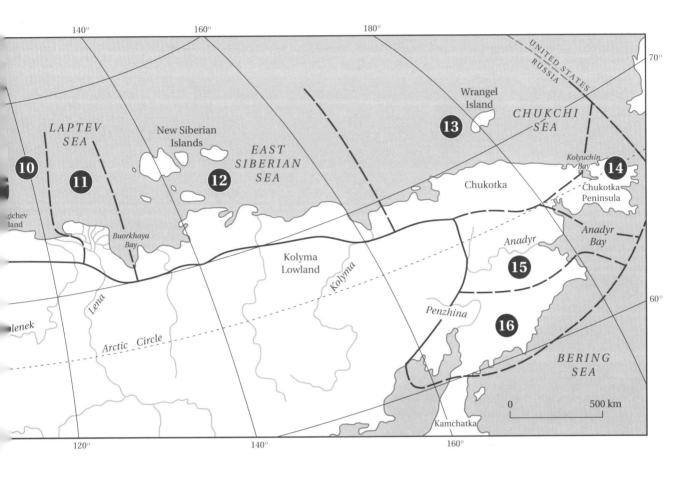

1. Murman District
2. Kanin-Pechora District
3. Polar Ural District
4. Yugorskiy District
5. Novaya Zemlya
6. Franz Josef Land
7. Ob-Tazovskiy District
8. Yenisey District
9. Taymyr District
10. Anabar-Olenek District
11. Lena District
12. Yana-Kolyma District
13. North Chukotka District
14. Beringian Chukotka District
15. Anadyr District
16. Koryak District

# FLORA OF THE RUSSIAN ARCTIC
VOLUME I

A.I. TOLMACHEV
*Russian Edition Editor*

J.G. PACKER
*English Edition Editor*

G.C.D. GRIFFITHS
*Translator*

# FLORA

*A Critical Review of the Vascular Plants Occurring in the Arctic Region of the Former Soviet Union*

## OF THE

## RUSSIAN

*Russian Edition Authors*

O.V. Rebristaya

A.K. Skvortsov

A.I. Tolmachev

N.N. Tsvelev

B.A. Yurtsev

## ARCTIC

The University of Alberta Press

VOLUME I

*Polypodiaceae—Gramineae*

First English edition published by
The University of Alberta Press
Athabasca Hall
Edmonton, Alberta
Canada T6G 2E8

Copyright © English edition by The University of Alberta Press 1995
Permission to translate and publish this edition of *Arkticheskaya Flora SSSR (Flora Arctica URSS)* was granted by the Severo-Zapadnoye Agentstvo po Avtorskim Pravam (SZAAP) on behalf of the original publishers, USSR Academy of Sciences, V.L. Komarov Botanical Institute.

Originally published as Arkticheskaya Flora SSSR (Flora Arctica URSS), Volumes I and II, by the USSR Academy of Sciences, V.L. Komarov Botanical Institute.

ISBN 0–88864–269-5

**Canadian Cataloguing in Publication Data**

Main entry under title:

Flora of the Russian Arctic

    Translation of: Arkticheskaya flora SSSR (Flora arctica URSS)
    Contents: Vol. 1. Polypodiaceae—Gramineae.
    Includes bibliographical references and index.
    ISBN 0-88864-269-5 (V.1)

    1. Botany—Russia, Northern. 2. Botany—Arctic regions.
I. Griffiths, Graham C.D. II. Packer, John G.
QK676.F66 1995      581.9472      C95–910815-7

All rights reserved.

No part of this publication may be produced, stored in a retrieval system, or transmitted in any forms or by any means, electronic, mechanical, photocopying, recording, or otherwise, without the prior permission of the copyright owner.

Printed on acid-free paper. ∞
Printed and bound in Canada by Best Book Manufacturers, Louiseville, Quebec.

COMMITTED TO THE DEVELOPMENT OF CULTURE AND THE ARTS

# Contents

- ix    Acknowledgements
- xi    Editor's Preface
- xv    Translator's Preface
- xix    Preface to Volume I of the Russian edition, Polypodiaceae-Butomaceae
- xxv    Preface to Volume II of the Russian edition, Gramineae
- xvii    Abbreviations Used in Citing Floristic and Systematic Literature

- 1    **FAMILY I / Polypodiaceae**—*True Ferns*
  - 4    GENUS 1 / Woodsia—*Woodsia*
  - 7    GENUS 2 / Cystopteris—*Bladder Fern*
  - 12    GENUS 3 / Dryopteris—*Shield Fern*
  - 17    GENUS 4 / Thelypteris—*Thelypteris*
  - 17    GENUS 5 / Gymnocarpium—*Oak Fern*
  - 19    GENUS 6 / Polystichum—*Holly Fern*
  - 19    GENUS 7 / Athyrium—*Lady Fern*
  - 21    GENUS 8 / Asplenium—*Spleenwort*
  - 21    GENUS 9 / Cryptogramma—*Rock Brake*
  - 22    GENUS 10 / Polypodium—*Polypody*

- 23    **FAMILY II / Ophioglossaceae**—*Adder's Tongue Family*
  - 23    GENUS 1 / Botrychium—*Moonwort*

- 25    **FAMILY III / Equisetaceae**—*Horsetails*
  - 25    GENUS 1 / Equisetum—*Horsetail*

- 37    **FAMILY IV / Lycopodiaceae**—*Club-Mosses*
  - 37    GENUS 1 / Lycopodium—*Club-Moss*

- 53    **FAMILY V / Selaginellaceae**—*Selaginella Family*
  - 53    GENUS 1 / Selaginella—*Selaginella, Little Club-Moss*

| | | |
|---|---|---|
| 57 | **FAMILY VI** / **Pinaceae**—*Pine Family* | |

| 57 | GENUS 1A / Abies—*Fir* |
|---|---|
| 58 | GENUS 1 / Picea—*Spruce* |
| 60 | GENUS 2 / Larix—*Larch* |
| 63 | GENUS 3 / Pinus—*Pine* |

| 65 | **FAMILY VII** / **Cupressaceae**—*Cypress Family* |
|---|---|

| 65 | GENUS 1 / Juniperus—*Juniper* |
|---|---|

| 67 | **FAMILY VIII** / **Sparganiaceae**—*Bur-Reed Family* |
|---|---|

| 67 | GENUS 1 / Sparganium—*Bur-Reed* |
|---|---|

| 71 | **FAMILY IX** / **Potamogetonaceae**—*Pondweed Family* |
|---|---|

| 71 | GENUS 1 / Potamogeton—*Pondweed* |
|---|---|
| 80 | GENUS 2 / Zostera—*Eel-Grass* |

| 81 | **FAMILY X** / **Juncaginaceae**—*Arrow-Grass Family* |
|---|---|

| 81 | GENUS 1 / Triglochin—*Arrow Grass* |
|---|---|
| 82 | GENUS 2 / Scheuchzeria—*Scheuchzeria* |

| 83 | **FAMILY XI** / **Alismataceae**—*Water-Plantain Family* |
|---|---|

| 83 | GENUS 1 / Alisma—*Water-Plantain* |
|---|---|

| 85 | **FAMILY XII** / **Butomaceae**—*Flowering Rush Family* |
|---|---|

| 85 | GENUS 1 / Butomus—*Flowering Rush* |
|---|---|

| 87 | **FAMILY XIII** / **Gramineae**—*Grasses* |
|---|---|

| 98 | GENUS 1 / Typhoides—*Reed Canary Grass* |
|---|---|
| 98 | GENUS 2 / Anthoxanthum—*Vernal-Grass* |
| 100 | GENUS 3 / Hierochloë—*Sweet Grass* |
| 104 | GENUS 4 / Milium—*Wood Millet* |
| 104 | GENUS 5 / Phleum—*Timothy* |
| 106 | GENUS 6 / Alopecurus—*Foxtail* |
| 115 | GENUS 7 / Arctagrostis—*Arctagrostis* |
| 119 | GENUS 8 / Agrostis—*Bent* |
| 127 | GENUS 9 / Calamagrostis—*Reed Grass* |
| 148 | GENUS 10 / Apera—*Silky Bent* |

- 148 GENUS 11 / Vahlodea—*Vahlodea*
- 150 GENUS 12 / Deschampsia—*Hair Grass*
- 164 GENUS 13 / Trisetum—*Trisetum*
- 171 GENUS 14 / Helictotrichon—*Oat Grass*
- 173 GENUS 15 / Beckmannia—*Slough Grass*
- 174 GENUS 16 / Phragmites—*Reed*
- 174 GENUS 17 / Molinia—*Moor Grass*
- 175 GENUS 18 / Koeleria—*June Grass*
- 179 GENUS 19 / Melica—*Melic*
- 179 GENUS 20 / Pleuropogon—*Semaphore Grass*
- 181 GENUS 21 / Dactylis—*Cocksfoot*
- 182 GENUS 22 / Poa—*Bluegrass*
- 224 GENUS 23 / Dupontia—*Dupontia*
- 229 GENUS 24 / Arctophila—*Arctophila*
- 231 GENUS 25 / Colpodium—*Colpodium*
- 233 GENUS 26 / Catabrosa—*Brook Grass*
- 234 GENUS 27 / Phippsia—*Phippsia*
- 237 GENUS 27A / Glyceria—*Manna Grass*
- 237 GENUS 28 / Puccinellia—*Alkali Grass*
- 264 GENUS 29 / Festuca—*Fescue*
- 278 GENUS 30 / Zerna—*Perennial Brome Grass*
- 283 GENUS 31 / Bromus—*Brome Grass*
- 284 GENUS 32 / Nardus—*Matgrass*
- 284 GENUS 33 / Roegneria—*Rhizomeless Wheat Grass*
- 299 GENUS 34 / Elytrigia—*Wheat Grass*
- 300 GENUS 35 / Leymus—*Wild Rye*
- 306 GENUS 36 / Hordeum—*Barley*

- 307 **APPENDIX I / Summary of Data on the Geographical Distribution of Vascular Plants of the Soviet Arctic**

  - 309 TABLE 1 / Distribution of Vascular Plants of the Soviet Arctic, *Polypodiaceae–Butomaceae*
  - 313 TABLE 2 / Distribution of Vascular Plants of the Soviet Arctic, *Gramineae*

- 319 Index of Plant Names

## ❧ Acknowledgements

THE TRANSLATION OF *Flora of the Russian Arctic* was initiated in 1988 when Dr. John Packer wrote to Norma Gutteridge, Director of the University of Alberta Press, regarding the feasibility of their publishing a translation of the then recently completed flora. With her favourable response, the process of securing the copyright to the translation rights was undertaken by the University of Alberta Press and accomplished in 1991.

Permission to translate and publish the 10-volume original *Arkticheskaya Flora SSSR (Flora Arctica URSS)* was granted by the Severo-Zapadnoye Agentstvo po Avtorskim Pravam (SZAAP) on behalf of the original publishers, the USSR Academy of Sciences, V.L. Komarov Botanical Institute. The Press acknowledges the assistance of B.A. Yurtsev of the V.L. Komarov Botanical Institute in obtaining these rights.

Initial funding for the translation and publication of the first volume was committed in 1993 by the University of Alberta Press in a farsighted and bold initiative by the Director that was crucial to the success of the enterprise.

The services of an expert translator were secured by contracting Dr. Graham Griffiths, whose past and present association with the University of Alberta in the Faculty of Science and knowledge of the Russian language made him a logical choice.

The Press also acknowledges the people involved in preparing the manuscript for publication, including Alan Brownoff, designer, Mary Mahoney-Robson, editor, Karen Chow and Heidi Betke, editorial assistants, Evelyn van der Heiden, graphics technician, and Cristina Munoz and Pat Mash of the Department of Botany at the University of Alberta who inputted the text.

In 1994 a successful application was made to the Selection Committee for Scientific Publication Grants, Natural Sciences and Engineering Research Council of Canada (NSERC) for a grant to assist in the publication of the first volume of *Flora of the Russian Arctic*. The University of Alberta Press would like to thank the Natural Sciences and Engineering Research Council of Canada for its committment to this project.

Finally, Dr. Keith Denford's involvement and support of this project since its inception is gratefully acknowledged.

# Editor's Preface

THE ARCTIC IS one of the world's major terrestrial biomes, a vast circumpolar ecosystem characterized by its distinct climate, soil conditions and biota. Compared with other biomes it is of relatively recent origin. Tertiary climatic deterioration, which was a significant factor in its development, led to extensive Pleistocene glaciation, particularly in North America and Europe. Even today sizable parts of the Arctic are still covered by ice.

Countries with arctic regions have taken a keen scientific interest in the Arctic especially over the past 50 years. As a consequence there is a vast amount of information available about its physical processes, ecology, plants and animals, both past and present. The present arctic flora is characterized by a very low overall species diversity, though many species have enormously wide distributions, small plant size, the virtual absence of annual species and generally high levels of both polyploidy and apomixis. Some of these phenomena have yet to receive satisfactory explanations and may not be susceptible to explanation in terms of current conventional thinking.

The work of documenting and describing arctic species and establishing their evolutionary relationships is an international enterprise that has been in progress for more than two hundred years. During this period floras for various parts of the Arctic have been published and there now exist contemporary floras or floras from the recent past for virtually all parts of the Arctic, including some for regions of the Russian Arctic, but there has never been a flora of the whole of the Russian Arctic. This has certainly impeded the work of arctic botanists both in Europe and probably even more so in North America. But with the publication of *Arkticheskaya Flora SSSR (Flora Arctica URSS)* the situation has changed dramatically.

Virtually any flora dealing exclusively with the Russian Arctic would be welcome but *Arkticheskaya Flora SSSR* is a special kind of flora. Not only does it provide a synthesis of taxonomic knowledge but, in detailed commentaries, it draws attention to unresolved problems. In many cases the discussions involve widely distributed species or complexes that are components of the European or North American Arctic as well, offering fresh insights for workers in these regions. The size of the area covered, the comprehensive content and the accomplished scholarship made *Arkticheskaya Flora SSSR* the most important flora ever published for any part of the Arctic. When it was begun it was a courageous undertaking; its completion in 1987 can only be regarded as an outstanding achievement of immense scientific significance.

*Arkticheskaya Flora SSSR* was written by taxonomic specialists at the V.L. Komarov Botanical Institute, St. Petersburg, under the editorship of A.I. Tolmachev and, following his death, by B.A. Yurtsev. The ten volumes of the flora were published from 1960 to 1987.

**Table P–1** Publication information on the original volumes of the Russian Edition, *Arkticheskaya Flora SSSR (Flora Arctica URSS)*

| Volume | Families | Pages | Imprimatur Date |
|---|---|---|---|
| I | Polypodiaceae—Butomaceae | 101 | November 23, 1960 |
| II | Gramineae | 272 | July 18, 1964 |
| III | Cyperaceae | 174 | January 14, 1966 |
| IV | Lemnaceae—Orchidaceae | 95 | April 5, 1963 |
| V | Salicaceae—Portulacaceae | 206 | December 15, 1966 |
| VI | Caryophyllaceae—Ranunculaceae | 246 | October 29, 1971 |
| VII | Papaveraceae—Cruciferae | 179 | September 2, 1975 |
| VIII (pt. 1) | Geraniaceae—Scrophulariaceae | 332 | December 12, 1980 |
| VIII (pt. 2) | Orobanchaceae—Plantaginaceae | 51 | June 9, 1983 |
| IX (pt. 1) | Droseraceae—Rosaceae | 332 | December 14, 1984 |
| IX (pt. 2) | Leguminosae | 187 | July 30, 1986 |
| X | Rubiaceae—Compositae | 410 | November 16, 1987 |

The flora, which uses the classification of Engler, treats some 360 genera, 1650 species and 220 infraspecific taxa, mostly subspecies. Included are descriptions of many new species and subspecies and numerous new combinations.

The number of species covered in *Arkticheskaya Flora SSSR* suggests a somewhat higher total for the whole of the Arctic than given in previously published figures. Polunin in his *Circumpolar Arctic Flora*, published in 1959, which he admits it is noncritical, recognizes 230 genera and 892 species in an Arctic defined more narrowly than that used in this flora. Löve and Löve in their *Cytotaxonomical Atlas of the Arctic Flora*, published in 1975, list 404 genera, 1629 species and 270 subspecies for much the same Arctic as used by *Arkticheskaya Flora SSSR*. However it should be observed that the 1650 species for the Russian Arctic is skewed considerably by the recognition of 52 *Taraxacum* species and 116 species of *Hieracium*.

As might be expected in a multi-authored work of this magnitude, different authors have different views on taxonomic concepts; some are more Komarovian than others. So, while some species are recognized that are generally not by western authors, the reverse is also true, as, for example, *Festuca baffinensis* Polunin, which is reduced to synonymy under *Festuca brachyphylla* Schultes, along with *Festuca hyperborea* Holmen. The mention of *Festuca* serves to remind us that the Gramineae volume was published in 1964 and that new discoveries have been made since then that are not included, a case in point being *Festuca brevissima*, described by Yurtsev.

With the publication of *Arkticheskaya Flora SSSR,* representing as it does a milestone in the history of arctic taxonomy, it would be lamentable if its full potential to advance our knowledge of the arctic flora went unrealized.

**TABLE P-2** Publication schedule for the English translation—*Flora of the Russian Arctic*

| Russian Edition | English Edition | Publication Date |
| --- | --- | --- |
| Volumes I – II | Volume I | 1995 |
| Volumes III – IV | Volume II | 1996 |
| Volumes V – VI | Volume III | 1996 |
| Volumes VII and IX (1) | Volume IV | 1997 |
| Volumes IX (2) and VIII (1 & 2) | Volume V | 1997–1998 |
| Volume X | Volume VI | 1998 |

However, there is no question that a lack of familiarity with the Russian language has been a serious obstacle to western botanists seeking information about taxa in the Russian Arctic and it has limited their entry to an abundant Russian literature. It has been a barrier to discourse. Regional taxonomies may be complete in themselves but circumpolar or widely distrtibuted taxa require a global perspective. For this reason, with a few notable exceptions, such studies by western taxonomists generally carry disclaimers, implicit or explicit, that the situation in the Russian Arctic is not clear, or remains to be worked out. Most arctic taxonomists have had to resort to this expedient at some time or other, and there is nothing wrong with it. Taxonomists of necessity accommodate themselves to the practicalities of a given situation. Clearly though, for botanists and for all biologists with an interest in the Arctic, to take full advantage of the benefits accruing from *Arkticheskaya Flora SSSR* a translation was necessary.

The English translation, *Flora of the Russian Arctic,* will be published in six volumes. The first two volumes include the ferns, gymnosperms and monocotyledons, the remaining four are devoted to the dicotyledons. The composition of the translated volumes and publication schedule are in Table P-2. The slightly revised order of publication for the translated volumes arranges the families in taxonomic order, something that was not possible in the original Russian volumes.

When biological systems merge that have evolved independently, or at least in partial isolation, exciting and scientifically interesting things begin to happen. The same surely can be said of information about biological systems. In the wake of the publication of *Flora of the Russian Arctic,* we may anticipate some absorbing taxonomic developments, the more so in view of the fact that its completion and translation could hardly have been more timely. For they have caught the tide of new technology now used daily in taxonomic research, which is transforming taxonomy and providing a better understanding of species and their relationships. This, combined with the circumstance that research projects of various degrees of magnitude abound in its pages, should assure us of a resurgent interest in the taxonomy of arctic vascular plants. Some of this work will be in the nature of tying up loose ends; there are lots of these and it is important to have this done. One also sees, not far off, major monographs emerging, some no doubt as cooperative enterprises involving arctic taxonomists from several countries. It is now possible to envisage monographs on such critical arctic genera as *Papaver, Draba,* and *Puccinellia,* something that was unheard of in the past.

JOHN G. PACKER
*English Edition Editor*

## Translator's Preface

In this translation of *Arkticheskaya Flora SSSR*, I have tried to preserve the flavour of the original edition. Some of the Russian authors (especially Tolmachev and Tsvelev) tend to write in much longer sentences than approved in current English style manuals. However, I do not think it the task of a translator to rewrite material merely to conform with current stylistic canons. I have followed the original sentence structure wherever possible, only breaking down or drastically reorganizing sentences when this seemed necessary to achieve clarity in English. Since this is a translation, not a revised edition, new nomenclatorial proposals are presented in the same form as in the original Russian edition, indicated by "sp. nova," "comb. nova," etc.

References to the Soviet Union (USSR) in the text naturally refer to the former Soviet Union within its postwar boundaries prior to its dissolution in 1991. With respect to the Arctic Region, the terms "Soviet Arctic" and "Russian Arctic" are equivalent since all arctic territory of the former Soviet Union lies within the present boundaries of the Russian Federation. However, references to distribution within the Soviet Union as a whole may include occurrence outside the present boundaries of the Russian Federation.

The transliteration system used for Russian personal and place names in this translation is the "popular system" that has evolved in English-language newspapers. It is the same as that used on maps produced by the National Geographic Society of the USA except for omission of the apostrophe used to represent the Cyrillic hard and soft signs. Such use of the apostrophe means nothing to English-speaking readers unfamiliar with the subtleties of Slavic pronunciation, and complicates printing for little advantage.

The following points will assist the reader with no knowledge of Russian with the pronunciation of the Russian words transliterated according to this system. The letter **y** does double service both as a vowel and consonant in this transliteration system (hence in words ending "**-yy**" the first **y** is to be read as a vowel). This vowel sound is pronounced more or less as in the French "feuille." While this double use of the letter **y** is perhaps a disadvantage of the system, the same double usage is present in English spelling. Differentiation of the consonantal **y** as **j** (according to German convention) would likely cause it to be mispronounced by unilingual English readers. Note that the vowel **i** is always to be pronounced as in "ravine," not as in "fine" (i.e., it represents the sound also represented in English by **double e**, as in "bee"). The Russian **e** is represented by **e** following a consonant, but by **ye** at the beginning of words and following a vowel. The consonant **ch** is to be pronounced as in "chalk," while the sound represented by the **ch** in "loch" is represented in transliterated Russian words by **kh**. The consonant **zh** represents the voiced equivalent of **sh**, a sound not normally used in English but roughly the same as the French pronunciation of **j**.

In rendering place names I have translated words with obvious English equivalents, such as bay for *guba* and *zaliv*, cape for *mys*, island for *ostrov*, lake for *ozero*, mountain for *gora*, peninsula for *poluostrov* and range for *khrebet*, rather than merely transliterated such words as on National Geographic maps. There is a general problem in rendering Russian place names in English arising from the highly inflected structure of the Russian language; this is whether to retain all the inflectional elements attached as suffices. I have adopted a pragmatic (though inconsistent) approach. In the case of major geographical features, I have used the shortened forms in common use on English maps (for example, "Barents Sea" rather than "Barentsovo Sea"). I have also used shortened forms in cases where the English reader can more readily relate these to a neighbouring town or river (for example, the "Kara Sea" rather than "Karskoye Sea" or the "Verkhoyansk Range" rather than "Verkhoyanskiy Range"). But for many localized features, which the reader may be able to locate only on Cyrillic maps, I have spelled out the fully inflected form in transliteration since that is the form to be found on such maps. In all cases I cite fully inflected place names in the nominative case, and wherever possible have checked the orthography against the 1986 edition of the *Atlas of the USSR*. There remains a possibility that a few errors may have been made in citing the nominative form of local names not included in the Atlas. It should be appreciated that many place names in the Russian North derive from aboriginal languages, and it is not always obvious (even to Russians) how they should be declined or whether they are declinable at all in Russian. The terms *kray* and *oblast* are transliterated when they refer to units of regional administration (such as *Primorskiy Kray* and *Leningrad Oblast*, since these terms are not exactly equivalent to any terms used for regional administrative units in western countries; but they are translated (as "territory" and "region") when not used in an administrative sense.

There are various discrepancies between the English transliterations of Russian personal names here used and the abbreviations of these names used in citing authorship of the Latin names of plants. This arises because the abbreviations for Russian authors given in standard lists generally follow German or Eastern European spelling conventions. I have decided to accept these discrepancies and to copy all abbreviations in the form given in the Russian edition. It would not be realistic to try to revise the abbreviations in this translation, since botanists are unlikely to accept such changes unless made in the context of publishing a new standardized list after widespread consultation. Examples of such discrepancies affecting the abbreviations for the name of authors of this Flora include Czer. for Cherepanov, Egor. for Yegorova, Jurtz. for Yurtsev and Tzvel. for Tsvelev.

A few unusual words for landform or vegetational types may need clarification. The term "placorn tundra" is used by Russian ecologists for level tundra with fine soil. *Baydzharakh* is a Yakutian word for small relict mounds enclosing polygonal ice blocks (normally arranged in groups or rows). These are not the same as pingos (*bulgunnyakhs* in Russian), which are larger isolated mounds resulting from the eruption of underground ice lenses. The term *khasyr* (presumably an aboriginal word, since I can not find it in any Russian dictionary) refers according to the context to small temporary lakes which support terrestrial (but flood-tolerant) vegetation on their beds. This description applies to what the Irish call a "turlough," so I have used this word in translation. I consider "barrens" to be an appropriate translation of the Russian term *goltsy* (the poorly vegetated high alpine zone of mountains), since the root of the Russian word means "bare." The Russian word *stlanik* is retained for vegetation dominated by cedar pine, a small tree or shrub with sprawling growth form. This term is sometimes translated as "krummholz," but not entirely appropriately since this growth form of cedar pine is genetically determined and does not require extreme environmental conditions.

Descriptive botanical terms have been checked in cases of doubt in the *Russian-English Botanical Dictionary* by P. Macrura (1982, Slavica Publishers Inc.).

With respect to the summaries of distribution in the North American Arctic, it should be noted that the authors often use the term "Labrador" to include arctic parts of Quebec (the Ungava Pensinsula, etc.), as well as Labrador in the present political sense.

The precise date of final approval for printing (the "imprimatur") is given at the end of each volume of the Russian edition. The date of effective publication in the sense of the International Code of Botanical Nomenclature, the date that the work became available, was presumably very shortly after that date. A list of the dates can be found in Table P-1 on page XII.

Maps in this English edition are numbered by the original Russian volume number and the corresponding original number of the map. For example, in this volume, Map II-4 is the fourth map in the Russian volume II on Gramineae.

I hope that his translation will facilitate further collaboration between Russian and western botanists in elucidating the fascinating present and past interrelationships within the circumpolar arctic flora.

G.C.D. Griffiths
*Translator*

# Preface to Volume I of the Russian Edition

*Arkticheskaya Flora SSSR*

*Polypodiaceae-Butomaceae*

---

THE PRESENT VOLUME represents the first part of a survey of the vascular plants of the arctic territory of the USSR, which will be published in the next few years. Only an insignificant number of the species included in the arctic flora, belonging mainly to families not very characteristic of the Arctic, is treated here. This volume of the Flora contains information on 67 plant species belonging to 25 genera and 12 families. In preliminary evaluation this is about 8% of the total species to be considered in our work. However, it seems justifiable to publish a restricted portion of the work because this will allow interested specialists to form opinions about the task we have undertaken and to express critical comments that will undoubtedly be useful for its continuation.

The series of principles the author bases his work on the flora of the Soviet Arctic have already been published.[1] This allows me not to dwell on the validation of decisions that have already been brought to the attention of the scientific community. But it seems appropriate to repeat here some of what has been said about the construction of the present work.

To the extent of accumulated knowledge, this flora represents a complete review of the wild vascular plants growing north of the polar limit of more or less closed forests, that is the floras of the forest-tundra ("northern forest-tundra" of some authors), true tundra and the so-called arctic deserts. The review includes all species of vascular plants occurring in the areas indicated, irrespective of the degree to which they are characteristic for these areas and the extent of their distribution. The arctic flora treated within such a scope will perhaps seem less distinctive and compact than if we had adopted a more rigorous principle of delimitation. But the task of any work of descriptive "inventory" consists of summarizing and describing that which objectively exists in nature, not in trying by the very choice of material described to anticipate certain scientific conclusions. The author fully takes into account that as elements of the flora (and more generally) different plant species are not of equal value. With specific reference to the Arctic, the importance of widely or practically universally distributed species and of species just penetrating the limits of the Arctic from the temperate zone is certainly not the same. But some importance attaches to the presence of any species within arctic limits, and consequently the attempt to "get rid of" species less characteristic of our area as "floral contaminants" would in practice be equivalent to consciously removing from the sphere of study, facts that ought to be studied and receive appropriate evaluation.

---

[1] A.I. Tolmachev. K Izucheniyu arkticheskoy flory SSSR [Towards the study of the arctic flora of the USSR]. Botan. Zhurn., vol. XLI, 1956, No. 6. Tolmachev was the sole author of Volume I of the Russian edition.

The arrangement of the entire contents of the Flora according to the system of A. Engler is dictated by purely practical considerations, on account of the need for convenient orientation of the reader to the content of any floristic survey. Disagreement with some of the phylogenetic considerations upon which the stated system is based does not, in my opinion, give one the right to replace it in descriptive floral works with other systems, since none of the latter has been elaborated in the required detail nor is free of inadequacies pertaining to the system which they have been proposed to replace. At any rate, while there exist today some more or less widely accepted treatments of plant phylogeny not agreeing with the Engler system, none of the newer systems constructed on the basis of these treatments has gained either general recognition as more valid than the Engler system or even appraisal as better than the systems proposed by other authors. In these circumstances refusal to follow the tradition built up in descriptive floristic works would not have advanced us one step forward, but would merely have displayed lack of concern for the interests of those who will use this published survey in their practical work.

As already indicated in the programmatic article mentioned, the author's treatment of the "species" category differs somewhat from the position of the majority of authors of the *Flora of the USSR*: geographical races connected to one another by transitional forms are not considered as separate species, but as subspecies of a single species. But in restoring use of the taxonomic category "subspecies," I am not at all calling for general "amalgamation" of species, that is for automatic rejection of differences which are relatively minor and sometimes difficult to recognize as differences between species (always providing that they are representative of the total complex of characters). The degree of difference between species may be nonuniform. Not every real distinction in nature lends itself to a corresponding distinction in our taxonomic practice with equal ease. But it is time to make a stand against irresponsible "elevation to the rank of species" of every geographical variant and any population that is to some degree distinctive. A substantive but rather narrow interpretation of the species is one thing; but "species fragmentation" under the flag of asserting that everything that merits the attention of the taxonomist is a species criterion while everything else may be totally disregarded (or "left to the disposal" of other specialists, such as geneticists), this is something else. To what extent the author's position on taxonomic questions is correct, can obviously only be judged on the basis of appraising the substantive results of our treatment of the arctic flora.

Much attention has been paid by us to the construction of *keys for plant identification,* the component of floristic surveys which in our opinion is underrated by many systematists. In constructing keys the author has tried to express the whole complex of characters able to serve in the correct recognition of each species, as far as possible independently of the phenological phase of the plant under study. While giving whole series of mutually opposing characters at the majority of steps in each key, we emphasize (by appropriate italicization) those characters which possess the greatest diagnostic importance or are more convenient for differentiating the species (or series of species) under comparison. We emphasize that the importance of characters is always evaluated with respect to the material with which we are dealing: if a given character is convenient as a means of orientation in arctic material, its use is upheld irrespective of to what extent it is valid in consideration of the relevant systematic group in a wider context.

In total the characteristics given in the keys for each species rather fully describe the appearance and the combination of diagnostic characters of the plants under consideration. Taking this into account, as well as the presence in literature published during recent decades of good comprehensive descriptions of the overwhelming majority of plants occurring in the territory of the USSR, the author has decided to refrain from repeating the morphological characteristics of species in the form of complete plant descriptions. We do not at all intend to deny the scientific value of such descriptions. But they should not be overvalued: basically duplicate descriptions which have already been published repeatedly bring little new to our understanding of the plant world. Furthermore they much increase the size of Floras. Possibly some colleagues will ask the question: where then will the reader find various data on the morphology of the species identified, if this is not included in the present Flora? But one should reject in advance as obviously unsound the demand that any floristic review should serve as the sole text for study of the plants of the relevant country. The need for detailed study inevitably has to be addressed for various scientific tasks, whether floristic, special systematic or of other character. And we do not intend that the worker in the Far North, having got his hands on this *Arkticheskaya Flora SSSR* will refrain from consulting the classical *Flora of West Siberia* or the just published detailed *Flora of Murmansk Oblast*, or finally the now almost complete *Flora of the USSR*. *Arkticheskaya Flora SSSR* is not at all intended to replace them. It only supplements them by treating the particular combination of plant species associated with the treeless expanses of the Far North, and by treating these in a rather different format from that in available floristic surveys.

While refraining from including detailed morphological species descriptions in the text of the Flora, we sometimes give rather comprehensive characterization of the variation of species and their component varieties, as well as certain commentaries with respect to the interrelationships between particular species and varieties of our flora. This kind of characterization seems especially appropriate within the framework of floristic surveys of the regional type, in connection with treatment of the substantive relationships existing within the confines of the particular region.

In passages dealing with particular species of our flora, numerous references to the literature must unavoidably be introduced. However, we emphasize that our survey does not pretend to transmit complete bibliographical information about the literary sources containing data on the relevant species. We also do not give a complete list of the synonymy of each species, restricting ourselves to the citation of those names under which the appropriate species is mentioned in works on the arctic flora. For the purpose of coordinating the data of our Flora with those of other survey works treating the Far North of the USSR, we include in all cases references to the treatment of the relevant species in the *Flora of the USSR*, *Flora of West Siberia* by P.N. Krylov, *Flora of the Northern Territory* by I.A. Perfilev and *Flora of Murmansk Oblast*, together with the incomplete *Flora of Yakutia* by V.A. Petrov and *Flora Arctica* by Gelert and

Ostenfeld. In relation to special works on the arctic flora, we restrict the references to works which are predominantly surveys of the flora of some part of the region under consideration. However, it would be burdensome (both in the sense of increasing the size of the publication and in relation to the reader's use of the survey data therein) to give references to all survey works devoted to one or other local flora. For instance, for Novaya Zemlya there are floristic reviews published at different times under the authorship of Trautvetter, Heuglin, Kjellman and Feilden. Reference to these would hardly have any practical importance, because a worker interested in deepening his knowledge of the flora of Novaya Zemlya need hardly concern himself with the study of these considerably outdated sources in view of the availability of Lynge's (1923) fundamental work, which combines all material of previous investigations together with the author's original data. But the availability of our more recent conspectus (1936) does not remove the need for reference also to Lynge's work, since the degree of detail of his treatment of the material preserves its importance as a basic text on the flora of Novaya Zemlya. Citations of works which do not have the character of floral surveys of some part of the Arctic, are made to the extent that these works contain factual information of definite importance for treating some species under consideration (not merely as locality sources). Therefore, lack of mention of some work in the literature list for a particular species should not be taken as evidence that there is no data at all on this species in the relevant work.

Together with references to literature on the flora of the Soviet Arctic and the flora of the USSR generally, we cite rather numerous references to literary sources on the flora of the Foreign Arctic. The need for this arises because the Arctic constitutes an integral botanical-geographical region and anyone who studies the flora of some part of this region must of necessity take an interest in data on the flora of the region as a whole. Of course, references to literature on the Foreign Arctic may be relatively brief. The availability in contemporary foreign literature (especially Scandinavian and Canadian) of a whole series of survey works on the flora of the Foreign Arctic (works of Hultén, Porsild, Polunin, etc.) has considerably eased our task. The rather heightened attention paid to literature on the Greenland flora (especially within this first volume of our Flora) is explained by the particularly significant successes achieved by study of the flora of that part of the Arctic Region during the last 25–30 years, and by the inclusion in many publications on Greenland of data on arctic plants having direct importance for our work.

Concerning the form of references to literary sources, in all cases the full surnames of the authors are cited (without abbreviations) in addition to abbreviations of the titles of the works to which reference is made (not of the periodical or serial publications in which the relevant works are printed). Therefore finding the relevant work should not present difficulties for the reader. At the end of our work we will give a detailed bibliographical list of all literature used and cited in all volumes of *Arkticheskaya Flora SSSR*.[2]

Data on conditions of growth and the associations of plants with particular vegetational communities are presented with as much detail as possible, at greater length than is customary in floristic surveys. The possibility of such treatment is facilitated by the relatively homogenous conditions of the region

---

[2] In the Russian edition such lists (or their addenda) appeared towards the ends of volumes 2–8, and no consolidated list for the whole *Arkticheskaya Flora SSSR* was in fact produced. The present English edition contains a consolidation of all these lists for the first time.

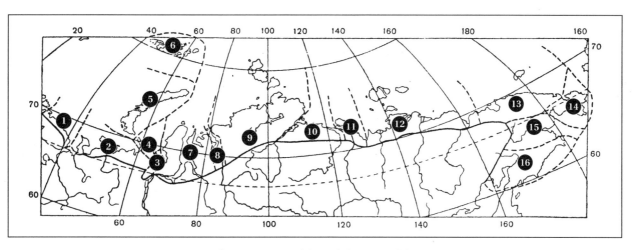

MAP I–1  Schematic map of the subdivision of the Soviet Arctic into districts.

whose flora is described, by the availability of relevant data in various works on the arctic flora, as well as by our rather numerous personal observations on the conditions of life of plants in the Arctic.

The geographical distribution of each plant is presented first in general form (zonal association of the species, continuous or fragmented character of its distribution), followed by a review of data on its distribution within the confines of the Soviet Arctic. We abstain from a complete listing of plant specimens studied as well as relevant literary sources. At the present stage of study of the arctic flora, this would be superfluous for the majority of species. A complete catalogue of localities would be too enormous, and the very numerous records referring mainly to better investigated parts of our region (such as, for example, the west coast of Novaya Zemlya) would divert attention from the no less important isolated records of discoveries in less studied parts of the Arctic. The presentation of such details remains appropriate in detailed floristic accounts for particular districts of the Arctic, but would not be justified in relation to the Arctic as a whole. Only in relation to species presently known from a very restricted number of sites does enumerating the known localities still make sense. For species really characteristic for the arctic flora and possessing a more or less extensive distribution in the Soviet Arctic, information on localities within its confines is given in summarized form on the distribution maps inserted in the text. The maps show only firm localities whose accuracy is not subject to any doubt for the species under consideration. On some maps the overall distribution is shown by hatching, and in certain cases by drawing distribution contours. For species widely distributed outside Arctic limits, the maps only present firm localities situated within the Arctic or in neighbouring parts of the forest zone. Only in relation to a few species do we depart from this principle and give a detailed representation of their distribution within the boundaries of the USSR as a whole. We emphasize that in those cases in which we indicate on our maps only particular localities without hatching and without drawing any distribution contour, all known localities falling within the boundary of the relevant map are accounted for. But there are few such species. In the text the review of the geographical distribution in the Arctic is given in a uniform sequence; from west to east, from Murman to the far north-east of Asia, without formal subdivision of the Soviet arctic territory into any districts.

After the review of data on the distribution of a species in the Soviet Arctic we present information on its foreign distribution within the confines of the Arctic Region, also the characteristics of its general distribution. Of course, this part of our data contains little that is original.

Finally, for convenient reference to the discovery of particular species in particular districts we have appended at the end of each volume of the Flora a series of tables giving in systematic form information on the distribution of the species treated therein both in the Soviet and Foreign Arctic. Their distribution in the Soviet Arctic is shown according to the districts whose characteristics are given in the article cited above and whose boundaries are schematically shown on the adjoining map (Map I–1).

In consideration of the poor accessibility of the text of our Flora to foreign readers, we give a list of the districts we have adopted also in the Latin language.[3] Therefore the tables should be accessible to everyone as reference material.

In publishing this first volume of our Flora I consider it my duty to thank my young collaborators O.V. Rebristaya, A.O. Tryasuchkin and B.A. Yurtsev for their diligent work in cataloging herbarium material of the arctic flora, drawing distribution maps and the constant help of many kinds which I have received from them in the course of completing my work.

*Leningrad, June 1958*

Postscript added in proof:

In the time which has passed since sending my manuscript to press, study of the arctic flora of various districts of our country has continued. Specifically, in the year 1958 new data were obtained on the flora of the far north of the Komi ASSR and part of the Nenetskiy National Okrug. Work during the year 1959 gave further additions to the flora of these districts. During these years our knowledge of the flora of eastern districts of Chukotka was also increased. Whatever of this new data was accessible to me prior to sending the manuscript to the printer was included in the basic text. In 1960 interesting new material was collected in a previously virtually unstudied part of Koryakia, in the district of the Bay of Korf. Appropriate additions were inserted in the text in proof. But it proved no longer possible to add the new localities to the distribution maps.

*Leningrad, September 1960*
A. Tolmachev

---

[3] In this English edition, only an English listing is given.

# Preface to Volume II of the Russian Edition

*Arkticheskaya Flora SSSR*

*Gramineae*

---

THE PRESENT VOLUME of the *Arkticheskaya Flora SSSR* (the second volume in the systematic sequence, the third in order of publication) is devoted to the family *Gramineae*, the family of flowering plants most richly represented in the Arctic Region. Among its members are numerous species which are very characteristic of the arctic flora and play an essential role in the formation of various plant communities. Many of them are widely distributed in the Arctic and function as characteristic components of its flora in most parts of our region. Some species extend here to very high latitudes, reaching the extreme northern limits of land. Among the arctic grasses are numerous species which appear to be endemic to the Arctic or whose distribution just extends beyond arctic limits (mainly in contiguous high alpine districts). Deserving of special attention is the presence in the Gramineae of endemic (or almost endemic) arctic genera, such as the oligotypic *Phippsia* and *Dupontia* and the monotypic *Arctophila*. This increases the importance of careful study of the grasses of the Arctic Region for understanding the history of the formation of its flora.

As in the composition of other families, the grasses occurring in the Arctic include many typically arctic-alpine species, partly arctic and partly alpine in their origin. A significant number of species have distributions of an overall boreal character, or are widespread in a whole series of zones in the extratropical part of the Northern Hemisphere. The majority of these species are associated in the Arctic with districts possessing milder climatic conditions: they penetrate the Arctic mainly in areas neighbouring the Atlantic or Pacific Oceans. To the contrary, typical arctic and arctic-alpine species are concentrated in greater numbers in districts characterized by stronger expression of continental climatic conditions. For this reason the richness of the floras in grass species characteristic of the Arctic appears especially pronounced in the eastern sector of the Eurasian Arctic and in north-west North America, where the role of grasses as the leading family in the composition of the floras is also most clearly expressed. Species of grasses characteristic of the Arctic but more narrowly distributed are native mainly to these sectors. Here the material of grasses demonstrates some connections between the arctic flora and cryoxerophytic floristic complexes of the interiors of the continents.

The grass family includes numerous genera (including some represented in the Arctic) whose systematic study is attended with considerable difficulties. The recognition of many species under field conditions is difficult for botanists without special expertise in the systematics of this family. On this account the material representing particular species in our herbaria is sometimes inadequate and fragmentary. Consequently our concepts of the Arctic distribution of many important grass species are still very far from complete. Strong variation in many species along with insufficiently complete material (often belonging to fortuitously distinctive populations) has in many cases led to overvaluation of differences characterizing particular populations and to inadequately founded description of new forms as of species rank. And in no less numerous cases concepts of the distribution of particular species have been distorted as a result of simple misidentifications which have been recorded in floristic literature.

In the present volume of the *Arkticheskaya Flora SSSR* data are presented on 36 genera and 162 species of grasses reliably reported within the confines of the territory covered by our Flora. The above numbers sufficiently demonstrate how essential a role grasses play in the formation of the arctic flora of Eurasia. Moreover it should be emphasized that a critical approach to material has forced the authors of the treatments of many genera to decline to recognize a considerable number of species described by various persons at various times from within the Soviet Arctic. The establishment of relevant conclusions and the need for more detailed clarification of certain other taxonomic questions has required authors to include in their text rather numerous and sometimes lengthy systematic commentaries, which we hope will be of interest to readers of the Flora.

Work on material for the present volume of the Flora was divided as follows:

N.N. Tsvelev composed the key for identification of genera and systematically treated the genera: *Agrostis, Alopecurus, Arctagrostis, Arctophila, Bromus, Calamagrostis, Colpodium, Deschampsia, Elytrigia, Koeleria, Leymus, Poa, Puccinellia, Roegneria, Vahlodea* and *Zerna*.

A.K. Skvortsov treated the genus *Festuca*.

O.V. Rebristaya treated the genus *Trisetum*.

B.A. Yurtsev treated the genera *Dupontia, Helictotrichon* and *Pleuropogon*.

The genera *Anthoxanthum, Apera, Beckmannia, Catabrosa, Dactylis, Hierochloë, Hordeum, Melica, Milium, Molinia, Nardus, Phippsia, Phleum, Phragmites* and *Typhoides* were treated by A.I. Tolmachev. He also wrote the general characterization of the family.

The ecological-geographical characteristics of species in the genera treated by A.K. Skvortsov and N.N. Tsvelev were expanded by B.A. Yurtsev.

Distribution maps were prepared by O.V. Rebristaya, N.V. Matveyeva and N.N. Kusheva with the assistance of V.V. Petrovskiy and A.E. Katenin.

A.I. Tolmachev
*Leningrad, July 1962*

# Abbreviations Used in Citing Floristic and Systematic Literature

THE LIST PRESENTED in this English edition is a consolidation of the lists and addenda published in volumes II-VIII of the Russian edition. Cyrillic titles are given both in transliteration and in English translation. The classification follows that of the last list on pages 36-40 of volume VIII(2) of the Russian edition.

## 1. Northern European Part of USSR

| | |
|---|---|
| Fl. Murm. | Flora Murmanskoy oblasti. [Flora of Murmansk Oblast]. Moscow; Leningrad, 1953, vol. I. 254 pp.; 1954, vol. II. 290 pp; 1956, vol. III. 450 pp.; 1959, vol. IV. 394 pp.; 1960, vol. V. 549 pp. |
| Fl. sev.-vost yevrop. ch. SSSR | Flora severo-vostoka yevropeyskoy chasti SSSR. [Flora of the North-East European part of the USSR]. Leningrad, 1974, vol. I. 272 pp.; 1976, vol. II. 305 pp.; 1976, vol. III. 295 pp.; 1977, vol. IV. 312 pp. |
| Fl. yevrop. ch. SSSR | Flora yevropeyskoy chasti SSSR. [Flora of the European part of the USSR]. Leningrad, 1974-1981, vols. 1–5. |
| Govorukhin, Fl. Urala | Govorukhin V.S. Flora Urala. Oprodelitel rasteniy, obitayushchikh na gorakh i v yevo predgoryakh ot beregov Karskovo morya do yuzhnykh predelov lesnoy zony. [Flora of the Urals. Key to plants living in the mountains and foothills from the shores of the Kara Sea to the southern limits of the forest zone]. Sverdlovsk, 1937. |
| Hanssen & Lid, Flow. pl. Franz. Josef L. | Hanssen O., Lid J. Flowering plants of Franz Josef Land. Skrifter om Svalbard og Ishavet, 1932, No. 39, pp. 1–42. |
| Igoshina, Fl. Urala | Igoshina K.N. Flora gornykh i ravninnykh tundr i redkolesiy Urala. [Flora of alpine and lowland tundras and open forests of the Urals]. In book: Rasteniya severa Sibiri i Dalnevo Vostoka. [Plants of Northern Siberia and the Far East]. Moscow; Leningrad, 1966, pp. 135–223. |
| Leskov, Fl. Malozem. tundry | Leskov A.I. Flora Malozemelskoy tundry. [Flora of the MalozemelskayaTundra]. Moscow; Leningrad, 1937. 106 pp. |
| Lynge, Vasc. pl. N.Z. | Lynge B. Vascular plants from Novaya Zemlya. Kristiania, 1923. 151 pp. |
| Perfilev, Fl. Sev. | Perfilev I.A. Flora Severnovo kraya. [Flora of the Northern Territory]. Arkhangelsk, 1934, part I. 160 pp.; 1936, parts II–III. 398 pp. |
| Rebristaya, Fl. Bolshezem. tundry | Rebristaya O.V. Flora vostoka Bolshezemelskoy tundry. [Flora of the Eastern Bolshezemelskaya Tundra]. Leningrad, 1977. 334 pp. |
| Ruprecht, Fl. samojed. cisur. | Ruprecht F.J. Flores samojedorum cisuralensium. In: Ruprecht F.J. Symbolae ad historiam et geographiam plantarum rossicarum. St. Petersburg, 1846, pp. 1–67. |

| | |
|---|---|
| Tolmatchev, Contr. fl. Vaig. | Tolmatchev A. Contributions to the flora of Vaigats and of mainland coast of the Yugor Straits. Tr. Botan. muzeya AN SSSR, 1926, vol. 19, pp. 121–154. |
| Tolmachev, Mat. fl. Mat. Shar | Tolmachev A.I. Materialy dlya flory rayona polyarnoy geofizicheskoy observatorii Matochkin Shar i sopredelnykh chastey Novoy Zemli. [Materials for the flora of the district of the Matochkin Shar Polar Geophysical Observatory and neighbouring parts of Novaya Zemlya]. Tr. Botan. muzeya AN SSSR, 1932, vol. 24, pp. 275–299; 1932, vol. 25, pp. 101–120. |
| Tolmachev, Obz. fl. N.Z. | Tolmachev A.I. Obzor flory Novoy Zemli. [Review of the flora of Novaya Zemlya]. Arctica, 1936, No. 4, pp. 143–178. |

**Additional Sources**

| | |
|---|---|
| Aleksandrova, Nov. dan. fl. Yuzh. o. N.Z. | Aleksandrova V.D. Novyye dannyye o flore Yuzhnovo ostrova Novoy Zemli. [New data on the flora of the South Island of Novaya Zemlya]. Biol. MOIP. Otd. biol., 1950, vol. IV, No. 4, pp. 76–85. |
| Andreyev, Mat. fl. Kanina | Andreyev V.N. Materialy k flore severnovo Kanina. [Materials towards the flora of Northern Kanin]. Tr. Botan. muzeya AN SSSR, 1931, vol. 23, pp. 148–196. |
| Andreyev, Rast. vost. Bolshezem. tundry | Andreyev V.N. Rastitelnost i prirodnye rayony vostochnoy chasti Bolshezemelskoy tundry. [Vegetation and natural districts of the eastern part of the Bolshezemelskaya Tundra]. Moscow; Leningrad, 1935. |
| Dahl & Hadač, Bidr. Spitzb. Fl. | Dahl O.C., Hadač E. Et bidrag til Spitzbergens flora. Norges Svalbards- og Ishavetsundersøkelser. Meddelelser, No. 63, Oslo, 1946. |
| Feilden, Fl. pl. N.Z. | Feilden H.W. The Flowering Plants of Novaya Zemlya. London, 1898. |
| Floderus, Nov. Sem. Salic. | Floderus B. Bidrag till kännedomen om Novaja Semljas Salices. Sv. bot. tidskr., VI, 1912. |
| Gorchakovskiy, Rastit. vysokogor. Urala | Gorchakovskiy P.L. Rastitelnyy mir vysokogoriy Urala. [Vegetational world of the high alpine Urals]. Moscow, 1975, pp. 82–120. |
| Holm, Nov. Zem. Veg. | Holm T. Novaia-Zemlia's Vegetation. Saerligt dens Phanerogamer. Copenhagen, 1885. |
| Katenin & al., Fl. Siv. Maski | Katenin A.E., Petrovskiy, V.V., Rebristaya O.V. Sosudistyye rasteniya. [Vascular plants]. In book: Ekologiya i biologiya rasteniy vostochno-yevropeyskoy lesotundry. [Ecology and biology of plants of the Eastern European forest-tundra]. Leningrad, 1970, pp. 37–48. |
| Kjellman & Lundström, Phanerogam. N.Z. Waig. | Kjellman F.R., Lundström A.N. Phanerogamen von Novaja-Semlja, Waigatsch und Chabarova. In book: Nordenskiöld A.E. Die wissenschaftlichen Ergebnisse der Vega-Expedition. Bd. I. Leipzig, 1883. |
| Lundström, Weiden Now. Sem. | Lundström, A.N. Kritische Bemerkungen über die Weiden Nowaja Semljas. Acta Reg. Soc. Sci. Upsal., 3, 1877. |
| Opred. rast. Komi | Opredelitel vysshikh rasteniy Komi ASSR. [Key to the higher plants of the Komi ASSR]. Moscow; Leningrad, 1962. |
| Perfilev, Mat. fl. N.Z. Kolg. | Perfilev I.A. Materialy k flore ostrovov Novoy Zemli i Kolguyeva. [Materials towards the flora of the islands of Novaya Zemlya and Kolguyev]. Arkhangelsk, 1928. 73 pp. |
| Sambuk, K fl. sev. yevrop. ch. SSSR | Sambuk, F.V. K flore severa yevropeyskoy chasti SSSR. [Towards the flora of the north of the European part of the USSR]. Zhurn. Russk. botan. obshch., vol XIV, No. 1, 1929. |

| | |
|---|---|
| Schrenk, Enum. pl. | Schrenk, A.G. Enumeratio plantarum in itinere per plages samojedorum cisuralensium per annum 1837 observatarum. In: Schrenk. Reise nach dem Nordosten des europäischen Russlands, Bd. 2, 1854. |
| Tolmachev, Fl. Kolg. | Tolmachev A.I. Floristicheskiye rezultaty Kolguyevskoy ekspeditsii Instituta po izucheniyu Severa. [Floristic results of the Kolguyev Expedition of the Institute for Study of the North]. Leningrad, 1930. 50 pp. (Tr. Polyarnoy Komissii; vol. 2). |
| Tolmachev, Fl. kr. sev. N.Z. | Tolmachev A.I. K flore kraynevo severa Novoy Zemli. [Towards the flora of the far north of Novaya Zemlya]. Izv. Glav. bot. sada, 1926, pp. 1–4. |
| Tolmachev, Fl. pober. Karsk. morya | Tolmachev A.I. K flore yugo-zapadnovo poberezhya Karskovo morya. [Towards the flora of the southwestern coast of the Kara Sea]. Botan. zhurn., 1937, vol. 22, No. 2, pp. 185–196. |
| Tolmachev, Mat. fl. yevr. arkt. ostr. | Tolmachev A.I. Materialy dlya flory yevropeyskikh arkticheskikh ostrovov. [Materials for the flora of the European Arctic Islands]. Zhurn. Rus. botan. obshch., 1931, vol. XVI, No. 56, pp. 459–472. |
| Tolmachev, Nov. dan. fl. Vayg. | Tolmachev A.I. Novyye dannyye o flore ostrova Vaygach. [New data on the flora of Vaygach Island]. Botan. zhurn., 1936, vol. 21, No. 1, pp. 88–91. |
| Tolmachev, Blyumental, Mat. fl. N.Z. | Tolmachev A.I., Blyumental I. Kh. Materialy dlya flory Novoy Zemli. [Materials for the flora of Novaya Zemlya]. Tr. Botan. muzeya AN SSSR, 1931, vol. 23, pp. 197–209. |
| Tolmachev, Tokarevskikh, Issled. rayona More-Yu | Tolmachev A.I., Tokarevskikh S.A. Issledovaniye rayona «lesnovo ostrova» u reki More-Yu v Bolshezemelskoy tundre. [Investigation of the "forest island" district on the River More-Yu in the Bolshezemelskaya Tundra]. Botan. zhurn., 1968, vol. 53, No. 4, pp. 560–566. |
| Trautvetter, Consp. fl. Now. Sem. | Trautvetter, E.R. Conspectus florae insularum Nowaja Semlja. Tr. SPb. bot. sada, I, 1871. |
| Vinogradova, Fl. Pym-va-shor | Vinogradova V.M. Flora rayona teplykh istochnikov Pym-va-shor v Bolshezemelskoy tundre. [Flora of the Pym-va-shor hot springs district in the Bolshezemelskaya Tundra]. Vestn. LGU, 1962, No. 9. Biol., vol. 2, pp. 22-34. |

## 2. Siberian Arctic

| | |
|---|---|
| Karavayev, Konsp. fl. Yak. | Karavayev, M.N. Konspekt flory Yakutii. [Conspectus of the flora of Yakutia]. Moscow; Leningrad, 1958. 190 pp. |
| Kjellman, Phanerog. sib. Nordk. | Kjellman F.R. Die Phanerogamenflora der sibirischen Nordküste. In: Nordenskiöld A.E. Die wissenschaftliche Ergebnisse der Vega-Expedition. Leipzig, 1883, Vol. I. pp. 94–139. |
| Krylov, Fl. Zap. Sib. | Krylov, P.N. Flora Zapadnoy Sibiri. [Flora of West Siberia]. Tomsk, 1927–1964, vols. I–XII. |
| Opred. rast. Yak. | Opredelitel vysshikh rasteniy Yakutii. [Key to the higher plants of Yakutia]. Novosibirsk, 1974. 533 pp. |
| Petrov, Fl. Yak. | Petrov V.A. Flora Yakutii [Flora of Yakutia], vol. I. Leningrad, 1930. |
| Schmidt, Fl. jeniss. | Schmidt F. Florula jenisseensis arctica. In: Wissenschaftliche Resultate der zur Aufsuchung eines angekündigten Mammuth-cadavers Expedition. St. Petersburg, 1872, pp. 73–133. |

| | |
|---|---|
| Tikhomirov, Petrovskiy, Yurtsev, Fl. Tiksi | Tikhomirov B.A., Petrovskiy V.V., Yurtsev B.A. Flora okrestnostey bukhty Tiksi (arkticheskaya Yakutiya). [Flora of the vicinity of the Bay of Tiksi (Arctic Yakutia)]. |
| Tolmachev, Fl. Taym. | Tolmachev A.I. Flora tsentralnoy chasti vostochnovo Taymyra. [Flora of the central part of East Taymyr]. Tr. Polyarnoy komissii, 1932, vol. 8, pp. 1–126; vol. 13, pp. 5–75; 1935, vol. 25, pp. 5–80. |
| Trautvetter, Fl. rip. Kolym. | Trautvetter E.R. Flora riparia Kolymensis. Acta Horti Petropol., 1877, vol. 5, pp. 495–574. |
| Trautvetter, Fl. taim. | Trautvetter E.R. Florula taimyrensis phaenogama. In: Phanerogame Pflanzen aus dem Hochnorden. St. Petersburg, 1847, pp. 17–64. |
| Trautvetter, Pl. Sib. bor. | Trautvetter E.R. Plantas Sibiriae borealis ab A. Czekanowski et F. Müller annis 1874 et 1875 lectas enumeravit. Acta Horti Petropol., 1877, vol. 5, p. 1-146. |
| Trautvetter, Syll. pl. Sib. bor.-or. | Trautvetter E.R. Syllabus plantarum Sibiriae boreali-orientalis a Dre Alex. Bunge fil. lectarum. Acta Horti Petropol., 1887, vol. 10, pp. 481–546. |

**Additional Sources**

| | |
|---|---|
| Aleksandrova, Fl. B. Lyakhovsk. | Aleksandrova V.D. Flora sosudistykh rasteniy ostrova Bolshovo Lyakhovskovo (Novosibirskiye ostrova). [Vascular plant flora of Bolshoy Lyakhovskiy Island (New Siberian Islands)]. Botan. zhurn., 1960, vol. 45, No. 11, pp. 1687–1693. |
| Drobov, Predst. sekts. *Ovinae* v Yakut. | Drobov, V.P. Predstaviteli sektsii *Ovinae* Fr. roda *Festuca* L. v Yakutskoy oblasti. [Representatives of section *Ovinae* of the genus *Festuca* L. in Yakutsk Oblast]. Petrograd, 1915. |
| Fl. Putorana | Flora Putorana. [Flora of Putorana]. Novosibirsk, 1976. 246 pp. |
| Fl. Stanov. nagorya | Vysokogornaya flora Stanovovo nagorya. [High alpine flora of the Stanovoye Highland]. Novosibirsk, 1972. 272 pp. |
| Hämet-Ahti, Cajand. vasc. pl. Lena R. | Hämet-Ahti L. A.-K. Cajander's vascular plant collection from the Lena River, Siberia, with his ecological and floristic notes. Ann. Bot. Fenn., 1970, vol. 7, pp. 255–324. |
| Korotkevich, Rastit. Sev. Zemli | Korotkevich E.S. Rastitelnost Severnoy Zemli. [Vegetation of Severnaya Zemlya]. Botan. zhurn., 1958, Vol. 43, No. 5, pp. 644–663. |
| Malyshev, Fl. Vost. Sayana | Malyshev L.I. Vysokogornaya flora Vostochnovo Sayana. [High alpine flora of the Eastern Sayan]. Moscow; Leningrad, 1965. 368 pp. |
| Middendorff, Gewächse Sibiriens | Middendorff, A.T. Die Gewächse Sibiriens. In: Middendorff, Sibirische Reise, Bd. IV, Theil 1, 1864. |
| Polozova, Tikhomirov, Rast. Tarei | Polozova T.G., Tikhomirov B.A. Sosudistyye rasteniya rayona Taymyrskovo statsionara (pravoberezhe Pyasiny bliz ustya Tarei. Zapadnyy Taymyr). [Vascular plants of the district of the Taymyr Station (right bank of the Pyasina near the mouth of the Tareya. West Taymyr)]. In book: Biogeotsenozy Taymyrskoy tundry i ikh produktivnost. [Biogeocenoses of the Taymyr tundra and their productivity]. Leningrad, 1971, pp. 161–184. |
| Scheutz, Pl. jeniss. | Scheutz N.J. Plantae vasculares jeniseenses. Kongl Svenska vetenskaps Akad. Handling., Bd. 22, No. 10, Stockholm, 1888. |
| Tikhomirov, Fl. Zap. Taym. | Tikhomirov B.A. K kharakteristike flory zapadnovo poberezhya Taymyra. [Towards the characterization of the flora of the west coast of Taymyr]. Petrozavodsk, 1948. 85 pp. (Tr. Karelofinskovo un-ta; vol. 2). |
| Tolmachev, Fl. o. Benneta | Tolmachev A.I. K flore ostrova Benneta. [Towards the flora of Bennet Island]. Bot. Zhurn., 44, 4, 1959. |

| | |
|---|---|
| Tolmachev, O fl. nakh. v tsentr. chasti Taym. | Tolmachev A.I. O neskolkikh neozhidannykh floristicheskikh nakhodkakh v tsentralnoy chasti Taymyrskovo polyostrova. [On some unexpected floristic discoveries in the central part of the Taymyr Peninsula]. DAN SSSR, ser. A, 1930, No. 5. |
| Tolmachev, Raspr. drev. porod | Tolmachev A.I. O rasprostranenii drevesnykh porod i severnoy granitse lesov v oblasti mezhdu Yeniseyem i Khatangoy. [On the distribution of tree species and the northern limit of forest in the region between the Yenisey and the Khatanga]. Tr. polyarnoy komissii, vol. 5, 1931. |
| Tolmachev, Rast. o. Sibiryakova | Tolmachev A.I. Obzor sosudistykh rasteniy ostrova Sibiryakova v Yeniseyskom zalive. [Review of the vascular plants of Sibiryakov Island in the Bay of Yenisey]. Tr. Botan. muzeya AN SSSR, 1931, vol. 23, pp. 211–218. |
| Tolmachev, Pyatkov, Obz. rast. Diksona | Tolmachev A.I., Pyatkov P.P. Obzor sosudistykh rasteniy ostrova Diksona. [Review of the vascular plants of Dikson Island]. Tr. Botan. muzeya AN SSSR, 1930, vol. 22, pp. 147–179. |
| Trautvetter, Fl. boganid. | Trautvetter E.R. Florula boganidensis phaenogama. In: Phanerogame Pflanzen aus dem Hochnorden. St. Petersburg, 1847, pp. 144–167. |
| Yurtsev, Fl. Suntar-Khayata | Yurtsev B.A. Flora Suntar-Khayata [Flora of Suntar-Khayata]. Leningrad, 1968, 235 pp. |

## 3. Far East, Chukotka

| | |
|---|---|
| Hultén, Fl. Kamtch. | Hultén E. Flora of Kamtchatka and the adjacent islands. Stockholm, 1927–1930, vols. I–IV. |
| Kjellman, Phanerog. as. K. Ber.-Str. | Kjellman F.R. Die Phanerogamenflora an der asiatischen Küste der Bering-Strasse. In: Nordenskiöld A.E. Die wissenschaftliche Ergebnisse der Vega-Expedition. Leipzig, 1883, vol. I, pp. 249–379. |
| Komarov, Fl. Kamch. | Komarov V.L. Flora poluostrova Kamchatki. [Flora of the Kamchatka Peninsula]. Leningrad, 1927-1930, vols. 1–3. |
| Opred. rast. Kamch. obl. | Opredelitel sosudistykh rasteniy Kamchatskoy oblasti. [Key to the vascular plants of Kamchatka Oblast]. Moscow, 1981, 411 pp. |
| Petrovskiy, Rast. o. Vrangelya | Petrovskiy V.V. Spisok sosudistykh rasteniy o. Vrangelya. [List of the vascular plants of Wrangel Island]. Botan. zhurn., 1973, vol. 58, No. 1, pp. 113–126. |
| Tikhomirov, Gavrilyuk, Fl. Bering. Chuk. | Tikhomirov B.A., Gavrilyuk V.A. K flore Beringovskovo poberezhya Chukotskovo poluostrova. [Towards the flora of the Beringian coast of the Chukotka Peninsula]. In book: Rasteniya severa Sibiri i Dalnevo Vostoka. [Plants of Northern Siberia and the Far East]. Moscow; Leningrad, 1966, pp. 58–79. |
| Trautvetter, Fl. Tschuk. | Trautvetter E.P. Flora terrae Tschuktschorum. St. Petersburg, 1878. 40 pp. (See also: Acta Horti Petropol., 1879, vol. 6, pp. 1–40). |
| Vasilev, Fl. Komand. ostr. | Vasilev V.N. Flora i paleogeografiya Komandorskykh ostrovov. [Flora and paleogeography of the Commander Islands]. Moscow; Leningrad, 1957. |
| Voroshilov, Opred. rast. D. Vost. | Voroshilov V.N. Opredelitel rasteniy sovetskovo Dalnevo Vostoka. [Key to plants of the Soviet Far East]. Moscow, 1982. 672 pp. |

**Additional Sources**

| | |
|---|---|
| Derviz-Sokolova, Fl. Dezhn. | Derviz-Sokolova T.G. Flora kraynevo vostoka Chukotskovo poluostrova (poselok Uelen–mys Dezhneva). [Flora of the far east of the Chukotka Peninsula (Uelen village–Cape Dezhnev)]. In book: Rasteniya severa Siberi i Dalnevo Vostoka. [Plants of Northern Siberia and the Far East]. Moscow; Leningrad, 1966, pp. 80–107. |
| Filin, Yurtsev, Rast. o. Ayon | Filin V.R., Yurtsev B.A. Sosudistyye rasteniya o. Ayon (Chaunskaya guba). [Vascular plants of Ayon Island (Chaun Bay)]. In book: Rasteniya severa Sibiri i Dalnevo Vostoka. [Plants of Northern Siberia and the Far East]. Moscow; Leningrad, 1966, pp. 44–57. |
| Floderus, Salic. Anadyr. | Floderus B. Salices peninsulae Anadyrensis. Arkiv f. bot., 25A, 10, 1933. |
| Floderus, Salix Kamtch. | Floderus B. On the Salix-flora of Kamtchatka. Arkiv f. bot., 20A, 6, 1926. |
| Fl. Sib. Daln. Vost. | Flora Sibiri i Dalnevo Vostoka. [Flora of Siberia and the Far East]. Moscow; Leningrad, 1966. |
| Kharkevich, Buch, Rast. Sev. Koryak. | Kharkevich S.S., Buch T.G. Sosudistyye rasteniya Severnoy Koryakii. [Vascular plants of Northern Koryakia]. Botan. zhurn., 1976, vol. 61, No. 8, pp. 1089–1102. |
| Khokhryakov, Fl. p-ova Taygonos | Khokhryakov A.P. K flore poluostrova Taygonos i severnovo poberezhya Gizhiginskoy guby. [Towards the flora of the Taygonos Peninsula and the north coast of the Bay of Gizhiga]. In book: Biologiya rasteniy i flora severa Dalnevo Vostoka. [Plant biology and flora of the Northern Far East]. Vladivostok, 1981, pp. 8–11. |
| Khokhryakov, Fl. r. Omolon | Khokhryakov A.P. K flore srednevo techeniya reki Omolon. [Towards the flora of the middle course of the River Omolon]. In book: Flora i rastitelnost Chukotki. [Flora and vegetation of Chukotka]. Vladivostok, 1978, pp. 53–75. |
| Khokhryakov, Mater. fl. yuzhn. ch. Magad. obl. | Khokhryakov A.P. Materialy k flore yuzhnoy chasti Magadanskoy oblasti. [Materials towards the flora of the southern part of Magadan Oblast]. In book: Flora i rastitelnost Magadanskoy oblasti. [Flora and vegetation of Magadan Oblast]. Vladivostok, 1976, pp. 3–36. |
| Khokhryakov, Yurtsev, Fl. Olsk. plato | Khokhryakov A.P., Yurtsev B.A. Flora Olskovo bazaltovovo plato (Kolymsko-Okhotskiy vodorazdel). [Flora of the Olskoye basalt plateau (Kolyma-Okhotsk watershed)]. Byul. MOIP. Otd. biol., 1974, vol. 79, No. 2, pp. 59–70. |
| Kitsing, Koroleva, Petrovskiy, Fl. b. Rodzhers | Kitsing L.I., Koroleva T.M., Petrovskiy V.V. Flora sosudistykh rasteniy okrestnostey bukhty Rodzhers (o. Vrangelya). [Vascular plant flora of the vicinity of Rodgers Bay (Wrangel Island)]. Bot. zhurn., 59, 7, 1974. |
| Kudo, Fl. Paramushir | Kudo Y. Flora of the island of Paramushir. J. Coll. Agric. Hokkaido Univ., 9, 2, 1922. |
| Kurtz, Fl. Tschuktsch. | Kurtz F. Die Flora der Tschuktschenhalbinsel. Nach den Sammlungen der Gebrüder Krause. Engler's Bot. Jahrb., 1895, Bd. 19, pp. 432–493. |
| Polezhayev & al., Fl. Bering. r-na | Polezhayev A.N., Khokhryakov A.P., Berkutenko A.N. K flore Beringovskovo rayona Magadanskoy oblasti. [Towards the flora of the Beringian district of Magadan Oblast]. Botan. zhurn., 1976, vol. 61, no. 8, pp. 1103–1110. |
| Schmidt, Reisen Amurl. | Schmidt Fr. Reisen im Amurlande und auf der Insel Sachalin. Mém. Acad. Sci. SPb., sér VIII, 12, 1, 1869. |
| Sugawara, Ill. fl. Saghal. | Sugawara S. Illustrated flora of Saghalien. II. 1939. Tokyo. |
| Vorobev, Mat. fl. Kuril. | Vorobev D.P. Materialy k flore Kurilskykh ostrovov. [Materials towards the flora of the Kurile Islands]. Tr. Dalnevost. filiala AN SSSR, ser. botan., vol. III(V), 1956. |
| Voroshilov, Fl. D. Vost. | Voroshilov V.N. Flora sovetskovo Dalnevo Vostoka. [Flora of the Soviet Far East]. Moscow, 1966, 476 pp. |

## 4. American Arctic

| | |
|---|---|
| Hultén, Comments Fl. Al. | Hultén E. Comments on the Flora of Alaska and Yukon. Stockholm, 1967. 147 pp. (Ark. Bot. Ser. 2, Bd. 7, Heft 1). |
| Hultén, Fl. Al. | Hultén E. Flora of Alaska and Yukon. Lund, 1941-1950, vols. I–X. |
| Hultén, Fl. Al. & neighb. terr. | Hultén E. Flora of Alaska and neighbouring territories. Stanford, 1968. 1008 pp. |
| Porsild, Ill. fl. Arct. Arch. | Porsild A.E. Illustrated flora of the Canadian Arctic Archipelago. Ottawa, 1957. 209 pp. (Nat. Mus. of Canada Bull. No. 146); 2nd edition, revised. Ottawa, 1964. |
| Porsild, Vasc. pl. W Can. Arch. | Porsild A.E. The vascular plants of the western Canadian Arctic Archipelago. Ottawa, 1955. 226 pp. (Nat. Mus. of Canada Bull. No. 135). |
| Porsild & Cody, Checklist pl. NW Canada | Porsild A.E., Cody W.J. Checklist of the vascular plants of the Continental Northwest Territories, Canada. Ottawa, 1968. 102 pp. |
| Porsild & Cody, Pl. continent. NW Canada | Porsild A.E., Cody W.J. Vascular plants of continental Northwest Territories, Canada. Ottawa, 1980. 667 pp. |
| Simmons, Survey phytogeogr. | Simmons H.G. A survey of the phytogeography of the Arctic American Archipelago with some notes about its exploration. Lund, 1913. 183 pp. |

**Additional Sources**

| | |
|---|---|
| Anderson, Fl. Al. | Anderson J.P. Flora of Alaska and adjacent parts of Canada. Ames, 1959. 724 pp. |
| Coville, Willows Alask. | Corille F.V. Willows of Alaska. Proceed. Washingt. Acad. Sci., 3, 1901. |
| Gjaerevoll, Bot. invest. centr. Alaska | Gjaerevoll O. Botanical investigations in central Alaska, especially in the White Mountains. Kgl. Norske Videnskabers Selskabs Skrifter, 1958, No. 5, pp. 1–74; 1963, No. 4, pp. 1–115; 1967, No. 10, pp. 1–63. |
| Gröntved, Vasc. pl. Arctic Amer. | Gröntved J. Vascular plants from Arctic North America. Report Fifth Thule exped. II, No. 1. Copenhagen, 1936. |
| Holm, Contr. morph. syn. geogr. distr. arct. pl. | Holm T. Contributions to the morphology, synonymy and geographical distribution of arctic plants. In: Report of the Canadian Arctic Expedition 1913–1918, vol. 5. Botany, part B. Ottawa, 1922. |
| Hultén, Fl. Aleut. Isl. | Hultén E. Flora of the Aleutian Islands and Westernmost Alaska Peninsula with notes on the flora of the Commander Islands. Stockholm, 1937. 397 pp.; 2nd edition, revised. Weinheim, 1960. 376 pp. |
| Hultén, Suppl. Fl. Al. | Hultén E. Supplement to Flora of Alaska and neighbouring territories. A study in the flora of Alaska and the Transberingian connection. Bot. Notis., 1973, vol. 126, No. 4, pp. 459–512. |
| Macoun & Holm, Vasc. pl. | Macoun J.M., Holm T. The vascular plants of the Arctic coast of America west of the 100th meridian. In: Report of the Canadian Arctic Expedition 1913–1918, vol. 5. Botany, part A. Ottawa, 1921. |
| Polunin, Bot. Can. E Arctic | Polunin N. Botany of the Canadian Eastern Arctic. National Museum of Canada, Bull. No. 97, Biological ser., No. 26, 1947. |
| Porsild, Bot. SE Yukon | Porsild A.E. Botany of Southeastern Yukon adjacent to the Canol Road. Bull. Nat. Mus. Canada, 151, 1951. |
| Porsild, Contrib. fl. Alaska | Porsild A.E. Contributions to the flora of Alaska. Rhodora, 41, 1939. |
| Porsild, Mat. fl. NW territ. | Porsild A.E. Materials for a flora of the continental Northwest territories of Canada. Sargentia, IV, 1943. |

| | |
|---|---|
| Raup, Bot. SW Mackenz. | Raup H.M. The botany of Southwestern Mackenzie. Sargentia, VI, 1947. |
| Raup, Willows Huds. | Raup H.M. The willows of the Hudson Bay region and the Labrador Peninsula. Sargentia, IV, 1943. |
| Raup, Willows W. Amer. | Raup H.M. The willows of boreal Western America. Contrib. Gray Herbar., CLXXXV, 1959. |
| Rydberg, Cespit. willows | Rydberg P.A. Cespitose willows of Arctic America. Bull. N.Y. Bot. Gard., I. 1899. |
| Schneider, Amer. willows | Schneider C. Notes on American willows. I, Bot. Gaz., LXVI, 2, 1918; II, ibid., LXVI, 4,1918; III, ibid., LXVII, 1, 1919; VI, J. Arnold Arboret., I, 1919, 67–97; VIII, ibid., 1919, 211–232; X, ibid., II, 1920, 65–90; XI, ibid., 1921, 185–204. |
| Simmons, Vasc. pl. Ellesm. | Simmons H.G. The vascular plants in the flora of Ellesmereland. Kristiania, 1906. |
| Welsh, Fl. Al. | Welsh S.L. Anderson's Flora of Alaska and adjacent parts of Canada. Provo, 1974. 724 pp. |
| Young, Fl. Lawr. Isl. | Young S.B. The vascular flora of St. Lawrence Island with special reference to floristic zonation in the arctic regions. Contrib. Gray Herbar. Harvard Univ., 1971, No. 201, pp. 11–115. |

## 5. Greenland

| | |
|---|---|
| Böcher & al., Fl. Greenl. | Böcher T.W., Holmen K., Jakobsen K. The flora of Greenland. 2nd edition. Copenhagen, 1968. 312 pp. |
| Böcher & al., Grønl. Fl. | Böcher T.W., Holmen K., Jakobsen K. Grønlands Flora. Copenhagen, 1957. |
| Holmen, Vasc. pl. Peary L. | Holmen K. The vascular plants of Peary Land, North Greenland. Meddl. om Grønl., 1957, Bd. 124, No. 9, pp. 1–149. |

**Additional Sources**

| | |
|---|---|
| Devold & Scholander, Fl. pl. SE Greenl. | Devold J., Scholander P.F. Flowering plants and ferns of Southeast Greenland. Skrifter om Svalbard og Ishavet, 1933, No. 56, pp. 1–209. |
| Floderus, Grönl. Salic. | Floderus B. Om Grönlands Salices. Meddl. om Grønl., 1923, 63. |
| Gelting, Vasc. pl. E Greenl. | Gelting P. Studies on the vascular plants of Eastern Greenland. Meddl. om Grønl., 1934, Bd. 101, No. 2, pp. 1–340. |
| Jorgensen & al., Flow. pl. Greenl. | Jorgensen C.A., Sørensen Th., Westergaard M. The flowering plants of Greenland. A taxonomical and cytological survey. Copenhagen, 1958. 172 pp. |
| Lagerkranz, Fl. W & E Greenl. | Lagerkranz J. Observations on the flora of West and East Greenland. Nova Acta Reg. Soc. Sci. Upsal., ser. 4, XIV, 6, 1950. |
| Lange, Consp. fl. groenl. | Lange J. M. C. Conspectus florae groenlandicae. Meddl. om Grønl., Bd. 3, 1880. |
| Ostenfeld, Fl. Greenl. | Ostenfeld C. H. The flora of Greenland and its origin. Det Kgl. Danske Videnskabernes Selsk. Biologiske Meddelelser, Bd. VI, 3, Copenhagen, 1926. |
| Polunin, Contrib. fl. SE Greenl. | Polunin N. Contribution to the flora and phytogeography of Southeastern Greenland. J. Linn. Soc., Bot., LII, 1943. |
| Porsild, Fl. Disko | Porsild M. The flora of Disko Island and the adjacent coast of West Greenland. Meddl. om Grønl., Bd 58, 1920. |
| Seidenfaden, Vasc. pl. SE Greenl. | Seidenfaden G. The vascular plants of Southeast Greenland from 60° 04' to 64° 30' N lat. Meddl. om Grønl., 1933, Bd. 106, No. 3, pp. 1–129. |

| | |
|---|---|
| Seidenfaden & Sørensen, Summary spec. E Greenl. | Seidenfaden G., Sørensen Th. Summary of all species found in Eastern Greenland. Meddl. om Grønl., Bd. 101, 4, 1937. |
| Seidenfaden & Sørensen, Vasc. pl. NE Greenl. | Seidenfaden G., Sørensen Th. The vascular plants of North-East Greenland from 74° 30' to 79°N lat. Meddl. om Grønl., Bd. 101, 4, 1937. |
| Sørensen, Revis. Greenl. sp. Puccinellia | Sørensen Th. A revision of the Greenland species of Puccinellia Parl. Copenhagen, 1953. |
| Sørensen, Vasc. pl. E Greenl. | Sørensen Th. The vascular plants of East Greenland from 71° 00' to 73° 30'. Meddl. om Grønl., 1933, Bd. 101, No. 3, pp. 1–177. |

## 6. Arctic Western Europe

| | |
|---|---|
| Fl. europ. | Flora europaea. Cambridge, 1964-1980, vols. 1–5. |
| Gröntved, Pterid. Spermatoph. Icel. | Gröntved J. The Pteridophyta and Spermatophyta of Iceland. Copenhagen, 1942. 427 pp. |
| Hultén, Atlas | Hultén E. Atlas of the distribution of vascular plants in NW Europe. Stockholm, 1950. 512 pp.; 2nd edition, revised, 1971. 531 pp. |
| Löve, Isl. Ferdafl. | Löve A. Islenzk Ferdaflora. Reykjavik, 1970. 428 pp. |
| Löve & Löve, Consp. Icel. fl. | Löve A., Löve D. Cytotaxonomical conspectus of the Icelandic flora. Acta Horti Gotoburg., 1956, vol. 20, No. 4, pp. 65–290. |
| Rønning, Svalb. fl. | Rønning O.I. Svalbards flora. Oslo, 1979. 128 pp. |
| Scholander, Vasc. pl. Svalb. | Scholander P.F. Vascular plants from northern Svalbard. Skrifter om Svalbard og Ishavet, 1934, No. 62, pp. 1–155. |

**Additional Sources**

| | |
|---|---|
| Andersson, Salic. Lappon. | Andersson N.J. Salices Lapponiae. 1845. Uppsala. |
| Benum, Fl. Troms | Benum P. Flora of Troms fylke. 1958. Tromsö. |
| Floderus, Salic. fennoscand. | Floderus B. Salicaceae fennoscandicae. Stockholm, 1931. |
| Fries, Mantissa | Fries E.M. Novitiarum florae Sueciae mantissa. I. 1832. Lund-Uppsala. |
| Hadač, Gefässpfl. Sassengeb. | Hadač E. Die Gefässpflanzen des «Sassengebietes» Westspitsbergen. Norges Svalbard- og Ishavetsundersøkelser. Skrifter, 87, Oslo, 1944. |
| Hadač, Hist. fl. Spitsb. | Hadač E. The history of the flora of Spitsbergen. Preslia, XXXII, 1960. |
| Hadač, Not. fl. Svalb. | Hadač E. Notulae ad floram Svalbardiae spectantes. Studia Bot. Cechica, vol. 5, 1942. |
| Hylander, Nord. Kärlväxtfl. | Hylander N. Nordisk Kärlväxtflora, I. Botan. Notiser, 1953, Heft 3. |
| Lagerberg & al., Pohj. luon. I | Lagerberg T., Kalela A., Väänänen H. Pohjolan luonnon kasvit. I. 1958. Helsinki. |
| Lid, Fl. Jan Mayen | Lid J. The flora of Jan Mayen. Oslo, 1964. 108 pp. (Norsk Polarinstitutt Skrifter; No. 130). |
| Lid, Norsk & Svensk fl. | Lid J. Norsk og Svensk flora. Oslo, 1974. 808 pp. |
| Lindman, Svensk fanerogamfl. | Lindman C.A.M. Svensk fanerogamflora. Utg. 2. 1926. Stockholm. |

| | |
|---|---|
| Rønning, Vasc. Fl. Bear Isl. | Rønning O.I. The vascular flora of Bear Island. Tromsø, 1959. 62 pp. (Acta Borealia A. Scientia; No. 15). |
| Wahlenberg, Fl. lappon. | Wahlenberg G. Flora lapponica. 1812. Berlin. |

## 7. General Sources

| | |
|---|---|
| Cherepanov, Rast. SSSR | Cherepanov S.K. Sosudistyye rasteniya SSSR. [Vascular plants of the USSR]. Leningrad, 1981. 509 pp. |
| Dorogostayskaya, Sorn. rast. Sev. | Dorogostayskaya E. Sornyye rasteniya Kraynevo Severa SSSR. [Weeds of the Far North of the USSR]. |
| Endem. vysokogor. rast. Sev. Azii | Endemichnyye vysokogornyye rasteniya Severnoy Azii. [Endemic high alpine plants of Northern Asia]. Novosibirsk, 1975. 336 pp. |
| Fl. SSSR | Flora SSSR. [Flora of the USSR]. Moscow; Leningrad, 1934–1960, vols. I–XXX. |
| Gelert & Ostenfeld, Fl. arct. | Gelert O., Ostenfeld C.H. Flora arctica. Part 1. Copenhagen, 1902. |
| Hultén, Amph-Atl. pl. | Hultén E. The Amphi-Atlantic plants and their phytogeographical connections. Stockholm, 1958. 340 pp. |
| Hultén, Circump. pl. | Hultén E. The circumpolar plants. Stockholm, 1962, pt. I; 1971, pt. II. |
| Löve & Löve, Cytotaxon. atlas arct. fl. | Löve A., Löve D. Cytotaxonomical atlas of the arctic flora. Vaduz, 1975. 598 pp. |
| Polunin, Circump. arct. fl. | Polunin N. Circumpolar Arctic Flora. Oxford, 1959. 515 pp. |

## 8. Miscellaneous

| | |
|---|---|
| Andersson, Monogr. Salic. | Andersson N.J. Monographia Salicum. 1867. Stockholm. |
| Andersson, Nordamer. Salic. | Andersson N.J. Bidrag till kännedomen om de i Nordamerika förekommande Salices. Öfversigtat K. Vet. Akad. Förhandl. XV, 1858. |
| Andersson, Salic. Japon. | Andersson N.J. Salices e Japonica. Mem. Amer. Acad., N.S., VI, 2, 1858. |
| Buchenau, Monogr. Junc. | Buchenau F. Monographia Juncacearum. Engler's Botan. Jahrb., Bd. 12. 1890. |
| Der. i kust. SSSR | Derevya i kustarniki SSSR. [Trees and shrubs of the USSR]. Vols. I–VI. Moscow-Leningrad, 1949–1962. |
| Dylis, Sib. listv. | Dylis N.V. Sibirskaya listvennitsa. [Siberian larch]. Mater. k pozn. fauny i flory SSSR, nov. ser., otd. botan., vol. 3. Moscow, 1947. |
| Fl. Az. Ros. | Flora Aziatskoy Rossii. [Flora of Asiatic Russia]. |
| Gilibert, Exerc. phytol. II | Gilibert J.E. Exercitia phytologica. II. 1792. Lyons. |
| Gorodkov, Obz. russk. osok | Gorodkov B.N. Obzor russkikh osok. [Review of Russian sedges]. Tr. Botan. muzeya AN SSSR, vol. XX, 1927. |
| Hagström, Crit. Res. Potamog. | Hagström J.O. Critical researches on the Potamogetons. K. Svenska Vetensk. Akad. Handl., Bd. 55, 5, 1916. |
| Hoffman, Hist. Salic. I | Hoffman G.F. Historia Salicum iconibus illustrata. I. 4. 1787. Leipzig. |

| | |
|---|---|
| Holmen & Mathiesen | Holmen K., Mathiesen H. Luzula Wahlenbergii in Greenland. Bot. tidsskr., Bd. 43, Heft 3, 1953. |
| Host, Salix | Host N.T. Salix. 1828. Vienna. |
| Keppen, Raspr. khv. der. | Keppen F. Geograficheskoye rasprostraneniye khvoynykh derev v yevropeyskoy Rossii i na Kavkaze. [Geographical distribution of coniferous trees in European Russia and the Caucasus]. St. Petersburg, 1885. |
| Kimura, Symb. Iteol. | Kimura A. Symbolae Iteologicae. IV, Sci. Repert. Tohoku Univ., Biol., XII, 1937; VI, ibid., XIII, 1938. |
| Krall & Viljasoo, Eestis pajud. | Krall H., Viljasoo L. Eestis Kasvavad pajud. 1965, Tartu. |
| Kükenthal, Cyper. Caricoid. | Kükenthal G. Cyperaceae-Caricoideae. In: Der Pflanzenreich, Heft 38, Leipzig, 1909. |
| Kükenthal, Cyper. Sibir. | Kükenthal G. Cyperaceae Sibiriae. Russk. botan. zhurn., 1911, Nos. 3–6. |
| Ledebour, Fl. alt. | Ledebour C.F. Flora altaica. IV. 1833. Berlin. |
| Ledebour, Ic. pl. fl. ross. | Ledebour C.F. Icones plantarum novarum vel imperfecte cognitarum floram rossicam, imprimis altaicam, illustrantes. V. 1834. Riga. |
| Linné fil., Supplem. | Linné C. fil. Supplementum plantarum systematis vegetabilium...1781. Braunschweig. |
| Moench, Meth. | Moench C. Methodus plantas horti botanici et agri Marburgensis a staminum situ describendi. 1794. Marburg. |
| Pallas, Fl. ross. | Pallas P.S. Flora rossica. I, 2. 1788. St. Petersburg. |
| Polunin, Real Arctic Pterid. | Polunin N. The Real Arctic and its Pteridophyta. American fern journ., vol. 41, No. 2, 1951. |
| Rasinsh, Ivy Latv. | Rasinsh A.P. Ivy Latviyskoy SSR. [Willows of the Latvian SSR]. Tr. Inst. biolog. AN Latv. SSR, 8, 1959. |
| Regel, Monogr. Betulac. | Regel E. Monographia Betulacearum hucusque cognitarum. 1861. Moscow. |
| Ruprecht, Distrib. crypt. vasc. | Ruprecht F.J. Distributio Cryptogamarum vascularium in Imperio Rossico. In: Ruprecht. Symbolae ad historiam et geographiam plantarum rossicarum. St. Petersburg, 1846. |
| Salisbury, Prodrom. stirp. horto Allerton | Salisbury R.A. Prodromus stirpium in horto ad Chapel Allerton vigentium. 1796. London. |
| Seemen, Salic. japon. | Seemen O. Salices japonicae. 1903. Berlin. |
| Skvortsov, Mat. iv. | Skvortsov A.K. Materialy po morfologii i sistematike ivovykh. [Materials for the morphology and systematics of Salicaceae]. I, Byull. MOIP, biol., LX, 3, 1959; II, ibid., LXI, 1, 1956; III, IV, Bot. mat. Gerb. Bot. inst. AN SSSR, XVIII, 1957; V, Sist. zamet. Gerb. Tomsk. univ., 1956, 79–80; VI, Bot. mat. Gerb. Bot. inst. AN SSSR, XIX, 1959; IX, ibid., XXI, 1961; X, Byull. MOIP, biol., XVI, 4, 1961; XI, Tr. MOIP, III, 1960. |
| Sukachev, Dendrologiya | Sukachev V.N. Dendrologiya s osnovami lesnoy geobotaniki. [Dendrology on the foundations of forest geobotany]. Leningrad, 1938. |
| Tikhomirov, Kedr. stlanik | Tikhomirov B.A. Kedrovyy stlanik, yevo biologiya i ispolzovaniye. [Cedar pine stlanik, its biology and utilization]. Mater. k pozn. fauny i flory SSSR, nov. ser., otd. botan., vol. 6. Moscow, 1949. |
| Tikhomirov, Proiskh. ass. kedr. stl. | Tikhomirov B.A. K proiskhozhdeniyu assotsiatsiy kedrovovo stlanika. [On the origin of the cedar pine stlanik association]. Mater. po istorii flory i rastitelnosti SSSR, sb. II. Moscow; Leningrad, 1946. |

| | |
|---|---|
| Tikhomirov, Raspr. papor. | Tikhomirov B.A. Rasprostraneniye paporotnikov v Sovetskoy Arktike. [Distribution of ferns in the Soviet Arctic]. Botan. mater. Gerbariya BIN, vol. 19, 1959. |
| Tolmachev, Ist. temnokhv. taygi | Tolmachev A.I. K istorii vozniknoveniya i razvitiya temnokhvoynoy taygi. [Towards the history of the origin and development of spruce-fir taiga]. Moscow; Leningrad, 1954. |
| Trautvetter, Incrementa | Trautvetter E.R. Incrementa florae phaenogamae rossicae. III. Acta Horti Petrop., IX, 1884. |
| Trautvetter, Salic. frigid. | Trautvetter E.R. De Salicibus frigidis Kochii. Nouv. Mém. Soc. Nat. Mosc., II, 1832. |
| Trautvetter, Salic. livon. | Trautvetter E.R. De Salicibus livonicis dissertatio. Nouv. Mém. Soc. Nat. Mosc., II, 1832. |
| Wimmer, Salic. europ. | Wimmer C.F.H. Salices europaeae. 1866. Bratislava. |
| Wolf, Mat. izuch. iv Yevr. Ross. | Wolf E. Materialy dlya izucheniya iv, rastushchikh diko v Yevropeyskoy Rossii. [Materials for the study of willows growing wild in European Russia]. Izv. SPb. Lesn. inst., 1900, 4–5. |

## FAMILY I
# Polypodiaceae R. Br.
**TRUE FERNS**

---

PERENNIAL HERBACEOUS PLANTS with prostrate stems (rhizomes) which are subterranean or partly exposed at the soil surface and at whose apices develop more or less divided leaves (fronds) frequently arranged in clusters. The degree and character of leaf division is very different in different genera and species. Sporangia develop either on the lower side of the normal leaves or on leaves which differ sharply from the others in appearance and structure.

The ferns growing in extratropical countries are found mainly in regions with relatively mild and persistently moist climate. In countries with hard winters their existence is mainly dependent on the presence of sufficiently thick snow cover which persists through the whole winter. Overwintering conditions are of critical importance for them. At the same time, certain species of ferns require only very limited summer heat, which allows them to grow at high latitude and in oceanic districts. However, in inland continental regions, despite the presence of a warmer summer, ferns are poorly represented and their distribution is narrowly localized.

*All* ferns growing in the Arctic are also native to the forest section of the temperate zone of the Northern Hemisphere. But they are rather sharply divisible into two groups according to their ecological associations and the character of their distribution. The first group includes true forest plants, distributed basically in the taiga zone or possessing a wider distribution (native also to the broad-leaved forest zone and sometimes to mountain forests at lower latitudes). They only locally exceed the northern limits of the taiga zone, occurring in the southern part of the tundra zone principally in regions with relatively mild oceanic climate (Atlantic or to a lesser degree Pacific districts of the Arctic). They are alien to the greater part of our region, that is to all high-latitudinal districts whose flora is typically arctic in all respects.

Other ferns associated (at least in the Arctic) with rocky habitats or growing in the vicinity of rocks, have a more extensive but not universal arctic distribution. This may be explained by taking into account not only contemporary ecological conditions but also peculiarities of the geological history of different arctic districts. Some members of this group (*Woodsia* and *Cystopteris* spp., *Dryopteris fragrans*) occur in localities at very high latitudes including districts with floras of high arctic type. All possess a more or less fragmented distribution, with particular localities sometimes very isolated from other parts of their range.

Irrespective of how characteristic they are for the arctic flora and the extent of their distribution within arctic limits, the species of ferns growing here do not display any tendency to form special arctic races and are represented in our region by the same forms which are distributed beyond its limits.

1. *Leaves of two different types:* vegetative (sterile) complexly divided into flat pinnules with scalloped margins, never bearing sporangia; fertile with pinnules whose revolute margins cover groups of confluent sori; from above these pinnules appear linear-elliptic with entire margins. Small plants (no higher than 20 cm) growing in boulder fields and rock crevices. ......... .................................... GENUS 9. **CRYPTOGRAMMA** R. BR.
- *All leaves uniform,* with flat segments on whose lower surfaces sporangia may develop. ................................................... 2

2. Rhizome *thick, obliquely ascending,* densely covered with remains of stipes of dead leaves. Leaves developing at tip of rhizome, *for the most part forming more or less dense cluster.* ................................. 3
- Rhizome *thin, slender, prostrate.* Leaves arising individually from its lateral offshoots or approaching its tip but arranged in row, not forming cluster. Leaf stipes always thin, for the most part slender. .................... 7

3. Leaves *rigid, leathery, overwintering in green state,* 15–40 cm long, with very short stipe. Leaf blade linear-lanceolate in general outline, divided into numerous, crowded, acuminate, mostly slightly falcate segments. These have sharply serrate margins with a triangular projection near the base on one side, narrowing to a spinose tip. Sori arranged in two rows parallel to the midrib of the segment, contiguous with one another at full development. ...... GENUS 6. **POLYSTICHUM** ROTH [**P. LONCHITIS** (L.) ROTH].
- Leaves less rigid, *developing at the beginning of summer and dying at the onset of cold.* Their structure other than described above. ............. 4

4. Low plants (leaves not more than 15 cm long) of stony and rocky places. Leaves pinnate, with segments divided into pinnules or rounded with crenate margins. *Sori at full development contiguous* with one another to such a degree that *they completely or almost cover the whole lower surface of the segments bearing them.* ....................................... 5
- *Plants predominantly taller* (leaf length 30–50 cm or more; only in one species 10–20 cm). Leaves 2–4 times pinnate. Sori arranged in two rows on the pinnules, where at full development they may cover a large part of the lower surface of a particular pinnule *but never form a complete covering over the lower surface of the segments,* in particular leaving a free strip along the midrib of the pinnule. ...................................... 6

5. *Leaf segments almost orbicular, with crenate margins,* with slightly tapering base. Leaf blade narrow, linear in general outline. Leaves bent downwards or almost prostrate on the substrate. ................................ ................................ GENUS 8. **ASPLENIUM** L. (**A. VIRIDE** HUDS.).
- *Leaf segments mostly divided into pinnules,* ovoid or oblong. Leaf blade oblong-lanceolate or linear in general outline. Leaves standing erect or ascending upwards. ........................... GENUS 1. **WOODSIA** R.BR.

6. *Sori oblique-elliptic or oblong, covered by semilunate indusia.* Lower part of leaf stipe bearing more or less numerous brownish scales; *upper part of stipe and rachis of leaf blade without scales.* Leaf blade oblong-elliptic in general outline, gradually narrowed and acuminate towards tip, moderately narrowed towards base. ................ GENUS 7. **ATHYRIUM** ROTH.
– *Sori orbicular, covered by orbicular-reniform indusia. Stipe along its whole length and rachis of leaf bearing brownish or ferruginous scales,* which sometimes form a rather thick covering. Leaf blade elliptic-lanceolate or ovoid-triangular. ......................... GENUS 3. **DRYOPTERIS** ADANS.

7. Leaf blade oblong-triangular, *divided into segments which are not completely separated from one another* but merge basally. Segments obtuse, with undulate or weakly dentate margins. Leaves *rather rigid,* dark green, *overwintering in green state.* Sori orbicular, rather large, arranged in two rows on lower surface of segments. .................................
............................ GENUS 10. **POLYPODIUM** L. (**P. VULGARE** L.).
– Leaf blade *divided into completely separate segments.* Leaves *tender, dying at the onset of cold.* .................................................. 8

8. Leaf blade in general outline broadly ovoid-triangular, *divided into three similar parts,* each with pinnate arrangement of segments. Lateral parts of leaf joined to its axis by long thin stipes, with distinct articulation at points of junction with main rachis. *Sori without indusia.* .....................
........................................ GENUS 5. **GYMNOCARPIUM** NEWM.
– Leaf blade normally bipinnate, sometimes incompletely pinnatifid on its distal part. The basal pair of segments can be larger and possess more complex subdivision than other pairs of segments, but remains basically similar to them. ..................................................... 9

9. Leaf blade *oblong-triangular* in general outline. Segments separate on basal part of leaf, *simply pinnatifid,* merging basally towards tip of leaf with subdivision scarcely indicated. *Basal pair of segments somewhat bent backwards* (towards base of stipe). *Sori without indusia.* ................
............ GENUS 4. **THELYPTERIS** SCHMIDEL [**TH. PHEGOPTERIS** (L.) SLOSS].
– Leaf blade *oblong or broadly triangular* in general outline, *2(–3) pinnate,* with separate segments often rather remote on basal part of leaf. *Sori with hyaline indusia* covering them from above, this later peeling and withering. ....................................... GENUS 2. **CYSTOPTERIS** BERNH.

## GENUS 1    Woodsia R.Br. — WOODSIA

SMALL FERNS WITH short thick rhizome, near whose tip the withered stipes of dead leaves are usually retained over a period of years, forming a peculiar "brush." Leaves arranged in a dense cluster at the tip of the rhizome. Densely crowded clusters can form small beds. Leaf blade rather thick, almost leathery, divided into numerous more or less crowded segments which are divided in their turn. Sori at full development fusing on the lower side of the segments into a continuous brownish covering reminiscent of felt.

Plants of rocky places, growing on cliffs, along rocky crests of tundra ridges and mounds, and among rock rubble. Arctic distribution very uneven. Absent not only from many lowland localities, but also from certain upland districts where conditions suiting the ecological requirements of *Woodsia* species doubtless exist (e.g. on Novaya Zemlya). Two of the three species represented in the arctic flora (*W. glabella, W. alpina*) are distributed in arctic districts at high latitude, while the third (*W. ilvensis*) penetrates the limits of the Arctic near its southern fringe. None of the species is a specifically arctic plant.

1. Leaf blade up to 6 cm long, *narrow and linear-lanceolate* in general outline. Stipe *very short, pale green* like the rachis, with a few pale scales. *Rachis and leaf segments glabrous* or only bearing isolated hairs. Segments *short*, oval-rhomboid or broadly triangular, most frequently incompletely divided into three lobes with dentate margins. Only bases of stipes of dead leaves retained, forming brush which only just projects above the rhizome. . . . . . . . . . . . . . . . . . . . . . . . . . . . . . . . . . . . . . . . . .1. **W. GLABELLA** R. BR.
– Leaf blade up to 10 cm long or more, *oblong-lanceolate*. Segments (at least the larger of them) *pinnatifid*. Stipe *yellowish brown* or *ruddy brown*, rather thick, ¼-⅔ as long as leaf blade, *distinctly rising above substrate*. Stipes of dead leaves forming rather high brush which is very conspicuous. . . . . . . . . . . . . . . . . . . . . . . . . . . . . . . . . . . . . . . . . . . . . . . . . . . . . . . . .2

2. Stipe *yellowish brown*, ¼-⅔ as long as leaf blade. Rachis and leaf blade *at first bearing narrow scales and hairs, later almost glabrous*. Blade *bright green*. Segments 8–12 pairs in total, with obtusely rounded tips, the larger with 2–3 pairs of lateral lobes. . . . . . . . . . . . . . .2. **W. ALPINA** (BOLTON) GRAY.
– Stipe *ruddy brown*, ½-⅔ as long as leaf blade. Stipe, rachis and lower surface of segments *densely covered with brown scales and long septate hairs*; upper surface of segments with scattered hairs. Leaf blade *dark green*, often brownish. Segments 8–20 pairs in total, the larger of them possessing *3–7 pairs of lateral lobes*. . . . . . . . . . . . . . . . . . . . .3. **W. ILVENSIS** (L.) R. BR.

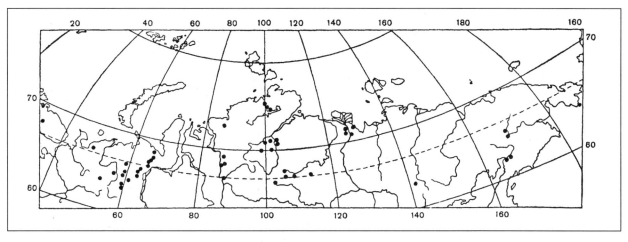

MAP I–2  Distribution of *Woodsia glabella* R. Br.

1. ***Woodsia glabella*** R. Br. in Richardson, Bot. App., Franklin Narrative (1823) 754; Ruprecht, Distrib. crypt. vasc. 123; Ledebour, Fl. Ross. IV, 511; Lange, Consp. fl. Groenl. 189; Scheutz, Pl. jeniss. 198; Simmons, Vasc. pl. Ellesm. 184; id., Survey Phytogeogr. 39; M. Porsild, Fl. Disko 25; Ostenfeld, Fl. Greenl. 55; Krylov, Fl. Zap. Sib. I, 14; Petrov, Fl. Yak. I, 5; Fomin, Fl. Sib. Daln.Vost. V, 16; id., Fl. SSSR I, 22; Devold & Scholander, Fl. pl. SE Greenl. 22; Sørensen, Vasc. pl. E Greenl. 21; Gelting, Vasc. pl. E Greenl. 30; Perfilev, Fl. Sev. I, 43; Leskov, Fl. Malozem. tundry 11; Gröntved, Vasc. pl. Arct. North Amer. 15; id., Pterid. Spermatoph. Icel. 112; Hultén, Fl. Al. I, 13; id., Atlas 13(51); Polunin, Bot. Can. E Arctic 28; Polunin, Real Arctic Pterid. 40; Poyarkova, Fl. Murm. I, 24; A.E. Porsild, Vasc. pl. W Canad. Arch. 69; id., Ill. Fl. Arct. Arch. 17; Holmen, Vasc. pl. Peary L. 42; Böcher & al., Ill. Fl. Greenl. 46; Tikhomirov, Raspr. papor. 611.

*W. ilvensis* var. *glabella* Trautv. — Gelert, Fl. arct. I, 7.

**Ill.:** Fl. arct. I, fig. 8; Fl. Murm. I, pl. 1,3; Porsild, Ill. Fl., fig 1c.[1]

Found on rocks and at the foot of rocks, on stony substrate or on patches of soil within boulder fields, mainly on south or southwest slopes, on sites sufficiently protected by snow cover during winter.[2] According to the reports of Tikhomirov for Taymyr and of Simmons for the north of the Canadian Archipelago, sites where *W. glabella* grows are characterized by a combination of protective snow cover in winter and conditions favourable for solar heating in summer. In Taymyr *W. glabella* sometimes grows in association with other larger plants whose tussocks (especially of *Sieversia glacialis*) serve to protect the small fern.

Normal sporulation (ripe spores are dispersed at the end of summer, the plant "dusts") is reported even at the most northern localities in the Soviet Arctic (Central Taymyr). Simmons' observations on Ellesmere Island indicate the association of *W. glabella* with massifs of the Archean group. Observations in the Soviet Arctic do not confirm that this association is obligatory (Tikhomirov).[3]

**Soviet Arctic.** West Murman (Pechenga); North Timan (very rare); Polar Ural (numerous localities, as far north as the basin of the River Kara); right bank of the Lower Yenisey (Khantayka, upper course of River Dudinka); lower course of River Pyasina; upper section of Lower Taymyra River and NW shore of Lake Taymyr; lower reaches of Olenek; lower reaches of Lena; Bay of Tiksi; SE Chukotka Peninsula (Provideniye Bay); upper part of Anadyr Basin; basin of River Penzhina. (Map I–2).

**Foreign Arctic.** West coast of Alaska; Canadian arctic coast; Labrador; Canadian Archipelago (on Ellesmere Island to 79°N); Greenland (north to Peary Land, south

---

[1] Entries are only given for figures which we recommend for use.

[2] According to observations in East Greenland (Sørensen), protection by snow cover is not absolutely necessary for overwintering of *W. glabella*.

[3] Evidently, in the given case we are faced with the generally not uncommon situation that the association of a particular species with particular (especially edaphic) conditions in some part of its range does not reflect absolute fidelity to them. In this case *W. glabella* is in fact restricted to places with a particular type of substrate in the north of the Canadian Archipelago; but apparently not because the presence of this substrate is *necessary* for its growth, but due to some other cause. In Taymyr a corresponding restriction to particular substrates has not at all been observed.

to 67°N on the west coast, to 61½°N on the east coast); Iceland; Spitsbergen; Arctic Scandinavia.

**Outside the Arctic.** Mountains of Central Europe, Northern Scandinavia and central part of Kola Peninsula; part of Pechora Basin near the Urals; Northern Ural; northern part of Central Siberian Plateau from the lower reaches of the Lower Tunguska to Norilsk and the western part of the Khatanga Basin, and to the upper course of the Olenek River (this extensive area of distribution scarcely delimited from the arctic part of the range); Verkhoyansk Range; mountains of Southern Siberia from the Altay to Northern Preamuria and the Lena-Okhotsk watershed; Sakhalin; Kamchatka; mountain forest districts of NW America.

2. ***Woodsia alpina*** (Bolton) Gray, Nat. Arrang. Brit. pl. II (1821), 17; Petrov, Fl. Yak. I, 7; Fomin, Fl. Sib. Daln. Vost. V, 14; id., Fl. SSSR I, 23; Devold & Scholander, Fl.pl. SE Greenl. 22; Hultén, Fl. Al. I, 13; id., Atlas 13 (50); Poyarkova, Fl. Murm. I, 22; Polunin, Bot. Can. E Arct. 28; id., Real Arctic Pterid. 40; A.E. Porsild, Ill. Fl. Arct. Arch. 17; Böcher & al., Ill. Fl. Greenl. 46; Tikhomirov, Raspr. papor. 599.

*Woodsia hyperborea* R. Br. — Ruprecht, Distrib. crypt. vasc. 121; Ledebour, Fl. Ross. IV, 511; Lange, Consp. fl. Groenl. 189.

*W. ilvensis* var. *alpina* Asch. et Graebn. — Gelert, Fl. arct. I, 7; Simmons, Vasc. pl. Ellesm. 183; id., Survey Phytogeogr. 39; M. Porsild, Fl. Disko, 25; Krylov, Fl. Zap. Sib. I, 13; Perfilev, Fl. Sev. I, 42; Gröntved, Pteridoph. Spermatoph. Icel. 113.

*W. ilvensis* — Ostenfeld, Fl. Greenl. 55 (pro parte).

*Acrostychum alpinum* Bolton, Fil. Brit. (1790) 76.

Ill.: Fl. arct. I, fig. 7; Fl. Murm. I, pl. I, 3; Porsild, Ill. Fl., fig. 1d.

Very rare plant with us. Reported in moist places on rock, at the outflows of hot springs on the Chukotka Peninsula, and on boulder fields in Koryakia.

**Soviet Arctic.** West Murman; Timan (Valsa); lower reaches of Lena (Kumakh-Surt); SE Chukotka Peninsula (at Senyavin Hot Springs); Bay of Korf.

**Foreign Arctic.** Beringian coast of Alaska; Labrador; Canadian Archipelago (on Ellesmere Island to 79°N); Greenland south of 74°N (rare); Iceland; Arctic Scandinavia.

**Outside the Arctic.** Northern Fennoscandia; mountains of Central and Western Europe (especially Great Britain); Urals (not significantly reaching Arctic limits); sporadically in a few places in the mountains of Southern Siberia; NW and Atlantic coasts of North America.

3. ***Woodsia ilvensis*** (L.) R. Br., Prodr. Fl. N. Hol. I (1810), 158; Ruprecht, Distrib. crypt. vasc. 120; Ledebour, Fl. Ross. IV, 510; Lange, Consp. fl. Groenl. 188; Scheutz, Pl. jeniss. 198; Ostenfeld, Fl. Greenl. 55 (pro parte); Fomin, Fl. Sib. Daln. Vost. V, 19; id., Fl. SSSR I, 23; Petrov, Fl. Yak. I, 6; Devold & Scholander, Fl. pl. SE Greenl. 23; Hultén, Fl. Al. I, 14; id., Atlas 13(52); Poyarkova, Fl. Murm. I, 21; Polunin, Bot. Can. E Arctic 27; id., Real Arctic Pterid. 41; A.E. Porsild, Ill. Fl. Arct. Arch. 17; Böcher & al., Ill. Fl. Greenl. 46; Tikhomirov, Raspr. papor. 615.

*W. ilvensis* var. *rufidula* (Michx.) Koch — Gelert, Fl. arct. I, 7; Gröntved, Pterid. Spermatoph. Icel. 113.

*W. ilvensis* ssp. *rufidula* Asch. & Graebn. — Krylov, Fl. Zap. Sib. I, 13.

*Acrostichum ilvense* L., Sp. pl. (1753) 1071.

Ill.: Fl. Sib. Daln. Vost. V, 20, 22; Fl. Murm. I, pl. I, 1; Porsild, Ill. Fl., fig. 1c.

Found mainly in gravelly areas, often near rocks, on crests and slopes of tundra ridges and mounds, in places moderately protected by snow cover. Sometimes forming small beds. Principally associated with outcrops of limestone and other calciferous rocks.

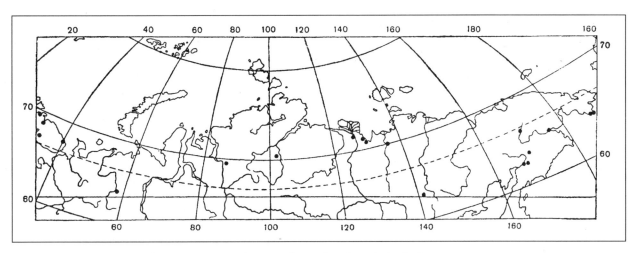

**Map I-3** Distribution of *Woodsia ilvensis* (L.) R. Br.

**Soviet Arctic.** Murman (western part and district of Iokanga); forest-tundra between Dudinka and Norilsk; lower reaches of Lena (Kumakh-Surt); district of the Bay of Tiksi; lower reaches of Yana; SE part of Chukotka Peninsula; Anadyr Basin; Penzhina Basin. (Map I-3).

**Foreign Arctic.** West coast of Alaska; Arctic Canada; Labrador; Canadian Archipelago (Baffin Island); Greenland north to 74° (reported as a common plant in SE Greenland; some records, especially for the west coast of Greenland, may refer to *W. alpina*); Iceland; Arctic Scandinavia.

**Outside the Arctic.** More or less mountainous districts of Northern, Central and Western Europe; Urals; northern part of Central Siberian Plateau; southern and central zone of Siberia from the Altay in the west to Sakhalin in the east, including high alpine districts. In the south reaching the Tien Shan, Korea and Central Japan. Isolated in the mountains of Asia Minor and Afghanistan. Forested regions of North America.

## ཀ GENUS 2   Cystopteris Bernh. — BLADDER FERN

THIS GENUS IS represented in the Arctic by three species, one of which *(C. montana)* is sharply distinguished from the others in morphological respects but scarcely crosses its borders and is essentially alien to the arctic flora. The two other species *(C. filix-fragilis, C. Dickieana)* are so close to one another that many investigators refrain from distinguishing them. But the latter point of view is apparently unfounded. Both these species are characteristic of the Arctic and widely distributed within its limits, reaching very high latitudes. Their distribution in different districts of the Arctic only manifests their dependence on the availability of suitable habitats. In some districts these ferns are common plants, but in others they are rarely met with and may be overlooked, especially in itinerant investigations, no doubt because of localization of their habitats. One of these species *(C. Dickieana)* shows a particular attraction to calciferous rock. The other *(C. filix-fragilis)* is apparently not closely tied to any chemical properties of the substrate. Its association with outcrops of particular rocks and absence from districts where others are developed is sometimes conditioned by differences in their physical properties, especially differences of cohesion and in the form of the fragments into which the rock breaks during weathering etc.

Both species characteristic of the Arctic are also extensively distributed outside its limits. One of them (*C. filix-fragilis*) possesses an almost cosmopolitan distribution and indeed is the most characteristic fern in the alpine flora of the temperate zone of the Northern Hemisphere. The other (*C. Dickieana*) is distributed outside the Arctic only in rather northern districts, mainly within Siberia, and the arctic part of its range appears fundamental.

1. Rhizome *thin, very extended.* Leaves arising from its offshoots, each of which bears a solitary leaf; consequently the individual leaves are rather distant from one another. Stipe *longer* than leaf blade. The latter *broad and ovoid-triangular* in general outline; *basal pair of segments considerably larger than the others and more strongly divided (2–3 pinnate)* ....... ............................................3. **C. MONTANA** (LAM.) DESV.
   
   – Rhizome not so long and *somewhat thickened close to its tip by numerous blackish remains of old stipes.* Stipe shorter (sometimes considerably so) than leaf blade, more rarely of equal length. Blade *extended, oblong-lanceolate in general outline,* with simply pinnate segments with crenate-dentate lobes. *Basal pair of segments of equal length to the succeeding pair* (sometimes slightly longer or slightly shorter), not noticeably differing from the others in degree of division; sometimes somewhat distant from the others ......................................................................2.

2. Stipe, rachis and leaf blade *bare,* more rarely (mainly on the stipe) with a few septate hairs. *Spores more or less densely covered with fine spinules* ... ...........................................1. **C. FILIX-FRAGILIS** (L.) BORB.
   
   – Stipe, rachis, segment stipes and leaf blade *bearing sparse brown* septate or unicellular capitate *hairs* (sometimes almost lacking). *Spores with obtusely tuberculate surface* .....................2. **C. DICKIEANA** SIMS.

   1. ***Cystopteris filix-fragilis*** (L.) Borbas, Balaton Fl. (1900) 314; Poyarkova, Fl. Murm. I, 25.
      *C. fragilis* (L.) Bernh. — Ruprecht, Distrib. crypt. vasc. 107; Ledebour, Fl. Ross. IV, 515; Lange, Consp. fl. Groenl. 188; Scheutz, Pl. jeniss. 198; Gelert, Fl. arct. I, 6; Simmons, Vasc. pl. Ellesm. 182 (pro parte); id., Survey Phytogeogr. 39; M. Porsild, Fl. Disko 24; Macoun & Holm, Vasc. pl. 7a; Lynge, Vasc. pl. N. Z. 13 (pro parte); Ostenfeld, Fl. Greenl. 54 (pro parte); Tolmatchev, Contr. Fl. Vaig. 139, 143 (? - possibly *C. Dickieana*); Krylov, Fl. Zap. Sib. I, 15; Petrov, Fl. Yak. I, 8; Fomin, Fl. Sib. Daln. Vost. V, 27; id., Fl. SSSR I, 24; Perfilev, Fl. Sev. I, 43; Devold & Scholander, Fl. pl. SE Greenl. 19; Sørensen, Vasc. pl. E Greenl. 20-21; Gelting, Vasc. pl. E Greenl. 29 (? - possibly *C. Dickieana*); Tolmachev, Obz. fl. N.Z. 149; id., Mater. fl. M. Shar. I, 277; Gröntved, Pterid. Spermatoph. Icel. 107; Hultén, Fl. Al. I, 16; id., Atlas, 12 (47); Polunin, Bot. Can. E Arctic 29; id., Real Arctic Pteridoph. 32; A.E. Porsild, Vasc. pl. W Can. Arch. 69; id., Ill. Fl. Arct. Arch. 18; Holmen, Vasc. pl. Peary L. 40; Böcher & al., Ill. Fl. Greenl. 48; Tikhomirov, Raspr. papor. 605.
      *C. fragilis* (L.) Bernh. s.l. — Polunin, Real Arctic Pterid. 32 (pro parte).
      **Ill.:** Petrov, Fl. Yak., fig. 10; Porsild, Ill. Fl., fig. 1a.

      One of the most widely distributed ferns in the Arctic. Growing in rocky places, rooting in crevices or at the foot of rocks, sometimes huddled among piles of large

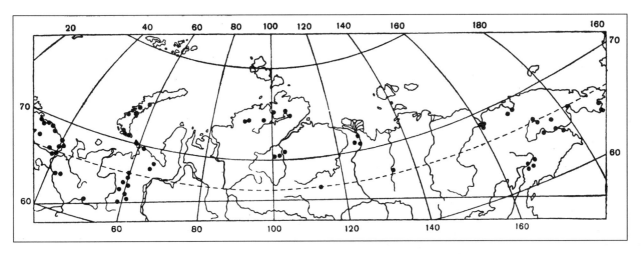

MAP I–4 Distribution of *Cystopteris filix-fragilis* (L.) Borb.

stone blocks and thus not avoiding rather well shaded places. Generally not rare, but for the most part encountered in small quantity. Sometimes forming small sparse beds on protected rocks in places well covered by snow in winter.

Associated only with districts of pronounced relief and sufficiently frequent repetition of rocky areas. It does not display fidelity to particular rocks under arctic conditions, but drops out wherever the rate of weathering and the shape of the rubble excludes the formation of stable habitats suitable for it.

Sporulating normally near the northern limit of its distribution (Central Taymyr, observation by Tikhomirov).

**Soviet Arctic.** Murman; upper reaches of River Usa; Polar Ural; (Vaygach?); Novaya Zemlya as far north as Mashigin Bay on the west coast and the Flott Peninsula on the east coast (that is to almost 75°N in both cases); mountainous districts of Central Taymyr (Byrranga Range west of Lake Taymyr; north shore of Lake Taymyr; banks of Lower Taymyra River not far from the outflow from the lake; reaching somewhat further North than 75°); between the Lena and Olenek, lower reaches of Lena; lower reaches of Kolyma; district of Bay of Chaun; SE Chukotka Peninsula; Anadyr River Basin; Penzhina River Basin. (Map I–4).

**Foreign Arctic.** West coast of Alaska; Arctic coast of Canada; Labrador; Canadian Archipelago (as far north as Northern Ellesmere Island); Greenland (reported on both coasts from the far south to the most northern points, but apparently the greater part of the records for the far north actually refer to *C. Dickieana*); Iceland; (Spitsbergen? — possibly really *C. Dickieana*); Arctic Scandinavia.

**Outside the Arctic.** Forested districts of Europe, Northern and Eastern Asia, North America; mountains of Europe, NW Africa and adjoining islands, Caucasus, the Middle East, Siberia, Central Asia; mountains of a series of tropical countries; subantarctic districts of South America; mountains of Tasmania and New Zealand; Kerguelen Island. In total possessing an almost cosmopolitan distribution, but everywhere confined to sites with cool or moderately warm, sufficiently moist climate. In mountainous districts reaching a very considerable elevation (in the mountains of Soviet Central Asia, for example, climbing to 3500–4000 m or higher), the commonest fern at alpine elevations in many mountainous districts.

2. ***Cystopteris Dickieana*** Sims. in Gard. Journ. (1848) 308; Krylov, Fl. Zap. Sib. I, 16; Sambuk, K fl. sev. yevrop. ch. SSSR; Petrov, Fl. Yak. I, 9; Fomin, Fl. Sib. Daln. Vost. V, 32; id., Fl. SSSR I, 25; Tolmachev & Blyumental, Mat. fl. N. Z. 198; Perfilev, Fl. Sev, I, 43; Tolmachev, Obz. fl. N. Z. 149; id., Fl. pober. Karsk. morya 186; Hadač, Not. fl.

Svalb. I; id., Gefässpfl. Spitzb. 9; Dahl & Hadač, Bidr. Spitsb. Fl. 7; Poyarkova, Fl. Murm. I, 26; Tikhomirov, Raspr. papor. 608.
*C. fragilis* — Simmons, Vasc. pl. Ellesm. 182 (pro parte); Lynge, Vasc. pl. N. Z. 13 (pro parte); many authors on Spitsbergen flora (pro parte?).
*C. fragilis* s.l. (incl. *C. Dickieana*) — Polunin, Real Arctic Pterid. 32 (pro parte).
*C. fragilis* ssp. *Dickieana* Hiyt. — Böcher & al., Ill. Fl. Greenl. 48.
*C. fragilis* var. *Dickieana* Moore.
*C. regia* var. *Dickieana* Milde.
Ill.: Fl. Sib. V, 32.

Apparently the sole fern species whose distribution type is *predominantly arctic*. Information available to date has established that it has a rather wide distribution outside arctic limits in Central and Eastern Siberia. But its known localities in temperate parts of the country are perhaps not very different from the arctic part of its range. A more or less extensive distribution of basically arctic plants within the taiga zone of Siberia where their growth is principally associated with rock exposures does not constitute anything exceptional. *Cystopteris Dickieana* is also found in the mountains of Scandinavia and Scotland, whose flora (like that of the Northern Ural) in many respects more closely approaches that of the arctic than that of high mountains in the true temperate zone. In the mountains of Central Europe, whose flora has been very fully studied, *C. Dickieana* has not been reported. Nor apparently is it found in the high alpine region of the Altay and Sayans, although somewhat further north (for example in the Kuznetsk Alatau and the basin of the Upper Lena) localities for it are known.

More reliable consideration of the distribution of *C. Dickieana* is hampered by the fact that for long the majority of investigators did not distinguish this species from the well known *C. filix-fragilis* (= *C. fragilis*). Thus it happened that especially in the Arctic *C. f.-fragilis* was for long considered the sole representative of its genus in the majority of districts. As a rule, arctic collections of *Cystopteris* were identified as *C. f.-fragilis* without detailed examination.

In more recent time, after the differences of *C. Dickieana* were paid due attention, very many plants in arctic material of *C. f.-fragilis* were discovered to really belong to *C. Dickieana*. Furthermore, it was reliably established that the characters of *C. Dickieana* do not originate spontaneously in different parts of the range of *C. filix-fragilis* and that, no matter how great the similarity of the two species in appearance, their distinction is not subject to doubt.

Review of material of *Cystopteris* from the Soviet Arctic has led to the conclusion that the ranges of *C. f.-fragilis* (s. str.) and *C. Dickieana overlap to a considerable degree*. In places (especially in Taymyr) *C. Dickieana* extends rather further north than does *C. f.-fragilis*. The same thing apparently occurs in Greenland. It cannot be excluded that on Spitsbergen maybe only *C. Dickieana* is found. On the other hand, on Novaya Zemlya the most northern finds belong as a rule to *C. f.-fragilis*.

In view of the considerable overlap of the ranges of *C. Dickieana* and *C. f.-fragilis*, the possibility that the former is a geographical race of the latter is out of the question. Furthermore, attention should be paid to A.V. Fomin's demonstration that the closer relationship of *C. Dickieana* is not with the latter species, but with *C. regia* (L.) Presl. growing in the mountains of Central Europe and the Caucasus. Is it possible that *C. Dickieana* is a differentiated northern race of the latter?

In places where the ranges of *C. Dickieana* and *C. f.-fragilis* overlap, plants which are difficult to identify and possibly of hybrid origin are sometimes reported. But from within arctic limits samples which cause real doubts regarding the correctness of their referal to one or the other of the stated species are not known to us.

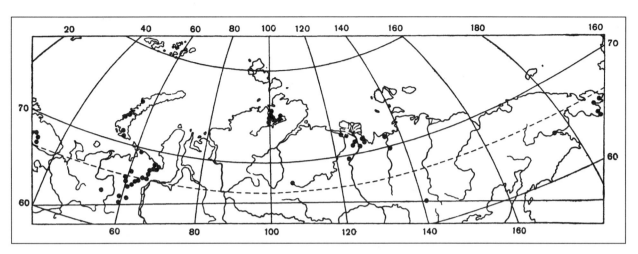

MAP I-5 Distribution of *Cystopteris Dickieana* Sims.

*Cystopteris Dickieana* (like *C. f.-fragilis*) occurs in the Arctic exclusively in districts with more or less pronounced relief and with considerable development of bedrock exposures. It grows in rocky places, most frequently at the foot of rocks on stony substrate. The majority of occurrences are restricted to outcrops of limestone or calciferous rock. It grows exclusively in places sufficiently protected by snow cover in winter, but avoids places with deep layers of snow where this lies persistently and where the soil at snowmelt simultaneously receives surplus moisture and cooling. The better heating of southern slopes is favourable for the development of *C. Dickieana*, but only in circumstances where it is not associated with lack of winter snow cover. Thus in very windy districts, such as at Amderma on the shore of the Kara Sea, where southwest winds blow snow off the southern slopes of tundra ridges, we find small beds of *C. Dickieana* always on the northeast side of low rocky crests where a certain depth of snow accumulates in winter without strong compaction.

Sporulating normally under arctic conditions as far as our most northern known localities (NW part of the South Island of Novaya Zemlya according to Lynge; Central Taymyr according to Tikhomirov).

**Soviet Arctic.** East Bolshezemelskaya Tundra (on Chernyshev Ridge near the River Adzva); Polar Ural (at many sites); Pay-Khoy (Amderma); Novaya Zemlya (South Island: Chernaya Bay in the far south, Propashchaya Bay at Kostin Shar, Bezymyannaya Bay and Gribov Bay in the north-west; North Island: Mashigin Bay). In Arctic Siberia on the north shore of Lake Taymyr; on the Lower Taymyra River (to 75°40'N); lower reaches of River Olenek; on the Olenekskaya Channel of the Lena Delta; between the Olenek and the Lena near the northern limit of forest; lower course of River Lena; district of Tiksi Bay; lower reaches of River Yana. In far NE Asia on the SE part of the Chukotka Peninsula and in the district of the mouth of the River Anadyr. (Map I–5).

**Foreign Arctic.** Canadian Archipelago (in the north known to exist on Ellesmere Island — Fram Harbour according to Hadač); Greenland (especially in more northern districts); Spitsbergen.

**Outside the Arctic.** Mountains of Scotland and Scandinavia; mountains of the central part of the Kola Peninsula; Northern Ural and vicinity (Pechora Basin); in rocky places in the forest zone of Siberia (on the Lower Tunguska, the upper reaches of the Olenek, in Central Yakutia and the Yana Basin; upper part of the Lena Basin). Extending south to the Kuznetsk Alatau and the northern foothills of the Sayans.

Details of the distribution insufficiently clarified. The available picture is evidently incomplete.

3. **Cystopteris montana** (Lam.) Desv., Prodr. (1827) 264; Ostenfeld, Fl. Greenl. 54; Krylov, Fl. Zap. Sib. I, 17; Fomin, Fl. Sib. Daln. Vost. V, 34; id., Fl. SSSR I, 25; Perfilev, Fl. Sev. I, 43; Leskov, Fl. Malozem. tundry 11; Hultén, Fl. Al. I, 18; id., Atlas 12 (48); Poyarkova, Fl. Murm. I, 28; Polunin, Real Arctic Pterid.; Böcher & al., Ill. Fl. Greenl. 47.

*C. montana* Link — Ruprecht, Distrib. crypt. vasc. 108; Ledebour, Fl. Ross. IV, 517; Scheutz, Pl. jeniss. 198.

*Polypodium montanum* Lam., Fl. Fr. I (1778), 23.

Ill.: Fl. Murm. I, pl. III.

Plant basically of the northern part of the forest zone of Eurasia and North America, growing principally on or near rocks in the shade of the forest canopy. It penetrates arctic limits in districts with more pronounced oceanic or moderately continental climatic traits. Here it retains an attraction to rocky places, but has less need of shade. On tundra sometimes encountered in mossy tall willow shrubbery. Everywhere relatively rare.

**Soviet Arctic.** Murman (Kildin Island, Ponoy); northern part of Malozemelskaya Tundra; SE Bolshezemelskaya Tundra; Polar Ural as far north as the Shchuchya River Basin. Almost everywhere on rocks. Right bank of the lower Yenisey within the limits of forest-tundra (shady mossy forest NE of Khantayka; Potapovskoye). Penzhina River Basin; Bay of Korf.

**Foreign Arctic.** West coast of Alaska; extreme SW Greenland; Arctic Scandinavia.

**Outside the Arctic.** Mountain forest districts of Western and Central Europe; mountains of Scandinavia; Kola Peninsula; sporadically in forests of the north of the European part of the USSR, especially in Karelia and in Arkhangelsk, Vologda and Novgorod Oblasts; more frequent in the Northern and Middle Urals, reaching the Southern Ural; sporadically in the forest zone of Western and Central Siberia, more frequently in Prebaikalia; reported on the Okhotsk Coast. Disjunctly from the main range in the Caucasus and Western Himalayas. In North America predominantly in inland mountain districts of the West, also near the Atlantic Coast.

In Northern Europe restriction of its occurrence to outcrops of rocks rich in lime has often been established.

## GENUS 3 — **Dryopteris** Adans. — SHIELD FERN

A WIDELY DISTRIBUTED genus containing a large number of species of rather diverse appearance. In general they are relatively large woodland plants, often playing an essential role in the composition of the lower strata of forest communities. A few species penetrate the limits of the Arctic in districts with relatively milder climate. They are confined here to protected places, warm enough during summer and well covered by deep and rather powdery snow during winter. Contrary to the remaining species, one species of *Dryopteris (D. fragrans)* is a plant characteristic of the Arctic occurring also in its districts at high latitudes. This species, common to forested regions of Northern Asia and North America, also makes do with open sites with little snow cover.

1. Leaf blade in general outline *oblong-lanceolate or almost elliptical, narrowing towards both ends*, rather thick, *dark green*. Stipe *very short, many times shorter than leaf blade* ........................................2.
- Leaf blade in general outline *oblong-triangular, with wide base*, narrowing towards tip, tender, *bright green. Stipe long,* no less than ⅖-½ as long as leaf blade, sometimes almost equal to it in length. ....................3.

2. *Large plant with leaves 50–70 cm long.* Leaf blade rather abruptly narrowed towards tip, *flat*, dark green above, *not very shiny*, lighter below, on lower surface (particularly on rachis) more or less densely clothed with light brown scales. Pinnules numerous, very similar to one another in size and shape. Sori as a rule *developing only on segments of upper half of leaf.* ........................................1. **D. FILIX-MAS** (L.) SCHOTT.
- *Plant of considerably smaller size, with leaves no more than 15–20 cm long, aromatic* through secretions of numerous glands. Leaf blade very gradually narrowing towards both ends, often slightly trough-shaped, densely covered basally by brown scales giving the whole plant a brownish tinge, glandular distally. Pinnules few, gradually diminishing in size from leaf rachis to tip of segment. *Sori developing more or less uniformly* on the segments *over the whole length of the leaf.* ........2. **D. FRAGRANS** (L.) SCHOTT.

3. Stipe thick, on its lower part up to 5 mm wide, *⅖-½ as long as leaf blade*, densely covered with lanceolate long-acuminate brown scales with darker stripe medially ......................3. **D. AUSTRIACA** (JACQ.) WOYNAR.
- Stipe rather thinner, *slightly shorter than (⅔-¾ as long as) leaf blade, sometimes equalling it in length*, on its lower part densely, above (as on rachis) sparsely clothed with light brown acuminate-ovoid *unicolorous* scales ..................................4. **D. SPINULOSA** (MUELL.) KTZE.

    **1. Dryopteris filix-mas** (L.) Schott, Gen. Filic. (1834) ad tab. 9; Ostenfeld, Fl. Greenl. 55; Krylov, Fl. Zap. Sib. I, 22; Fomin, Fl. Sib. Daln. Vost. V, 54; id., Fl. SSSR I, 36; Perfilev, Fl. Sev. I, 44; Devold & Scholander, Fl. pl. SE Greenl. 19; Gröntved, Pterid. Spermatoph. Icel. 108; Hultén, Atlas 14 (55); Poyarkova, Fl. Murm. I, 31; Polunin, Real Arctic Pterid. 39; Böcher & al., Ill. Fl. Greenl. 50.

  *Aspidium filix-mas* (L.) Sw.— Ruprecht, Distrib. crypt. vasc. 103; Gelert, Fl. arct. I, 5.
  *Polystichum filix-mas* Roth — Ledebour, Fl. Ross. IV, 514
  *Lastraea filix-mas* Presl — Lange, Consp. fl. Groenl. 187.
  *Polypodium filix-mas* L., Sp. pl. (1753) 36.
  Ill.: Fl. Murm. I, pl. IV.

    Forest plant characteristic of coniferous, mixed and deciduous forests of the temperate zone, principally in regions with more or less pronounced oceanic or moderately continental climate. Just penetrating the limits of the Arctic, where it grows on protected southern slopes which are warm enough in summer but lie beneath deep snow cover in winter.

  **Soviet Arctic.** Murman (Kildin Island, northern forest boundary south of Murmansk).
  **Foreign Arctic.** Greenland (SW and extreme SE); Iceland; Arctic Scandinavia.

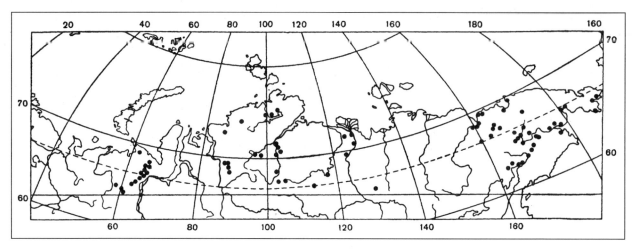

MAP I–6  Distribution of *Dryopteris fragrans* (L.) Schott.

**Outside the Arctic.** Forested regions of the greater part of Europe (absent from the extreme northeast!); Caucasus; SE West Siberia; sporadically in the south of East Siberia; mountain forests of Kazakhstan and northern parts of Central Asia; Eastern North America. Replaced in the Far East (Preamuria, Sakhalin, Northern China, Japan) by the closely related *D. crassirhizoma* Nakai (not recognized as of species rank by everyone). Other closely related species or races in the Himalayas, in mountains of tropical Asia and South America, on islands in the Indian and Pacific Oceans, and in South Africa.

2. ***Dryopteris fragrans*** (L.) Schott, Gen. Filic. (1834) ad tab. 9; Simmons, Survey Phytogeogr. 40; M. Porsild, Fl. Disko 23; Macoun & Holm, Vasc. pl. 7a; Ostenfeld, Fl. Greenl. 55; Krylov, Fl. Zap. Sib. I, 23; Petrov, Fl. Yak. I, 13; Fomin, Fl. Sib. Daln. Vost. V, 61; id., Fl. SSSR I, 38; Perfilev, Fl. Sev. I, 44; Gröntved, Vasc. pl. Arct. N. Am. 15; Hultén, Fl. Al. I, 23; id., Atlas 14 (56); Polunin, Bot. Can. E Arct. 31; id., Real Arctic Pterid. 39; A.E. Porsild, Vasc. pl. W Can. Arch. 70; id., Ill. Fl. Arct. Arch. 18; Böcher & al., Ill. Fl. Greenl. 49; Tikhomirov, Raspr. papor. 601.

*Aspidium fragrans* Sw. — Ruprecht, Distrib. crypt. vasc. 103; Gelert, Fl. arct. I, 5; Simmons, Vasc. pl. Ellesm. 182.

*Polystichum fragrans* Ldb., Fl. Ross. IV, 514; Schmidt, Fl. jeniss. 130; Scheutz, Pl. jeniss. 198.

*Lastraea fragrans* Presl — Lange, Consp. fl. Groenl. 186.

*Polypodium fragrans* L., Sp. pl. (1753) 1089.

**Ill.:** Fl. arct. I; Porsild, Ill. Fl., fig. 16.

Characteristic fern of rocky places in the Arctic and the northern forest zone. It grows in rock crevices, on rock ledges, on gravelly slopes near rock, and on boulder fields. Under arctic conditions it shows a clear preference for well warmed southern slopes. On these it sometimes forms beds in places sheltered by neighbouring cliffs. On rocks it grows as separate tufts. Concealment by snow cover through the winter is favourable for the development of *D. fragrans*, but not an essential condition for its survival (it can also grow on cliffs without snow cover). Living parts of the plant are protected to a considerable degree by a covering of old dead leaves retained over the course of several years. A protective role is apparently also played by the resinous secretions of the numerous glands, which are the source of the odour given off by the fern (rather strong on warm sunny days).

*Dryopteris fragrans* does not display close association with particular rock formations, at least not in Northern Siberia. A series of investigators have reported it as a common plant on outcrops of diabase and other igneous rocks. In the north of the Canadian Archipelago it is characteristic of regions with outcrops of Archean gneiss, but also occurs on outcrops of limestone and other rocks.

The leaves of *D. fragrans* overwinter in green condition, dying in summer as the young leaves develop and subsequently remaining on the plant for several years. Abundant sporulation is observed as far as the extreme limits of the species' distribution in the Arctic. According to Tikhomirov's observations in the central part of Taymyr, the plants are conspicuously "dusted" with spores at the end of summer.

**Soviet Arctic.** Polar Ural (numerous localities along the whole range); Pay-Khoy (rare); Transyenisey forest-tundra (Khantayka, Norilsk); Taymyr on the southern slopes of the Byrranga Range (on the River Pyasina where it cuts through the range; NW shores of Lake Taymyr; on the Bay of Yamu-Baykur; reported also on the lower course of the Taymyra River shortly above its entry into Lake Taymyr); between the rivers Olenek and Lena; lower reaches of the Kolyma and further east on the shore of the East Siberian Sea; district of the Bay of Chaun; (Wrangel Island?); SE coast of Chukotka Peninsula; Anadyr Basin; Penzhina Basin; Bay of Korf. (Map I–6).

**Foreign Arctic.** West coast of Alaska; arctic coast of Canada; Labrador; Canadian Archipelago (reaching 79°N on Ellesmere Island); Greenland (almost to 80°N in the northwest; south to 64°44' on the west coast; in East Greenland only on Scoresby Sound, about 71°N); Arctic Scandinavia (single locality in the extreme NW of Finnish Lapland).

**Outside the Arctic.** In Siberia principally at subarctic latitudes; Northern Ural south to 62°; Central Siberian Plateau (district of Norilsk, northern edge of the plateau in the Khatanga Basin, upper and middle reaches of the Olenek; lower and middle reaches of the Lower Tunguska; south to the Podkamennaya Tunguska and the SW part of the Vilyuy Basin); Yenisey Ridge; Verkhoyansk Range. Further south rather common in the Stanovoy Range, on the Vitimskoye Highland, and in Transbaikalia. Rare in the Sayans and the Tannu-Ola Range, very rare in Altay. In the Far East in the mountains of Northern Preamuria; ranging south to the Sikhote-Alin, and penetrating NE China and Northern Korea; occurring on Sakhalin, Northern Japan and as a great rarity on Kamchatka. In North America found in inland districts of Alaska and Western Canada, and in the forest zone of Central and Eastern Canada and the NE United States.

3. ***Dryopteris austriaca*** (Jacq.) Woynar in Vierteljahrschr. Nat. Ges. Zürich LX (1919), 339; Fomin, Fl. SSSR I, 41; Hultén, Fl. Al. I, 21; id., Atlas 14 (53); Gröntved, Pterid. Spermatoph. Icel. 108; Poyarkova, Fl. Murm. I, 34; Polunin, Real Arctic Pterid. 39.

*D. dilatata* A. Gray — M. Porsild, Fl. Disko, 24; Ostenfeld, Fl. Greenl. 55; Fomin, Fl. Sib. Daln. Vost. V, 70; Böcher & al., Ill. Fl. Greenl. 50.

*D. spinulosa* ssp. *dilatata* Aschers. — Krylov, Fl. Zap. Sib. I, 26; Perfilev, Fl. Sev. I, 44.

*Aspidium spinulosum* ssp. *dilatatum* Roep. — Gelert, Fl. arctica, I, 6.

*Aspidium dilatatum* Sw. — Ruprecht, Distrib. crypt. vasc. 106.

*Lastraea spinulosa* β *intermedia* Milde — Lange, Consp. fl. Greenl. 187.

*Dryopteris spinulosa* (Muell.), s.l. — Devold & Scholander, Fl. pl. SE Greenl. 21 (?).

*Polypodium austriacum* Jacq., Obs. I (1764), 45.

Widely distributed forest fern just penetrating arctic limits. Here it grows in willow shrubbery and among thickets of dwarf birch (*Betula nana*), as well as in "island" forest of the forest-tundra.

**Soviet Arctic.** Murman (more or less ubiquitous); South Kanin; Timanskaya forest-tundra (isolated localities); Bay of Korf in Koryakia.

**Foreign Arctic.** West coast of Alaska; Greenland (on west coast north to Disko Island, in the south-east to north of 62°); Iceland; Arctic Scandinavia.

**Outside the Arctic.** Widely distributed in the forest zone and mountain forests of more southern regions of Europe, Asia and North America (south to Asia Minor, the

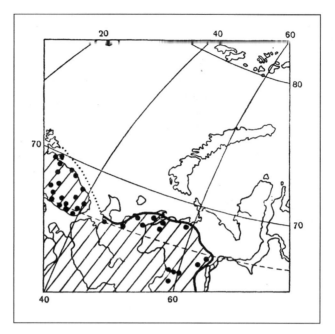

MAP I-7  Distribution of *Dryopteris spinulosa* (Muell.) Ktze.

Caucasus, the Himalayas, Central Japan, California, and Virginia). In inland continental regions absent from the northern part of the taiga zone.

4. ***Dryopteris spinulosa*** (Muell.) O. Ktze., Rev. gen. pl. II (1891), 813; Fomin, Fl. SSSR I, 32; Leskov, Fl. Malozem. tundry 12; Poyarkova, Fl. Murm. I, 32.
*Aspidium spinulosum* Sw. — Ruprecht, Distrib. crypt. vasc. 105.
*D. spinulosa* ssp. *euspinulosa* Asch. — Krylov, Fl. Zap. Sib. I, 26; Perfilev, Fl. Sev. I, 44.
*D. euspinulosa* (Diels) Fomin, Fl. Sib. Daln. Vost. V, 66.
*Polystichum spinulosum* Ldb., Fl. Ross. IV (pro parte).
*Polypodium spinulosum* Muell., Fl. Friedrichst. (1767) 193.
Ill.: Fl. Murm. I, pl. V.

Forest fern principally associated with countries bordering the Atlantic Ocean. Penetrating the Arctic in Northern Europe. It grows here in willow shrubbery on the slopes of tundra mounds and near streams, also in protected openings within shrub thickets where snow cover is long persistent. Restricted to sites with considerable depth of snow cover and relatively rich soil.

**Soviet Arctic.** Murman (more or less ubiquitous); southern part of Kanin; Timanskaya and Malozemelskaya Tundras (common in the forest-tundra zone, rarer in the north); west margin of the Bolshezemelskaya Tundra [right bank of the Pechora slightly north of 68°]. (Map I–7).

**Foreign Arctic.** Arctic Scandinavia.

**Outside the Arctic.** Forest regions of Europe, West Siberia and Eastern North America. Reaching almost 65°N on the west slopes of the Urals (Mount Sablya), not above 60°N in West Siberia.

## GENUS 4    Thelypteris Schmidel — THELYPTERIS

1. ***Thelypteris phegopteris*** (L.) Sloss in Rydberg, Fl. Rocky Mts. (1917) 1043; Poyarkova, Fl. Murm. I, 36.
    *Polypodium phegopteris* L., Sp. pl. (1753) 1089; Ruprecht, Distrib. crypt. vasc. 52; Ledebour, Fl. Ross. IV, 509; Lange, Consp. fl. Groenl. 185; Scheutz, Pl. jeniss. 198.
    *Aspidium dryopteris* (L.) Baumg. — Gelert, Fl. arct. I, 4.
    *Dryopteris phegopteris* (L.) C. Chr. — Ostenfeld, Fl. Greenl. 55; Krylov, Fl. Zap. Sib. I, 29; Fomin, Fl. Sib. Daln. Vost. V, 85; id., Fl. SSSR I, 44; Devold & Scholander, Fl. pl. SE Greenl. 20; Hultén, Fl. Al. I, 28; id., Atlas 15 (60); Gröntved, Pterid. Spermatoph. Icel. 110; Polunin, Bot. Can. E. Arct. 31.
    *Lastraea phegopteris* (L.) Borg. — Böcher & al., Ill. Fl. Greenl. 49.
    *Phegopteris polypodioides* Fée — Perfilev, Fl. Sev. I, 45.
    **Ill.:** Fl. Murm. I, pl. VI.

    Forest fern widely distributed in the temperate zone of the Northern Hemisphere, marginally penetrating the Arctic in regions bordering the Atlantic Ocean and in the Pacific part of North America. On forest islands of the forest-tundra (especially in birch woods), among shrubs on rocks, in all cases in sites well protected by snow cover in winter.

    **Soviet Arctic.** Murman, more or less ubiquitous. East of the White Sea not reaching the polar limit of forest.

    **Foreign Arctic.** Beringian coast of Alaska; Labrador; Greenland south of the Arctic Circle; Iceland; Arctic Scandinavia.

    **Outside the Arctic.** Forest (including mountain forest) districts of Europe, Asia and North America, south to the Pyrenees, Transcaucasus, Northern Asia Minor, the Himalayas, Northern Korea, Central Japan, NW United States and central states of the Eastern United States.

## GENUS 5    Gymnocarpium Newm. — OAK FERN

SMALL FOREST FERNS with tender leaf blade distinctly divided into three roughly equal parts, each of which resembles a separate triangular pinnate leaf in shape. Two species penetrate arctic limits in areas bordering the Atlantic and Pacific Oceans.

1. Leaf blade spreading almost horizontally, with flat segments. *All three parts of it of roughly the same size* and divided to the same degree. *Stipe and rachis of leaf not glandular.* .............1. **G. DRYOPTERIS** (L.) NEWM.
– Leaf blade for the most part directed obliquely upwards. Segments often somewhat recurved and therefore appearing narrower. *Apical (axillary) third of leaf blade always distinctly larger than lateral*, with more numerous segments. *Upper part of stipe and rachis of leaf covered with fine yellowish glands.* .....................2. **G. CONTINENTALE** (PETR.) POJARK.

1. ***Gymnocarpium dryopteris*** (L.) Newm. in Phytologist 4 (1851), 371; Poyarkova, Fl. Murm. I, 38.
    *Polypodium dryopteris* L., Sp. pl. (1753) 1093; Ruprecht, Distrib. crypt. vasc. 52; Ledebour, Fl. Ross. IV, 509; Lange, Consp. fl. Groenl. 185; Scheutz, Pl. jeniss. 198.

*Aspidium dryopteris* (L.) Baumg. — Gelert, Fl. arct. I, 4.
*Phegopteris dryopteris* (L.) Fée — Perfilev, Fl. Sev. I, 45.
*Lastraea dryopteris* (L.) Borg. — Böcher & al., Ill. Fl. Greenl. 49.
*Dryopteris pulchella* (Salisb.) Hayek — Ostenfeld, Fl. Greenl. 55; Krylov, Fl. Zap. Sib. I, 28.
*Dryopteris Linnaeana* C. Chr. — M. Porsild, Fl. Disko, 23; Fomin, Fl. Sib. Daln. Vost. V, 78; id., Fl. SSSR I, 43; Devold & Scholander, Fl. pl. SE Greenl. 20; Leskov, Fl. Malozem. tundry 12; Hultén, Fl. Al. I, 25; id., Atlas 15 (57); Gröntved, Pterid. Spermat. Icel. 109.
*Dryopteris disjuncta* (Ledb.) Mort. — Polunin, Real Arctic Pterid. 39.
Ill.: Fl. Murm. I, pl. VII.

    Forest (basically taiga) fern, penetrating the limits of the Arctic in its Atlantic portion and near the Bering Strait. Found in the Arctic in birch and spruce woods of the forest-tundra, in shrub thickets, below protective rocks, and in well warmed places (sometimes at hot springs). Exclusively associated with sites which are well protected by deep snow cover in winter. Sporulation more or less suppressed at the northern limit of its distribution.

**Soviet Arctic.** Murman (more or less everywhere); South Kanin; south shore of Cheshskaya Bay; southern districts of the Timanskaya and Malozemelskaya Tundras; lower reaches of the Pechora (within the limits of forest-tundra). Further east within the limits of forest-tundra in the Pechora Basin, on the lower reaches of the Ob (Salekhard), and the lower reaches of the Yenisey (north to $68\frac{1}{2}°$; east of the Yenisey the boundary of the range shifts abruptly to the south); SE Chukotka Peninsula.

**Foreign Arctic.** West coast of Alaska (Nome); Labrador; West Greenland north to Disko Island; East Greenland north to Angmagssalik; Iceland; Arctic Scandinavia.

**Outside the Arctic.** Forested districts of Western and Central Europe; mountains of Central and Southern Europe; Caucasus; forest zone of Siberia and the Far East (south to the Altay, Northern Mongolia, NE China, and Central Japan); mountains of Central Kazakhstan; Himalayas; temperate northern North America.

2. **Gymnocarpium continentale** (Petr.) Pojark., Tr. Tadzh. filiala AN SSSR 22 (1950), 10.
*Dryopteris pulchella* var. *continentalis* Petrov, Fl. Yak. I (1930), 14.
*Dryopteris continentalis* Petrov, ibid. 15-16; Fomin, Fl. SSSR I, 43.
*Dryopteris Robertiana* (Hoffm.) Chr. — Fomin, Fl. Sib. Daln. Vost. V, 80 (pro parte); Hultén, Fl. Al. I, 29.
Ill.: Petrov, Fl. Yak. I, fig. 12 (p. 15).

    Species close to *G. dryopteris* and especially to the European *G. Robertianum* (Hoffm.). Occurring mainly in mountainous districts of Northern Asia and Western North America, for the most part on rocks. Reported especially on outcrops of limestone and other calciferous rocks. Found in the arctic on rocky slopes.

**Soviet Arctic.** Lower reaches of Lena (within the forest-tundra). Bay of Korf (Koryakia). Near the edge of our region on the northern part of the Central Siberian Plateau, on the middle course of the River Medvezhya, tributary of the Kheta (Khatanga Basin), about 70°N.

**Outside the Arctic.** Forested region of Siberia east of the Yenisey (reported from the basins of the Vilyuy and the Aldan, the upper course of the River Olenek, the middle course of the River Yana, that is in general north of the Siberian part of the range of *G. dryopteris*). In the Far East, especially in Sakhalin; Alaska.

## GENUS 6 — **Polystichum** Roth — HOLLY FERN

1. ***Polystichum lonchitis*** (L.) Roth in Roem. Mag. 2 (1799), 106; Ruprecht, Distrib. crypt. vasc. 106; M. Porsild, Fl. Disko 24; Ostenfeld, Fl. Greenl. 55; Krylov, Fl. Zap. Sib. I, 30; Fomin, Fl. Sib. Daln. Vost. V, 89; id., Fl. SSSR I, 46; Perfilev, Fl. Sev. I, 45; Devold & Scholander, Fl. pl. SE Greenl. 21; Gröntved, Pterid. Spermat. Icel. 112; Hultén, Fl. Al. I, 32; id., Atlas, 17 (66); Poyarkova, Fl. Murm. I, 40; Polunin, Real Arctic Pterid. 40; Böcher & al., Ill. Fl. Greenl. 50.

   *Aspidium lonchitis* (L.) Sw. — Ledebour, Fl. Ross. IV, 512; Lange, Consp. fl. Groenl. 186; Gelert, Fl. arct. I, 6.

   *Polypodium lonchitis* L., Sp. pl. (1753) 1088.

   Ill.: Fl. Murm. I, pl. VIII.

   Very unusual fern with stiff leathery leaves which overwinter in green condition and live for several years. Basically associated with mountain forests of the northern forest zone of Eurasia and North America, but climbing to above treeline in the mountains of Fennoscandia and the Urals. Growing in districts with rather moist climate throughout the whole year.

   Just penetrating arctic limits. Confined to sites well covered with snow through the winter.

   **Soviet Arctic.** Rybachiy Peninsula in Murman.

   **Foreign Arctic.** Greenland (to almost 70°N on the west, to 67°N on the east coast); Iceland; Arctic Scandinavia.

   **Outside the Arctic.** Distributed disjunctly: mountains of Scandinavia, Northern Great Britain, Central and Southern Europe; Caucasus; mountains of Asia Minor; in the north of the European part of the USSR in mountains of the central and southwestern part of the Kola Peninsula (rather common!), and in the Northern and Prepolar Urals (betwen 60° and 65°N); Tien Shan, Altay, Sayans, mountains of Prebaikalia (far from everywhere); Western Himalayas; Sakhalin Island; Northern Kurile Islands; Commander and Aleutian Islands; Southern Alaska; mountain forest districts of Western and Eastern North America.

## GENUS 7 — **Athyrium** Roth — LADY FERN

FOREST FERNS DIFFERING in size and external appearance, often large. Two large-leaved species with elongate pinnate leaves (*A. filix-femina* and *A. alpestre*) penetrate arctic limits (and extend above treeline in the mountains) at sites which are most favourable with respect both to general climatic and narrowly local conditions. Both species are very similar to one another.

1. Apical portions of leaf segments with sharp teeth. Sori oblong, normally covered by well developed indusia. ..........1. **A. FILIX-FEMINA** (L.) ROTH.
– Apical portions of leaf segments blunt-toothed. Sori almost orbicular, with poorly developed or no indusia. .............2. **A. ALPESTRE** (HOPPE) RYL.

1. ***Athyrium filix-femina*** (L.) Roth, Tent. fl. Germ. III (1800), 65; Ruprecht, Distrib. crypt. vasc. 108; Krylov, Fl. Zap. Sib. I, 33; Petrov, Fl. Yak. I, 18; Fomin, Fl. Sib. Daln. Vost. V, 106; id., Fl. SSSR I, 53; Perfilev, Fl. Sev. I, 45; Gröntved, Pterid. Spermat. Icel. 104; Hultén, Atlas 12 (46); Poyarkova, Fl. Murm. I, 44.

   *A. filix-femina* ssp. *cyclosorum* (Rupr.) Chr. — Hultén, Fl. Al. I, 34.

*Asplenium filix-femina* Bernh. — Ledebour, Fl. Ross. IV, 518; Scheutz, Pl. jeniss. 199.
*Polypodium filix-femina* L., Sp. pl. (1753) 1090.

Widely distributed forest plant of the temperate zone of the Northern Hemisphere, encountered mainly in coniferous forests, also in birch woods. Just penetrating the limits of the Arctic, where it grows in birch groves of the forest-tundra and in willow shrubbery.

**Soviet Arctic.** Murman (district of Kola Bay, lower reaches of River Voronya, Iokanga, Terskiy Shore); western part of Bolshezemelskaya Tundra (slopes of Salindey-Musyur).

**Foreign Arctic.** Bering coast of Alaska; Iceland; Arctic Scandinavia.

**Outside the Arctic.** Widely distributed in forests of Europe, Siberia, the Far East and North America. Mountain forests of Central and Southern Europe, NW Africa, Asia Minor, Caucasus, Northern Iran, Northern Tien Shan, and Himalayas. Northern China; Northern Japan. In the mountains of North America south to the Mexican frontier. Closely related species in the mountains of a series of tropical countries. Represented by different varieties or geographical races in different parts of its wide range. It is possible that some of these in fact represent distinct species.

2. ***Athyrium alpestre*** (Hoppe) Rylands in Moore, Ferns of Gr. Brit. I, tab. 7 (1857); Gelert, Fl. arct. I, 4; Ostenfeld, Fl. Greenl. 54; Krylov, Fl. Zap. I, 34; Fomin, Fl. Sib. i Daln. Vost. V, 116; id., Fl. SSSR I, 57; Perfilev, Fl. Sev. I, 45; Devold & Scholander, Fl. pl. SE Greenl. 18; Gröntved, Pterid. Spermatoph. Icel. 103; Poyarkova, Fl. Murm. I, 46; Hultén, Atlas 11 (44); Polunin, Real Arctic Pterid. 38; Böcher & al., Ill. Fl. Greenl. 48; Tikhomirov, Raspr. papor. 597.

*Athyrium alpestre* var. *americanum* Butt. — Hultén, Fl. Al. I, 34.
*Polypodium alpestre* Hoppe — Lange, Consp. fl. Groenl. 186.
*Aspidium alpestre* Hoppe, Neue Bot. Taschenb. (1805) 216.

Ill.: Fl. Murm. I, pl. IX; Fl. Sib. i Daln. Vost. V, 117.

Mainly a mountain forest species, but extending in the mountains also above the altitudinal limit of forest. Here, as in the Arctic, it is found in shrub thickets and especially on rocky slopes, near streams, often in places where snow long persists. In extreme NE Asia near the outflows of hot springs. On open areas (without trees or shrubs) sometimes forming dense beds.

**Soviet Arctic.** Murman (more or less everywhere); Kanin (north to Cape Kanin). Not reported from the tundras near the Pechora. Not reaching arctic limits in the Urals, although frequently reported above treeline in the Sablya Massif and the Lyapin River Basin. Certainly absent from Arctic Siberia. In extreme NE Asia on the SE part of the Chukotka Peninsula near outflows of hot springs, near the Bay of St. Lawrence and near Provideniye Bay.

**Foreign Arctic.** South Greenland (north to 66° on the west, to 63½° on the east coast); Iceland; Arctic Scandinavia.

**Outside the Arctic.** Scandinavia; mountains of Northern Great Britain and Central Europe; Caucasus; Urals; mountains of Southern Siberia (from the Altay to Prebaikalia); Kamchatka; extreme east of North America. In western North America represented by var. *americanum* Butt. (which the Greenland plants also approach in a series of characters), and on the Gaspé Peninsula by var. *gaspense*.

### GENUS 8 — **Asplenium** L. — SPLEENWORT

1. ***Asplenium viride*** Huds., Fl. Angl. (1762) 383; Ruprecht, Distrib. crypt. vasc. 112; Ledebour, Fl. Ross. IV, 521; Lange, Consp. fl. Groenl. 305; Scheutz, Pl. jeniss. 199; Gelert, Fl. arct. I, 8; Krylov, Fl. Zap. Sib. I, 38; Fomin, Fl. Sib. Daln. Vost. V, 143; id., Fl. SSSR I, 65; Perfilev, Fl. Sev. I, 46; Devold & Scholander, Fl. pl. SE Greenl. 18; Hultén, Fl. Al. I, 37; id., Atlas 11 (43); Poyarkova, Fl. Murm. I, 49; Polunin, Real Arctic Pterid. 38; Böcher & al., Ill. Fl. Greenl. 51.
   **Ill.**: Fl. Murm. I, pl. XII, 2.
   
   Small fern growing in rock crevices, mainly in the forest section of the temperate zone, just penetrating arctic limits.
   
   **Soviet Arctic.** Murman (in the basin of the River Voronya; considerably more frequently encountered in mountains of the central and southwestern parts of the Kola Peninsula); Polar Ural (in the district of Yeletskiy Pass, the Ray-Iz Mountains and further south); right bank of the Yenisey at Khantayka.
   
   **Foreign Arctic.** Alaska (Nome); Greenland (in the southwest north to 62°, on the east coast between 63° and 66°N); Arctic Scandinavia.
   
   **Outside the Arctic.** Widely but far from universally distributed in more or less mountainous forested districts of temperate Europe, Asia and North America.

### GENUS 9 — **Cryptogramma** R.Br. — ROCK BRAKE

MOUNTAIN FERNS GROWING for the most part on coarse boulder fields, also on rocks. Not encountered everywhere, but can be common where they occur. Two species widely distributed in the forest section of the north temperate zone penetrate arctic limits.

Species of *Cryptogramma* are sharply distinguished from the various other ferns growing in the Arctic by the presence of two different types of leaves: sterile leaves fulfilling only assimilative functions, and fertile leaves possessing narrower pinnules whose revolute margins cover dense accumulations of sori.

1. Leaves crowded in dense cluster, strongly subdivided into very large number of pinnules; each segment (primary subdivision) divided into 26–60 pinnules in total. Leaf stipes green. ............... 1. C. CRISPA (L.) R. BR.
-  Leaves somewhat distant from one another, considerably less subdivided; each segment divided into no more than 10–15 pinnules. Leaf stipes reddish brown. ............................2. C. STELLERI (GMEL.) PRANTL.

1. ***Cryptogramma crispa*** (L.) R. Br. in Richardson, Bot. App., Franklin Narrative (1823) 767; Krylov, Fl. Zap. Sib. I, 41; Fomin, Fl. Sib. Daln. Vost. V, 164; id., Fl. SSSR I, 77; Perfilev, Fl. Sev. I, 46; Gröntved, Pterid. Spermatoph. Icel. 106; Hultén, Atlas 9 (33); Poyarkova, Fl. Murm. I, 50.
   *Allosorus crispus* Bernh. — Ruprecht, Distrib. crypt. vasc. 115; Ledebour, Fl. Ross. IV, 525.
   *Osmunda crispa* L., Sp. pl. (1753) 1067.
   **Ill.**: Fl. Sib. Daln. Vost. V, 165; Fl. Murm. I, pl. XI.
      On rocks and boulder fields.
   **Soviet Arctic.** Murman (Rybachiy Peninsula, Kildin Island).

**Foreign Arctic.** Iceland; Arctic Scandinavia.
**Outside the Arctic.** Mountains of Scandinavia, Great Britain, Central and Southern Europe, Asia Minor and the Caucasus; Northern Ural (north to 65°).

2. ***Cryptogramma Stelleri*** (Gmel.) Prantl in Bot. Jahrb. III (1882), 413; Petrov, Fl. Yak. I, 19; Fomin, Fl. Sib. Daln. Vost. V, 172; id., Fl. SSSR I, 78; Perfilev, Fl. Sev. I, 46; Hultén, Fl. Al. I, 41; Polunin, Real Arctic Pterid. 38.
*Allosorus Stelleri* Rupr., Distrib. crypt. vasc. 115.
*Allosorus gracilis* Presl — Scheutz, Pl. jeniss. 199.
*Pteris Stelleri* S.G. Gmel. in Nov. Comment. Acad. Petrop. 12 (1768) 519.

In rock fissures.

**Soviet Arctic.** Right bank of the Yenisey in the forest-tundra zone (Khantayka); lower reaches of Lena (Kumakh-Surt, Ayakit, etc.; rather common); Penzhina River Basin.
**Outside the Arctic.** Northern Ural (from 58° to 65°N); right bank of the Yenisey within the limits of the forest zone; mountains of South Siberia (Kuznetskiy Alatau, Altay, Sayans, mountains of Prebaikalia); Dzhugdzhur Range and hills on the Okhotsk Coast; Verkhoyansk Range. Disjunctly in the Himalayas and Japan. NW North America; Rocky Mountains south to the State of Colorado; Eastern Canada and United States.

## GENUS 10    Polypodium L. — POLYPODY

1. ***Polypodium vulgare*** L., Sp. pl. (1753) 1085; Ruprecht, Distrib. crypt. vasc. 118; Ledebour, Fl. Ross. IV, 508; Krylov, Fl. Zap. Sib. I, 45; Fomin, Fl. Sib. Daln. Vost. V, 181; id., Fl. SSSR, I, 85; Perfilev, Fl. Sev. I, 46; Gröntved, Pterid. Spermatoph. Icel. 111; Hultén, Atlas 17 (67); Poyarkova, Fl. Murm. I, 53.
**Ill.:** Fl. Murm. I, pl. XII, 1.

Basically a forest fern, often growing on shady mossy rocks. Penetrating arctic limits in Atlantic districts of Europe. Leaves overwintering in green condition.

**Soviet Arctic.** Murman (more or less ubiquitous).
**Foreign Arctic.** Iceland; Arctic Scandinavia.
**Outside the Arctic.** Forest (mountain forest at more southern latitudes) districts of Europe (south to the Mediterranean mountains), including the west half of the European part of the USSR; Northern (as far north as the Telpos-Iz Mountains) and Middle Urals; Crimea; Caucasus; mountains of NW Africa and neighbouring islands; mountains of Asia Minor and Iran. In Siberia in the Kuznetskiy Alatau, Altay and adjoining territory, and at one site in the Western Sayan. Reported far from its basic range in the Hawaiian Islands, South Africa and Kerguelen Island.

In Western North America replaced by the race *P. vulgare* ssp. *occidentale* (Hook.) Hult., which does not penetrate arctic limits. In East Siberia and the Far East as well as Eastern North America, there is a closely related species, *P. virginianum* L., which nowhere reaches 60°N.

## FAMILY II

# Ophioglossaceae R. Br.

**ADDER'S TONGUE FAMILY**

---

### GENUS 1 — Botrychium Sw. — MOONWORT

SMALL HERBACEOUS PERENNIALS with more or less erect underground stem (length 1–4 cm), with thick unbranched pale roots. The above-ground part of the plant consists of a fleshy vertical rachis ("stalk") divided above a third of the total height of the plant into two segments: a flat, more or less fleshy bright or pale green leaf somewhat divergent from the rachis, and an erect continuation of the rachis which is paniculately subdivided above (dense like a bunch of grapes when young) and bears on its branches numerous spherical sporangia.

The rather few species of this genus are distributed in rather moist regions of the tropical and temperate zones. In the northern forest zone some of them are distributed as far as its northern limit. Isolated species penetrate the limits of the Arctic. Of the two species reported for the Soviet Arctic, one (*B. lunaria*) is in general very widely distributed. The other (*B. boreale*) is more rarely encountered, and possesses a subarctic distribution. Irrespective of their degree of abundance, species of *Botrychium* nowhere play an essential role in the formation of vegetational communities.

1. Whole plant 3–15 cm high. Sterile leaf blade arising from about the middle of the above ground rachis. Blade oblong, up to 6 cm long, 2–3 times longer than its width, *divided into several (up to 8) pairs of uniform broad semilunate or reniform lobes* whose margins slightly overlap one another. Above the origin of the leaf blade the rachis continues straight upwards and branches repeatedly, forming an *elongate erect-standing brush (almost like a bunch of grapes)* which bears numerous sporangia on its branches. Entire plant, with the exception of the bright-green glossy leaf blade, pale green, almost colorless at its base. Sporiferous brush yellowish.
   ....................................................1. B. LUNARIA (L.) SW.
– Sterile leaf blade arising from the upper half of the above ground rachis (at about two-thirds of the total height of the plant). Blade *broadly ovoid-triangular, with pinnately lobed segments* gradually diminishing in size from base to tip. Sporiferous part of plant *diffusely paniculate*. Otherwise similar to the preceding species. ..................2. B. BOREALE (FR.) MILDE.

   *1. Botrychium lunaria* (L.) Sw. in Schrad. Journ. Bot. II (1801), 110; Ledebour, Fl. Ross. IV, 504; Ruprecht, Distrib. crypt. vasc. 101; Lange, Consp. fl. Groenl. 190; Scheutz, Pl. jeniss. 197; Gelert, Fl. arct. I, 2; Krylov, Fl. Zap. Sib. I, 3; Petrov, Fl. Yak. I, 23; Tolmachev, Fl. Kolg. 12, 39; Porsild, Fl. Disko, 25; Ostenfeld, Fl. Greenl. 54; Fomin, Fl. Sib. Daln. Vost. V, 206; id., Fl. SSSR I, 98; Perfilev, Fl. Sev. I, 48; Devold & Scholander, Fl. pl. SE Greenl. 17; Sørensen, Vasc. pl. SE Greenl. 20; Leskov, Fl.

Malozem. tundry 12; Gröntved, Pterid. Spermatoph. Icel. 101; Hultén, Fl. Al. I, 48; id., Atlas 6 (24); Poyarkova, Fl. Murm. I, 56; Böcher & al., Ill. Fl. Greenl. 44.
*Osmunda lunaria* L., Sp. pl. (1753) 1064.
**Ill.:** Fl. arct., fig. 1; Fl. Murm. I, pl. XIII, 1.

Species widely distributed in forest regions of the temperate zone, in the north of its range rather significantly penetrating arctic limits.

Growing on grassy slopes of tundra mounds and ridges, most often on loamy soil. Also found in shrub thickets with moss carpet, and at the edges of forest-tundra groves. Not rare in the European North, but for the most part encountered in insignificant quantity. Normal development of the reproductive parts is reported more or less everywhere.

**Soviet Arctic.** Murman (from the national frontier to Teriberka); Kanin; Timanskaya and Malozemelskaya Tundras (frequent); Kolguyev Island; Bolshezemelskaya Tundra (more or less ubiquitous); Polar Ural (rather rare); Pay-Khoy; lower reaches of Ob. Not found in the greater part of Arctic Siberia. In NE Asia reported on the lower reaches of the Kolyma and in the Penzhina Basin.

**Foreign Arctic.** Alaska (Beringian coast); West Greenland north to 69½° and East Greenland almost to 72°N; Iceland; Arctic Scandinavia.

**Outside the Arctic.** Found in the greater part of Europe, in the Caucasus, in the forest zone of West Siberia, in the Yenisey Basin north to 66°, in the southern half of East Siberia (north to Central Yakutia), on the coasts of the Okhotsk Sea, and on Kamchatka. Further south, to the frontiers of Mongolia and in the Far East as far as Northern China and Central Japan. Mountainous districts of the Middle East and Central Asia, Himalayas. Forested regions of North America (Alaska, Canada, western, northern and central zones of the USA). Also found in the Southern Hemisphere (Southern South America, New Zealand and SE Australia).

2. ***Botrychium boreale*** (Fr.) Milde in Bot. Zeit. (1857) 478, 880; Scheutz, Pl. jeniss. 197; Gelert, Fl. arct. I, 2; Ostenfeld, Fl. Greenl. 54; Krylov, Fl. Zap. Sib. I, 5; Fomin, Fl. Sib. Daln. Vost. V, 203; id., Fl. SSSR I, 97; Devold & Scholander, Fl. pl. SE Greenl. 17; Perfilev, Fl. Sev. I, 48; Hultén, Fl. Al. I, 47; id., Atlas 6 (22); Poyarkova, Fl. Murm. I, 56; Böcher & al., Ill. Fl. Greenl. 45.

*B. lunaria* var *boreale* Fr., Herb. norm. 16, 85; Hultén, Fl. Al. I, 47.

**Ill.:** Fl. arct. I, fig. 2; Fl. Murm. I, pl. XIII, 2.

Subarctic species with fragmentary distribution in the northern part of the forest zone and penetrating the southern fringe of the tundra zone. Here it occurs as a decidedly rare plant. It is reported from mixed herb meadows, on well-drained slopes, sometimes among shrubs. Normally on sandy soil.

**Soviet Arctic.** Murman (series of reliable localities from the district of Kola Bay in the west to Iokanga in the east); lower reaches of Pechora (dunes colonized by birch near Naryan-Mar); the Vangurey Hills in the west part of the Bolshezemelskaya Tundra (grassy slopes); upper reaches of River Pym-Va-Shor (tributary of the Adzva) and upper course of River Usa; on the River Pyderata near the extremity of the Polar Ural (grassy slopes); lower reaches of the Yenisey (Dudinka); SE Chukotka Peninsula (herb beds in the vicinity of Chaplino Hot Springs); Bay of Korf.

**Foreign Arctic.** Extreme SW and SE Greenland.

**Outside the Arctic.** Northern Fennoscandia; Arkhangelsk Oblast; Northern Preamuria; Kamchatka; Alaska.

Apparently, the extremely fragmentary picture of the range of this species is due to a significant degree to its rarity and consequent absence from herbaria. But the fragmentation of its range is real.

## FAMILY III
# Equisetaceae L. C. Rich.
**HORSETAILS**

---

IN THE CONTEMPORARY flora this family is represented by the single genus *Equisetum*, consisting in total of up to 30 species which are distributed from tropical to arctic latitudes and whose occurrence is associated with rather moist habitats.

---

### GENUS 1
## Equisetum L. — HORSETAIL

PERENNIAL HERBACEOUS PLANTS with creeping branching rhizomes which often deeply penetrate the soil. Stems jointed, rather rigid due to their rich retention of silica, more or less ribbed externally, often scabrous, hollow inside, annual or overwintering in green state (perennial). Branches (not developed in all species) arising from the stem at the nodes, in the majority of cases forming whorls. Leaves weakly developed, adnate to one another, forming sheaths which conceal the nodes of the stem and branches. Free tips of leaves (often conspicuous through their dark colour, often with hyaline margins) taking the form of teeth of the sheath. Sporangia in oblong spikes developing at the tips of the stems. In certain species there is a differentiation of the stems into two types: vegetative green stems fulfilling assimilative functions and fertile stems without green pigment bearing the spikes.

The majority of species in this genus possess wide distributions. This applies also to the species growing in the Arctic. The greater part of the species represented in the arctic flora are distributed predominantly in the temperate zone and only penetrate the Arctic on the northern edge of their range, occurring in those parts of the Arctic where conditions are rather temperate. Specifically arctic species are not found in this genus, but at least three of the 8 species occurring in the Arctic are characteristic of the arctic flora. One of these (*E. variegatum*) is a predominantly arctic-alpine plant in the general nature of its distribution (in mountainous parts of the temperate zone, but not confined to high mountains). The closely related *E. scirpoides* is widely distributed in the northern forest zone, but has a more restricted distribution in the Arctic. The commonest species in the Arctic, *E. arvense*, has a very wide distribution. This species shows a tendency towards differentiation of an arctic race, *E. arvense* ssp. *boreale*, to which at least the basic bulk of arctic material of *E. arvense* should be referred. However, this has not achieved complete differentiation.

1. *Stems overwintering in green state, rigid,* distinctly ribbed, not flattening during drying, unbranched or with a few branches not forming whorls,

rather dark, dull green. Sporiferous spikes relatively short, always developing on the usual green stems. ........................................2.
- *Stems annual*, dying in winter, *relatively soft*, readily flattening during drying. Branches (if present) numerous, arranged in whorls. Plants more or less pale or bright green. Sporiferous spikes most frequently oblong. In some species they develop on special pale brown stems devoid of green pigment which later (after sporulation) either transform into the usual green stems or die. ...............................................4.

2. *Large* (up to 30 cm high) plants. *Stems erect*, unbranched, *thick* (about 0.5 cm in width), *circular in general outline, with numerous* (more than 10, often 15–20) *weakly prominent* ribs, *strongly scabrous*, in cross-section with broad circular central cavity. Sheaths dark, closely appressed to stem, with *numerous narrow* acute teeth. Spike ovoid, about 1 cm long, equal to stem in thickness. .......................................1. E. HIEMALE L.
- *Small* (not more than 15–20 cm high, often less than 10 cm) plants *with slender* (less than 2 mm in width) erect or flexuous stems (occasionally somewhat branching on their lower part). *Ribs few, sharply prominent.* Central cavity narrow or absent. Sheaths with *a few broad* teeth with finely acuminate tips. Spikes small (up to 0.5 cm long), thicker than stems. ....3.

3. Stems usually more or less crowded, forming virtual clusters, *erect or slightly bent, 1.0–1.5 mm in width, with 4–6 sharply prominent ribs, each of which is bisected by a longitudinal furrow* (so that in cross-section the stem appears 8–12 ridged with the ridges approximated in pairs). Central cavity of stem circular in cross-section, not broad but wider than vallecular cavities. Spikes up to 5 mm long, slightly thicker than stems. ..........
...........................................2. E. VARIEGATUM SCHL.
- Stems frequently prostrate on their lower part, often *bent, arched-recurved, thin (less than 1 mm in width)*, usually *6-ridged*, in cross-section with three elongate vallecular cavities. Central cavity not developed. Spikes 3–4 mm long, considerably thicker than the stems so that the stem tips appear clavate when spikelets are present. ....3. E. SCIRPOIDES MICHX.

4. All stems *erect*, green, *rather thick* (3–5 mm in width), *up to half a metre high*, unbranched or bearing whorls of thin, rather short branches. Spikes oblong-oval, dark, 1.2–2.0 cm long, developing on well developed green stems, slightly raised above their tips. Aquatic plant, sometimes growing on intermittently desiccated banks of waterbodies. ......4. E. LIMOSUM L.
- Green stems erect or flexuous, thinner (1.5–3.0 mm in width). In some species spikes developing on pale brownish stems which later turn green or die. Small (no more than 20–25 cm high) plants growing on dry land. .5.

5. Stems green, sometimes bearing sporiferous spikes at their tips. .......6.
- Stems without green pigment (pale, brownish, fleshy) bearing sporiferous spikes at their tips. ................................................10.

6. Stems erect, bright (or pale) green, up to 25 cm high and 3 mm wide.

Branches *long and thin, with many branchlets* (as if disheveled), for the most part *arched-recurved downwards* so that the whorl as a whole is somewhat convex above. Sheaths divided into 2–3 broad lobes formed from fused teeth. Green stems may carry dead sporiferous spikelets at their tips. .................................................6. **E. SILVATICUM** L.

– Branches without or with weak development of branchlets. ............7.

7. Stems always erect, pale green, rather rigid, up to 20 cm high, 1.5–2.0 mm wide, with scabrous ribs on their upper part. *Whorls not developed below middle third* (often only on upper half) of stem. *All branches of more or less uniform length, not divided into branchlets, projecting horizontally* from stem or leaning slightly arcuately. Sheaths closely appressed to stem, with narrow teeth. ......................................7. **E. PRATENSE** EHRH.

— Stems erect or flexuous. Branches *directed obliquely upwards*, often considerably longer in the lower whorls than in the upper. ................8.

8. Plant *dull green*, often glaucous. Stems more or less erect, with mircotuberculate or transversely rugose ribs. Branches *simple, directed obliquely upwards, with their tips somewhat incurved towards stem.* Sheaths with rather broad teeth. Sporiferous spikes oblong-cylindrical, 1.5–2.0 cm long, developing at the tips of the green stems and larger branches and able to persist long after sporulation ......................5. **E. PALUSTRE** L.

– Plant *bright green*. Stems erect, or *obliquely ascending*, or *arched-recurved on their lower part and then prostrate*, up to 15 (more rarely 20) cm high, 1.5–2.5 mm wide, with a small number of ribs (smooth below, obtusely tuberculate above). *Branches directed obliquely upwards*, shorter nearer tip of stem; unbranched tip of stem often projecting considerably above them. *On prostrate stems position of branches strongly asymmetrical*, all being turned more or less upwards. Sheaths with triangular, rather well spaced teeth. ..........................(8. **E. ARVENSE** L., S.L.) — 9.

9. Stems *erect or obliquely ascending, or almost erect at first then genuflected, drooping towards the ground*. Branching always considerable. Branches 4–5 (rarely 3) ridged; on erect or obliquely ascending stems the branches are more or less erect, but on drooping stems the branches curve on the lower (ground-facing) side of the stem so that *they all* (irrespective of their origin) *have more or less upwardly directed tips* (away from the ground). . .
...........................8. **E. ARVENSE** L., S. STR. (various varieties)

– Stems predominantly *prostrate*, more rarely more or less erect. Branching often weak, principally developed on lower whorls. Branches normally *3-ridged, thin*, in prostrate specimens *sometimes somewhat arcuately recurved from the stem*. Frequently stems almost unbranched. .........
........................8a. **E. ARVENSE** SSP. **BOREALE** (BONG.) RUPR.

10. Fertile stems developing at the beginning of summer, pale and brownish, somewhat thicker than the sterile. After sporulation they turn green and develop green branches like those of the sterile stems. ..............11.

– Fertile stems developing at the beginning of summer considerably thicker than the sterile, fleshy, pale, light brown, with large sheaths. Spike oval-cylindrical, on rather long stalk. After sporulation all or the majority of the fertile stems die. A minority remain, developing (mainly from the base) a small number of green branches. . . . . . . . . . .(8. **E. ARVENSE** L., S.L.) — 12.

11. Sheaths *large* (2–3 cm long), *with a few broad teeth*. Spikes almost cylindrical, 1.0–2.5 cm long, on rather long stalk. Branches beginning to form branchlets soon after their appearance (while still short) . . . . . . . . . . . . . .
. . . . . . . . . . . . . . . . . . . . . . . . . . . . . . . . . . . . . . .6. **E. SILVATICUM** L.

– Sheaths not more than 1.5 cm long, *with narrow unfused teeth*. Spikelets oblong-oval, 1.0–1.5 cm long, on short stalk. Branches appearing after sporulation not forming branchlets. . . . . . . . . . . . . .7. **E. PRATENSE** EHRH.

12. Fertile stems *thick, with a considerable number of relatively short internodes*, a considerable portion of whose length is covered by the sheaths. After sporulation all stems quickly die. *Spikes rather large*, 1.5–2.0 cm long.
. . . . . . . . . . . . . . . . . . . . . . . . . . . . . . . . . . . . . . .8. **E. ARVENSE** L. (S. STR.)

– Fertile stems low but *relatively thin, with a small number of extended internodes*, only covered by sheaths near the nodes. *Spikes small*, for the most part about 1 cm (sometimes up to 1.5 cm) long. Dying of the fertile stems after sporulation is often delayed, and then a *few thin drooping green branches develop from the lower stem nodes*. . . . . . . . . . . . . . . . . . . . . .
. . . . . . . . . . . . . . . . . . . . . . . .8a. **E. ARVENSE** SSP. **BOREALE** (BONG.) RUPR.

1. *Equisetum hiemale* L., Sp. pl. (1753) 1062; Ruprecht, Distrib. crypt. vasc. 93; Gelert, Fl. arct. I, 9; Ostenfeld, Fl. Greenl. 54; Krylov, Fl. Zap. Sib. I, 57; Petrov, Fl. Yak. I, 32; Ilin, Fl. SSSR I, 110; Perfilev, Fl. Sev. I, 50; Gröntved, Pterid. Spermatoph. Icel. 92; Hultén, Atlas 4(13); Selivanova-Gorodkova, Fl. Murm. I, 68; Karavayev, Konsp. fl. Yak. 41.

   **Ill.:** Fl. Murm. I, pl. XIX.

   Basically a plant of the northern forest zone, growing mainly in river valleys, beneath the forest canopy. Just penetrating arctic limits.

   **Soviet Arctic.** West Murman (Pechenga); Kildin Island.

   **Foreign Arctic.** Extreme SW Greenland; Iceland; Arctic Scandinavia.

   **Outside the Arctic.** Widely distributed in the forest zone of Europe, Asia and North America. Reaching furthest north in Fennoscandia. In NE Europe not reaching the Arctic Circle, in Siberia not further north than 62–63°. Southern distributional limits: Southern Europe, southern edge of the forest zone of the USSR, Northern Mongolia, NE China, Japan.

2. *Equisetum variegatum* Schleich., Catal. pl. Helvet. (1807) 27; Weber & Mohr, Bot. Tasch. (1807) 60, 447; Ruprecht, Distrib. crypt. vasc. 94; Lange, Consp. fl. Groenl. 191; Scheutz, Pl. jeniss. 196; Gelert, Fl. arct. I, 9; Simmons, Survey Phytogeogr. 40; M. Porsild, Fl. Disko 26; Lynge, Vasc. pl. N. Z. 14; Ostenfeld, Fl. Greenl. 54; Krylov, Fl. Zap. Sib. I, 58; Petrov, Fl. Yak. I, 33; Tolmachev, Fl. Taym. I, 91; id., Obz. fl. N. Z. 149; Ilin, Fl. SSSR I, 111; Perfilev, Fl. Sev. I, 50; Seidenfaden & Sørensen, Vasc. pl. NE Greenl. 28, 170; Gröntved, Pterid. Spermatoph. Icel. 95; Hultén, Fl. Al. I, 59; id., Atlas 5(20); Selivanova-Gorodkova, Fl. Murm. I, 70; Polunin, Bot. Can. E Arctic 34; id., Real Arctic Pterid. 42; A.E. Porsild, Vasc. pl. W Can. Arch. 70; id., Ill. Fl. Arct.

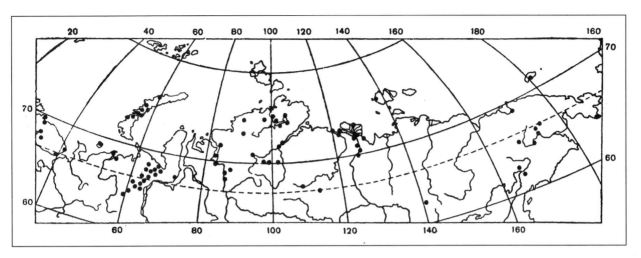

Map I-8  Distribution of *Equisetum variegatum* Schleich.

Arch. 19; Holmen, Vasc. pl. Peary L. 40; Böcher & al., Ill. Fl. Greenl. 42; Karavayev, Konsp. fl. Yak. 41.

**Ill.:** Fl. Murm. I, pl. XX, 1; Porsild, Ill. Fl., fig. 2d, e.

Found in places with ample but not excessive moisture, mainly where turf formation is incomplete, on stony and clayey slopes (sometimes on steep rocky slopes), also in springfed fens and in moist meadows near streams. More rarely observed in thickly moss-covered sites. Widely distributed in the Arctic, mainly in its mountainous districts.

**Soviet Arctic.** Murman (more or less ubiquitous); Malozemelskaya Tundra (isolated localities); Kolguyev Island (rare); Bolshezemlskaya Tundra (common on the eastern margin); Polar Ural (very common); Novaya Zemlya (common, reaching 76°N). Rare in the West Siberian Arctic. Common on the lower reaches of the Yenisey and in Taymyr, in the northwest of the latter reaching beyond 76°N; arctic coast of Yakutia; New Siberian Islands; Chukotka (apparently rare); Wrangel Island; Anadyr and Penzhina Basins; Bay of Korf. (Map I–8).

**Foreign Arctic.** Western and arctic coasts of Alaska, NW Canada, Labrador, Canadian Archipelago (on Ellesmere Island almost reaching 82°N); all Greenland (in the north reported beyond 82°); Iceland; Spitsbergen and Bear Island; Arctic Scandinavia.

**Outside the Arctic.** Mainly in mountainous districts, sporadically in northern parts of the forest zone. Fennoscandia, mountains of Central Europe; locally in the north of the European part of the USSR; Urals; Central Siberian Plateau; Verkhoyansk Range; Altay, Sayans, mountains of Prebaikalia and Northern Preamuria; Stanovoy Range, Sakhalin, Kurile Islands, Kamchatka; Caucasus; Northern Mongolia (mountains); mountains of Western China; mountainous districts and taiga region of North America.

**3. *Equisetum scirpoides*** Michx., Fl. bor. amer. II(1803), 281; Ruprecht, Distrib. crypt. vasc. 94; Schmidt, Fl. jeniss. 130; Lange, Consp. fl. Groenl. 191; Scheutz, Pl. jeniss. 196; Gelert, Fl. arct. I, 9; M. Porsild, Fl. Disko 26; Lynge, Vasc. pl. N. Z. 14; Ostenfeld, Fl. Greenl. 54; Krylov, Fl. Zap. Sib. I, 58; Petrov, Fl. Yak. I, 34; Tolmachev, Fl. rez. Kolguyev. eksp. 13, 39; id., Obz. fl. N. Z. 149; Ilin, Fl. SSSR I, 111; Perfilev, Fl. Sev. I, 50; Leskov, Fl. Malozem. tundry 14; Hultén, Fl. Al. I, 57; id., Atlas 5 (17); Polunin, Bot. Can. E Arctic 36; id., Real Arctic Pterid. 41; Selivanova-Gorodkova, Fl. Murm. I, 72; A.E. Porsild, Vasc. pl. W Can. Arch. 70; id., Ill. Fl. Arct. Arch. 19; Böcher & al., Ill. Fl. Greenl. 41; Karavayev, Konsp. fl. Yak. 41.

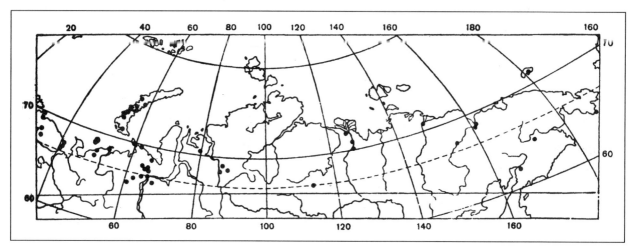

MAP I–9   Distribution of *Equisetum scirpoides* Michx.

Ill.: Fl. Murm. I, pl. XX, 2; Porsild, Ill. Fl., fig. 2 f,g.

In mossy, moderately moist areas of tundra not exposed during wintertime. Usually encountered in small quantity. Growing to a considerable degree buried in the moss carpet. Widely distributed in the Arctic and probably more common than so far established, since easily overlooked during cursory collections in the course of itinerant work.

**Soviet Arctic.** Murman (relatively rare, mainly in West Murman); Malozemelskaya Tundra (more or less common); Bolshezemelskaya Tundra (not everywhere); Kolguyev and Vaygach Islands; Pay-Khoy and Polar Ural; Novaya Zemlya north to 75°; Priobskaya Tundra; Gydanskaya Tundra; lower reaches of Yenisey; lower reaches of Lena, Indigirka and Kolyma; Wrangel Island; Anadyr and Penzhina Basins. (Map I–9).

**Foreign Arctic.** Alaska, north coast of Canada, Labrador; southern fringe of Canadian Archipelago; West coast of Greenland, from the southern extremity to 70–71° N; Spitsbergen (reaching almost 80°N); Arctic Scandinavia.

**Outside the Arctic.** Widely distributed in the temperate zone of the Northern Hemisphere, growing basically in mossy coniferous forests; Fennoscandia, northern zone of European part of USSR, almost all the forested region of Siberia south to the Sayans, Northern Preamuria and Sakhalin; forested zone of North America.

4. ***Equisetum limosum*** L., Sp. pl. (1753) 1062; Scheutz, Pl. jeniss. 196; Gelert, Fl. arct. I, 11; Petrov, Fl. Yak. I, 31; Hultén, Fl. Al. I, 54; Tolmachev, Fl. rez. Kolguyev. eksp. 13, 39; Selivanova-Gorodkova, Fl. Murm. I, 68. — **Swamp Horsetail.**

*E. fluviatile simplex* — Ruprecht, Distrib. crypt. vasc. 92.

*E. fluviatile* L. — Gröntved, Pterid. Spermatoph. Icel. 92; Hultén, Atlas 3(12); Polunin, Real Arctic Pterid. 41.

*E. heleocharis* Ehrh. — Krylov, Fl. Zap. Sib. I, 55; Ilin, Fl. SSSR I, 108; Perfilev, Fl. Sev. I, 49; Leskov, Fl. Malozem. tundry 13.

*E. heleocharis* var. *limosum* — Andreyev, Mat. fl. Kanina 154.

Ill.: Fl. Murm. I, pl. XVIII.

Growing along channels between lakes, and in shallow shoreline parts of waterbodies. Forming dense beds. In the Far North represented mainly by an unbranched form, f. *Linnaeanum* Döll (= *E. limosum* s. str.). As a plant basically associated with the temperate zone, confined to the southern fringes of the Arctic.

**Soviet Arctic.** Murman (frequent), Kanin (everywhere), Timanskaya and Malozemelskaya Tundras (common); Kolguyev Island; Bolshezemelskaya Tundra

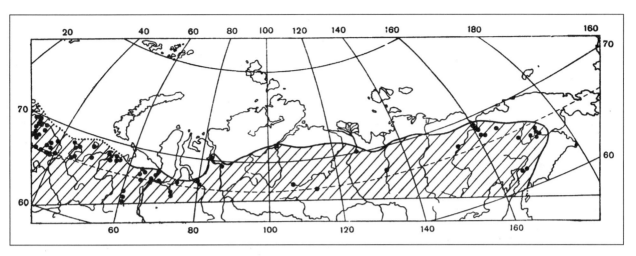

MAP I–10  Distribution of *Equisetum limosum* L.

(west and northwest, in the east only in the Usa Basin); Polar Ural; Priobskaya Tundra, lower reaches of Nadym and Nyda, lower reaches of Taz; lower reaches of Yenisey north to Nikandrovskiy Island (70°20'N). On lower reaches of Lena at Kyusyur (forest-tundra), on the Kolyma almost at its mouth. Greater part of Anadyr Basin, Penzhina Basin, Bay of Korf. (Map I–10).

**Foreign Arctic.** West coast of Alaska; Iceland; Arctic Scandinavia.

**Outside the Arctic.** Almost all Europe, Caucasus, Asia Minor, all of Siberia and the Soviet Far East, Northern Kazakhstan, Northern Mongolia, NE China, Korea, Northern Japan, forest zone of North America.

5. ***Equisetum palustre*** L., Sp. pl. (1753) 1061; Ruprecht, Distrib. crypt. vasc. 91; Scheutz, Pl. jeniss. 195; Gelert, Fl. arct. I, 11; Krylov, Fl. Zap. Sib. I, 54; Petrov, Fl. Yak. I, 30; Tolmachev, Fl. rez. Kolguyev. eksp. 39; id., Nov. dan. fl. Vayg. 81; Ilin, Fl. SSSR I, 108; Perfilev, Fl. Sev. I, 49; Leskov, Fl. Malozem. tundry 13; Gröntved, Pterid. Spermatoph. Icel. 92; Hultén, Fl. Al. I, 55; id., Atlas 4 (14); Selivanova-Gorodkova, Fl. Murm. I, 66; Karavayev, Konsp. fl. Yak. 41. — **Marsh Horsetail.**

Ill.: Fl. Murm. I, pl. XVII.

Distributed basically in the temperate zone, but with quite considerable extension into the Arctic in the western half of Eurasia. Grows in moist places on the banks of channels and streams (sometimes entering the water), in very moist shoreline willow thickets, sometimes on moist tundra with mossy shrubbery.

**Soviet Arctic.** Murman, Kanin, Timanskaya and Malozemelskaya Tundras; Kolguyev Island; West and East Bolshezemelskaya Tundra; Polar Ural (to its northern extremity); southern Vaygach Island; Priobskaya Tundra; lower reaches of Yenisey north to 69°30'. In Arctic Yakutia reported from the lower reaches of the Lena north to 71°30'. Between the Yenisey and the Lena recorded so far only for the forest zone. East of the Lena not reaching the boundaries of the Arctic. (Map I–11).

**Foreign Arctic.** West coast of Alaska; Iceland; Arctic Scandinavia.

**Outside the Arctic.** Greater part of Europe, Caucasus, Northern Asia Minor, Siberia, Northern Kazakhstan, Northern Mongolia, the Far East from Kamchatka in the north to Korea and Central Japan in the south; forest zone of North America.

6. ***Equisetum silvaticum*** L., Sp. pl. (1753) 1061; Ruprecht, Distrib. crypt. vasc. 91; Lange, Consp. fl. Groenl. 193; Scheutz, Pl. jeniss. 195; Gelert, Fl. arct. I, 10; M. Porsild, Fl. Disko 27; Ostenfeld, Fl. Greenl. 54; Krylov, Fl. Zap. Sib. I, 52; Petrov, Fl. Yak. I, 26; Ilin, Fl. SSSR I, 107; Perfilev, Fl. Sev. I, 49; Leskov, Fl. Malozem. tundry 14;

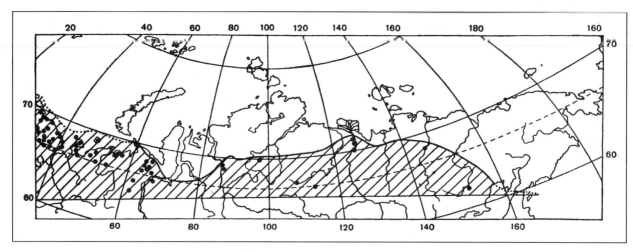

MAP I–11 Distribution of *Equisetum palustre* L.

Gröntved, Pterid. Spermatoph. Icel. 94; Hultén. Fl. Al. I, 57; id., Atlas 5 (18); Selivanova-Gorodkova, Fl. Murm. I, 64; Polunin, Real Arctic Pterid. 42; Böcher & al., Ill. Fl. Greenl. 41; Karavayev, Konsp. fl. Yak. 41. — **Wood Horsetail.**
**Ill.:** Fl. Murm. I, pl. XVI.

On forest islands and within shrubby birch woods of the forest-tundra, beneath the canopy of tundra shrubs, on the banks of lakes and streams, especially on readily warmed sandy soil. Within arctic limits only in the more temperate areas, especially in the European North.

**Soviet Arctic.** Murman (more or less ubiquitous); Kanin; Timanskaya and Malozemelskaya Tundras (common); southern fringe of Bolshezemelskaya Tundra; southern part of Polar Ural; forest-tundra on the shores of the Ob Sound (north to the Nyda); lower reaches of Yenisey north to Dudinka (69°24′N). East of the Yenisey the range boundary considerably recedes from the northern boundary of forest. In the Far East in the Penzhina Basin and on the Bay of Korf. There is a record for this species growing in Chukotka, but without details of localities. (Map I–12).

**Foreign Arctic.** Alaskan coast south of the Bering Strait; Labrador; West Greenland (north to 70°); Iceland; Arctic Scandinavia.

**Outside the Arctic.** Widely distributed in the forest region of Europe (including mountainous districts) and the Caucasus, in the temperate part of Siberia (in Yakutia only in the northwest and in southern districts), and in the whole forest region of the Far East south to Northern Korea; temperate north of North America.

7. ***Equisetum pratense*** Ehrh., Hannover. Magaz. 9 (1784), 138; Ruprecht, Distrib. crypt. vasc. 90; Krylov, Fl. Zap. Sib. I, 52; Petrov, Fl. Yak. I, 26; Tolmachev, Fl. rez. Kolguyev. eksp. 39; Andreyev, Mater. fl. Kanina 154; Ilin, Fl. SSSR I, 104; Perfilev, Fl. Sev. I, 49; Leskov, Fl. Malozem. tundry 13; Gröntved, Pterid. Spermat. Icel. 93; Hultén, Fl. Al. I, 56; id., Atlas 4 (15); Selivanova-Gorodkova, Fl. Murm. I, 62; Karavayev, Konsp. fl. Yak. 41. — **Meadow Horsetail.**

In riverine willow thickets, and on meadow (mainly southfacing) slopes in the forest-tundra and the southern part of the true tundra.

**Soviet Arctic.** Murman (more or less ubiquitous); Kanin (except the most northern part); Timanskaya and Malozemelskaya Tundras; southern Kolguyev Island (rare); Bolshezemelskaya Tundra; upper reaches of the River Usa and the Polar Ural; Priobskaya Tundra; SE Yamal; on the Yenisey north to Dudinka (69°24′N). For Arctic Yakutia there is only one old record (mouth of Lena, St. Matthew Island, 7

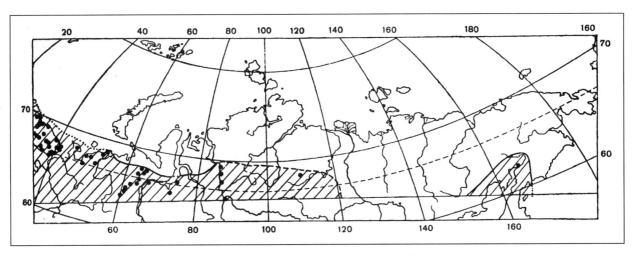

MAP I–12  Distribution of *Equisetum silvaticum* L.

VIII 1883, Priklonskiy), not confirmed by later collections. In the Far East the basins of the Penzhina and the Anadyr (except the portion near its mouth), also the district of the Bay of Korf. (Map I–13).

**Foreign Arctic.** West coast of Alaska; Iceland; Arctic Scandinavia.

**Outside the Arctic.** Widely distributed in Northern and Central Europe, in the whole forest zone of the European part of the USSR, Siberia and the Far East, south to the northern part of the Ukraine, Northern Kazakhstan, Northern Mongolia and NE China; forest region of North America.

8. ***Equisetum arvense*** L., Sp. pl. (1753) 1061; Ruprecht, Distrib. crypt. vasc. 87; Ledebour, Fl. Ross. IV, 486; Schmidt, Fl. jeniss. 130; Scheutz, Pl. jeniss. 195; Lange, Consp. fl. Groenl. 191; Gelert, Fl. arct. I, 10; Simmons, Vasc. pl. Ellesm. 180; id., Survey Phytogeogr. 40; M. Porsild, Fl. Disko 27; Holm, Contr. Morph. Syn. Geogr. arct. pl. 68; Lynge, Vasc. pl. N. Z. 13; Ostenfeld, Fl. Greenl. 54; Krylov, Fl. Zap. Sib. I, 51; Petrov, Fl. Yak. I, 27; Tolmachev, Fl. rez. Kolguyev. eksp. 39; id., Fl. Taym. I, 91; id., Obz. fl. N. Z. 149; Seidenfaden, Vasc. pl. SE Greenl. 42; Devold & Scholander, Fl. Pl. SE Greenl. 16; Scholander, Vasc. pl. Svalb. 15; Ilin, Fl. SSSR I, 103; Perfilev, Fl. Sev. I, 49; Seidenfaden & Sørensen, Vasc. pl. NE Greenl. 27, 170; Gröntved, Pterid. Spermatoph. Icel. 91; Hultén, Fl. Al. I, 51; id., Atlas 3 (11); Polunin, Bot. Can. E Arct. 33; id., Real Arctic Pterid. 41; Selivanova-Gorodkova, Fl. Murm. I, 61; A.E. Porsild, Vasc. pl. W Can. Arch. 70; id., Ill. Fl. Arct. Arch. 19; Holmen, Vasc. pl. Peary L. 39; Böcher & al., Ill. Fl. Greenl. 41; Karavayev, Konsp. fl. Yak. 41. — **Field Horsetail.**

**Ill.:** Fl. Murm. I, pl. XIV; Porsild, Ill. Fl., fig. 2a,b,c.

The commonest and most widely distributed species of horsetail in the Arctic. The exceptionally great polymorphism of *E. arvense* has long attracted the attention of botanists. This has been reflected in description of a considerable number of varieties, as well as in attempts to segregate separate forms of species rank from *E. arvense* as a whole. An attempt to systematize the diversity of forms of *E. arvense* s.l. was made in his time by Ruprecht (1845), who gave descriptions of four "microspecies" subordinate to *E. arvense* [*E. boreale* Bong. (1831), *E. alpestre* Wahlb. (1812), *E. campestre* Schulz (1819) and *E. arcticum* Rupr.]. But this treatment to a high degree represented the nature of variation in *E. arvense* s.l., rather than the racial differentiation. Many investigators of the northern flora have established the existence of diverse forms of *E. arvense* within the confines of the floras studied by them, but as a rule have refrained from more categorical evaluation of their taxonomic significance. Apparently, the resolution of relevant questions is

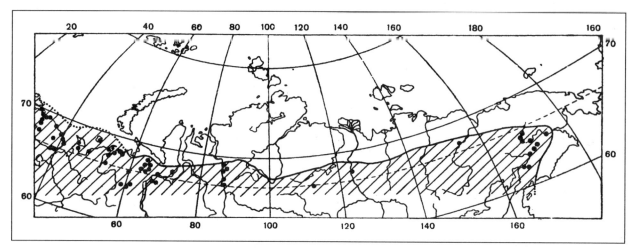

MAP I–13  Distribution of *Equisetum pratense* Ehrh.

complicated by the presence of parallelism between individual variation (directly dependent on the conditions of growth of individual specimens) and variation of a deeper character testifying to the presence of racial differentiation within *E. arvense* s.l. Hultén (1940) has paid particular attention to the possible existence of such variation in connection with his review of data for *E. arvense* in Alaska. He indicates that there is a possibility that *E. boreale* Bong. can be treated there as a race, but refrains from categorical conclusions. Perhaps the most clearly elucidated enquiry into the existence of *E. boreale* as a specific "geographical variety" of *E. arvense* is in the works of E. Hadač devoted to the flora of Spitsbergen (Hadač, 1944; Dahl & Hadač, 1946). After studying arctic material of *E. arvense* s.l. from Spitsbergen and elsewhere, this author reached the conclusion that in high arctic districts there occurs a form of this collective species which is relatively stable in its characters, differing from plants found in temperate latitudes by a whole complex of characters and to the greatest degree agreeing with the characterization of *E. boreale* Bong.

Review of our arctic material compels us to support Hadač's conclusions. In arctic districts of the USSR such as Novaya Zemlya, Vaygach, Taymyr, Arctic Yakutia and Chukotka, apparently there grows a *single race* ("geographical variety") of *E. arvense*, rather variable in form like other races but possessing a certain complex of stable characters. Perhaps the most substantial of its differences from *E. arvense* s. str. is the more or less regularly *incomplete die-off of fertile stems*, which develop a small number of *green branches* after sporulation. The vegetative stems of this form are almost always prostrate. The fertile stems are distinguished (after just a little growth) by relatively extended internodes. *Spikes considerably shorter* than in *E. arvense* from more southern districts. Branches almost always *3-ridged* (in other forms of *E. arvense*, as a rule 4- or sometimes 5-ridged). It is interesting that the last character is retained also by large specimens of this race growing in the south of the tundra zone (e.g., on the Bolshezemelskaya Tundra) and even beyond its limits (e.g., in Alaska).

While having established the distinctness of the arctic race of *E. arvense*, we do not consider it possible to separate it from the stated species on account of incomplete stability of some of its characters, and also because in more temperate parts of the tundra region a clear mixing of the characters of "*E. boreale*" and other forms of *E. arvense* s.l. has been observed. In these circumstances we consider it under the following name:

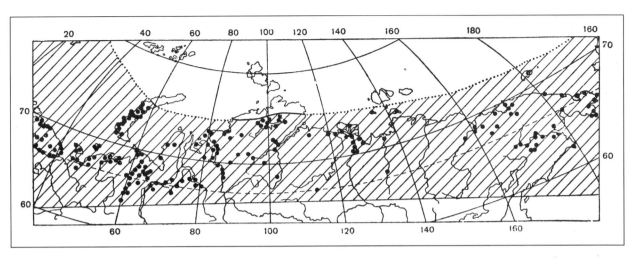

MAP I–14  Distribution of *Equisetum arvense* L.

**8a. *Equisetum arvense* subsp. *boreale*** (Bong.) Rupr., Distrib. crypt. vasc. (1845) 87 [under the name *E. arvense* L. β *E. boreale* Bong.]. — *E. boreale* Bong. in Mém. Ac. Sc. Pétersb. IV, 2 (1832), 174. — *E. arvense* var. *boreale* Milde, Sporenpfl. (1865) 98.

*E. arvense* var. *boreale* — Lange, Consp. fl. Groenl. 192; Tolmatchev, Contr. Fl. Vaigats 125; Petrov, Fl. Yak. I, 28; Andreyev, Mater. fl. Kanina 154; Tolmachev & Pyatkov, Obz. rast. o. Diksona 152; Tolmachev, Fl. rez. Kolguyev. eksp. 12; id., Mater. fl. Mat. Shar 278; Hadač, Gefässpfl. Sassengeb. 10; Dahl & Hadač, Bidrag Spitsb. Fl. 7; Böcher & al., Ill. Fl. Greenl. 41.

*E. arvense* ε *E. arcticum* Rupr., Distrib. crypt. vasc. (1845) 89 (pro maxima parte); *E. arvense* var. *arctica* — Trautvetter, Syllab. pl. Sibir. bor.-orient. (1888) 65; Petrov, Fl. Yak. I, 29.

To this race must be unconditionally referred *all* material of *E. arvense* s.l. from Novaya Zemlya, Vaygach, the shores of the Kara Sea, Taymyr, Arctic Yakutia and adjacent islands, Chukotka and Koryakia. In Murman, Kanin, the tundras near the Pechora, and some parts of the lower reaches of the Yenisey the boundaries between spp. *boreale* and *E. arvense* sensu stricto are obliterated.

Most often encountered on sandbars or gravelbars on the shores of rivers, as well as on banks not subject to surf action on protected portions of sea shores. Common on shoreline slopes and bluffs formed by friable deposits, especially in rather eroded areas with incomplete turf formation. Common on stabilized alluvial areas with meadowlike vegetation, and in riverine willow thickets of the more southern parts of the tundra zone. In the forest-tundra also frequent in areas with sparse tree cover.

Often encountered in considerable quantity, growing for the most part in patches, unevenly distributed over the surface of the general area of habitation. On sandbars often growing apart from other plants, sometimes forming small but not very dense, pure beds. At other sites normally growing intermingled with other plants.

The development of fertile stems has been observed both in more temperate and extreme districts of the Arctic, namely in Taymyr and the European arctic islands. However the spores far from always ripen, and the strongly developed vegetative reproduction of *E. arvense* is of decisive importance for the maintenance of the species. Associated with this is the frequently reported communal growth of this horsetail.

In comparison with other horsetails characteristic of the Arctic, this species is emphatically distinguished by the unbroken nature of its distribution.

**Soviet Arctic.** Murman (everywhere); Kanin, Timanskaya and Malozemelskaya Tundras (ubiquitous); Kolguyev, Bolshezemelskaya Tundra, Polar Ural, Pay-Khoy (everywhere), Vaygach, Novaya Zemlya north to 75°; Yamal, Obsko-Tazovskiy Peninsula, Gydanskaya Tundra; lower reaches of Yenisey, shores and islands of the Bay of Yenisey; Taymyr, north to the mouth of the Pyasina, the southern slopes of the Byrranga Range and the shores of Lake Taymyr; Arctic Yakutia from the lower reaches of the Anabar and Olenek to the mouth of the Kolyma; New Siberian Islands (Stolbovyy and Bolshoy Lyakhovskiy Islands); Wrangel Island; Polar and Beringian coasts of Chukotka; Anadyr Basin; coast of Koryakia; Penzhina Basin. (Map I–14).

**Foreign Arctic.** Arctic Alaska and Canada; Labrador; Canadian Arctic Archipelago (to 82°N on Ellesmere Island); all Greenland (reported in Peary Land above 82°30'N); Iceland; Jan Mayen; Bear Island; Spitsbergen (reaching beyond 80°N); Arctic Scandinavia.

**Outside the Arctic.** Throughout the northern forest region of Europe, Asia and America, further south only in moister districts. In the south reaching Southern Europe, the northern part of Asia Minor, the mountains of Central Asia, the Himalayas, China, and Southern Japan. In North America south to California and North Carolina. Canary Islands, NW Africa. Isolated in the Southern Hemisphere in the Cape Region.

Subspecies *boreale* locally penetrates the limits of the forest region in the Urals, Northern Yakutia and North America. The overwhelming bulk of plants from the temperate north belong to other forms.

## FAMILY IV

# Lycopodiaceae L.C. Rich.

**CLUB-MOSSES**

---

**GENUS 1**     **Lycopodium** L. — CLUB-MOSS

PERENNIAL HERBACEOUS (but with more or less rigid stems) evergreen plants, with small simple elongate (most often lanceolate) leaves. In some species leaves scalelike, adnate to each other and to stem. Spores developing in reniform or semicircular sporangia located in the axils either of the normal leaves or of shortened acuminate-ovoid pale coloured sporophylls which are aggregated in long spikes situated at the tips of branches.

    Club-mosses are basically forest plants. The existence of the majority of them is associated with conditions of moderate warmth and stable, rather high humidity. In the Northern Hemisphere they are found mainly in the taiga zone and in mountain forests of more southern zones, avoiding hot and dry regions. At the same time protection of the places where they grow by snow cover during winter seems to be an indispensable condition for their existence here. In regions with milder and snowy winters they extend considerably further northwards than in regions possessing a more continental climate. The species encountered in the Arctic are not exclusive to it, but penetrate its limits from the northern forest region. One species *(Lycopodium alpinum)* is arctic-alpine. But it is characteristic of this species that it grows only in more temperate parts of the Arctic, predominantly occupied by floras of subarctic type. At the same time it rather significantly penetrates the northern part of the forest zone. *Lycopodium pungens* approaches the arctic-alpine type in the nature of its distribution, being common to the Arctic and to a series of high alpine districts of the temperate zone. But in view of its considerable distribution in northern parts of the forest zone, it is more correct to consider this a *subarctic-alpine* species. Its closest relative *L. annotinum* is a typical forest plant, just penetrating arctic limits on their southern fringe. Rather different relationships occur between the almost cosmopolitan *L. clavatum* and its northern race, *L. clavatum* ssp. *monostachyon*. Typical *L. clavatum* is associated in the Northern Hemisphere with the taiga and accompanying communities. On the fringe of the Arctic it is reported only for a few localities and is essentially alien to the arctic flora. In the northern region of its distribution it is replaced by a special race, ssp. *monostachyon*, which occurs also in the mountains of Southern Siberia. This race *(L. cl. monostachyon)* must be considered to have a distribution of subarctic character: the greater part of its range is found within the taiga zone (especially its northern part), but in West and extreme NE Eurasia it penetrates arctic limits and is rather characteristic of a series of more temperate districts.

    The most extensively distributed species in the Arctic is the generally widespread forest species, *L. selago*. This enters the composition of floras of typically arctic or even high arctic type as the sole representative of its genus. In many arctic districts it is represented by one or two (very rarely three) of the forms occurring also in the taiga zone. But in a considerable part of Arctic Siberia we meet with a special, specifically arctic race, *Lycopodium selago* ssp. *arcticum*, which has achieved a considerable degree of differentiation. This race, which just penetrates the region of dominance of open larch forest, is actually *endemic to the Arctic*.

1. *Stems erect or bent on their lower part, ascending,* almost from their base *evenly dichotomously divided* into branches which are identical to them in structure. Short roots arising from base of stem. *No long prostrate stems.* Branches erect or slightly bent, dichotomously forked in their turn. Whole plant often having the form of a dense bush tapered basally, expanded above. *Sporangia developing in axils of normal leaves,* mainly on upper part of plant. Reproduction also accomplished with the aid of *special gemmae,* developing as small protuberances in the leaf axils, then falling. . . . . . . . . . . . . . . . . . . . . . . . . . . . . . . . . . [SUBGENUS **UROSTACHYA**; L. SELAGO L., S.L.] — 2.
– *Stems long, prostrate,* frequently shallowly buried in moss carpet or soil, rooting, resembling rhizomes. From them arise numerous branches directed vertically or obliquely upwards, which in their turn are (strongly or weakly) branched and more densely leaved than the prostrate stems. The "bushes" formed in several species as a result of branching of lateral branches are joined to one another by means of the prostrate stem (not always immediately obvious, but easily extracted from the soil on account of its considerable strength), and appear externally like small separate plants. *Sporangia concentrated in oblong pale spikes,* which develop (1–3 in number) at the tips of the vertical branches and are sometimes elevated above them on long peduncles, bearing weakly developed and widely spaced pale leaves. . . . . . . . . . . . . . . . . . [SUBGENUS **RHOPALOSTACHYA**] — 3.

2. *Plant light yellowish green* (sometimes almost yellow), 4–11 cm high. Branches more or less densely crowded, together forming a small dense "bush." *Lower leaves on stem and branches lanceolate,* up to 4–5 mm long, obliquely divergent; upper leaves (from about one-third of total height of plant) *closely appressed, short, ovoid-lanceolate or lanceolate, thickish, fleshy,* about 2.5 mm long and 0.75 mm wide. Thickness of branches together with leaves (on upper half of plant) 3–5 mm. Tips of branches obtusely rounded. Sporangia developing in small quantity or absent. *Gemmae usually numerous,* especially on upper part of plant, conspicuous, very prominent; the leaves shielding them are rather prominent and the branches in corresponding places appear distinctly thickened. . . . . . . . . . . . . . . . . . . . . . . . . . . . . . . . . . . . 1a. **L. SELAGO** SSP. **ARCTICUM** (GROSSH.) TOLM.
– *Plant bright or light green,* sometimes yellowish, normally larger, 4–20 cm high (may be taller in the forest region and in oceanic districts of the Foreign Arctic). Leaves *approximately uniform along whole extent of stem and branches, lanceolate,* (4)5–9 mm long, 0.8–1.2 mm wide. Thickness of branches together with leaves 4–10 mm, possibly more than 10 mm on plants with projecting leaves. *Leaves more or less appressed to one another*

*and to stem, or obliquely or horizontally divergent.* Tips of branches (when not thickened due to the presence of gemmae) appearing acuminate (on account of the narrowly acuminate form of the upwardly directed leaves). Sporangia often developed in great quantity and very obvious. Gemmae often more densely accumulated at the tips of branches, not very prominent on their sides in forms with divergent leaves but forming obvious protuberances in forms with appressed leaves. . . . . . .1. **L. SELAGO** L., S. STR.
**VARIATION:**

α. Leaves more or less *closely appressed to stem* (often except for the lower part of the stem where they are obliquely divergent), on account of which the total thickness of branches together with leaves is not great (4–5 mm). Plant *pale green,* often slightly yellowish. Sporangia most often few. Gemmae usually present in considerable quantity and very obvious. Total height of plant not more than 10–13 cm (may be taller in forest zone). . . . . . . . . . . . . . . . . . . . . . . . . . . . . . . . . . . . . . . . . . . . . . . . .VAR. **APPRESSUM** DESV.

β. Leaves *directed obliquely upwards, not appressed,* often narrowly lanceolate. Plant *bright green,* up to 20 cm high. Sporangia developing in profusion. Gemmae often few and relatively inconspicuous. . . . . . . . . . . . . . . . . . . . . . . . . . . . . . . . . . . . . . . . . . . . . . . . . . . . . . . . . . . . . . . . . .VAR. **LAXUM** DESV.

γ. Leaves *more or less horizontally spreading along whole extent of stem and branches,* sometimes slightly inclined downwards. Plant *intensively bright-green,* normally with friable branches, not forming dense "bushes." Sporangia abundant. Gemmae inconspicuous due to the prominence of the large leaves shielding them. . . . . . . . . . . . . . . . . . . . .VAR. **PATENS** DESV.

3. Leaves *lanceolate, free from base,* directed obliquely upwards (sometimes loosely appressed to branches) or spreading horizontally. Thickness of branches together with leaves 5–10 mm. . . . . . . . . . . . . . . . . . . . . . . . . .4.
– Leaves firm, leathery, *scalelike, closely appressed to each other and adnate to each other and to the branch for up to half* (or more) *of their length.* Branches often flattened. Their thickness together with leaves no more than 4 mm. . . . . . . . . . . . . . . . . . . . . . . . . . . . . . . . . . . . . . . . . . . . . . . . . . . .8.

4. Leaves lanceolate, with entire margins (sometimes finely dentate below branches), directed obliquely upwards and rather closely appressed to branches, *gradually attenuate towards end and transformed apically into long translucent awn.* Spikes on peduncles. . . . . .[**L. CLAVATUM** L., S.L.] — 5.
– Leaves *abruptly narrowed at end, with acute tip.* Spikes sessile. . . . . . . . .6.

5. *Spikes 2–3(4) in number,* situated in close group on common *peduncle* (this branched on its uppermost part and *2–3 times as long as the spikes), narrowly cylindrical, of uniform thickness* along their whole length, firm, 1.5–3.5(4) cm long, 2.5–3.5 mm wide. Tips of leaves often freely divergent. Thickness of branches together with leaves 4–8 mm. Branches *without very obvious girdles.* . . . . . . . . . . . . . . . . . . . . . . . . . .4. **L. CLAVATUM** L., S. STR.
– *Spikes solitary, cylindrical or ovoid-cylindrical* (thicker basally than distally), 0.8–2.0 cm long, 2.5–3.0(3.5) mm wide. *Peduncle about equal to spike in length* or slightly shorter, sometimes slightly longer. Branches with very

dense arrangement of leaves whose tips are usually somewhat incurved, possessing *clearly visible girdles*. Thickness of branches together with leaves (2.5)3–4(5) mm. .................................................
................4. **L. CLAVATUM** SSP. **MONOSTACHYON** (GREV. & HOOK.) SEL.

6. Plant *bright green*. Leaves on creeping part of stem arranged densely. Leaves on branches *spreading horizontally or inclined slightly downwards*, sometimes directed obliquely upwards but not closely appressed to one another, on their distal part with serrate margins. *Branchlets without or with inconspicuous girdles*. Spike large, more than 2 cm long. ............
...........................................................2. **L. ANNOTINUM** L.
– Leaves on creeping part of stem arranged more or less sparsely, on branches either *directed obliquely upwards and loosely appressed to one another* or *obliquely divergent (sometimes spreading almost horizontally), with entire margins. Branchlets with distinctly visible girdles*, sometimes moniliformly constricted. Spike up to 1.5 cm long. ....................7.

7. Leaves *more or less appressed to branches or obliquely divergent, narrow*, with attenuate prickly tip. Whole plant pale, yellowish green. ............
...........................................................3. **L. PUNGENS** LA PYL.
– Leaves *spreading almost horizontally, broad, oval-lanceolate*, abruptly narrowed to subulate tip. ....................**L. SUBARCTICUM** V. VASS.[1]

8. Plant *dark green*. Prostrate stem (often more or less hidden in soil or among stones) bearing well spaced vertical branches which are repeatedly forked, with *erect closely approximated branchlets forming dense clusters* ("bushes") of a total height of 4–10 cm. Branchlets *not* or scarcely *flattened*. Leaves thickish, adnate approximately to half their length, ovoid-lanceolate, 2.0–2.5 mm long. Spikes *sessile, solitary*, rather thick, often friable. ...............................................................7. **L. ALPINUM** L.
– Branches often arising obliquely from the prostrate stem, *broadly dichotomously forked, with perceptibly flattened branchlets spreading like a fan*. Total height of branches up to 20 cm. *Spikes on long peduncles, narrowly cylindrical*, solitary or arranged in groups of 2–3, firm. .................
...........................................[**L. COMPLANATUM** L. (S.L.)] — 9.

9. *Branchlets narrow*, 1.5–1.8 mm wide, gathered in dense clusters. Plant *glaucous green*. ..........................6. **L. TRISTACHYUM** PURSH.
– *Branchlets broader*, 2–4 mm wide, very noticeably flattened, more interruptedly distributed, shining bright green above, rather dark below (but not glaucous). ..........5. **L. COMPLANATUM** L., S. STR. (**L. ANCEPS** WALLR.).

**1. *Lycopodium selago*** L., Sp. pl. (1753), 1102; Ruprecht, Distrib. crypt. vasc. 95; Ledebour, Fl. Ross. IV, 496; Schmidt, Fl. jeniss. 130; Scheutz, Pl. jeniss. 196; Lange, Consp. fl. Groenl. 183; Gelert, Fl. arct. I, 12; Simmons, Vasc. pl. Ellesm. 179; id., Survey Phytogeogr. 41; M. Porsild, Fl. Disko 27; Lynge, Vasc. pl. N. Z. 15; Perfilev, Mater. fl. N. Z. Kolg. 50; Ostenfeld, Fl. Greenl. 54; Tolmatchev, Contr. Fl. Vaig. 125, 143; Krylov, Fl. Zap. Sib. I, 60; Tolmachev, Fl. rez. Kolguyev. eksp. 13, 39; Andreyev, Mater. fl. Kanina 155; Devold & Scholander, Fl. pl. SE Greenl. 15; Seidenfaden,

---

[1] A species in need of critical study described from the Commander Islands but recorded by the author of the description also for Murman (without precise reference to specimens). So far there are no reliable records for the Soviet Arctic. In view of the undoubted peculiarity of the type from the Commander Islands and in consideration of V.N. Vasilev's records of the existence of plants identified by him as *L. subarcticum* in Murman, we have included *L. subarcticum* Vass. in the key in order to draw attention to the possibility of its discovery in some district of the Arctic. This seemed all the more desirable because this recently described species is not included in available comprehensive works treating the northern flora. Plants similar to the type from the Commander Islands are present among material of *L. annotinum* s.l. from the Kanin Peninsula. I refrain from evaluating the taxonomic status of *L. subarcticum* at this time.

Vasc. pl. SE Greenl. 43; Ilin Fl. SSSR I, 114; Perfilev, Fl. Sev. I, 50; Seidenfaden & Sørensen, Vasc. pl. NE Greenl. 28; Gröntved, Pterid. Spermatoph. Icel. 98; Hultén, Fl. Al. I, 71; id., Atlas 2(7); Polunin, Botany Can. E Arctic 37; id., Real Arctic Pterid. 42; Kuzeneva, Fl. Murm. I, 76; A.E. Porsild, Vasc. pl. W Can. Arch. 71; id., Ill. Fl. Arct. Arch. 20; Böcher & al., Ill. Fl. Greenl. 36; Karavayev, Konsp. fl. Yak. 41.

*L. appressum* (Desv.) Petr. — Ilin, Fl. SSSR I, 115 (pro parte); Tolmachev, Obz. fl N.Z. 149; Leskov, Fl. Malozem. tundry 15.

Ill.: Fl. Murm. I, pl. XXI; Porsild, Ill. Fl., fig. 2h.

Growing in protected places, on slopes, in small tundra depressions, invariably in places sufficiently protected by snow cover in winter. On Novaya Zemlya reported for lichen tundras, together with *Dryas octopetala* and lichens. Further south found on mossy tundras and on meadowlike slopes, often growing on peaty soil, sometimes accompanied by cloudberry, the lichen *Ochrolechia tartarea* and other plants characteristic of drier parts of hummocky tundra. In the south of the tundra zone and in the forest-tundra occurring frequently beneath shrub canopy *(Betula nana,* occasional large bushes of *B. tortuosa,* certain willow species). Here it can also be found in shady mossy spruce woods.

Represented by different varieties dependent on the district and the nature of the habitat. In the extreme north of its range (Novaya Zemlya, Vaygach, Kolguyev, Yamal, Gydanskaya Tundra) the low-growing appressed-leaved var. *appressum* is exclusively encountered, which has occasioned a series of authors to conclude that this completely replaces the true *L. selago* in the Arctic. Indeed this form becomes the sole representative of the species wherever the habitats associated with the growth of other varieties disappear.

In shrubby tundras of the more temperate districts of the Arctic (e.g. the Kanin Peninsula) and in the forest-tundra, complete transitions from var. *appressum* to *var. laxum* occur. In Murman var. *laxum* most often predominates. The var. *patens* is considerably more rarely encountered within the Arctic, being entirely confined to shady forested habitats and therefore restricted in its distribution to "island" forests of the forest-tundra. In some circumstances both extreme types (var. *appressum* and var. *patens*) can be found in topographically very close habitats. For instance, in the basin of the River Kuya (western edge of the Bolshezemelskaya Tundra) we observed in dry shrubby tundra on the plateau typical var. *appressum,* while not at all far away in a shady spruce wood in the river valley grow large specimens of var. *patens.* In such cases the impression is created that var. *appressum* is completely distinct from "typical" *L. selago.* But observations in other districts (Kanin, Murman, relatively temperate districts of Greenland) show the existence of a whole gamut of transitions between varieties of *L. selago* when suitable habitats are present.

In its extreme expression var. *appressum* is very similar to *L. selago* ssp. *arcticum,* and in strictly quantitative comparison the differences between them are less than those between var. *appressum* and other varieties of *L. selago* s. str. in their typical expression. The most stable difference between *L. s. arcticum* and extreme forms of var. *appressum* is the shortened form and greater fleshiness of leaves on the upper half of the plant. In the former the difference in leaf form between the lower (lower third or quarter) and upper parts of the plant is rather striking; but in var. *appressum* this difference is weakly expressed and the leaves along the whole extent of the branches retain a narrower lanceolate form. In combining on the same plant both closely appressed leaves and leaves which are noticeably divergent from the stem (low on the stems), *L. s. arcticum* approaches *L. s.* var. *dubium* Sanio intermediate between var. *appressum* and other varieties of "typical" *L. selago.*

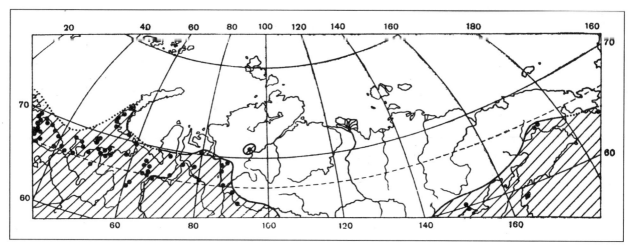

MAP I–15  Distribution of *Lycopodium selago* L. s. str.

Normal sporulation is observed everywhere in var. *laxum* and var. *patens*. In var. *appressum* it is often more or less suppressed and the vegetative gemmae have assumed basic importance as the means of reproduction.

**Soviet Arctic.** Murman (everywhere); Kanin Peninsula, Timanskaya and Malozemelskaya Tundras (everywhere as far as the arctic coast); Bolshezemelskaya Tundra; Kolguyev Island (more or less everywhere, but not in great quantity); Polar Ural (frequent); Pay-Khoy and Vaygach Island (rare and local); on Novaya Zemlya reported for the South Island (so far only on the west coast) and for the extreme SW part of the North Island as a rare plant; in Arctic Siberia in the southern half of Yamal, the lower reaches of the Taz and the Gydanskaya Tundra; in the lower reaches of the Yenisey extending north to Dudinka, reported near Norilsk in the Pyasina Basin, in the Khatanga Basin in open woodland transitional to tundra on the Volochanka. Replaced by *L. s. arcticum* at the mouth of the Yenisey, in Taymyr, Arctic Yakutia and the greater part of Chukotka. Appearing anew in the southeast of the Chukotka Peninsula, the Anadyr Basin and the Bay of Korf. (Map I–15).

**Foreign Arctic.**[2] Alaska (west and north coasts), Labrador, Canadian Arctic Archipelago (on Ellesmere Island to 81°40'N), Greenland (to 80°N in the North-West and to 76°45'N on the East coast), Iceland, Spitsbergen, Arctic Scandinavia.

**Outside the Arctic.** Widely distributed in forest regions of the temperate zone of the Northern Hemisphere both in Eurasia and North America. Restricted to mountains in more southern districts. In relatively northern high alpine districts represented exclusively by var. *appressum*, which is also found in mountainous districts far from the Arctic (Alps, Caucasus, etc.) and in the northern forest zone in peat bogs and sometimes on rocks.

In Eastern Asia (Primorskiy Kray, Sakhalin, Kurile Islands, China, Japan) replaced by the closely related species *L. chinense* Chr., which apparently also occurs in Southern Alaska. A series of other closely related species replace *L. selago* in the mountains of South America, the Hawaiian Islands, Madeira, the Azores and a series of extratropical districts of the Southern Hemisphere.

**1a. *Lycopodium selago* subsp. *arcticum*** (Grossh.) Tolm. in Not. Syst. Herb. Inst. Bot. (1960), 39.

*L. arcticum* Grossh., Fl. Kavk. I (1939) 374 (nomen!).

*L. appressum* Petrov (non Desv.!), Fl. Yak. I, 37, pro maxima parte; Ilin, Fl. SSSR I, 115, pro parte; Karavayev, Konsp. fl. Yak. 42.

---

[2] With respect to the Canadian Archipelago, northern districts of Greenland and Spitsbergen, there is uncertainly whether data on the distribution of *L. selago* available in the literature (with or without the reservation that the plants belong to "f. *appressa*") should be referred to *L. selago* s. str, rather than to *L. s. arcticum*. The latter (particularly with respect to plants from the Canadian Archipelago) appears correct. I admit that all Greenlandic plants seen by me belong to *L. selago* s. str., moreover plants from the west coast collected below 72°54'N should be referred to var. *laxum*. But I have not seen material from more northern localities. It is undisputed that at least part of the Spitsbergen plants belong to *L. s. arcticum*.

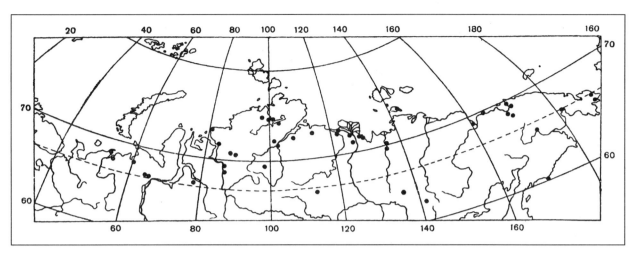

MAP I–16  Distribution of *Lycopodium selago arcticum* (Grossh.) Tolm.

*L. selago* L. — Tolmachev & Pyatkov, Obz. rast. o. Diksona 152; Tolmachev, Fl. Taym. I, 91; *L. selago* var. *appressa* Desv. — Trautvetter, Pl. Sibir. bor. 145; id., Syll. pl. Sibir. bor.-orient. 65.

**Ill.:** Tolmachev in Not. Syst. (1960) 40, 41.

In mossy areas, on *Dryas*-lichen tundra, among thickets of *Cassiope tetragona* and *Ledum decumbens*, most frequently on sloping sites with sandy loam or gravelly soil, sometimes in small depressions. Associated with sites sufficiently protected by snow cover during wintertime and more or less well drained. Growing for the most part in small quantity but not a rarity (especially in Arctic Yakutia). For the most part encountered in small colonies in portions of tundra which are isolated from one another. Sporulation normally not observed. Reproducing with the aid of abundant development of gemmae.

A truly arctic race extending only to mountainous districts of subarctic Siberia.

**Soviet Arctic.** Basic region of distribution extending from the lower reaches of the Yenisey and the Bay of Yenisey in the west (Dikson Island, Golchikha, Dudinka, south as far as Khantayka), through Taymyr (River Pyasina below 71°N, lower course of River Dudypta, Central Taymyr; River Mamontovaya in the Lower Taymyra Basin, 75°17'N; Botling Island and the Bay of Ozhidaniye in the NW part of Lake Taymyr; lower reaches of River Yamu-Tarid), the northern part of the Khatanga Basin (River Novaya), the Popigaya River Basin, the lower reaches of the Anabar River, the lower reaches of the Olenek and Lena (on the Gusinaya Channel of the River Olenek; Stannakh-Khocho on the shore of Olenek Bay; Chay-Tumus on the Olenekskaya Channel of the Lena Delta; Tyriya River between the Olenek and the Lena; Osuk near the mouths of the Lena; Bay of Neyelov; Tiksi Bay), the lower course of the River Yana (Dzhanky, Magyl); Bolshoy Lyakhovskiy Island; lower reaches of Kolyma and the coast to the east of it (Pokhodskoye, River Lelendey, B. Baranov rock), the district of the Bay of Chaun (Pevek, Ust-Chaun, etc.), Chukotkan coast of Arctic Ocean (River Chegitun) to the Bering Strait (Lawrence Bay) in the east; shore of the Bay of Krest. Locally in the Anadyr Basin (mouth of River Belaya, Medvezhi Hills between the village of Markovskoye and the Yeropol), on the Koryak coast (Capes Goven and Olyutorskiy). (Map I–16).

West of the Yenisey scattered localities embedded in the range of *L. selago* s. str.: lower reaches of the River Bolshoy Pur; Cape Kruglyy on Tazovskaya Bay (doubtful specimen); Cape Khayen-Sale on Yamal; middle course of River Khanma in the Priobskaya Tundra and its sources in the Polar Ural; upper reaches of River

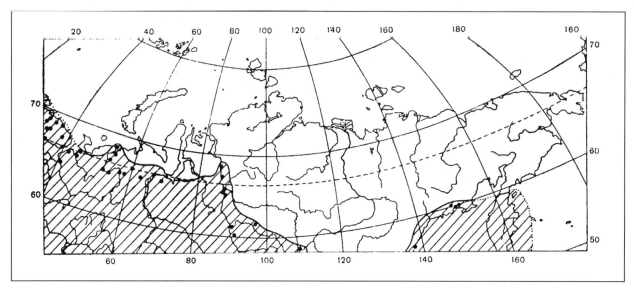

**MAP I–17**  Distribution of *Lycopodium annotinum* L.

[3] Apparently the Canadian Archipelago is the source of a specimen preserved in the Leningrad herbarium with label notation "ex itin. Parry" without collection locality. Simmons records *L. selago* for the Melville Peninsula, Baffin Island, Devon Island and from rather numerous points on Ellesmere Island where the species extends north to Discovery Bay (81°40'N). Porsild (1955) additionally records the presence of *L. selago* on Victoria Island. I have not seen material from Ellesmere Island. All Greenlandic plants seen by me undoubtedly belong to *L. selago* s. str., but I have not seen plants from the NW coast of Greenland nor from East Greenland north of the 70th parallel (i.e. from districts where the occurrence of *L. s. arcticum* is theoretically most likely). The presence of *L. s. arcticum* on Spitsbergen indirectly indicates the probability of its occurrence in Greenland, in so far as *L. selago* var. *appressum* grows on Novaya Zemlya; thus the question of whether the Spitsbergen localities of *L. s. arcticum* are connected to its basic range through Novaya Zemlya is answered negatively.

Apparently, in Arctic America *L. selago* is distributed partly disjunctively. It is

Adzva and the district of the River Yushinoy (on the Pechora) in the Bolshezemelskaya Tundra; east of Chernaya Hill in the Malozemelskaya Tundra.

**Foreign Arctic.**[3] Spitsbergen [district of Ice Fjord, Bear Valley (Björndalen); King's Bay]; northern districts of Greenland? (Canadian Arctic Archipelago).

**Outside the Arctic.** Reported only from high alpine parts of the Verkhoyansk Range (Tompo River Basin), on the northern edge of the Central Siberian Plateau (alpine tundra in the upper reaches of the River Savida, tributary of the Kotuy), on lowland at the Kheta-Khatanga junction (vicinity of the village of Khatanga and a marshy depression in the northern foothills of the plateau), and in the district of the Olenek-Vilyuy watershed (Daaldyn). Records of occurrence in mountains of the Lena-Okhotsk and Lena-Amur watersheds under the name *L. appressum* Petr. (cf.: Petrov, Fl. Yak. I, 38) should be referred to *L. selago* var. *appressum*.

**2. Lycopodium annotinum** L., Sp. pl. (1753) 1103; Ruprecht, Distrib. crypt. vasc. 96; Ledebour, Fl. Ross. IV, 497; Scheutz, Pl. jeniss. 196; Lange, Consp. fl. Groenl. 183 (α *vulgare*); Gelert, Fl. arct. I, 12 (pro minima parte); M. Porsild, Fl. Disko, 28 (pro parte); Ostenfeld, Fl. Greenl. 54 (pro parte); Krylov, Fl. Zap. Sib. I, 61; Petrov, Fl. Yak. I, 40; Devold & Scholander, Fl. pl. SE Greenl. 15; Ilin, Fl. SSSR I, 117; Perfilev, Fl. Sev. I, 51; Leskov, Fl. Malozem. tundry 15; Gröntved, Pterid. Spermatoph. Icel. 96 (pro parte); Hultén, Fl. Al. I, 63 (pro parte); id., Atlas 1(2); Kuzeneva, Fl. Murm. I, 78; Polunin, Real Arctic Pterid. 42 (pro parte); Karavayev, Konsp. fl. Yak. 42.

**Ill.:** Fl. Murm. I, pl. XXII, 1.

In its typical form well distinguished from the generally similar *L. pungens*. The differences are fully evident also in areas where the ranges of the two species overlap. For instance, in the forest-tundra near the Pechora one can find *typical L. annotinum* and *typical L. pungens* at a distance of no more than a few hundred metres from one another, the former as a plant of shady "island" spruce woods, the latter on elevated, partially shrub-covered areas of tundra. However, in certain circumstances specimens which are transitional or more precisely of doubtful specific identity are also observed. A hybrid origin of such plants is not excluded in the considerable area of overlap of the ranges of the two species. Moreover, in habitats which are rather moist and shady *L. pungens* can apparently assume a form in several respects closer to *L. annotinum* than the typical form: in these plants the not reported on the arctic coast between Alaska and the Melville Peninsula, and its localities on islands of the Canadian Archipelago are thus separated from the basic continental part of the range. This is additional evidence of the possible connection of these localities with the Arctic-Siberian race *(L. s. arcticum)*. We note that in East Siberia the range of *L. s. arcticum* (rather common in Arctic Yakutia) is over a considerable distance quite separate from the range of the taiga-dwelling *L. selago* s. str.; connections between them are recorded only through high alpine localities within the Verkhoyansk Range and the mountain system of the Stanovoy Range.

leaves are broader than in the type and somewhat spreading from the rachis of the branches. But otherwise the characters of *L. pungens* are revealed in them on careful examination.

*Lycopodium annotinum* is a *forest*, basically *taiga*, plant. In the Arctic it is found mainly in the forest-tundra, especially beneath the canopy of "island" spruce woods or spruce-larch groves in Siberia. At some distance from treeline it is confined to tall willow shrubbery in valleys.

**Soviet Arctic.** Murman; southern part of Kanin Peninsula, riverine willow thickets in the Northern Malozemelskaya Tundra; forest-tundra east of the Pechora; tundra on the lower reaches of the Ob; shore of Ob Sound north of Nyda; lower reaches of Yenisey north to Dudinka. East of the Yenisey not reaching the northern limit of forest. (Map I–17).

**Foreign Arctic.** West coast of Alaska; Greenland, on the west coast at least to 60°N (data on the distribution of *L. annotinum* and *L. pungens* not clearly distinguished), in the east apparently to 66–67°N; Iceland; Arctic Scandinavia.

**Outside the Arctic.** Widely distributed in the taiga zone and partly in more southerly located deciduous forests. Forested districts of Western, Central and Northern Europe, in the European part of the USSR south to the forest-steppe zone; West Siberia from the northern limit of forest to the southern fringe of the forest zone; east of the Yenisey mainly in temperate northern districts with sufficient development of coniferous forest, in Yakutia only in the far south (upper part of Aldan Basin); in the mountains of Southern Siberia only within the forest zone; in the Soviet Far East on the Okhotsk coast, Sakhalin, the Kurile Islands and Kamchatka (commoner here than *L. pungens*). South of the USSR in Northern Mongolia, mountainous districts of China, the Himalayas, Northern and Central Japan. Widely distributed in the forest zone of North America.

3. **Lycopodium pungens** La Pyl. in Mém. Soc. Linn. Paris VI (1827), 182; Petrov, Fl. Yak. I, 39; Andreyev, Mater. fl. Kanina 155; Ilin, Fl. SSSR I, 117; Perfilev, Fl. Sev. I, 51; Leskov, Fl. Malozem. tundry 15; Kuzeneva, Fl. Murm. I, 80; Karavayev, Konsp. fl. Yak. 42.

*L. annotinum* var. *pungens* Desv. — Gelert, Fl. arct. I, 12; M. Porsild, Fl. Disko 28; Seidenfaden, Vasc. pl. SE Greenl. 42; Devold & Scholander, Fl. pl. SE Greenl. 14; Sørensen, Vasc. pl. E Greenl. 19; Gröntved, Pterid. Spermatoph. Icel. 96; A.E. Porsild, Ill. Fl. Arct. Arch. 19; Böcher & al., Ill. Fl. Greenl. 36.

*L. annotinum* var. *alpestre* Hartm. — Lange, Consp. fl. Groenl. 184; Scheutz, Pl. jeniss. 196; Polunin, Bot. Can. E Arctic 38.

*L. annotinum* — Simmons, Survey Phytogeogr. 41; Tolmachev, Fl. rez. Kolguyev. eksp. 39.

Ill.: Petrov, Fl. Yak. I, fig. 22; Porsild, Ill. Fl., fig. 2i.

Subarctic species characteristic of the northern fringe of the taiga zone (where it is found principally in more or less open habitats) and of the southern, more temperate part of the tundra zone. Within the Arctic it is found on the slopes of sandy ridges and hills, and on flat ground among tall but not too dense shrubs (in the European North especially in areas with tall shrubs of *Betula tortuosa*; in the Far East among *B. Middendorffii*, as well as in open thickets of *Alnus fruticosa* and stands of cedar pine, *Pinus pumila*). Frequent as an inhabitant of dry areas with peaty soil. Confined to places sufficiently protected by snow cover during wintertime.

Normal sporulation reported more or less everywhere within the tundra part of the range.

**Soviet Arctic.** Murman (everywhere); southern half of Kanin Peninsula; Timanskaya and Malozemelskaya Tundras; Kolguyev Island; Bolshezemelskaya Tundra except for the coastal zone; Polar Ural; lower reaches of Ob and SE coast of Yamal; shores

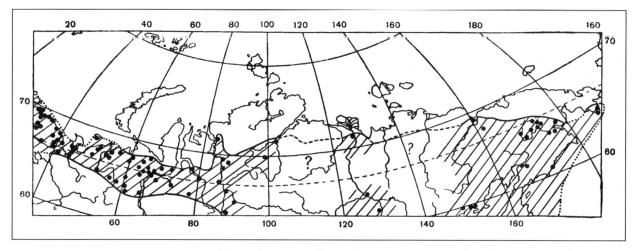

MAP I–18 Distribution of *Lycopodium pungens* La Pyl.

of Tazovskaya Bay, lower reaches of Taz; lower reaches of Yenisey north to its mouth (Sopochnaya Korga). Further east the distribution is extremely inadequately clarified. Reported in the Pyasina and Khatanga Basins on the northern edge of the Central Siberian Plateau and near the village of Khatanga, not found in Taymyr; occurring in the lower reaches of the Lena (Kumakh-Surt), at the mouth of the Kolyma (Pokhodskoye) and in forest-tundra adjacent to its lower reaches; SE Chukotka Peninsula; at many sites in the Anadyr Basin and in Koryakia. (Map I–18).

**Foreign Arctic.** West coast of Alaska; West Greenland north to 72° 50′, East Greenland north to 72°45′ (evidently common further south, but details of distribution unclear because records of *L. pungens* have usually not been distinguished from those of *L. annotinum*); Iceland; Arctic Scandinavia.

**Outside the Arctic.** Extreme NE Scandinavia ("Lapland"); Kola Peninsula; the northeast of the European part of the USSR (south to the Solovetskiye Islands, the Zimniy shore of the White Sea, and the middle course of the River Pechora); Northern West Siberian Lowland (south to 64° N); on the Yenisey from the southern part of the tundra zone south to 63°N. East of the Yenisey reported on the Lower Tunguska; in Central and Southern Yakutia, near Baykal (mainly in mountains), here and there in the Sayans, in mountainous districts of Dahuria, and in the Stanovoy Range. Occurring on the western and northern coasts and on islands of the Sea of Okhotsk; on Sakhalin in the mountains; on Kamchatka considerably rarer than *L. annotinum*.

Distribution outside the USSR difficult to estimate because records of *L. pungens* and *L. annotinum* have usually not been distinguished. *Lycopodium pungens* certainly occurs in inland districts of Alaska and in mountainous districts and part of the forest zone of Canada.

**4. Lycopodium clavatum** L. subsp. **monostachyon** (Grev. et Hook.) Sel.; *L. clavatum* var. *monostachyon* Grev. et Hook. in Hooker, Bot. Misc. 2 (1831), 375; Hultén, Fl. Al. I, 67; Böcher & al., Ill. Fl. Greenl. 36.

*L. clavatum* L. — Ruprecht, Distrib. crypt. vasc. 97 (pro parte); Ledebour, Fl. Ross. IV, 499 (pro parte); Lange, Consp. fl. Groenl. 184; Gelert, Fl. arct. I, 12; Ostenfeld, Fl. Greenl. 54; Krylov, Fl. Zap. Sib. I, 62 (pro parte); Petrov. Fl. Yak. I, 41; Perfilev, Fl. Sev. I, 51 (pro parte); Leskov, Fl. Malozem. tundry 15; Karavayev, Konsp. fl. Yak. 42; Hultén, Atlas 1(3).

*L. lagopus* (Laest.) Zinserl. — Kuzeneva, Fl. Murm. I, 80.

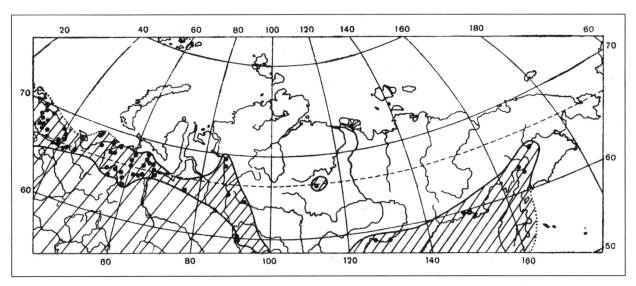

**MAP I-19** Distribution of *Lycopodium clavatum monostachyon* (Grev. et Hook.) Sel.

Ill.: Fl. Murm. I, pl. XXXIII, 3.

Found in dry places, on well-drained tundra slopes, in the forest-tundra among open stands of tall birch shrubbery (*Betula tortuosa*) and in isolated pieces of forest. In the extreme north of its range (e.g. on Kolguyev Island, on the Malozemelskaya Tundra, on the NW Bolshezemelskaya Tundra, at the mouth of the Taz, in the district of Dudinka on the Yenisey) in most cases only sterile individuals develop, clear evidence of a certain reduction in the vitality of the species at the northern edge of its range. Normal reproduction is most often reported in the vicinity of the forest boundary, that is in plants growing in open woodland of the forest-tundra (spruce stands on the River Usa, the district of Salekhard on the Ob, open forest near Khantayka on the Yenisey).

**Soviet Arctic**. Murman; Kanin Peninsula (except extreme north); Malozemelskaya Tundra; Kolguyev Island; Bolshezemelskaya Tundra north to the Vangurey Hills and the Khalmer-Yu River; Polar Ural south from the upper reaches of the Sob (slightly north of 67°N); lower reaches of Ob (Salekhard district); the mouth of the Taz; the lower reaches of the Yenisey north to Dudinka. Further east only in the Penzhina Basin and the southwestern part of the Anadyr Basin. (Map I–19).

**Foreign Arctic**. Bering coast of Alaska; extreme SW Greenland; Arctic Scandinavia (except for the most northerly projecting coastal areas).

**Outside the Arctic**. *Lycopodium clavatum* is widely distributed in the forest region of Europe, Asia and North America. It is also found in mountain forests of Japan, China, the Himalayas and several tropical regions (Indonesia, tropical Africa, etc.), where it is represented by distinct forms.

The *typical* form of *L. clavatum* L. is reported in the Soviet Arctic only in forest-tundra, in the district of Salekhard (bare mainland bank of River Poluy; 30 VI 1913, Pole and Rozhdestvenskiy).

*Lycopodium clavatum* L. is a widely distributed, almost cosmopolitan species (but everywhere confined to forested districts with more or less cool climate). In the Arctic it is represented by its northern race, ssp. *monostachyon* (Grev. & Hook.) Sel., distinguished from the type by the solitary, somewhat shortened sporiferous spikes borne on short peduncles, by the smaller leaves appressed to the branches, and by the presence on the branches of distinct girdles marking the annual growth. This race *replaces* typical *L. clavatum* everywhere on the northern fringe of its range, occupying a zone of fluctuating width (2-5° on the meridian). The significant distinctness of *L. cl. monostachyon* is underscored by the almost complete lack of plants combining characters of this race and the typical form. Plants[4] with solitary, relatively long spikes on peduncles equalling or exceeding them in length may be considered transitional. *Series* of specimens may be considered transitional, when some of the specimens have solitary spikes, others spikes in twos or threes. It is quite possible that sometimes we are dealing (in the herbarium!) with samples which actually represent parts of the same plant, separated from a common stem. All such cases are restricted to points of contact between the ranges of *L. clavatum* s. str. and *L. cl. monostachyon*, moreover no true overlap of their ranges has been observed (with the possible exception of Kamchatka, where on account of the mountainous nature of the country the pattern of distribution of different forms of plants is very much complicated). This is the typical pattern for mutually exclusive subspecies of a single species.

The general character of the range of *L. cl. monostachyon* is subarctic, a large part of it being situated within the limits of the northern forest zone. In East Siberia the range does not cross these limits into the Arctic, but this occurs, on the one hand, in the west starting from the lower reaches of the Yenisey (especially in the European North), and on the other hand in the Anadyr and Penzhina Basins. The southern limit of the distribution of *L. cl. monostachyon* (at the same time the northern limit for *L. clavatum* s. str.) passes through the central part of Karelia, crosses the east shore of the White Sea north of Arkhangelsk, encloses the lower part of the Pechora Basin north of the 64th parallel (apparently nowhere touching the basin of the Severnaya Dvina), shifts southwards in the Urals to the latitude of the sources of the Pechora but on the Ob climbs north to Salekhard (on the Arctic Circle!), after that assuming a general southeastern direction. On the Yenisey the typical form of *L. clavatum* is distributed north to the mouth of the Podkamennaya Tunguska. More northern known localities belong to *L. cl. monostachyon*. This race grows on the heights of the Yenisey Ridge, locally in the Northern Angara region, on the shores of the northern part of Baykal and in the basin of the Upper Angara River, in South Yakutia (Aldan Basin), and in the region of the Lena-Amur watershed. On the shore of the Sea of Okhotsk the southern boundary of the range of *L. cl. monostachyon* exits near Ayan (where both this and the typical form can be found) and encloses Feklistov Island in the Shantar group of islands. On Kamchatka this race occurs mainly in more northern districts and is absent from the Commander Islands (but *L. clavatum* s. str. occurs on Bering Island).

South of the boundary indicated the typical (many-spiked) form of *L. clavatum* is exclusively found everywhere, but in the *high alpine* Sayans and Altay (where *L. clavatum* is generally rare) ssp. *monostachyon* is locally reported again. The range of this race thus has a subarctic-alpine character. In Prebaikalia the northern and southern (high alpine) parts of the range are practically contiguous.

5. **Lycopodium complanatum** L., Sp. pl. (1753) 1104; Ruprecht, Distrib. crypt. vasc. 97; Ledebour, Fl. Ross. IV, 499; Scheutz, Pl. jeniss. 197; Krylov, Fl. Zap. Sib. I, 64; Hultén, Fl. Al. I, 68.

*L. anceps* Wallr. (1840) — Petrov, Fl. Yak. I, 42; Ilin, Fl. SSSR I, 121; Kuzeneva, Fl. Murm. I, 81; Karavayev, Konsp. fl. Yak. 42.

*L. complanatum* var. *anceps* — Krylov, Fl. Zap. Sib. I, 64; Perfilev, Fl. Sev. I, 51; Leskov, Fl. Malozem. tundry 15; Hultén, Atlas 1(4).

Typically a forest plant, penetrating the limits of the tundra zone only on its very margin. In forest-tundra confined to areas of open forest on gentle, well-drained slopes. Sometimes found on sand, in euthermal areas.

**Soviet Arctic.** Murman (more or less everywhere); southern part of Kanin Peninsula; shore of Cheshskaya Bay and the Timanskaya Tundra (locally); SW part of Malozemelskaya Tundra. East of the Pechora only in more southern parts of the

---

[4] We note that for one of the basic distinguishing characters of *L. clavatum* s.str. and *L. cl. monostachyon* (the number of spikes) an intermediate condition is excluded by the very nature of the character.

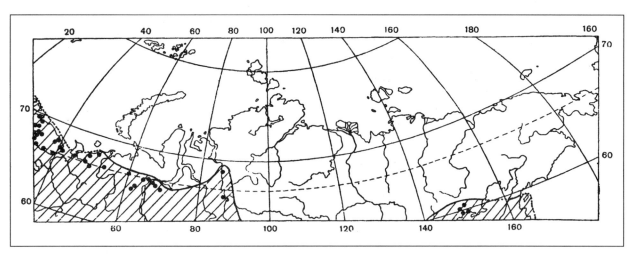

MAP 1-20  Distribution of *Lycopodium complanatum* L.

forest-tundra (on the River Shapkina; lower course of the Adzva). In the Urals not north of the Arctic Circle. On the lower reaches of the Ob at Salekhard, also on the River Poluy; not reported for the shores of Ob Sound. On the lower reaches of the Yenisey within the forest-tundra (Khantayka). East of the Yenisey the boundary of the range shifts into the depths of the taiga zone. (Map 20).

**Foreign Arctic.** Beringian coast of Alaska; locally in Arctic Scandinavia.

**Outside the Arctic.** Widely distributed in forested regions of temperate Eurasia and North America. Replaced by closely related species in Western Europe, the more southern parts of Eastern North America, the island of Madeira and in mountains of a series of tropical countries. The general character of its distribution in Europe and North America underscores the poor adaptation of the species to the conditions of countries with typically oceanic climate.

6. ***Lycopodium tristachyum*** Pursh, Fl. Am. Sept. (1814) 653; Ilin, Fl. SSSR I, 121; Kuzeneva, Fl. Murm. I, 81.

   *L. sabinaefolium* Willd. — Ruprecht, Distrib. crypt. vasc. 98.

   *L. chamaecyparissus* A. Br. in Mutel, Fl. Fr. IV (1837), 192; Lange, Consp. fl. Groenl. 184.

   *L. complanatum* var. *chamaecyparissus* A. Br. — M. Porsild, Fl. Disko 28.

   *L. complanatum* ssp. *chamaecyparissus* Döll. — Krylov, Fl. Zap. Sib. I, 64; Perfilev, Fl. Sev. I, 51.

   *L. complanatum* var. — Ostenfeld, Fl. Greenl. 54.

   *L. complanatum* (*L. chamaecyparissus*) — Gelert, Fl. arct. I, 13.

   Forest zone of Europe and Atlantic North America. In the Arctic in dry areas of open forest and among tundra shrubs.

**Soviet Arctic.** Found once in Murman (Iokanga).

**Foreign Arctic.** West Greenland from the southern extremity to Disko Island (69¼°N); East Greenland (Angmagssalik); Arctic Scandinavia.

**Outside the Arctic.** Western and parts of northeastern districts of the forested part of the European territory of the USSR; Northern and Western Europe; mountains of Southern Europe and Asia Minor; eastern districts of Canada and the United States.

7. ***Lycopodium alpinum*** L., Sp. pl. (1753) 1104; Ruprecht, Distrib. crypt. vasc. 98; Ledebour, Fl. Ross. IV, 498; Scheutz, Pl. jeniss. 197; Lange, Consp. fl. Groenl. 184; Gelert, Fl. arct. I, 13; M. Porsild, Fl. Disko 28; Ostenfeld, Fl. Greenl. 54; Krylov, Fl.

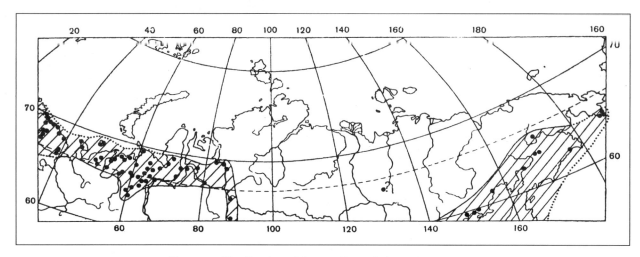

MAP I–21  Distribution of *Lycopodium alpinum* L.

Zap. Sib. I, 65; Petrov, Fl. Yak. I, 43; Tolmachev, Fl. rez. Kolguyev. eksp. 13, 39; Andreyev, Mater. fl. Kanina 155; Devold & Scholander, Fl. pl. SE Greenl. 14; Ilin, Fl. SSSR I, 122; Perfilev, Fl. Sev. I, 51; Leskov, Fl. Malozem. tundry 14; Gröntved, Pterid. Spermatoph. Icel. 96; Hultén, Fl. Al. I, 62; id., Atlas 1(1); Polunin, Real Arctic Pteridoph. 42; Kuzeneva, Fl. Murm. I, 82; Karavayev, Konsp. fl. Yak. 42; Böcher & al., Ill. Fl. Greenl. 37.

Arctic-alpine species rather widely distributed in the Arctic and developing well there, but not penetrating high arctic latitudes. To the contrary, it extends locally rather considerably beyond arctic limits southwards into northern parts of the forest zone.

In dry tundra (shrubby, lichen or shrubby-herbaceous), especially on slopes, on sandy or gravelly substrate. Encountered on slopes and crests of tundra mounds and ridges, and in unflooded places near rivers and streams. Always associated with well-drained areas which are protected to greater or lesser degree by snow cover in winter. In mountainous districts often among stones, in boulder fields on mountain slopes, sometimes on rock. In the forest-tundra found in dry open forest.

In most districts developing normally and often sporulating profusely. Sometimes found only in a vegetative state near the northern limit of its distribution.

**Soviet Arctic.** Murman (everywhere); Kanin (mainly the northern part); Timanskaya and Malozemelskaya Tundras; Kolguyev Island; Bolshezemelskaya Tundra, Pay-Khoy, Polar Ural; lower reaches of Ob, South-Central Yamal; Obsko-Tazovskiy Peninsula; lower reaches of Yenisey to 69½°N; SE Chukotka; Anadyr Basin; Koryakia. (Map I–21).

Not reported from Taymyr, Arctic Yakutia, Northern Chukotka and arctic islands other than Kolguyev.

**Foreign Arctic.** West coast of Alaska; Baffin Island; West Greenland north to 70°40', East Greenland north to 71°; Iceland; Arctic Scandinavia.

**Outside the Arctic.** Moderately extending into taiga districts of Fennoscandia, including the south of the Kola Peninsula and the north of Karelia; ubiquitous in the mountains of Scandinavia. In the northern part of the forest zone in the Pechora Basin. In the Urals descending south to 59°N, additionally isolated in Taganay. In the Siberian Mountains in the Altay, Sayans, near Baykal, in Transbaikalia, the Stanovoy Range, Dzhugdzhur, and the Yenisey Ridge. Additionally in the northern part of the forest zone in Siberia beyond the Yenisey (district of Turukhansk on the

Lower Tunguska). Kamchatka, high mountains of Sakhalin and other districts of the Far East. Mountains of Central and Southern Europe, Caucasus, mountains of Asia Minor. Mountains of NE China and Northern Japan. Alaska. In the mountains of Western North America south to the U.S. frontier, in Eastern North America south to the Gaspé Peninsula.

## FAMILY V

# Selaginellaceae Metten

**SELAGINELLA FAMILY**

---

**GENUS 1**     **Selaginella** Spring. — SELAGINELLA, LITTLE CLUB-MOSS

SMALL LOWLY PLANTS, in general appearance reminiscent of a small low-growing club-moss or moss. Stems prostrate, branched, forming small tufts. Sporangia situated in axils of leaves, associated in poorly differentiated spikes at the tips of the branches. They are of two types: reniform microsporangia in which the numerous small prickly microspores develop, and macrosporangia enclosing 3-4 large, almost globose macrospores.

*Selaginella* is basically a genus of tropical mountain forests. In temperate latitudes of the Northern Hemisphere it is represented by a small number of species, two of which (*S. selaginoides* and *S. sibirica*, very much different from one another and belonging to different sections) penetrate arctic limits. The first of these possesses a fragmented distribution of a generally subarctic-subalpine character. In Europe it is more or less continuously distributed in the northern forest and southern tundra zones, and separated from its northern range in the mountains of Central Europe and the Caucasus. In Siberia and the Far East, where its distribution is still inadequately studied, it is found mainly outside the Arctic in mountainous districts. The second species is a typical rock plant of the forest zone of East Siberia, in mountains climbing to the barrens, and in the north of its range extending north of the northern limit of forest without any change in the nature of its habitats.

1. Plant *light green*, with soft leaves, in appearance rather reminiscent of a moss. Leaves *spreading* from stem, directed obliquely upwards, ovoid-lanceolate, acuminate, up to 3–4 mm long, *with acuminate denticles on their margins*. Spikes at tips of branches, 1–3 cm long, about 0.5 cm wide. (In moist places in the temperate part of the European Arctic and in the lower reaches of the Yenisey). .............1. S. SELAGINOIDES (L.) LINK.
– Plant *grey-green*, in appearance reminiscent of a small, densely branched club-moss, forming a rather dense tuft. Branchlets short (up to 3 cm), slender (together with leaves up to 2 mm wide). Leaves *appressed to branches, closely imbricate*, greyish, narrow, *linear-lanceolate with entire margins*, 1.0–1.5 mm long, at tips with long notched awn. Spikes about 1 cm long, 1.5 mm wide. (In dry, rocky or stony places in the East Siberian and Far Eastern Arctic. ..................2. S. SIBIRICA (MILDE) HIERON.

    *1. **Selaginella selaginoides** (L.) Link, Fil. Horti Berol. (1841) 158; Gelert, Fl. arct. I, 13; Ostenfeld, Fl. Greenl. 55; Krylov, Fl. Zap. Sib. I, 67; Petrov, Fl. Yak. I, 46; Ilin, Fl. SSSR I, 124; Devold & Scholander, Fl. pl. SE Greenl. 15; Gröntved, Pterid.

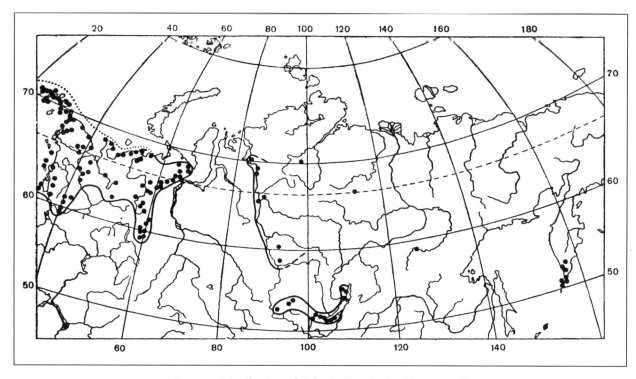

MAP I–22 Distribution of *Selaginella selaginoides* (L.) Link.

Spermatoph. Icel. 98; Hultén, Fl. Al. I, 74; id., Atlas 2(8); Kuzeneva, Fl. Murm. I, 84; Leskov, Fl. Malozem. tundry 15; Böcher & al., Ill. Fl. Greenl. 38.

*S. spinosa* P. Beauv. — Ledebour, Fl. Ross. IV, 501; Ruprecht, Distrib. Crypt. vasc. 99; Lange, Consp. fl. Groenl. 183; Perfilev, Fl. Sev. I, 51.

*S. spinulosa* A. Br. — Scheutz, Pl. jeniss. 197.

*Lycopodium selaginoides* L., Sp. pl. (1753) 1101.

**Ill.**: Fl. Murm. I, pl. XXV.

Growing in moist places on stream banks, on mossy hummocks, sometimes beneath a canopy of shrubs. Also on slopes with sandy loam or stony soil, and in the forest-tundra on the edges of "island" forest. Gravitates to sites with adequate ground moisture in the presence of drainage. Sufficient winter snow cover is, so it appears, an indispensible condition. Often forming small loose tufts.

**Soviet Arctic.** Murman (more or less ubiquitous); northern part of Kanin (common); Timanskya and Malozemelskaya Tundras, west, central and eastern parts of Bolshezemelskaya Tundra (apparently not uncommon); Polar Ural north to Mount Minisey (68½°N; common); lower reaches of Yenisey north to 70½° (Potapovskoye, Dudinka, Tolstyy Nos). (Map I–22).

**Foreign Arctic.** Beringian coast of Alaska (Nome), SW (north to 65°) and SE (between 63 and 64°) Greenland, Iceland, Arctic Scandinavia.

**Outside the Arctic.** Atlantic districts of Northern Europe, the greater part of Scandinavia, northern and central districts of Finland; the north of the European part of the USSR south to Estonia (disjunct), the north of Leningrad Oblast, and the district of Vologda.[1] In the Urals south to 68°N. Isolated in the mountains of Central Europe (Pyrenees, Alps, Carpathians) and the Great Caucasus. Absent from West Siberia. On the Yenisey and east of it reported on the Yenisey Ridge, near Turukhansk, on the lower course of the Lower Tunguska, on the northern edge of the Central Siberian Plateau (in the western part of the Khatanga Basin), and on the Olenek-Vilyuy watershed. More or less common on the southern part of Baykal

---

[1] In old literary sources there are records of the occurrence of *S. selaginoides* on the Volga, Oka and Kama, specifically near the cities of Gorkiy (Lower Novgorod) and Kazan (cf. Ruprecht, l.c.).

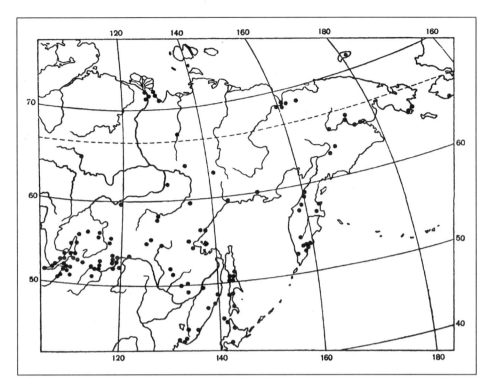

MAP I–23   Distribution of *Selaginella sibirica* (Milde) Hieron.

and within the Eastern Sayan. Rarer in more western parts of the Sayan Mountains. Rare in Northern Baykal, on the upper Lena and Kirenga, and in the upper reaches of the Aldan. Isolated in the far south of Kamchatka and on Paramushir Island. Northern Japan. Temperate part of the taiga zone of North America.

The fragmentation of the range is doubtless partly attributable to natural causes (disjunct distribution: absence from the Altay and the West Siberian Lowland being an undisputed fact), partly due to the inconspicuousness of the plant as a result of which it is often missed during collecting.

2. ***Selaginella sibirica*** (Milde) Hieron., Hedwigia, 39 (1900), 290; Petrov, Fl. Yak. I, 44; Ilin, Fl. SSSR I, 124; Hultén, Fl. Al. I, 75.

*S. rupestris* (L.) Spring — Ruprecht, Distrib. crypt. vasc. 98; Gelert, Fl. arct. I, 14; Krylov, Fl. Zap. Sib. I, 67.

*S. rupestris* f. *sibirica* Milde, Filic. Eur. et Atl. (1867) 262.

Growing in rock crevices, on gravelly slopes and boulder fields, in stony tundra on the crests of uplands, invariably on relatively dry sites. In distinction from *S. selaginoides*, not avoiding sites which lack snow cover in the course of the winter.

**Soviet Arctic.** Lower reaches of Lena (Kumakh-Surt, Tuora-Sis Range opposite Chekurovka, River Sietchan); shore of Buorkhaya Bay (Tiksi Bay, etc.); lower reaches of Kolyma and adjacent tundra (River Tymyveyem); Wrangel Island; southeastern part of Chukotka Peninsula; Anadyr Basin; eastern part of the Penzhina River Basin; on the Bay of Korf. (Map I–23).

**Foreign Arctic.** Beringian coast of Alaska.

**Outside the Arctic.** Widely distributed and common in many places in mountainous districts of East Siberia (shores of Baykal, all Transbaikalia, Southern and Central Yakutia southwards and eastwards from the River Lena, the Verkhoyansk Range and Verkhoyansk Depression) and of the Far East (northern part of Amur Basin,

Primorskiy Kray, Sakhalin, Hokkaido, Okhotsk coast, Kamchatka Peninsula except for extreme south). In North America in inland districts of Alaska and NW Canada.

In the southwestern mountains and the east of North America replaced by other closely related species or races of the collective species *S. rupestris* (L.) Spring. This collective species is also reported in the mountains of China and in the Himalayas. There are records for South America and South Africa.

Records in the literature of the occurrence of *S. rupestris* (s.l.) in the Altay and Urals have not been confirmed by most recent data.

## FAMILY VI
# Pinaceae Lindl.
**PINE FAMILY**

---

VAST GROUP OF arboraceous plants (a few species assuming a shrubby form, mainly under extreme environmental conditions). Distributed predominantly in temperate and subtropical latitudes of the Northern Hemisphere. Isolated northern species of several genera marginally penetrate the limits of the Arctic.

1. Shoots all long, directly bearing leaves (needles) which are arranged individually (not forming bunches). Needles short, dark green, not falling in winter. Cones oblong, with broad flat scales. .................... 2.
- Shoots of two types; long shoots bearing short secondary shoots which bear leaves (needles) arranged in pairs or up to 5 or more, forming veritable bunches. ................................................. 3.

2. Needles soft, flat, with blunt, very slightly bifurcate tip. Cones standing erect, with broad reniform scales, shattering upon ripening, leaving their woody axes (which terminate in a slight thickening) standing erect on the twigs of the tree. ............................. [1a. **ABIES** MILL. — **FIR**].
- Needles firm, tetragonal, acuminate apically. Cones pendulous, their scales either ovoid (narrowing apically) or broader with rounded margin. Scales closely appressed to one another until full ripening of the seeds, then spreading so that the seeds fall freely. Cones falling whole after release of seeds. ........................... 1. **PICEA** DIETR. — **SPRUCE**.

3. Leaves (needles) soft, light green, turning yellow and falling in autumn, arranged on short shoots in bunches of 20–40. Cones reddish or greenish at first, yellowish or brownish at ripening, with thin flat scales. ................................................. 2. **LARIX** MILL. — **LARCH**.
- Leaves (needles) dark or glaucous green, persisting through several years, arranged on short shoots in pairs or in bunches of up to 5. Cones more or less dark green at first, later dark grey or dark brown, with thick angulate woody scales. ........................... 3. **PINUS** (TOURN.) L. — **PINE**.

---

## GENUS 1A
# Abies Mill. — FIR

*Abies sibirica* Ldb., Fl. Alt. IV (1833), 202; Middendorff, Gewächse Sibiriens 548; Keppen, Raspr. khr. der. 399; Krylov, Fl. Zap. Sib. I, 71; Komarov, Fl. SSSR I, 139; Perfilev, Fl. Sev. I, 52; Ya. Vasilev, Der. i kyst. SSSR I, 68. — **Siberian Fir**.
*Pinus sibirica* Ldb. — Scheutz, Pl. jeniss. 193.

Forest tree (able to adopt a stlanik form at the limits of its distribution), distributed from the Northern Dvina Basin in the west to the western part of the Lena Basin in the east. Normally not reaching forest boundaries. In NE Europe not crossing north of the Arctic Circle. Extending further northwards along the Yenisey, where it is found locally between 67 and 68°N. An isolated specimen of fir, evidently resulting from fortuitous seed transport, was found by Lundström in August 1875 on Cape Sopochnaya Korga near the mouth of the Yenisey (71°53'N). This plant has long ceased to exist, but we mention the stated record since the possibility of transport of fir seeds to arctic latitudes at the present time cannot be excluded.

## GENUS 1 — **Picea** Dietr. — SPRUCE

THERE ARE RECORDS of the growth of two species on the southern fringe of the Soviet Arctic.

1. Cone scales with sharply pointed or obtusely truncate tips. Twigs glabrous or sparsely pubescent. . . . . . . . . . . .1. **P. EXCELSA** LINK. — EUROPEAN SPRUCE
– Cone scales with rounded tip. Young twigs densely pubescent. . . . . . . . . . .
. . . . . . . . . . . . . . . . . . . . . . . . . . . . . . .2. **P. OBOVATA** LDB. — SIBERIAN SPRUCE

   **1. *Picea excelsa*** Link in Linnaea XI (1841), 517; Keppen, Raspr. khv. der. 268; Komarov, Fl. SSSR I, 144; Perfilev, Fl. Sev. I, 53; Sukachev, Dendrol. 103; Ya. Vasilev, Der. i kust. SSSR I, 130 — **European Spruce.**
   *Picea abies* (L.) Karst. — Hultén, Atlas 19 (73).
   Forest tree reaching the polar limit of the distribution of forests in the Far North.
   **Soviet Arctic.** Reported for West Murman (Pechenga, Southern Rybachiy Peninsula, district of Kola Bay) and on the Terskiy Shore of the Kola Peninsula (Orlov district). Records for the northern boundary of forest in the Mezen district are doubtful.
   **Foreign Arctic.** Arctic Scandinavia (in the polar forest fringe and forming restricted "island" forests).
   **Outside the Arctic.** Scandinavia, Finland, northern half of the European part of the USSR, east about to Mezen and the central part of the basin of the Northern Dvina. Mountains of Central Europe.
   Certain authors (Orlova, Fl. Murm. I) dispute the fact of the occurrence of *P. excelsa* in Murmansk Oblast, referring relevant records almost entirely to *P. obovata* (or in much lesser proportion to *P. fennica* Rgl.). It is difficult to judge to what extent this is sound without undertaking a complete revision of the material which serves as the basis for establishing the wide distribution of *P. excelsa* in the Soviet part of Fennoscandia, especially the material studied by Scandinavian authors. This task exceeds the scope of our investigation. Accordingly we have decided not to exclude *P. excelsa* from the list of plants represented in the flora of the Murman forest-tundra.

   **2. *Picea obovata*** Ldb., Fl. Alt. IV (1833), 201; Middendorff, Gew. Sibiriens 541; Krylov, Fl. Zap. Sib. I, 74: Petrov, Fl. Yak. I, 49; Tolmachev, Raspr. drev. porod 4; Andreyev, Mater. fl. Kanina 115; Komarov, Fl. SSSR I, 145; Perfilev, Fl. Sev. I, 53; Leskov, Fl. Malozem. tundry 16; Sukachev, Dendrol. 120; Ya. Vasilev, Der. i kyst. SSSR I, 135;

Orlova, Fl. Murm. I, 90 (pro parte ?); Karavayev, Konsp. fl. Yak. 45. — **Siberian Spruce**.

*Abies obovata* Rupr., Fl. samoj. cisur. (1846) 56; Schmidt, Fl. jeniss. 120.
*Pinus orientalis* Ldb. — Scheutz, Pl. jeniss. 194.
*Abies orientalis* Schrenk, Enum. plant. 526.
*Picea excelsa* Keppen, Raspr. khv. der. 268 (pro parte).
*P. abies* ssp. *obovata* Hultén, Atlas 19 (75).

Widely distributed forest-forming tree of the north and east of the European part of the USSR and a large part of Siberia. Within Fennoscandia penetrating the forest-tundra only in the form of isolated trees, exceeded in distribution northwards by other trees, especially birch. In the region between the White Sea and the Urals, to the contrary, this is the most northern tree, almost everywhere forming the northern fringe of closed forests and the greater part of the forest "islands" of the forest-tundra. East of the Urals it is exceeded in distribution northwards by larch. In the forest-tundra it occurs either as an admixture in open larch forests or surrounded by them in the form of small stands in river valleys. Occasional saplings reach true tundra, growing in more protected sites with favourable exposure. East of the Yenisey the gap between the northern boundaries of the ranges of larch and spruce increases and spruce does not reach the northern limit of forest.

**Soviet Arctic**. Isolated trees in the forest-tundra of Murman. As forest-former at the northern edge of closed forests and in "island" forests of Southern and Central Kanin, and in the Timanskaya and Malozemelskaya Tundras. In the Bolshezemelskaya Tundra mainly in valleys of tributaries of the Lower Pechora and of right-bank tributaries of the Usa, forming separate forest islands sometimes at a considerable distance from the region of closed forest growth. One of the most northern spruce groves is the "forest island" at almost 68°N on the river More-Yu (Khaypudyra) which enters Khaypudyrskaya Bay of the Barents Sea. Near the Urals the northern boundary of spruce shifts somewhat southwards. In the West Siberian forest-tundra, for example in Southern Yamal, spruce is found mainly in valley forests surrounded by open larch forests. On the Yenisey the boundary of the range of spruce advances almost to 69°30'N. Near Dudinka spruce occurs in the form of isolated trees on the southern slopes of tundra mounds. In the Pyasina Basin between 69°20' and 69°30'N it is found more abundantly than on the Yenisey and is represented by larger trees, but still grows as an admixture in larch groves. There is no reliable information on the distribution of spruce in the Khatanga Basin. In the Olenek Basin spruce is widely distributed and is recorded on the lower course of that river below 70°20'N, within the forest zone. Further east the boundary of the range is at first gradually then (on the right bank of the Lena) abruptly displaced to the south. Spruce is absent from the basins of the Yana, Indigirka, Kolyma and Anadyr.

**Foreign Arctic**. Not occurring.

**Outside the Arctic**. Widely distributed in the forest zone of the east of the European part of the USSR and Siberia. In mountains south to the Dzhungarskiy Alatau, the Tannu-Ola Range, Northern Mongolia and Northern Preamuria.

## GENUS 2    Larix Mill. — LARCH

REPRESENTED IN THE Soviet Arctic by two species which form hybrid populations where their ranges meet.

1. Cones of more or less ovoid shape, 2–3 cm long. Their scales broadly ovoid or almost circular, with more or less curved margin, slightly bent inwards at tip ("spoon-shaped"), dull, with a considerable part of their external surface densely clothed with ferruginous pile. Number of scales (with rare exceptions) greater than 20. ........ 1. **L. SIBIRICA** LDB. — **SIBERIAN LARCH**
Represented in the Far North by two races: *L. sibirica* ssp. *rossica* (Rgl.) Sukacz. and *L. sibirica* var. *polaris* Dyl.

– Ripe cones usually widely spreading, not more than 2 cm long (usually about 1.5 cm). Scales spatulate, with straightly truncated or excavated margin which is not bent inwards, bright, ferruginous, shining, without external pile. Number of scales less than 20. ...........................
............................... 2. **L. DAHURICA** TURCZ. — **DAHURIAN LARCH**
A form apparently of hybrid origin (*L. dahurica* × *sibirica* = *L. Czekanowskii* Sz.) is distinguished by the outwardly bent margins of the cone scales and sometimes by weakly developed ferruginous pile on the outside of the scales.

    *1. Larix sibirica* Ldb., Fl. Alt. IV (1833), 204; Middendorff, Gew. Sibir. 527; Schmidt, Fl. jeniss. 120; Krylov, Fl. Zap. Sib. I, 75; Petrov, Fl. Yak. I, 54; Tolmachev, Raspr. drev. porod 5; Komarov, Fl. SSSR I, 155; Perfilev, Fl. Sev. I, 54; Leskov, Fl. Malozem. tundry 16; Sukachev, Dendrol. 212; Dylis, Sib. listven.; Ukhanov, Der. i kust. SSSR I, 165; Orlova, Fl. Murm. I, 92; Hultén, Atlas 18 (72); Karavayev, Konsp. fl. Yak. 45. — **Siberian Larch.**
*Pinus Ledebourii* Endl. (1847) — Ledebour, Fl. Ross. III, 672; Scheutz, Pl. jeniss. 194.
*Abies Ledebourii* Rupr., Fl. samoj. cisural. (1846) 56.
*Larix Ledebourii* — Schrenk, Enum. plant. 527.
*Larix sibirica* oec. *rossica* Sukacz., Dendrol. 218.
*Larix decidua* var. rossica Rgl., Gartenflora (1871).
*L. sibirica* f. *rossica* Szafer.
*Larix Sukaczewii* Dyl. in Dokl. AN SSSR (1945) 489; id., Sib. listven. (1947) 489; Ukhanov, Der. i kust. SSSR, I, 170.

    · One of the basic forest-forming trees in Western and Central Siberia. In the north-east and east of the European part of the USSR mainly contributing to the formation of tree stands of mixed composition. From the Polar Ural to the region of the upper reaches of the Pyasina, it forms the fringe of closed forests and "island" forests of the forest-tundra.

**Soviet Arctic.** Isolated locality in forest-tundra on the Terskiy Shore of the Kola Peninsula (opposite Sosnovets Island). On the forest boundary in South Kanin near Malaya Nes. Forest-tundra of Timan between the rivers Indiga and Velt. Lower reaches of the Pechora in the Naryan-Mar district and slightly further north, lower part of the basin of the River Kuya a tributary of the lower Pechora (mainly on sands, forming open forests); isolated in the basin of the River Pay-Yaga (Chernaya) in the northwest of the Bolshezemelskaya Tundra. Absent from near the forest bundary over a considerable distance beyond the Pechora. Appearing again in the foothills of the Polar Ural and reaching 68°N on its eastern slopes (on the River Pyderata). Further east forming open forests (becoming scattered trees northwards) in the southern part of Yamal, on the Obsko-Tazovskiy Peninsula, and on the right bank of Tazovskaya Bay. On the Yenisey occurring as scattered trees more or less regularly to the latitude of the rivermouth port (Ust-Port at about 69°40'N) and forming open forests to approximately 69°20'N, formerly continuous to the village of Dudinka. Once found (in 1914) near the mouth of the Yenisey at the mouth of Chayka Creek (71°51'N!). East of the Yenisey common in forest-tundra between Dudinka and Norilsk. Forming forests and open forests in the Norilsk Depression. Isolated stand ("Koyev Wood") on the west shore of Lake Pyasino. In the district of the Vvedenskiy Station (Limka) on the Pyasina only hybrids with *L. dahurica* have been reported. Evidently the range boundary shifts abruptly southwards from the NE margin of the Norilsk Depression (over the whole expanse from the region of open forest between the rivers Pyasina and Khatanga as far south as the Lower Tunguska concurrence between the ranges of *L. sibirica* and *L. dahurica* has not so far been reliably demonstrated).

**Foreign Arctic**. Not occurring.

**Outside the Arctic**. NE European part of the USSR, Western and Central Siberia, east to the southern part of the Vilyuy Basin, the Lena at the mouth of the Vitim, and western districts of Transbaikalia. South (in the mountains) reaching Tarbagatay, Saur and Northern Mongolia.

*Larix sibirica* in the wide sense is a rather polymorphic species. This is reflected in longstanding attempts to divide it into geographical varieties or races, especially to contrast the form growing in NE Europe (*Larix archangelica* Laws., *L. decidua* var. *rossica* Rgl., *L. rossica* Rgl.) with the Siberian (or more narrowly Altaic) type. The name "*rossica*" as designating the European race or variety of *L. sibirica* has been quite widely used in the literature. N.V. Dylis (1945) advocated recognition of the specific distinctness of the "Northern Russian" larch and proposed for it the new name *L. S*ukaczewii Dyl.

The data presented by N.V. Dylis in his monograph (1947) force one to recognize that treatment of the European race of Siberian Larch as a separate species is taxonomically exaggerated. When the distinction is made by comparing the Northern European form with the *Altaic* type of *L. sibirica*, the possibility of distinguishing them as separate (although close) species appears real. But if account is taken of the total diversity of forms of *L. sibirica* (s. str.) characterized in that work, this impression disappears. In particular, the West Siberian forest-tundra form *L. sibirica* var. *polaris* Dyl., described there (page 75) and treated as a northern variety of typical *L. sibirica*, possesses characters agreeing with the characterization given on page 71 for *Larix Sukaczewii* ("seed scales large, woody...distinctly turned inwards at their tips") not with *L. sibirica*, to which the author expresses no doubts regarding the attribution of the stated variety. The factual data presented in the monograph clearly do not agree with the author's thesis of relatively little variation in *L. Sukaczewii*.

In view of these considerations we prefer to consider the Northern European form of larch as one of the races of Siberian larch, adopting for it the longstanding designation *Larix sibirica* ssp. *rossica* (Rgl.) Sukacz. To this race must be referred all larch growing on the expanse from the shores of the White Sea (Sosnovets, Nes) to the Polar Ural. In the Urals it is reported here and there in the basin of the River Sob. The great bulk of larch beyond the Urals, in particular all the most northern populations, belong to *L. sibirica* var. *polaris* Dyl.

2. ***Larix dahurica*** Turcz. in Bull. Soc. Nat. Mosc. (1838) 101; Trautvetter, Fl. boganid. 148; Middendorff, Gewächse Sibiriens 527; Gelert, Fl. arct. I, 16; Petrov, Fl. Yak. I, 55: Tolmachev, Raspr. drev. porod 8; Komarov, Fl. SSSR I, 156; Sukachev, Dendrol.

221; Ukhanov, Der. i kust. SSSR I, 171; Karavayev, Konsp. fl. Yak. 45. — **Dahurian Larch.**

*Pinus dahurica* Fisch. — Ledebour, Fl. Ross. III, 673; Trautvetter, Pl. Sib. bor. 111.

Occurring both in tree form (low-growing at arctic boundaries, reaching a height of 3–4 m at an age of 150–200 years, but considerably taller-growing at a short distance from tundra) and as stlanik, individually or in small groups, forming open forests or denser groves in protected sites. Fruiting abundantly in forest-tundra. Satisfactory regeneration has been observed on the whole northern limit of its distribution. The stlanik form is usually found in mountainous sites.

At the polar limit of it distribution keenly responsive to changes of slope exposure (even at very gentle inclination) and the degree of protection from winds (especially the north winds which blow in summer). Not fastidious about soil conditions, but avoiding very boggy sites. In tetragonal bogs in the southern part of the tundra zone growing only on elevated areas along the fissures which break up the surface of the bog.

**Soviet Arctic.** From the banks of the River Pyasina near its outfall from Lake Pyasino, east all the way to the Anadyr Basin. Near the Pyasina reaching approximately 70°30'N. Further east its distributional limit gradually shifts northwards. In the expanse between the foot of the Central Siberian Plateau and the valley of the river Dudypta there is formed a belt of open larch forest which gradually broadens eastwards. Open forest does not reach the Dudypta itself, but scattered trees occur on the right bank of the river. In the Kheta Basin larch forms more or less dense groves on the Kheta itself and on the lower courses of its tributaries the Volochanka and the Boganida, but is absent from the region of their sources. Further east the boundary of the continuous range of larch is almost coincident with the watershed boundary between small left-bank tributaries of the lower Kheta and Khatanga and rivers flowing into a major tributary of the latter, the Novaya River. North of the watershed scattered trees are met with for a certain distance. In the Khatanga Valley the boundary advances north-east to somewhat beyond the mouth of the Novaya River. In the valley of the Novaya River between 72°30' and 72°40'N is situated the most northern forest "island," Arymas, and above this on the river course below 72°40'N grow the most northern larch specimens. Occasional dwarf specimens of larch also occur somewhat south-east of the stated locality on the upper course of the Novaya River.

East of the Khatanga the northern boundary of the range of larch shifts somewhat southwards, but the furthest localities are situated north of the 72nd parallel all the way to the Lena Valley. On the lower course of the Olenek larch reaches a latitude close to 72°30'N. On the Lena the furthest forest "island" is located on the island of Tit-Ary (72°N). After the felling of all larger trees, satisfactory regeneration of larch in the form of an open forest of saplings has been observed there. North of Tit-Ary larch is found in small quantity in protected sites on the right bank of the Lena as far as 72°10(15)'N. East of the Lena the northern boundary of the range of larch has been traced only very approximately. In comparison with the stretch from the Khatanga to the Lena, it is here displaced considerably to the south. In general it proceeds between 70 and 71°N, shifting northwards in the valleys of major rivers. East of the mouth of the Kolyma the range boundary withdraws more abruptly from the coast and descends to the south, enclosing the upper part of the Anadyr Basin and exiting southwestwards into the Sea of Okhotsk.

**Foreign Arctic.** Not occurring.

**Outside the Arctic.** Distributed through the greater part of the expanse of East Siberia, from the basin of the Lower Tunguska, the upper reaches of the Podkamennaya Tunguska, the northern shores of Baykal and the Yablonovyy Range east to the shores of the Pacific Ocean. In the south widespread in the Amur Basin and on Sakhalin.

Within the limits of its range, *L. dahurica* is represented by different forms, some of which have been described under different species names: in particular, *L. Cajanderi* Mayr described from Yakutia which should maybe be considered a northeastern race of *L. dahurica*. In view of the considerable overlap (which was doubtless still greater in the recent geological past) of the ranges of *L. dahurica* and *L. sibirica*, the formation of hybrids between them is in no way exceptional and has already given rise to rather numerous populations of hybrid origin. It is possible that the results of hybridization are manifested in the polymorphism of *L. dahurica*, especially in the more western parts of its range. A form of undisputed hybrid origin described under the name *L. Czekanowskii* Szafer (1918) enjoys a considerable distribution in the overlap zone of the ranges of *L. dahurica* and *L. sibirica*. Characters of *L. dahurica* generally predominate in its structure. A characteristic difference from both parent species is the outward bending of the margins of the seed scales. Larches with these characters are particularly common in the area between the Pyasina and the Khatanga.

Since the question of racial differentiation within *Larix dahurica* can in principle only be decided by revising all material of this species, not only plants from the polar fringe of its range, we choose to abstain from this consideration, treating *L. dahurica* as a species in a wide sense.

## ❧ GENUS 3  **Pinus** (Tourn.) L. — PINE

REPRESENTED IN THE Soviet Arctic by two species, one of which (*P. silvestris* ssp. *lapponica*) penetrates the fringe of the Arctic in the most western part of our country, while the other (*P. pumila*) enjoys a considerable distribution in the far north-east.

1. Straight-trunked or contorted tree, able to adopt a stlanik form under extreme conditions. Needles *stiff, relatively short* (up to 4 cm long), arranged in pairs on short twigs, glaucous green. Cones ovoid, more or less acute, shining green when young, grey when mature, opening widely when the seeds are fully ripe. Seeds small, adorned with wings. . . . . . . . . .
. . . . . . . . . . . . . . . . . . . . .1. **P. SILVESTRIS** SSP. **LAPPONICA** FR. — **LAPLAND PINE**
- Small tree bending towards the ground, semiprostrate or prostrate, sometimes very lowly. *Needles soft, long* (4–7 cm long), dark green with glaucous stripes, arranged in bunches *of 5 together*. Cones large, obtusely ovoid, dark green at first, later dark brown, weakly splitting open when the seeds are ripe. *Seeds large* with thick, dark brown husk, *without wings* ("cedar nuts"). . . . . . . . . . . . . . . . . . . . . . . . .2. **P. PUMILA** (PALL.) RGL. — **CEDAR PINE**.

1. **Pinus silvestris** subsp. **lapponica** Fries (pro var.); Sukachev, Dendrol. 176. — **Lapland Pine**.
    *P. lapponica* Mayr, Fremdl. Wald- und Parkb. (1906) 348; Orlova, Fl. Murm. I, 93.
    *P. silvestris* L. (pro parte) — Middendorff, Gew. Sibiriens 527; Keppen, Raspr. khv. der. 38; Komarov, Fl. SSSR I, 167; Perfilev, Fl. Sev. I, 53; Maleyev, Der. i kust. SSSR I, 53; Hultén, Atlas 18 (71).
        Forest-forming tree of the northern part of the taiga zone of Fennoscandia. Forming more or less sparse groves at the northern boundary of forest in

Murmansk Oblast. Of scattered occurrence in forest-tundra, absent from pure tundra.

**Soviet Arctic.** Murman (in forest-tundra and on the northern fringe of closed forests).

**Foreign Arctic.** Extreme north of Fennoscandia.

**Outside the Arctic.** Northern parts of the taiga zone of Fennoscandia (including the south of Murmansk Oblast and the north of the Karelian ASSR). Replaced further south by typical *P. silvestris* L. The overall distribution of *P. silvestris* extends from Northern Great Britain in the west to the Lena and Amur Basins in the east, from northern parts of the taiga zone in the north to the steppe zone in the south. East of the White Sea not penetrating the limits of true forest-tundra.[1]

2. **Pinus pumila** (Pall.) Rgl., Ind. Sem. Horti Petropol. (1858) 23; Middendorff, Gew. Sibiriens 560; Petrov, Fl. Yak. I, 64; Komarov, Fl. SSSR I, 164; Sukachev, Dendrol. 570; Tikhomirov, Proiskh. assots. kedr. stl. (1946); id., Kedr. stlanik (1949); Maleyev, Der. i kust. SSSR I, 196; Karavayev, Konsp. fl. Yak. 45. — **Cedar Pine.**

*P. cembra pumila* Pall. — Gelert, Fl. arct. I, 15.

Small tree or shrub with trunks bending towards the ground and branches raising themselves obliquely upwards. At sites subject to strong wind gusts and only feebly protected by snow cover in winter, adopting a totally lowly form, appressed to the ground. Under favourable conditions forming closed thickets which can cover considerable areas.

Basically a plant of the mountains and sea coasts of East Siberia and the Far East.

**Soviet Arctic.** On the lower reaches of the Lena penetrating the limits of forest-tundra, distributed on the slopes of the hills on the right bank of the river almost to 71°N. On the Yana and Indigirka reaching somewhat short of the northern forest boundary. On the lower reaches of the Kolyma at the forest boundary. Of abundant occurrence in the Anadyr Basin, the mountains of Koryakia, the Penzhina Basin and the shores of the northern part of the Sea of Okhotsk.

**Foreign Arctic.** Not occurring. The record in the literature of the growth of cedar pine stlanik on Kotzebue Sound in Alaska was evidently based on a misunderstanding (confusion in labelling?). Current data indicate the absence of *P. pumila* from Alaska.

**Outside the Arctic.** In East Siberia east from the Eastern Sayans and the shores of Baykal, the middle course of the Aldan, the lower reaches of the Vilyuy and the valley of the lower Lena all the way to the shores of the Bering, Okhotsk and Japan Seas. Distributed south to the mountains of Northern Mongolia, the hills of the left-bank part of the Amur Basin, and near the sea to the mountains of North Korea and Central Japan.

---

[1] The records in "Flora of the USSR" (I, page 168) of the growth of pine on the Yenisey at 69° and on the upper course of the Pyasina north of 70°N are based on some kind of misunderstanding. In reality the northern boundary of the range of pine crosses the Yenisey near the 66th parallel and proceeds further east, invariably remaining south of the Arctic Circle. In the Pyasina Basin pine is certainly absent.

## FAMILY VII

# Cupressaceae F.W. Neger

### CYPRESS FAMILY

---

VAST FAMILY WIDELY distributed in temperate and subtropical latitudes and in mountains of the tropical zone of both hemispheres. One species of one genus (*Juniperus*) penetrates arctic limits.

---

### GENUS 1

## Juniperus L. — JUNIPER

1. ***Juniperus sibirica*** Burgsd., Anleit., 2. Aufl. (1790), 127, 128; Petrov, Fl. Yak. I, 71; Komarov, Fl. SSSR I, 181; Maleyev, Der. i kust. SSSR I, 349; Orlova, Fl. Murm. I, 96: Karavayev, Konsp. fl. Yak. 46. — **Siberian Juniper.**
    *J. nana* Willd. — Keppen, Raspr. khv. der. 464; Andreyev, Mater. fl. Kanina 155; Perfilev, Fl. Sev. I, 45; Leskov, Fl. Malozem. tundry 16.
    *J. alpina* Clus. — Lange, Consp. fl. Groenl. 182.
    *J. Niemannii* Wolf — Perfilev, Fl. Sev. I, 54.
    *J. communis* var. *nana* — Schrenk, Enum. pl. 525; Krylov, Fl. Zap. Sib. I, 84.
    *J. communis* var. *depressa* — Perfilev, Mater. fl. N. Z. Kolg. 50.
    *J. communis* f. *montana* — Hultén, Fl. Al. I, 89; Böcher & al., Ill. Fl. Greenl. 53.
    *J. communis* L. — Ruprecht, Fl. samoj. cisur. 55; Schmidt, Fl. jeniss. 119; Scheutz, Pl. jeniss. 195; Gelert, Fl. arct. I, 16; Ostenfeld, Fl. Greenl. 55; Tolmachev, Fl. rez. Kolguyev. eksp. 39; Leskov, Fl. Malozem. tundry 16; Hultén, Atlas 18 (69), pro parte; ? Gröntved, Pterid. Spermatoph. Icel. 114.

    Small shrub, often appressed to the ground, relatively common in the taiga zone of Siberia and the European North. Locally common in the forest fringe. Penetrating rather far into the tundra zone in the European and West Siberian sectors of the Arctic. Growing mainly in dry sandy places, commonly on sloping well-drained areas which are adequately protected by snow cover in winter. Sometimes found in considerable quantity, forming small thickets.

    **Soviet Arctic.** Murman (ubiquitous), the whole of Kanin, Timanskaya and Malozemelskaya Tundras; lower reaches of Pechora (to its mouth), Bolshezemelskaya Tundra north to the shores of the Sea of Pechora and the foothills of Pay-Khoy (on the River Korotaikha); Kara River Basin, Polar Ural to its northern extremity; tundra of the lower reaches of the Ob; on the Yenisey as far as Tolstyy Nos (a little north of 70°). In extreme NE Asia in western and southern parts of the Anadyr Basin, in the Penzhina Basin and on the Beringian coast of Koryakia. Over the expanse from the Yenisey to the Anadyr Basin so far not found in the Arctic proper. It is reported from the northern edge of the Central Siberian Plateau, from subarctic districts of Yakutia and from the Kolyma Basin (on the River Bolshoy Anyuy). Krylov's record of its growing on the Khatanga below 71°45'N needs checking. Reports of the growth of juniper in Southern Novaya Zemlya have not been confirmed by subsequent data.

    **Foreign Arctic.** West coast of Alaska; West and East Greenland north to 69°(?); Iceland (?); Arctic Scandinavia.

    **Outside the Arctic.** Widely distributed in Siberia and the Far East, and in the north of the European part of the USSR. Penetrating far to the south by ascending mountains. Present in mountains of Japan and Korea, Mongolia, Southern Siberia, and Central Asia. In Europe widely distributed in Scandinavia and recorded for the mountains of Central Europe. Reported for the Caucasus and Asia Minor.

It is difficult to write an accurate description of the range of *J. sibirica* because of obscurity regarding the question of its delimitation from *J. communis* and the correct identity of other low-growing forms (e.g. *J. nana* Willd.). While the Central European high alpine *J. nana* when transplanted to the plains "behaves" like *J. communis*, gradually losing its differences from the latter, it is undisputed that the Siberian and Far Eastern *J. sibirica* does not manifest this tendency, but retains a prostrate shrubby growth form in relatively far southern districts (e.g. on the south of Sakhalin Island). The question of the systematic position of low-growing forms of juniper of the *J. communis*-type needs further investigation. Accordingly, we have decided not to assert whether the prostrate juniper of Greenland and Iceland should be treated as *J. sibirica* or as one of the forms of *J. communis* s. str.

## FAMILY VIII

# Sparganiaceae Engl.

BUR-REED FAMILY

---

### GENUS 1 — Sparganium L. — BUR-REED

PLANTS WITH MORE or less fleshy herbaceous stems and simple slender ribbonlike leaves, mostly bright green. Growing in wet places, predominantly in shallow water but also on well moistened shores of waterbodies. Stems and leaves floating in many species, in others standing erect. Small greenish flowers assembled in spherical heads, either sessile or distant from the stem on rather long peduncles. Some of the heads forming a simple or weakly branched inflorescence on the upper part of the stem.

The species of this genus, few in total number, are distributed predominantly in the temperate zone of the Northern Hemisphere. A considerable portion of them are represented in the flora of the USSR. Isolated species occur in the Southern Hemisphere (in Australia and New Zealand).

Certain species widespread in the northern temperate zone penetrate arctic limits. They are confined here to areas not very far removed from the northern limits of forests. One species (*S. hyperboreum* Laest.) is typically subarctic: it is widely distributed in the relatively temperate part of the Arctic and the northern part of the forest zone, but absent from more southern districts of the latter.

1. Inflorescence consisting of a few (2–3–5) female and a few (2–4–6) male heads. Heads relatively large, the female 7–11 (6–13) mm in diameter at flowering time, 15–18 mm in fruit. Styles long, always very conspicuous. . . . . . . . . . . . . . . . . . . . . . . . . . . . . . . . . . . . . . . . . . . . . . . . . . . . . . . . .2.
   – Inflorescence consisting of a few (usually 2–3) female and a single male head. Heads small, the female 3.5–5.0 mm in diameter at flowering time, 7–12 mm in fruit. Styles shortened. Leaves always very narrow. Plants of very small dimensions . . . . . . . . . . . . . . . . . . . . . . . . . . . . . . . . . . . . . . . . . . .3.

2. Stems standing erect (very rarely floating), rather thick. Leaves flatly triple-edged, with rather thick, sharply projecting midrib (often convoluted in dry form), relatively broad (3–10 mm wide). Inflorescence large, elongate, with 3–5 female and 4–6 male heads. Beak long, of approximately equal length to achene . . . . . . . . . . . . . . . . . . . . . . . . . . . . . . . .1. **S. SIMPLEX** HUDS.
   – Stems floating (very rarely standing erect), always rather slender and weak. Leaves narrow (2–5 mm wide), semicylindrical or flatly ribbonlike, with thin non-projecting midrib (not convoluted in dried leaves). Inflorescence shortened, with 2–3 female and 2–3 male heads. Beak approximately half as long as achene . . . . . . . . . . . . .2. **S. AFFINE** SCHNITZL.

3. Stems mainly floating, in erect-standing individuals often much shortened (up to 10 cm high). Leaves 1.5–5.0 mm wide, thin, shorter than stem. Male head in inflorescence separated from upper female by an obvious internode. Achene gradually narrowing to rather short but well defined beak ... .................................................. .3. **S. MINIMUM** HILL.

– Stems standing erect, with small number of internodes. Leaves long, rather thick, very narrow (1–3 mm wide). Inflorescence with a few unevenly spaced female heads and single male head closely contiguous with upper female. Achene abruptly narrowing apically, with very short, often scarcely developed beak .................... .4. **S. HYPERBOREUM** LAEST.

1. ***Sparganium simplex*** Huds. Fl. Angl., ed. 2 (1778), 401; Ledebour, Fl. Ross. IV, 4 (pro maxima parte); Scheutz, Pl. jeniss. 162; Rotert, Fl. Az. Ros. I, 30; Krylov, Fl. Zap. Sib. I, 99: Yuzepchuk, Fl. SSSR I, 223; Perfilev, Fl. Sev. I, 55; Kuzeneva, Fl. Murm. I, 100; Hultén, Atlas 22 (85).

    Ill.: Fl. Murm. I, pl. XXX, 1.

    Characteristic plant of the temperate zone of the USSR. Growing on the margins of standing waterbodies, and on banks of rivers and oxbows, often in considerable quantity. As a rule not passing north beyond the limits of the forest zone. Only rarely and locally reported in the forest-tundra.

    **Soviet Arctic.** Reported in West Murman in the district of Pechenga, in forest-tundra on the southern fringe of the Kanin Peninsula (Nes) and on the shore of Tazovskaya Bay near the mouth of the River Taz.

    **Foreign Arctic.** Not reported.

    **Outside the Arctic.** Greater part of Europe, including the European part of the USSR (on the lower Pechora almost at 67°N); Caucasus; Siberia (on the Ob north to 65°, on the Yenisey to the Arctic Circle, in Yakutia to the middle course of the Vilyuy); North and East Kazakhstan; temperate Far East (Preamuria, Sakhalin, NE China, Northern Korea); Northern India; forested regions of North America. Reported also in Australia.

2. ***Sparganium affine*** Schnitzl, Typhac. (1845) 27; Scheutz, Pl. jeniss. 162; Ostenfeld, Fl. arct. I, 18 (pro parte); id., Fl. Greenl. 69; Rotert, Fl. Az. Ros. I, 32; Petrov, Fl. Yak. I, 82; Perfilev, Fl. Sev. I, 55; Yuzepchuk, Fl. SSSR I, 224; Seidenfaden & Sørensen, Vasc. pl. NE Greenl. 180; Gröntved, Pterid. Spermat. Icel. 114; Kuzeneva, Fl. Murm. I, 102.

    *S. angustifolium* Michx. — Hultén, Atlas 20 (78); Böcher & al., Ill. Fl. Greenl. 307.

    Ill.: Fl. Murm. I, pl. XXXI, 2.

    Common plant of waterbodies of the northern forest zone. As a rule not passing north beyond its limits. There are records for the tundra of West Murman and for forest-tundra of the lower reaches of the Yenisey and Lena.

    **Soviet Arctic.** West Murman (Rybachiy Peninsula, Pechenga); lower reaches of Pechora (Naryan-Mar); lower reaches of Yenisey (Luzino; cited by Scheutz with some doubt on the basis of sterile material); lower reaches of Lena (vicinity of Kyusyur village, about 70°50′N; flood pond in valley of River Ebetem; 31 VII 1935, Gorodkov and Tikhomirov; well developed flowering plant!).

    **Foreign Arctic.** Greenland as far as 68°N; Iceland; Arctic Scandinavia. Replaced in Alaska by the closely related species *S. angustifolium* Michx.

    **Outside the Arctic.** North and north-west of the European part of the USSR, Northern and Central Europe; Kamchatka (?).

    Both records for the Asian Arctic deserve much attention in view of their disjunction from the basic European range of the species, as well as from the likewise disjunct Kamchatka localities. Perhaps they indicate a more extensive eastwards distribution of the species also at temperate latitudes.

3. ***Sparganium minimum*** Hill, Brit. Herb. (1756) 507; Scheutz, Pl. jeniss. 162; Ostenfeld, Fl. arct. I, 17; Rotert, Fl. Az. Ros. I, 34; Krylov, Fl. Zap. Sib. I, 100; Petrov, Fl. Yak. I, 85; Yuzepchuk, Fl. SSSR I, 225; Perfilev, Fl. Sev. I, 55; Leskov, Fl. Malozem. tundry 17; Hultén, Fl. Al. I, 94; id., Atlas 21 (82); Gröntved, Pterid. Spermat. Icel. 116; Kuzeneva, Fl. Murm. I, 103.

Growing in ponds within peat bogs. Rarely reported in the Arctic, only at a short distance from the forest boundary.

**Soviet Arctic.** Timanskaya Tundra (Indiga Basin); West Bolshezemelskaya Tundra (district of Vangurey, River Kolva); shore of Tazovskaya Bay (doubtful specimens); Anadyr River Basin (isolated localities, particularly Lake Krasnoye); district of the Bay of Penzhina.

**Foreign Arctic.** West coast of Greenland north to 64°30'; Iceland; Arctic Scandinavia.

**Outside the Arctic.** Northern and Central Europe; Transcaucasus; West Siberian Lowland north to 64°, Yenisey Basin north to the Arctic Circle (Kureyka); SW part of East Siberia proper; Kamchatka; temperate districts of North America. Records in the literature of occurrence in NE Yakutia (Yana Basin) actually refer to *S. hyperboreum*.

4. ***Sparganium hyperboreum*** Laest. apud Beurl., Öfvers. Vet.-Akad. Förhandl. 1852 (1853), 192; Scheutz, Pl. jeniss. 162; Lange, Consp. fl. Groenl. 116; Rotert, Fl. Az. Ros. I, 35; Ostenfeld, Fl. Greenl. 69; Krylov, Fl. Zap. Sib. I, 101; Petrov, Fl. Yak. I, 84; Yuzepchuk, Fl. SSSR I, 226; Perfilev, Fl. Sev. I, 56; Leskov, Fl. Malozem. tundry 17; Hultén, Fl. Al. I, 93; id., Atlas 21 (81); Gröntved, Pterid. Spermat. Icel. 115; Kuzeneva, Fl. Murm. I, 103; Böcher & al., Ill. Fl. Greenl. 307.

*S. submulticum* (Hartm.) Neum. — Ostenfeld, Fl. arct. I, 18; M. Porsild, Fl. Disko 29.

Ill.: Petrov, Fl. Yak. I, fig. 31.

The only species of *Sparganium* relatively widely distributed in the Arctic. Growing on the shores of tundra lakes, both in the water and in moist places. Not avoiding the edges of small overgrown waterbodies. Encountered also on streams through lowland tundra.

**Soviet Arctic.** Murman (not uncommon, especially along the coast of the Barents Sea); Kanin Peninsula (in the north in the Nottey River Basin); Timanskaya and Malozemelskaya Tundras (Yeney, Adzva, Vorkuta, Khalmer-Yu); tundra east of the River Kara; forest-tundra on the right bank of the lower Ob (Poluy River Basin); lower reaches of Yenisey (Dudinka); lower reaches of Lena (Kyusyur); lower course of Yana (Muntaya); NW Chukotka (Chaun district); Anadyr Basin (found often, numerous localities); district of the Bay of Penzhina.

**Foreign Arctic.** West coast of Alaska; Labrador; Greenland (on the west coast north to 68°40'); Iceland; Arctic Scandinavia.

**Outside the Arctic.** Northern Europe; temperate and subarctic north of Siberia and the Far East; Northern North America.

## FAMILY IX
# Potamogetonaceae Engl.
**PONDWEED FAMILY**

---

AQUATIC PLANTS, FOUND in a completely submersed state or partly floating on the water surface with their inflorescences emergent above it. Able to grow both in standing waterbodies and in rivers (in the latter, if the current is slow). Some of the species are saltwater plants, growing in shallow waters on sea coasts.

1.  Leaves relatively broad (2–5 times as long as wide) or very narrow, linear, sometimes almost filiform, but in such cases relatively short (up to 10–15 cm long). Flowers with perianth, assembled in a series of whorls which form an interrupted or dense spikelike inflorescence on a common peduncle. Plants of fresh (occasionally brackish) waterbodies. . . . . . . . . . . .
    . . . . . . . . . . . . . . . . . . . . . . . . . . . . . .GENUS 1. POTAMOGETON L. — PONDWEED.
–   Leaves narrow, linear, very long (for the most part 0.5–1.0 m long, 2–9 mm wide). Flowers without perianth, arranged in strongly compressed spikelets. Marine plants growing in salt water. . . . . . . . . . . . . . . . . . . . . . . . . .
    . . . . . . . . . . . . . . . . . . . . . . . . . . . . . . . .GENUS 2. ZOSTERA L. — EEL-GRASS

---

## GENUS 1
# Potamogeton L. — PONDWEED

AQUATIC PLANTS WITH floating stems and leaves, with their inflorescences emergent above the water surface at flowering time. In some species the stems and leaves are completely submersed, in others the leaves partly float on the water surface. In their shape and denser consistency these floating leaves differ more or less sharply from the submersed leaves. Leaves always simple, of narrowly linear to broadly oval shape. Flowers unattractive, greenish, assembled in whorls which may be arranged more or less separately from one another on a long peduncle or may be aggregated and together form a dense spikelike inflorescence; in such cases the peduncle is usually distinctly thickened.

Growing at shallow and moderate depths in standing bodies of fresh water and slowly flowing rivers and streams, sometimes forming beds. Certain species can also grow in brackish waterbodies.

Vast genus possessing an almost cosmopolitan distribution. Particularly well represented in the temperate zone of the Northern Hemisphere. In the Arctic the growth potential of *Potamogeton* species is restricted by the coldness of the majority of waterbodies and their thawing out for too short a time. The majority of the species reported within arctic limits are plants widespread in the temperate north which do not penetrate very far into the Arctic. The northern distributional limits of the majority of species has in general only been established approximately. Apparently, the valleys of the large rivers from the south serve as routes for species of *Potamogeton* to penetrate the Arctic, and along these valleys the northern boundaries of their ranges bend northwards. This is easily explained by the relatively warm water of the large northern rivers even in their lower reaches, thanks to which far more favourable conditions for the growth of *Potamogeton* are created in their floodplains than in surrounding tundra, and no doubt partly also because the rivers provide a more or less regular transport of fragments of *Potamogeton* to the Far North from more temperate regions situated further south. A specially far advance to the north of a whole series of *Potamogeton* species is reported, in particular, in the Yenisey valley.

In contrast with the majority of the species represented in the arctic flora whose arctic localities represent extreme northern points of their generally extensive (non-arctic) ranges, a few species have so far been found exclusively within arctic limits. Such, in particular, is the peculiar *P. subretusus*, found in relative abundance at certain localities in the lower reaches of the Yenisey. Another species so far known only from there, *P. subsibiricus*, should perhaps be considered a northwestern race of the closely related *P. sibiricus* described from the Vilyuy Basin.

The Arctic American *P. Porsildorum*, found so far at a few points from the shore of the Bering Strait to Hudson Bay, also belongs here. One endemic species, *P. groenlandicus*, close to and replacing the widespread *P. pusillus*, is peculiar to Greenland.

It must be emphasized that, in view of the very poor investigation of the aquatic flora of the Arctic and Subarctic, we must bear in mind the possibilities both that some temperate northern species so far not reported within arctic limits will be discovered there, and that species known to us only from within arctic limits will in future also be found in the temperate north.

The systematics of the genus *Potamogeton* is much complicated by the abundance of hybrids between the different species. However, far northern material consists predominantly of plants whose specific identity does not occasion any doubts. On the other hand, another difficulty for the study of *Potamogeton* is felt particularly acutely in the North. This is the incompleteness of knowledge of their distribution. This is attributable partly to its fragmentation (arising from the restriction of their occurrence to areas with conditions outstandingly favourable for the growth of aquatic plants), and partly to the fact that many investigators of the northern flora paid only minimal attention to the collection of aquatic plants.

1. Entire plant submersed. Leaves always narrowly linear, with *well developed sheath*. Inflorescence elongate, on slender peduncle, consisting of *small, more or less separated whorls*. (SUBGENUS **COLEOGETON** RAUNK.) — 2.
– Leaves without or with short sheath, of diverse form, in some species all submersed, in others partly submersed partly floating. Inflorescence on more or less thick peduncle, *consolidated, spikelike, with whorls of flowers more or less closely contiguous with one another.* .......................
...........................(SUBGENUS **EUPOTAMOGETON** RAUNK.) — 5.

2. Leaves up to 10 cm long, very narrow, filiform, blunt, many *with shallow notch at tip*. Inflorescence consisting of numerous (6–8) small whorls which are considerably and more or less uniformly separated .......... ...........................................2. **P. SUBRETUSUS** HAGSTR.
– Leaves acuminate or blunt, but always *without notch at tip*. ............3.

3. Leaves very delicate, blunt or acute at tip, very narrowly filiform, *with single median nerve* from which fine lateral nerves run towards the leaf margins at a right or acute angle. Leaf sheaths with brownish fringe, *on their lower part with fused margins*. Whorls of inflorescence *few* (3–5), widely separated. .......................................1. **P. FILIFORMIS** PERS.
– Leaves somewhat thicker. *Besides the basic median nerve, there are two or more additional longitudinal nerves parallel to it*. Leaf sheaths *split to the base*, rolled into a tube. ..............................................4.

4. Leaf tips *rounded*. Leaf sheaths strongly developed, broad, enclosing 3–4 (more rarely 2) branchlets. Whorls of inflorescence *numerous* (mostly 8), *uniformly separated*. ...........................3. **P. VAGINATUS** TURCZ.
– Leaf tips *acuminate*. Leaf sheaths mostly narrow, enclosing up to 2 branchlets. Whorls of inflorescence (mostly 5) *nonuniformly separated*. .. ................................................4. **P. PECTINATUS** L.

5. *Small, totally submersed* plants with short and narrow, linear (up to 2 mm wide) leaves. Peduncles short and thin. ..............................6.
– Leaves all submersed or partly floating on the water surface; in the latter case the shape and consistency of floating and submerged leaves is different. Leaves always relatively broad, more or less oblong (from lanceolate-taeniate to ovoid or almost orbicular in shape). Peduncles often long, frequently thickened. ................................................7.

6. Stem with few branches. Leaves 4–6 cm long and about 2 mm wide, rounded or slightly acuminate at their tips, with *numerous* (13–17) *longitudinal nerves*. Inflorescence consisting of 3–4 aggregated whorls. ....... .........................................5. **P. SUBSIBIRICUS** HAGSTR.
– Stems mostly strongly branched on their upper part, with short internodes. Leaves 1.5–5.0 cm long, up to 1.5 mm wide, mostly with acute tips, *with three longitudinal nerves*. Inflorescence consisting of 3–6 aggregated whorls. ...................................................6. **P. PUSILLUS** L.

7. Leaves of two types: submerged leaves narrower, delicate, often with slightly undulate surface, quickly desiccating and readily shrivelling in air, *tapering at base, sessile or with short petiole*; floating leaves broader and flatter, relatively thick, often leathery, always petiolate (if floating leaves not developed, the plant may be recognized by the form of the base of the submersed leaves and often by the presence of a petiole on them). .....8.
– *All submersed leaves* sessile, *with base more or less clasping stem*. .......13.

8. All leaves (floating and submersed) with entire margins. ............9.
– Submersed leaves undulate on their margins, with small denticles. ....12.

9. Stems often reddish. Leaves dark green, mostly *with reddish tinge*. Petioles of floating leaves relatively *short* (often only ⅓ - ½ as long as leaf blade). Stipules about 6 cm long, red-brown. ................................10.
– Stems and leaves *without reddish tinge*. Floating leaves *with long petioles*, of approximately the same length as the blade, occasionally somewhat longer than it. Stipules falling early, not red-brown. ..................11.

10. Submersed leaves lanceolate, 6–7 times as long as their width; floating leaves oblong-obovate, tapering at base into short petiole. Inflorescence of uniform density. ........................................7. **P. ALPINUS** BALB.
– Leaves of narrower shape, submersed leaves linear-lanceolate, 10 or more times as long as their width; floating leaves narrowly elliptic (their length approximately 5 times as great as their width), with very short petioles. Lower whorls of inflorescence often somewhat separated. ...............
........................7a. **P. ALPINUS** SUBSP. **TENUIFOLIUS** (RAF.) HULT.

11. Submersed leaves normally developed, lanceolate or ovoid-lanceolate, tapering at both ends, about 10 cm long. Floating leaves ovoid or ovoid-lanceolate, about 7 cm long and 3 cm wide, tapering at base. Inflorescence dense, 2-3 cm long. .................................**P. DIGYNUS** WALL.
– Submersed leaves *linear, up to 50 cm long, with undeveloped leaf blade*, reduced to the level of phyllodia. Floating leaves ovoid or elliptic, *with rounded or cordate base*, often larger (can reach 10 cm long and 4–5 cm wide). Stipules large (up to 10 cm long), falling early. Inflorescence dense, cylindrical, large (up to 4–5 cm long). .....................8. **P. NATANS** L.

12. Stems *thin, strongly branched*. Leaves mostly of two types: submersed leaves linear-lanceolate, tapering at both ends, with acute tips; floating leaves small, thin, mostly elliptic. Stipules small. Inflorescence dense, 2.5–5.0 cm long. ......................................9. **P. GRAMINEUS** L.
– Stems *thick* (3–4 mm wide), often very long, branched to varying degree. Leaves *only submersed*, elliptic or lanceolate, *large* (can reach 20 cm long and 4 cm wide), tapering at base, acuminate or blunt at tip, *yellowish green*. Inflorescence dense, up to 6 cm long. .................**P. LUCENS** L.

13. Leaves elongate, *ovoid-lanceolate*, up to 15 cm long, *4–6 times as long as their width*, with rounded or weakly clasping base. Stipules large (1.5–6.0 cm long), thick, long retained, yellowish. Inflorescence dense, 3–6 cm long, *on very long* (up to 20 cm or more) *peduncle*. .....................
.............................................10. **P. PRAELONGUS** WULF.
– Leaves *relatively short and broad*, from oblong-ovoid to almost circular, up to 10 cm long, *1½ - 2½ times as long as their width, with deeply cordate clasping base*, undulate on margin. Stipules short, whitish, falling early. Inflorescence rather dense, 1.5–2.5 cm long, *on short* (about 5 cm long) *peduncle*. ........................................11. **P. PERFOLIATUS** L.

1. ***Potamogeton filiformis*** Pers., Syn. pl. I (1805) 152; Hagström, Crit. Res. Potamog. 14; Ostenfeld, Fl. arct. I, 21; id., Fl. Greenl. 69; M. Porsild, Fl. Disko 31; Krylov, Fl. Zap. Sib. I, 114; Petrov, Fl. Yak. I, 90; Yuzepchuk, Fl. SSSR I, 236; Perfilev, Fl. Sev. I, 59; Seidenfaden & Sørensen, Vasc. pl. NE Greenl. 180; Gröntved, Pterid. Spermat. Icel. 119; Hultén, Fl. Al. I, 99; id., Atlas 24 (94); Kuzeneva, Fl. Murm. I, 107; Böcher & al., Ill. Fl. Greenl. 304.

    *P. marinus* — Ledebour, Fl. Ross. IV, 31; Lange, Consp. fl. Groenl. 117.

    Ill.: Fl. Murm. I, pl. XXXII, 1.

    Growing in small lakes and streams, sometimes in brackish waterbodies near the sea shore.

    **Soviet Arctic.** Murman (Rybachiy Peninsula; Kharlovka in East Murman); NW Chukotka [mouth of Chaun River, "freshwater" (i.e. brackish? — A.T.) lake; 3 IX 1938, Yakovlev].

    **Foreign Arctic.** West coast of Alaska, north coast of Canada, Labrador, Southern Baffin Island, West and East Greenland (reported in East Greenland as far as 74°25'N, the most northerly of all known localities for *Potamogeton*), Iceland, Arctic Scandinavia.

    **Outside the Arctic.** NW Europe (east to the Kola Peninsula, the shores of the White Sea, Leningrad Oblast, and the upper reaches of the Volga); isolated on the lower reaches of the Pechora (just below 67°N) and in the Central Volga district; Caucasus; Central Asia, South Siberia, temperate Far East (China, Northern Japan, Sakhalin, Kamchatka). In North America locally in Alaska, Canada and the northern USA.

    The distribution of *P. filiformis* is very peculiar. In extratropical Asia the species is confined to relatively southern districts. In particular, it has not so far been reported in Siberia inside the true taiga zone, occurring only in forest-steppe districts on its southern fringe. Its basic European range (emphatically Atlantic!) is separated from its Asian range. In North America the distribution of *P. filiformis* is also interrupted, but is here characterized by extensive penetration of the species into far northern districts. All truly arctic and the greater part of the Alaskan plants are distinguished by American systematists as the special variety (more correctly subspecies), var. *borealis* (Raf.) St. John (= *P. borealis* Raf., 1808; ? = *P. filiformis* f. *polaris* Hagstr., l.c., p. 17). Very probably it is proper to refer also our Chukotkan plants to this variety or subspecies [*P. filiformis* subsp. *borealis* (Raf.)].

2. ***Potamogeton subretusus*** Hagström, Crit. Res. Potamog. (1916) 30; Petrov, Fl. Yak. I, 91; Yuzepchuk, Fl. SSSR I, 238.

    *P. pectinatus* — Scheutz, Pl. jeniss. 164 (pro parte).

    Ill.: Hagström, l.c., fig. 10 (p. 31).

    Found in lakes and oxbows, in shallow places (about 1 m depth or less), on the lower reaches of the Yenisey and adjacent tundra. Forming beds in shallow water.

    **Soviet Arctic.** Lower reaches of Yenisey (Nikandrovskiy Island, 70°20'N; Brenner, 1876; Malo-Brekhovskiy Island, 70°50'N; Salberg, 1876); Boganidskoye Lake on the eastern fringe of the basin of the River Dudinka, a right-bank tributary of the Yenisey, 69°22'N (Tolmachev, 1932).

    Not found in the **Foreign Arctic** or **outside arctic limits**.

    Up to the present time *P. subretusus* has been found only at the three listed points lying within a distance of no more than 150–200 km. Two of the localities are situated in the Yenisey valley proper, one beyond its limits in a depression intermediate between the valley of a tributary of the Yenisey (the River Dudinka) and the basin of the River Pyasina. At least at the last site, the excellent development both of the vegetative and reproductive organs of the plant and its abundant growth are striking to behold. This is evidence of the high adaptation of *P. subretusus* to conditions of the temperate Arctic, especially if one takes into account that

Boganidskoye Lake is situated outside the sphere of influence of the warm water of the River Yenisey.

It is difficult to judge to what extent the available data on the distribution of *P. subretusus* accurately present a picture of its range. If we start from the relative closeness of *P. subretusus* to *P. vaginatus* Turcz., as indicated by Hagström, then it is possible that the former species should be considered an arctic race of the latter. But whether their ranges ever adjoin is unclear, because the localities for *P. vaginatus* in the lower reaches of the Yenisey are so far absolutely disjunct from its basic range. It is also noteworthy that "true" *P. vaginatus* is found in Arctic America, not any replacement form.

The presence of a notch in the tip of the leaf in *P. subretusus* is apparently a fully real distinguishing character of this species. It is certainly not the result of fortuitous change in a dead state. This character is well expressed in plants collected by me in 1932.

3. ***Potamogeton vaginatus*** Turcz., Fl. baic.-dahur. (1854) 66; Hagström, Crit. Res. Potamog. 32; Petrov, Fl. Yak. I, 91; Yuzepchuk, Fl. SSSR I, 238; Hultén, Fl. Al. I, 104; id., Atlas 27 (108).

*P. pectinatus* — Scheutz, Pl. jeniss. 164 (pro parte).

Plants of standing waterbodies of the forest zone of East Siberia and North America. Occurring sporadically in the Arctic, on the lower reaches of large rivers.

**Soviet Arctic.** Lower reaches of Yenisey (Tolstyy Nos, Nikandrovskiy Island).

**Foreign Arctic.** Mouths of the Yukon and Mackenzie Rivers.

**Outside the Arctic.** The range of the species is divided into two main parts: Siberian and American. In East Siberia *P. vaginatus* is found especially in the Angara region and Transbaikalia, where it crosses into Northern Mongolia. On the Lena it reaches the borders of Yakutia, but is not reported further north. In North America it occurs in Alaska, Canada and the northern USA. The European range of the species occupies an absolutely isolated position, being restricted to the coasts of the Gulf of Bothnia; isolated localities are also known in the Karelian ASSR. Arctic localities of the species must evidently be associated with transport of fragments by large rivers to their lower reaches.

Plants from the coasts of the Gulf of Bothnia represent, in the opinion of certain authors, the hybrid *P. pectinatus* × *vaginatus* (= *P. bottnicus* Hagstr.). Plants from the Karelian localities also belong to this.

4. ***Potamogeton pectinatus*** L., Sp. pl. (1753) 127; Ledebour, Fl. Ross. IV, 30; Scheutz, Pl. jeniss. 164 (pro parte); Hagström, Crit. Res. Potamog. 39; Krylov, Fl. Zap. Sib. I, 113; Petrov, Fl. Yak. I, 92; Yuzepchuk, Fl. SSSR I, 239; Perfilev, Fl. Sev. I, 59; Leskov, Fl. Malozem. tundry 17; Kuzeneva, Fl. Murm, I, 107; Hultén, Atlas 26 (102).

In rivers and lakes at shallow depth. Not avoiding bodies of brackish water.

**Soviet Arctic.** South Kanin, Timanskaya and Malozemelskaya Tundras; lower reaches of Pechora; Karskaya Tundra (on the Kara River); lower reaches of Yenisey. Near the arctic boundary on the River Khatanga, about 72°N.

**Foreign Arctic.** Arctic Scandinavia (in Norway).

**Outside the Arctic.** Widely distributed in the temperate north of Europe, Asia and America. Of uneven occurrence in subarctic districts. The total range of the species is almost cosmopolitan.

5. ***Potamogeton subsibiricus*** Hagstr., Crit. Res. Potamog. 84; Petrov, Fl. Yak. I, 94; Yuzepchuk, Fl. SSSR I, 243.

? *P. pusillus* — Scheutz, Pl. jeniss. 163.

Ill.: Hagström, l. c., fig. 33 (p. 84).

Soviet Arctic. Lower reaches of Yenisey (Dudinka, Nikandrovskiy Island); Chukotka Peninsula (Lawrence Bay).
Not found in the **Foreign Arctic** or **outside the Arctic**.

Apparently close to *P. sibiricus* A. Benn. described from Yakutia (valley of the Vilyuy River) and remaining, like the present species, one of the least known representatives of its genus. Hultén indicates that *P. subsibiricus* is very close to the American *P. Porsildorum* Fern., distributed from Alaska to Hudson Bay.

6. ***Potamogeton pusillus*** L., Sp. pl. (1753) 127; Ledebour, Fl. Ross. IV, 29; Ostenfeld, Fl. arct. I, 20 (pro parte); Hagström, Crit. Res. Potamog. 121; Krylov, Fl. Zap. Sib. I, 112; Petrov, Fl. Yak. I, 95; Yuzepchuk, Fl. SSSR I, 247; Perfilev, Fl. Sev. I, 59; Leskov, Fl. Malozem. tundry 17; Gröntved, Pterid. Spermat. Icel. 123; Hultén, Fl. Al. I, 103; id., Atlas 27 (105); Kuzeneva, Fl. Murm. I, 108; Karavayev, Konsp. fl. Yak. 45.

Ill.: Hagström, l.c., fig. 54 and 55 (pp. 122-123); Fl. Murm. I, pl. XXXII, 2.

Mainly in small bodies of standing water.

Soviet Arctic. Murman (isolated points near the Norwegian frontier, Iokanga, forest-tundra south of the Ponoy); Timanskaya Tundra (Cape Barmin); lower reaches of Pechora (Naryan-Mar district). Records for Arctic Siberia (Scheutz, Pl. jeniss. 163) should be referred to another species (*P. subsibiricus* Hagstr.).

Foreign Arctic. Iceland, Arctic Scandinavia. Replaced in Greenland by the closely related endemic species *P. groenlandicus* Hagstr.

Outside the Arctic. Almost all of Europe, the European part of the USSR including Crimea and the Caucasus, Siberia, Central Asia, Iran, Northern India, East Asia, North and South America, North and South Africa.

Plants from the Lower Pechora differ from typical *P. pusillus* in the more extended internodes and generally rather larger size. Their leaves are broader than in typical *P. pusillus* and mostly possess five longitudinal nerves. In general they somewhat approach the Siberian *P. subsibiricus*.

7. ***Potamogeton alpinus*** Balb. in Mem. Ac. Sc. Turin (1804) 329; Ostenfeld, Fl. Arct. I, 19; id., Fl. Greenl. 69; Hagström, Crit. Res. Potamog. 141; Krylov, Fl. Zap. Sib. I, 106; Yuzepchuk, Fl. SSSR I, 252; Perfilev, Fl. Sev. I, 57; Gröntved, Pterid. Spermat. Icel. 116; Hultén, Atlas 23 (90); Kuzeneva, Fl. Murm. I, 108.

*P. rufescens* Schrad. — Scheutz, Pl. jeniss. 163; Lange, Consp. Fl. Groenl. 116.

? *P. salicifolia* Wolfg. — Scheutz, Pl. jeniss. 163.

Ill.: Hagström, l.c., fig. 63 (p. 142); Fl. Murm. I, pl. XXXIII.

In lakes and oxbows, in shallow or moderately deep water.

Soviet Arctic. Murman (South Rybachiy Peninsula, Murmansk district, forest-tundra near Ponoy River); Kanin Peninsula (middle course of the River Sess-Yaga); lower reaches of Pechora (Naryan-Mar); lower reaches of Ob (near Salekhard); lower reaches of Yenisey (Zaostrovskiy Island, vicinity of Dudinka) and tundra to the east of it (pond in the depression of Boganidskoye Lake in the basin of the River Dudinka, a tributary of the Yenisey).

Foreign Arctic. Iceland, Arctic Scandinavia.

Outside the Arctic. Northern and Central Europe, Caucasus, West and Central Siberia, NE Kazakhstan.

The hybrid *P. alpinus* × *pusillus* (= *P. lanceolatus* Sm.) occurs on the lower reaches of the Pechora (near Naryan-Mar).

7a. ***P. alpinus*** subsp. ***tenuifolius*** (Raf.) Hultén, Fl. Aleut. (1937) 65; id., Fl. Al. I, 98; Böcher & al., Ill. Fl. Greenl. 306.

*P. tenuifolius* Raf., Med. Repos. Lex. 3, II (1811), 409; Yuzepchuk, Fl. SSSR I, 253; Karavayev, Konsp. fl. Yak. 45.

*P. alpinus* Balb. — M. Porsild, Fl. Disko 30; Ostenfeld, Fl. Arct. I, 9; id., Fl. Greenl. 69; Petrov, Fl. Yak. I, 96.

In lakes and oxbows.

**Soviet Arctic.** Lower reaches of Lena (vicinity of Kyusyur village, flood pond in valley of Ebetem River; Gorodkov and Tikhomirov, 1935); Penzhina River Basin (oxbow in valley of Penzhina near the mouth of the Slovutnaya River; Gorodkov and Tikhomirov, 1932); district of the Bay of Korf, in lakes.

**Foreign Arctic.** West coast of Alaska, Labrador, West Greenland (north to 67°).

**Outside the Arctic.** NE Asia (Yakutia, Preamuria, Kamchatka, Sakhalin, North Japan); forest region of Alaska and Canada, northern USA.

8. ***Potamogeton natans*** L., Sp. pl. (1753) 126; Ledebour, Fl. Ross. IV, 23; Hagström, Crit. Res. Potamog. 191; Krylov, Fl. Zap. Sib. I, 105; Petrov, Fl. Yak, I, 97; Yuzepchuk, Fl. SSSR I, 255; Perfilev, Fl. Sev. I, 57; Gröntved, Pterid. Spermat. Icel. 120; Hultén, Atlas 25 (98).

Common and widely distributed plant of lakes and oxbows in the northern forest zone. Growing both in shallows and in relatively deep water. Just reaching arctic boundaries.

**Soviet Arctic.** In the basin of the Poluy River, a right-bank tributary of the lower Ob, in a stream (Leskov, 1932).

**Foreign Arctic.** Iceland, Arctic Scandinavia.

**Outside the Arctic.** Greater part of Europe, Caucasus, Siberia south from the lower reaches of the Ob, from 65–66°N on the Yenisey and from the southern fringe of the Lena Basin. In the South reaching North Africa, Iran, Afghanistan, Northern India, Northern China, and Japan. Temperate part of North America. Disjunctly in Southern Africa.

9. ***Potamogeton gramineus*** L., Sp. pl. (1753) 127; Ledebour, Fl. Ross. IV, 25; Scheutz, Pl. jeniss. 163; Ostenfeld, Fl. Arct. I, 19; id., Fl. Greenl. 69; Hagström, Crit. Res. Potamog. 204; M. Porsild, Fl. Disko 30; Krylov, Fl. Zap. Sib. I, 109; Perfilev, Fl. Sev. I, 57; Leskov, Fl. Malozem. tundry 17; Gröntved, Pterid. Spermat. Icel. 119; Hultén, Fl. Al. I, 100; id., Atlas 24 (96).

*P. heterophyllus* Schreb. — Lange, Consp. fl. Groenl. 117; Petrov, Fl. Yak. I, 98; Yuzepchuk, Fl. SSSR I, 256; Kuzeneva, Fl. Murm. I, 112.

Ill.: Fl. Murm. I, pl. XXXIV.

In lakes and channels with slow-moving water.

**Soviet Arctic.** Murman (Rybachiy Peninsula, Pechenga, valley of Ponoy River within the forest-tundra); Timanskaya Tundra (Indiga River Basin); upper course of River Usa (within forest-tundra); lower reaches of Ob (near the Arctic Circle).

**Foreign Arctic.** West coast of Alaska; Greenland (north to almost 70° on the west coast); Iceland; Arctic Scandinavia.

**Outside the Arctic.** Europe except southern fringe (on the Pechora reported near 67°N); Caucasus; forest zone of Siberia and the Far East; from the Arctic Circle on the Ob and Yenisey, the middle course of the River Olenek, and Kamchatka in the north to Kazakhstan, Pamir, NE China, Korea and Central Japan in the south; greater part of forest region of North America.

10. ***Potamogeton praelongus*** Wulf. in Roemer, Arch. Bot. III, st. 3 (1805), 331; Ledebour, Fl. Ross. IV, 27; Scheutz, Pl. jeniss. 163; Ostenfeld, Fl. Arct. I, 20; Hagström, Crit. Res. Potamog. 250; Krylov, Fl. Zap. Sib. I, 107; Petrov, Fl. Yak. I, 100; Yuzepchuk, Fl. SSSR I, 259; Perfilev, Fl. Sev. I, 5; Leskov, Fl. Malozem. tundry 17; Gröntved, Pterid. Spermat. Icel. 123; Hultén, Fl. Al. I, 103; id., Atlas 26 (104); Kuzeneva, Fl. Murm. I, 114.

? *P. salicifolia* Wolfg. — Scheutz, Pl. jeniss. 163.

Ill.: Fl. Murm. I, pl. XXXV.
   In lakes, including relatively deep places.
Soviet Arctic. Murman (Pechenga, coast SE of Svyatoy Nos); middle part of Kanin Peninsula (lake on floodplain of River Sess-Yaga); Malozemelskaya Tundra (Lake Saundey); Priobskaya Tundra; lower reaches of Yenisey (Nikandrovskiy Island; Zaostrovskoye?).
Foreign Arctic. District of mouth of Mackenzie River; Labrador; Iceland; Arctic Scandinavia.
Outside the Arctic. Greater part of Europe, Caucasus, Siberia from the fringe of the Arctic in the west and Central Yakutia and Kamchatka in the east as far as the southern frontiers; Northern China, Japan. Widely distributed in North America.

11. *Potamogeton perfoliatus* L., Sp. pl. (1753) 126; Ledebour, Fl. Ross. IV, 27; Trautvetter, Fl. rip. Kolym. 67; Scheutz, Pl. jeniss. 163; Ostenfeld, Fl. arct. I, 20; Hagström, Crit. Res. Potamog. 254; Krylov, Fl. Zap. Sib. I, 106; Petrov, Fl. Yak. I, 101; Yuzepchuk, Fl. SSSR I, 260; Perfilev, Fl. Sev. I, 58; Leskov, Fl. Malozem. tundry 17; Gröntved, Pterid. Spermat. Icel. 121; Hultén, Fl. Al. I, 101; id., Atlas 26 (103); Kuzeneva, Fl. Murm. I, 114.
Ill.: Fl. Murm. I, pl. XXXVI.
   In standing waterbodies and slowly flowing rivers, partly at rather considerable depth.
Soviet Arctic. Murman (Pechenga, forest-tundra in the basins of the Iokanga and Ponoy); Timanskaya Tundra (Indiga, Pulskoye Lake); Malozemelskaya Tundra; lower reaches of Pechora, Western Bolshezemelskaya Tundra (vicinity of Naryan-Mar, upper course of River Shapkina); lower reaches of Yenisey (Vershininskoye, Dudinka, Tolstyy Nos, Sopochnyy Island); lower reaches of Kolyma (Panteleyevo).
Foreign Arctic. West coast of Alaska (Kotzebue Sound); Iceland; Arctic Scandinavia.
Outside the Arctic. Greater part of Europe, Caucasus, forest region of Siberia (more or less ubiquitous); in Asia south to Iran, India and China; Alaska and Yukon in North America.
   Replaced by *P. perfoliatus* subsp. *Richardsonii* (Benn.) Hult. (= *P. Richardsonii* Fern.) on the Pacific coast of Alaska, in Canada and the USA.

**DOUBTFUL SPECIES**

*Potamogeton lucens* L., Sp. pl. (1753) 126; Ledebour, Fl. Ross. IV, 26; Hagström, Crit. Res. Potamog. 233; Krylov, Fl. Zap. Sib. I, 108; Petrov, Fl. Yak. I, 99: Yuzepchuk, Fl. SSSR I, 257; Perfilev, Fl. Sev. I, 57; Hultén, Atlas 27 (97).
? *P. salicifolia* Wolfg. — Scheutz, Pl. jeniss. 163.
   Species widely distributed at temperate latitudes in Europe and North America, normally not reaching the northern limits of the forest zone. Apparently as an exception penetrating arctic limits along the large Siberian rivers.
Soviet Arctic. Recorded by Krylov for the Poluy River near Salekhard, and by Scheutz (under the name *P. salicifolia*) for the lower reaches of the Yenisey (Zaostrovskoye below 69°45'N). I have not seen arctic plants.
Foreign Arctic. Not reported.
Outside the Arctic. Widely distributed in temperate and partly subtropical latitudes of the Northern Hemisphere. In Siberia reliable localities are restricted to southern parts of the taiga zone, which casts doubt on the accuracy of the identification of the plants from the lower reaches of the Ob and Yenisey. It is possible that these plants (similar to what has been shown with respect to "*P. lucens*" from the Vilyuy Basin) will prove to belong to *P. praelongus*. Hagström, when revising the *Potamogeton* material from the lower reaches of the Yenisey worked on by Scheutz, does not indicate the presence of *P. lucens* in this district; but he also does

not mention the plants collected by Arnell at Zaostrovskoye in his treatment of *P. praelongus*, under which he includes *P. salicifolius* Wolfg. as a synonym.

**Potamogeton digynus** Wall., Catal. (1828), no. 5177; Yuzepchuk, Fl. SSSR I, 253; Karavayev, Konsp. Fl. Yak. 46.

Species of southern and eastern Asia, recorded in the literature for the lower reaches of the Lena. There is no corresponding herbarium material in our holdings. The growth in Arctic Yakutia of this almost tropical species, whose reliable localities in the USSR are confined to Primorskiy Kray, is most improbable. It is possible that *P. gramineus*, reliably known from subarctic Yakutia (Olenek Basin), or *P. alpinus* ssp. *tenuifoius*, of which there is material from the lower reaches of the Lena, were mistaken for *P. digynus*.

## GENUS 2    Zostera L. — EEL-GRASS

MARINE HERBS, GROWING in a submersed state in moderately deep (from 1 to 10 m) water at sites protected from strong surf. One of the few species penetrates districts on the fringes of the Arctic.

1. ***Zostera marina*** L., Sp. pl. (1753) 968; Ledebour, Fl. Ross. IV, 20; Lange, Consp. fl. Groenl. 117; Ostenfeld, Fl. arct. I, 18; id., Fl. Greenl. 69; Yuzepchuk, Fl. SSSR I, 266; Perfilev, Fl. Sev. I, 59; Gröntved, Pterid. Spermat. Icel. 125; Hultén, Fl. Al. I, 95; id., Atlas 22 (87); Kuzeneva, Fl. Murm. I, 118; Böcher & al., Ill. Fl. Greenl. 306.

Ill.: Fl. Murm. I, pl. XXXVII.

On sea shores at shallow depth at protected sites.

**Soviet Arctic.** Murman (west coast of Rybachiy Peninsula, Terskiy Shore between the Arctic Circle and the mouth of the Ponoy); coast of Koryakia (Bay of Korf).

**Foreign Arctic.** West coast of Alaska south of the Bering Strait; West Greenland between 64 and 65°N; Iceland; Arctic Scandinavia.

**Outside the Arctic.** Widely distributed on the shores of northern parts of the Atlantic and Pacific Oceans, also on the Baltic, Mediterranean and Black Seas.

## FAMILY X

# Juncaginaceae Lindl.

**ARROW-GRASS FAMILY**

---

### GENUS 1 — Triglochin L. — ARROW GRASS

REPRESENTED IN THE Arctic by two species confined to its relatively temperate parts and possessing a wide distribution in the temperate zone of the Northern Hemisphere (south to the subtropics).

Plants with narrow, linear or awlshaped leaves and long spikelike inflorescences. Growing in moist silty places on river banks and partly in saline areas near the sea shore.

1. Stem thick, about 3 mm wide, bearing dense inflorescence (up to 20 cm long) with very numerous flowers on relatively short pedicels. Leaves fleshy, up to 2 mm wide, forming dense basal tuft. Carpels and stigmas 6 ............................................................1. **T. MARITIMUM** L.
- Stem thin, up to 1 mm wide. Inflorescence sparse, with few (up to 10) flowers on thin, obliquely divergent pedicels of length almost equal to that of the flowers. Leaves thin, up to 1 mm wide, few. Carpels and stigmas 3 ............................................................2. **T. PALUSTRE** L.

   1. ***Triglochin maritimum*** L., Sp. pl. (1753) 339; Ledebour, Fl. Ross. IV, 35; Ruprecht, Fl. samoj. cisur. 57; Schrenk, Enum. plant. 527; Trautvetter, Pl. Sib. bor. 113; id., Fl. kolym. 561; Ostenfeld, Fl. arct. I, 21; Krylov, Fl. Zap. Sib. I, 119; Petrov, Fl. Yak. I, 104; Andreyev, Mater. fl. Kanina 155; Perfilev, Fl. Sev. I, 60; Fedchenko, Fl. SSSR I, 276; Leskov, Fl. Malozem. tundry 18; Gröntved, Pterid. Spermat. Icel. 126; Hultén, Fl. Al. I, 105; id., Atlas 29 (116); Kuzeneva, Fl. Murm. I, 122.

   Found in saline sites with silty soil, directly on the sea shore (at sites protected from surf) but especially in saltmarshes in river mouths in the zone of tidal flooding ("tamtsy," "estuarine saltmarshes") within the European Arctic. Outside arctic limits the association with sea coasts is not obligatory.

   **Soviet Arctic.** Murman (on the entire coast, not uncommon); shores of Kanin Peninsula and Cheshskaya Bay; coast of Timanskaya Tundra. After a considerable gap reported for the mouth of the More-Yu River (Khaypudyra) in the middle part of the Bolshezemelskaya Tundra. In Siberia found at the polar limit of forest between the lower Lena and the Olenek (at a considerable distance from the sea). In the Khatanga and Kolyma Basins apparently not significantly reaching the boundaries of the forest zone[1]; SE Chukotka Peninsula; shore of Bay of Korf.

   **Foreign Arctic.** Labrador, Iceland, Arctic Scandinavia.

   **Outside the Arctic.** Widely distributed in the northern temperate zone: in Northern Europe mainly on sea coasts (more rarely in marshes away from the sea, for example on the middle course of the Mezen River), further south and in Siberia predominantly in moist places of the forest-steppe and steppe zones; in Yakutia ranging far to the north. Southern limit of distribution in subtropical latitudes. Disjunctly in the extreme south of South America.

---

[1] Ostenfeld's record of the growth of *T. maritimum* at the mouth of the Kolyma is evidently erroneous. Trautvetter, to whom Ostenfeld refers, recorded *T. maritimum* only for the middle course of the Kolyma (Sredne-Kolymsk district). There is material from this district in the Leningrad herbarium.

2. ***Triglochin palustre*** L., Sp. pl. (1753) 338; Ledebour, Fl. Ross. IV, 35; Ruprecht, Fl. samoj. cisur. 57; Lange, Consp. fl. Groenl. 121; Ostenfeld, Fl. arct. 21; id., Fl. Greenl. 69; M. Porsild, Fl. Disko 31; Krylov, Fl. Zap. Sib. I, 118; Petrov, Fl. Yak. I, 103; Andreyev, Mater. fl. Kanina 156; Devold & Scholander, Fl. pl. SE Greenl. 106; Fedchenko, Fl. SSSR I, 277; Perfilev, Fl. Sev. I, 60; Leskov, Fl. Malozem. tundry 18; Seidenfaden & Sørensen, Vasc. pl. NE Greenl. 180; Gröntved, Pterid. Spermat. Icel. 127; Hultén, Fl. Al. I, 106; id., Atlas 30 (117); Kuzeneva, Fl. Murm. I, 121; Böcher & al., Ill. Fl. Greenl. 306.

Ill.: Fl. Murm. I, pl. XXXVIII, 1.

Found in moist places devoid of continuous plant cover, on river banks, at river mouths and frequently near the sea shore, on slightly saline soil. Sometimes encountered side-by-side with *T. maritimum*. Generally more widely distributed.

**Soviet Arctic.** Murman (more or less ubiquitous); Kanin Peninsula north to Kanin Nos; Timanskaya and Malozemelskaya Tundras; Bolshezemelskaya Tundra north to Varandey and Khaypudyrskaya Bay; Karskaya Tundra near the Polar Ural. East of the Ob reported within arctic limits only in the extreme east of the USSR (SE part of Chukotka Peninsula and shore of Bay of Korf).

**Foreign Arctic.** West coast of Alaska; Labrador; West Greenland north to 71°; East Greenland north to 73°50' (scattered localities); Iceland; Arctic Scandinavia.

**Outside the Arctic.** Widely distributed in Europe and Asia, south to the Mediterranean Sea, Asia Minor, Afghanistan, the Himalayas, Tibet, Northern China, Northern Korea, and Northern Japan; North America. Disjunctly in the extreme south of South America. The northern range boundary in Siberia advances in the basins of the Ob and Yenisey to between 60 and 63°N, and in Yakutia (Olenek Basin) shifts to the Arctic Circle.

## GENUS 2  **Scheuchzeria** L. — SCHEUCHZERIA

REPRESENTED IN THE Arctic by the sole species, which just penetrates arctic limits from the northern forest zone.

1. ***Scheuchzeria palustris*** L., Sp. pl. (1753) 338; Ledebour, Fl. Ross. IV, 37; Krylov. Fl. Zap. Sib. I, 120; Perfilev, Fl. Sev. I, 60; Fedchenko, Fl. SSSR I, 278; Kuzeneva, Fl. Murm. I, 122; Hultén, Atlas 30 (118).

Ill.: Fl. Murm. I, pl. XXXIX.

Characteristic plant of very moist mossy bogs of the forest zone. In NW Europe occasionally penetrating the forest-tundra.

**Soviet Arctic.** Rarely in forest-tundra of the Kola Peninsula. East of the White Sea nowhere reported at the northern limit of forest.

**Foreign Arctic.** Arctic Scandinavia.

**Outside the Arctic.** Widely distributed in the forest region of Europe, in the middle and southern zone of Siberia, in the temperate part of the Far East, and in temperate northern districts of North America. Represented in America by a special race, *Sch. p.* ssp. *americana* (Fern.) Hult.

## FAMILY XI

# Alismataceae DC.

### WATER-PLANTAIN FAMILY

### GENUS 1     Alisma L. — WATER-PLANTAIN

1. ***Alisma plantago-aquatica*** L., Sp. pl. (1753) 343; Yuzepchuk, Fl. SSSR I, 280; Kuzeneva, Fl. Murm. I, 126; Hultén, Atlas 31 (122).

   *A. plantago* L. — Ruprecht, Fl. samoj. cisur. 57; Scheutz, Pl. jeniss. 164; Krylov, Fl. Zap. Sib. I, 121; Perfilev, Fl. Sev. I, 60.

   *A. latifolium* Gilib. — Petrov, Fl. Yak. I, 105.

   Ill.: Fl. Murm. I, pl. XL.

   Found in wet places on the shores of waterbodies and at the edge of sites inundated during floods, mainly in the temperate zone.

   **Soviet Arctic.** Rarely in Murman (specifically Pechenga). East of the White Sea not reaching the polar limit of forest. Reported on the Ob to 64°N (Berezov), and on the Yenisey to 66°N (Turukhansk district). From Yakutia there is a specimen with imprecise indication "lower reaches of River Lena" collected by Bychkov in 1896; probably it was collected in the north of the forest zone, but it may have originated from forest-tundra.

   **Foreign Arctic.** Not reported.

   **Outside the Arctic.** Widely distributed in the temperate zone of Eurasia, south to the region of the Mediterranean Sea, the Caucasus and Central Asia.

## FAMILY XII

# Butomaceae S.F. Gray

### FLOWERING RUSH FAMILY

### GENUS 1    Butomus L. — FLOWERING RUSH

*1. **Butomus umbellatus** L., Sp. pl. (1753) 372; Ledebour, Fl. Ross. IV, 43; Krylov, Fl. Zap. Sib. I, 127; Fedchenko, Fl. SSSR I, 292; Perfilev, Fl. Sev. I, 61; Leskov, Fl. Malozem. tundry 13; Kuzeneva, Fl. Murm. I, 129; Hultén, Atlas 32 (127).

Aquatic plant growing near the shores of rivers and lakes, mainly within the limits of the northern forest zone. Just penetrating arctic limits in the NE European part of the USSR.

**Soviet Arctic.** Single reliable locality at the mouth of the Mutnaya River (Indiga River Basin, Timanskaya Tundra), on the shore and in the water (Dedov, 1928).

**Foreign Arctic.** Not reported.

**Outside the Arctic.** Greater part of Europe, middle and southern zones of Siberia, Preamuria, Northern Mongolia, Northern China; Central and Western Asia, south to Afghanistan and Northern India.

## FAMILY XIII

# Gramineae Juss.

### GRASSES

---

ONE OF THE LARGEST families of flowering plants, including seven thousand species and possessing a cosmopolitan distribution. Richly represented in all holarctic floras and invariably occupying one of the leading positions in their composition.

Grasses everywhere play a very conspicuous role in the composition of the arctic flora, occupying along with *Cyperaceae, Compositae, Cruciferae* and *Caryophyllaceae* one of the leading positions according to number of species. Frequently they stand in first place among these families. Grasses are not only well represented in arctic floras as a whole, but also invariably contribute to the composition of the flora in those parts of the Arctic at the highest latitudes.

In the flora of the Soviet Arctic, the grass family is represented by a total of 36 genera and 162 species. A portion of the latter exhibit considerable polymorphism, which has been reflected in the description of numerous forms under separate species names.

A rather considerable number of the grasses occurring in the Arctic are not characteristic of it and are actually foreign to the arctic flora. But the bulk of the representatives of the family growing in the Arctic enter into the composition of its flora as an organic component. Diminution of the number of grass species in the course of transition to higher latitudes in the Arctic takes place slowly, and certain species persist in the composition of the flora right to the furthest northern outposts of land.

The distribution of grasses in particular sectors of the Arctic is not uniform. In districts with more oceanic climate, there are rather fewer grasses truly characteristic of the Arctic but its limits are here often penetrated by species distributed mainly in the temperate zone and constituting a "casual" component of the arctic flora.

A considerable number of genera of grasses are characteristic of the Arctic; the rich representation of this family in the arctic flora is invariably a function of the relatively high diversity of its generic composition, not of a large number of species in any particular genus. The genera most richly represented in the Arctic are *Poa, Puccinellia, Calamagrostis* and *Festuca*. However, the list of characteristic arctic plants also includes members of certain genera poor in arctic species or entirely oligotypic, such as (for example) species of *Hierochloë, Arctagrostis* and certain others.

Among the arctic grasses there are many endemic or almost endemic species, of which a considerable portion are widely distributed. Many grasses characteristic of the Arctic possess arctic-alpine distributions differing in details.

Very significant is the presence in the Arctic not only of endemic species, but also endemic genera of grasses. Especially deserving mention here are the oligotypic arctic genera *Dupontia* and *Phippsia*,\* and the monotypic subarctic genus *Arctophila*. The existence of these genera, as well as the considerable degree of differentiation of certain arctic species of other genera, attests to the great antiquity of the development of grasses under arctic conditions and, evidently, to the membership of a considerable number of representatives of the family in the ancient autochthonous core of the arctic flora.

The ecological associations of arctic grasses are diverse. The more characteristic arctic species include plants of open stony tundras sometimes almost devoid of snow in wintertime (*Poa abbreviata, Hierochloë alpina, Festuca brachyphylla*), plants of deeply snow-covered sites permanently cooled by meltwater and with severely restricted duration of the growing season (especially the genus *Phippsia*), and plants of moderately moist clayey or clayey-stony tundras moderately protected by snow in winter (*Arctagrostis latifolia, Alopecurus alpinus, Deschampsia alpina*, etc.). There are also plants typical of marshy tundras of varying degrees of wetness (*Dupontia, Calamagrostis, Hierochloë pauciflora*) and halophytes of sea shores (species of *Puccinellia*). Certain species take part in the formation of meadowlike communities, mainly on slopes (for example, certain species of *Poa* and *Festuca*). Finally, some species (*Pleuropogon Sabinii, Arctophila fulva*) are non-obligatory hydrophytes.

The species which are not specifically arctic include, on the one hand, a group of meso-hygrophytes distributed mainly in oceanic districts of the temperate Arctic and native to meadow or littoral habitats; also, a group of species growing on dry, more or less stony, poorly vegetated substrates which are genetically connected with mountain and steppe landscapes of inland continental regions. These plants occur mainly in harsher continental districts of the Arctic with little snow cover.

Noteworthy as a special biological property of a whole series of arctic grasses is the phenomenon of vivipary. In the majority of species this phenomenon is observed to be facultative; but in several cases the viviparous forms have become predominant (or locally completely replace the non-viviparous type), and sometimes, finally, the species may be represented only by the viviparous form (*Deschampsia alpina*).

1. Inflorescence a one-sided, very narrow spike with solitary sessile spikelets arranged in 2 rows on its triangular rachis; spikelets narrowly lanceolate, 1-flowered, usually appressed to rachis of spike; *glumes absent*; style *1, with filiform stigma*. Perennial greyish green plant, forming dense tufts; culms above base without nodes; leaf blades very narrow, setiformly folded lengthwise. . . . . . . . . . . . . . . . . . . . . . . . . . . . . . . . . . . . . . . . . GENUS 32. **NARDUS** L.
– Inflorescence otherwise; if spikelike, then *2 well developed glumes always present* at base of spikelets; stigmas *always* 2, sessile or on more or less long styles. . . . . . . . . . . . . . . . . . . . . . . . . . . . . . . . . . . . . . . . . . . . . . . . . . . . . . . .2.

2. Spikelets 8–15 mm long, *more or less nodding*, 3–10 in number, aggregated on short (1–4 mm) slender pedicels into *one-sided racemose* inflorescence, each with 6–14 florets; glumes significantly shorter than lemmas, obtuse,

---

\* Formally the genus *Phippsia* cannot be considered endemic, because isolated colonies of its species occur disjunctly from their main ranges in the mountains of Southern Norway and the Western United States.

often more or less lobed; lemmas obovate, dark violet, scabrous with rather long acicules, with 7 more or less outstanding nerves, obtuse at tip, with broad membranous border; paleas *with 2–4 long awnlike appendages* arising from their keels significantly below their tip. Perennial plant with subterranean stolons; leaf blades 1–3 mm wide, flat or folded lengthwise. .
. . . . . . . . . . . . . . . . . . . . . . . . . . . . . . . . . . . . . .GENUS 20. **PLEUROPOGON** R. BR.

– Plant of different appearance; keels of paleas *always without awnlike appendages.* . . . . . . . . . . . . . . . . . . . . . . . . . . . . . . . . . . . . . . . . . . . . . . . . . . . . . .3.

3. Inflorescence a *true spike:* spikelets *sessile or almost sessile* (on short thick pedicels up to 1 mm long), arranged singly or in groups of 2–3 in regular longitudinal rows on 2 sides of common rachis of inflorescence. . . . . . . . .4.
– Inflorescence *paniculate or racemose*, frequently very dense, spikelike but with spikelets never arranged in regular longitudinal rows on 2 sides of unbranched common rachis of inflorescence . . . . . . . . . . . . . . . . . . . . . . . .7.

4. Spikelets arranged *in groups of 2–3* on common rachis of spike, but lateral spikelets in groups of 3 may be imperfect (then containing 1 staminate floret or completely lacking florets); glumes *lanceolate-subulate or subulate*, with 1–5 not always conspicuous nerves, usually more or less displaced to one side of spikelet. . . . . . . . . . . . . . . . . . . . . . . . . . . . . . . . . . . . . . . . . . . . . . . .5.
– Spikelets arranged *singly* on common rachis of spike; glumes *lanceolate or oblong-lanceolate*, with 3–5 conspicuous nerves, usually not displaced to one side of spikelet. . . . . . . . . . . . . . . . . . . . . . . . . . . . . . . . . . . . . . . . . . . . . . . . .6

5. Rachis of spike *without joints*, not disarticulating during fruiting or in dry state; spikelets all uniform, sessile, arranged in groups of 2–3 on rachis of spike, each with 2–5 aggregated bisexual florets; lemmas *without or with very short (up to 3 mm) awns*; glumes lanceolate-subulate, usually with more or less conspicuous nerves. Perennial plant with long subterranean stolons (rhizomes). . . . . . . . . . . . . . . . . . . . . . . . . . . . . . . . . . . . . . . .35. **LEYMUS** HOCHST.
– Rachis of spike *with joints*, when in fruit or in dry state readily disarticulating; spikelets arranged in groups of 3 on rachis of spike; middle spikelet in each group of 3 sessile with 1 bisexual floret, the 2 lateral on very short (up to 1 mm) but conspicuous pedicels, usually smaller with 1 staminate floret or completely lacking florets, strongly reduced; lemmas *with more or less long awns*; glumes subulate, without conspicuous nerves. Perennial or annual plant, forming more or less dense tufts, without subterranean stolons. . . . . . . . . . . . . . . . . . . . . . . . . . . . . . . . . . . . . . . . . . . . . . . . . . . . .36. **HORDEUM** L.

6. Glumes over almost whole surface but especially on nerves *scabrous with small acicules, more rarely pubescent*, during fruiting usually long persistent on rachis of spike; lemmas smooth, scabrous or pubescent, basally *with rather long* (usually about 0.4 mm), *obtusely triangular and short-pubescent* (at high magnification) *callus*; anthers broadly linear, *1–2.5 mm long*. Caryopses slightly concave on ventral side. Perennial plant, *usually* forming more or less dense tufts, *without subterranean stolons.* . . . . . . . . . .
. . . . . . . . . . . . . . . . . . . . . . . . . . . . . . . . . . . . . . . . . . . .33. **ROEGNERIA** C. KOCH.

– Glumes *bare and smooth* or only *more or less scabrous distally on* keel and lateral nerves, during fruiting often falling together with lemmas and paleas; lemmas bare and only more or less scabrous distally on keel, basally *with shorter* (usually about 0.2 mm), *broadly rounded, bare or almost bare callus*; anthers linear, *3–5 mm long*. Caryopses with deep groove on ventral side. Perennial plant, *often with long subterranean stolons* (rhizomes). . . . . . . . . . . . . . . . . . . . . . . . . . . . . . . . . . . . . . . . . .34. **ELYTRIGIA** DESV.

7. Ligule appearing as very narrow membranous strip, *modified almost from base into series of more or less long hairs* which are always conspicuous without magnification. . . . . . . . . . . . . . . . . . . . . . . . . . . . . . . . . . . . . . . . . . . . . . .8.
– Ligule membranous, sometimes appearing as very narrow (scarcely evident) membranous strip, but *never modified into series of hairs*; however, the margin of the ligule may bear very minute cilia (visible only at high magnification), whose length is always somewhat less than the length of the membranous part of the ligule. . . . . . . . . . . . . . . . . . . . . . . . . . . . . . . . .9.

8. Spikelets *7–12 mm long*, aggregated in more or less dense paniculate inflorescence, each with 3–7 florets; rachis of spikelet *clothed with long hairs*; glumes lanceolate, membranous, strongly unequal; lemmas *narrowly lanceolate, extended at tip into long subulate mucro*. Rather tall perennial plant with long subterranean stolons (rhizomes); culms *with numerous nodes;* leaf blades greyish green, *0.8–3 cm wide* . . . .16. **PHRAGMITES** ADANS.
– Spikelets *4–8 mm long*, aggregated in condensed (sometimes almost spikelike) paniculate inflorescence, each with 2–5 florets; rachis of spikelet *bare*; glumes broadly lanceolate, shorter than spikelet; lemmas *lanceolate*, with 3–5 weak nerves. Perennial plant, usually forming more or less dense tufts; culms *above base without nodes*; leaf blades *up to 1 cm wide* . . . . . . . . . . . . . . . . . . . . . . . . . . . . . . . . . . . . . . . . . . . . . . . . . . . . .17. **MOLINIA** SCHRANK.

9. Spikelets orbicular-ovoid, strongly laterally compressed, 2–3 mm long, containing 1 or 2 florets, *arranged* on very short pedicels *in 2 closely approximated rows in one-sided oblong spikes*, which are in turn aggregated into a linear, likewise one-sided, spikelike panicle; glumes equal in length to spikelet, each of same shape and size, keeled, somewhat swollen; lemmas almost completely concealed by glumes, lanceolate, keeled, with 5 inconspicuous nerves; during fruiting the spikelets fall together with the glumes. Rather tall perennial plant. . . . . . . . . . . . . . .15. **BECKMANNIA** HOST.
– Inflorescence otherwise; if dense and spikelike, then spikelets *never arranged in 2 closely approximated rows on its lateral branchlets*. . . . . . .10.

10. Spikelets 1.5–3.5 mm long, aggregated in condensed or more or less diffuse paniculate inflorescence with completely bare and smooth branchlets, with 1 or 2 widely spaced florets; glumes from ovoid to almost orbicular, *usually less than half as long as lemmas, sometimes completely absent*, rarely upper glume slightly more than half as long as lemma (then rachis of spikelet and lemmas and paleas completely bare and smooth); lemmas

obtuse, awnless, *with 3 conspicuous more or less raised nerves* but with weakly expressed keel, bare or more or less pubescent; 1–3 stamens. . . .11.

– Combination of characters otherwise; spikelets often (but far from always) larger or with greater numbers of florets; both glumes or at least upper glume *significantly longer than half length of lemmas*, rarely glumes more or less equal to half length of lemma or even slightly shorter, but then lemmas wi*th 5–7 more or less uniformly developed nerves*; stamens 3, rarely 2. . . . . . . . . . . . . . . . . . . . . . . . . . . . . . . . . . . . . . . . . . . . . . . . . . . . . . . .12.

11. Spikelets *2.5–3.5 mm long*, aggregated in more or less diffuse paniculate inflorescence, with 1–2 florets; glumes *always present, relatively large* (upper sometimes more than half as long as lemma); lemmas and paleas bare and smooth; stamens *3*; anthers *1–2 mm long*. Plant with subterranean or submerged stolons, usually not forming tufts; leaf blades flat, 2–6 mm wide. . . . . . . . . . . . . . . . . . . . . . . . . . . . . . . .26. **CATABROSA** BEAUV.

– Spikelets *1.5–2.5 mm long*, aggregated in more or less condensed, more rarely diffuse, paniculate inflorescence, usually with 1 (more rarely 2) florets; glumes *very small, often absent*; lemmas more or less pubescent especially on nerves, more rarely bare; *1–2* stamens; anthers *0.4–0.8 mm long*. Small (often dwarf) perennial plant, usually forming small tufts; leaf blades flat or folded lengthwise, usually 1–3 mm wide. . .27. **PHIPPSIA** R. BR.

12. Spikelets 3–7 mm long, aggregated in spikelike or more or less diffuse paniculate (sometimes racemose) inflorescence; glumes herbaceous-membranous, usually almost equal to spikelet in length; lowest 2 lemmas *usually of greater thickness* (often almost leathery), bearing in their axils 1 *staminate floret with 3 stamens or completely lacking florets*; central floret (third or sole in spikelet) *bisexual, with 2 stamens*, surrounded by awnless *membranous* lemma and palea. Perennial plant possessing, especially in dry state, strong aromatic odour of dry hay (due to presence of coumarin). . . . . . . . . . . . . . . . . . . . . . . . . . . . . . . . . . . . . . . . . . . . . . . . . . . . . . . .13.

– Structure of spikelets otherwise: spikelets with 1 or more numerous florets sequentially arranged on more or less elongate rachis of spikelet; lemmas *of similar thickness* on all florets in spikelet, usually differing only in size (with exception of genus *Typhoides*); *all* developed florets *bisexual, always with 3 stamens*. Plant without aromatic odour. . . . . . . . . . . . . . . . . . . . . . .14.

13. Spikelets 5–7 mm long, *lanceolate, with 1 bisexual floret*, aggregated in loose spikelike panicle with very short (up to 7–8 mm long) branchlets; lower of the 2 outer glumes ½-⅔ as long as upper; next 2 floret scales (in origin lemmas of reduced lower florets of spikelet) just over half as long as spikelet, rather copiously pubescent, with more or less long, often geniculate awn on their back. . . . . . . . . . . . . . . . . . . . . . . . . . .2. **ANTHOXANTHUM** L.

— Spikelets 3–6 mm long, *ovoid*, more or less shining, aggregated in condensed or more less diffuse paniculate or racemose inflorescence; *3 florets in each spikelet*, the central bisexual (with pistil and 2 stamens), the lateral staminate (each with 3 stamens); glumes equal to spikelet and slightly

longer than lemmas, membranous; lemmas of lateral staminate florets leathery, often more or less pubescent, often with short awn near tip. . . . . .
.................................................3. **HIEROCHLOË** R. BR.

14. All spikelets in inflorescence *with only 1 floret*; rachis of spikelet *usually not prolonged* above base of this single floret, *more rarely prolonged in form of small pubescent or bare rachilla not bearing floret scales (lemma and palea) at tip* (only in cases of abnormal development in some species of *Calamagrostis* can isolated spikelets in the panicle possess rudimentary floret scales at the tip of the long-pubescent rachilla, the remnant of the rachis of the spikelet). .............................................15.
– All spikelets in inflorescence bearing *2 or more florets*; the second floret may sometimes be imperfectly developed, but in that case the rachis of the spikelet is *always considerably prolonged* above the base of the first floret and *bears at its tip the normally developed or more or less reduced lemma and palea of the second floret.* .............................22.

15. Spikelets 4–5 mm long, awnless, laterally compressed, aggregated in very dense, almost spikelike paniculate inflorescence with strongly abbreviated branchlets; glumes equal to spikelet in length, each of almost same length, with sharp keel on back; in the axils of both glumes (or of only one of them) *are situated an additional 2 (or 1) significantly smaller, narrowly linear pubescent scales (inner glumes)*, being in origin the lemmas of the reduced lower florets of the spikelet; floret scales (lemmas and paleas) almost leathery, shorter than glumes. Rather tall perennial plant with long subterranean stolons and broad (4–10 mm) leaf blades. ...............
.................................................1. **TYPHOIDES** MOENCH.
– Spikelet *always with only 2 glumes and 2 (more rarely 1) floret scales.* . . .16.

16. Numerous, strongly laterally compressed spikelets 2.5–6 mm long aggregated in very dense spikelike inflorescence *of truly cylindrical, short-cylindrical or ovoid shape* ("tail"); glumes each of same shape and size, equal to spikelet in length, carinate with strongly prominent keel.. ..............17.
– Spikelets aggregated in more or less diffuse or condensed paniculate inflorescences, sometimes almost spikelike with strongly abbreviated branchlets, *but always more or less lobate, not possessing truly cylindrical or ovoid shape.* ........................................................18.

17. Glumes *with their margins fused near base,* more or less covered with rather long hairs, *without awn at tip but sometimes with very short mucro*; lemmas *usually with awn* arising from their back, rarely awnless; paleas *absent.* Perennial plant, forming more or less dense tufts or with long subterranean stolons. .................................6. **ALOPECURUS** L.
– Glumes *free from base,* usually with short inconspicuous hairs but often pectinate-ciliate on keel, *with mucro or short awn at tip*; lemmas always awnless; paleas *shorter than lemmas, with 2 nerves.* Perennial plant, forming more or less dense tufts. ............................5. **PHLEUM** L.

18. *Annual* plant 20–80 cm high. Spikelets about 2.5 mm long, aggregated in a more or less lax paniculate inflorescence; lemmas *with awn 4–7 mm long, arising from just below their tip and exserted far beyond spikelet.* . . . . . . . . . . . . . . . . . . . . . . . . . . . . . . . . . . . . . . . . . . . . . . . . . . . . . . . 10. **APERA** ADANS.
   – *Perennial* plant. Lemmas *awnless or with awns arising usually from their back, more rarely from their tip but then significantly shorter.* . . . . . . . . . . 19.

19. Spikelets *strongly laterally compressed*, 2.5–7 mm long, aggregated in a condensed (more rarely more or less diffuse) paniculate inflorescence often with strongly abbreviated branchlets; glumes from broadly lanceolate to ovoid, *almost always shorter than lemmas*; lemmas *equal to spikelets in length*, lanceolate or lanceolate-ovoid, *with strongly raised sharp keel* and inconspicuous lateral nerves, *on basal half over whole surface scabrous with rather long acicules*; paleas slightly shorter than lemmas. Perennial plant with subterranean stolons and rather broad, almost always flat leaf blades. . . . . . . . . . . . . . . . . . . . . . . . . . . 7. **ARCTAGROSTIS** GRISEB.
   – Spikelets *weakly laterally compressed* or even flattened on back; glumes *always equal to spikelet in length*; lemmas *usually shorter than spikelets*, more rarely equalling them in length, *without raised keel on back or with weakly indicated keel*, over whole outer surface (except callus) *bare and smooth or more or less scabrous, but then mainly on distal part of lemmas.* . . . . . . . . . . . . . . . . . . . . . . . . . . . . . . . . . . . . . . . . . . . . . . . . . . . . . . . . 20.

20. Spikelets 2–9 mm long, aggregated in condensed (often almost spikelike) or more or less diffuse paniculate inflorescence; lemmas almost always awned, basally (on callus) *always with tufts of more or less long hairs which are equal to, longer than, or not less than 1/6 - 1/3 as long as the lemmas*; at the base of the single floret there is *almost always* a well expressed remnant of the rachis of the spikelet in the form of a long-pubescent rachilla appressed to the outer surface of the palea; rarely this is absent, but then the callus hairs are very long, equalling or exceeding the lemma in length. . . . . . . . . . . . . . . . . . . . . . . . . 9. **CALAMAGROSTIS** ADANS.
   – Callus of lemmas *bare or with very short hairs* (1/10 - 1/8 as long as the lemmas); remnant of rachis of spikelet *always absent*. . . . . . . . . . . . . . . . . . . . . 21.

21. Glumes membranous, *without keel on back*; lemmas *ovoid, awnless, almost leathery, strongly indurate during fruiting*; paleas *slightly shorter than lemmas*, also leathery; spikelets 2.5–3.5 mm long, aggregated in lax effuse panicle. Leaf blades 3–12 mm wide. . . . . . . . . . . . . . . . . . . . 4. **MILIUM** L.
   — Glumes membranous, *with more or less expressed keel on back*; lemmas *broadly lanceolate, awned or awnless, membranous, not indurate during fruiting*; paleas *usually significantly shorter than lemmas (1/8 - 2/3 as long), sometimes absent*; spikelets 1.5–3 mm long, aggregated in more or less diffuse or condensed paniculate inflorescence. Leaf blades usually narrower. . . . . . . . . . . . . . . . . . . . . . . . . . . . . . . . . . . . . . . . . . . . . . . . . . . . . 8. **AGROSTIS** L.

22. Lemmas laterally compressed, carinate, awnless (but sometimes with mucro at tip up to 1 mm long), with 3–7 nerves of which the median is

raised in the form of a *sharp keel conspicuous along the whole length of the lemma*, but lateral nerves usually inconspicuous. .....................23.
- Lemmas rounded on back, *without keel or with keel weakly indicated distally*, rarely (genus *Trisetum*) with well expressed keel but then lemmas with long awns. .................................................26.

23. Panicles very dense, spikelike, with strongly abbreviated branchlets (longer of them ⅛-¼ as long as the whole panicle), which *together with the common rachis of the inflorescence are copiously covered with short hairs*; spikelets 3–6 mm long, with 2–4 florets, distinctly laterally compressed; lemmas usually slightly shorter than glumes, lanceolate, more or less pubescent or bare over almost whole surface, acuminate or obtuse at tip, with 3 nerves of which the lateral are scarcely evident, with very short stiff hairs on callus. Perennial plant, forming tufts or with short subterranean stolons. ........................................18. **KOELERIA** PERS.
- Panicle more or less diffuse or condensed (sometimes spikelike), with branchlets *bare and smooth or scabrous with acicules, but always without hairs*. ........................................................24.

24. Numerous spikelets 5–7 mm long grouped in bunches at tips of more or less abbreviated branches and secondary branchlets *on one side* of more or less lobed or almost spikelike inflorescence, each with 2–5 florets; branches arising successively and *singly* from only 2 sides of triangular common rachis of panicle; glumes and lemmas *lanceolate, long-acuminate at tip, often with mucro or short* (to 1 mm) *awn*. Rather tall perennial plant; leaf sheaths closed to middle or higher, on lower leaves flattened and keeled. ...........................................21. **DACTYLIS** L.
- Inflorescence otherwise, sometimes very dense and spikelike, but *not one-sided*; its lower branchlets *usually arranged in twos to fives per node, rarely singly*; lemmas *lanceolate to oblong-lanceolate, usually blunt or obtuse at tip*. ........................................................25.

25. Spikelets 2.5–7 mm long, with 2–5 florets or sometimes modified into vegetative buds (vivipary), aggregated in diffuse or more or less condensed (to almost spikelike) paniculate inflorescences with smooth or more or less scabrous branchlets; lemmas more or less pubescent, more rarely bare, with bare or pubescent callus, *often with sharply differentiated tuft of long crinkly hairs on their back but almost always* (except *Poa eminens*) *without hairs between bases of keel and marginal nerves*; paleas *pubescent on keel or more or less scabrous with* (sometimes very *short*) *acicules*; segments of rachis of spikelet short, with well expressed joints. Perennial plant, forming tufts or with long subterranean stolons. ....................22. **POA** L.
— Spikelets usually 4.5–9 mm long, with 2–5 florets, always normally developed, aggregated in diffuse paniculate inflorescence with completely smooth branchlets; lemmas with broadly hyaline margins, on basal half especially on nerves with rather copious and long pubescence, with distal part of callus (*including that adjoining internerves*) also copiously covered

with rather long crinkly hairs which do *not*, however, *form a sharply differentiated tuft of long crinkly hairs on its back*; paleas *bare and smooth on keels*; segments of rachis of spikelet elongate, with relatively weakly developed joints. Perennial plant with subterranean stolons; leaf blades without acicules on margin, greyish green above. . . . . . . . . . . . .25. **COLPODIUM** TRIN.

26. Lemmas *broadly lanceolate to oblong-ovoid, always blunt at tip, awnless, with 7–9 conspicuous raised nerves*, bare but usually more or less scabrous with acicules. Perennial plant with long subterranean stolons/rhizomes, not forming dense tufts; leaf sheaths closed for almost whole length (fused at margins). . . . . . . . . . . . . . . . . . . . . . . . . . . . . . . . . . . . . . . . . . . . . . . . . . .27.
– Lemmas *with 3–5 usually weak nerves, more rarely with 7–9 also inconspicuous nerves but then lemmas of different shape, often awned.* . . . . . .28.

27. Spikelets 6–9 mm long, containing (*besides the 1–2 lower bisexual florets*) *a club-shaped appendage formed by abortive lemmas* situated at the tip of the rachis of the spikelet; inflorescence a lax one-sided raceme or panicle with relatively few (5–18), more or less nodding spikelets on short thin pedicels; glumes *slightly shorter than spikelets, with 3–5 nerves.* . . . . . . . . . .
. . . . . . . . . . . . . . . . . . . . . . . . . . . . . . . . . . . . . . . . . . . . . .19. **MELICA** L.
— Spikelets 3–8 mm long, *aggregated in rather lax* paniculate inflorescence, *with 2–8 florets*; glumes *significantly shorter than spikelets, with 1 nerve.* . .
. . . . . . . . . . . . . . . . . . . . . . . . . . . . . . . . . . . . . . . . . . . . . .27a. **GLYCERIA** R. BR.

28. Spikelets (excluding awns) *1–2.5 cm long*, aggregated in condensed or more or less diffuse panicle or raceme with relatively few spikelets, with 4–10 florets; lemmas oblong-ovoid or oblong-lanceolate, 7–15 mm long (excluding awns), with 5–9 inconspicuous nerves and usually with straight or weakly bent awn arising near its tip, rarely awnless; leaf sheaths *closed for almost whole length* (fused at margins); leaf blades usually flat and rather broad, more rarely loosely folded lengthwise. . . . . . . . . . . . . . . . . . .29.
– Spikelets usually smaller, *up to 1 cm long, more rarely larger* (to 2 cm long) but *then* leaf sheaths *closed from base to less than half their length* (usually to ⅛-⅓). . . . . . . . . . . . . . . . . . . . . . . . . . . . . . . . . . . . . . . . . . . . . . . . . .30.

29. *Annual* plant. Lower glume *with 3–5*, upper *with 5–7 nerves.* . . . . . . . . . . . . .
. . . . . . . . . . . . . . . . . . . . . . . . . . . . . . . . . . . . . . . . . .31. **BROMUS** L.
– *Perennial* plant. Lower glume *with 1*, upper *with 3 nerves.* . . . . . . . . . . . . . . .
. . . . . . . . . . . . . . . . . . . . . . . . . . . . . . . . . . . . . . . . .30. **ZERNA** PANZ.

30. Lemmas *with more or less geniculate, more rarely straight awns arising from their back always distinctly below their tip*, bare but usually with more or less long stiff hairs on callus; awns rarely completely reduced (in one species of *Deschampsia* which forms dense tufts without subterranean stolons), but then callus of lemmas with numerous rather long hairs (⅕-⅓ as long as lemmas). . . . . . . . . . . . . . . . . . . . . . . . . . . . . . . . . . . . .31.

– Lemmas *awnless or with straight awns arising from their tip* (i.e. tip of lemma directly transformed into awn), bare or more or less pubescent. ...................................................................34.

31. Spikelets *1–2 cm* long, aggregated in racemose or paniculate inflorescence *with relatively few spikelets*; 2–5 florets in spikelet; lemmas on back with long awn which is more or less geniculate and twisted on its lower part. Perennial plant, forming rather dense tufts without subterranean stolons. . .............................................14. **HELICTOTRICHON** BESS.
– Spikelets *smaller and usually more numerous*. .......................32.

32. Spikelets 4–6 mm long, aggregated in more or less diffuse panicle with relatively few spikelets, always with 2 florets; glumes equal to spikelet in length and completely covering both florets; lemmas *broadly ovoid, ⅖-½ as long as glumes*, with short, more or less geniculate awn (not exserted from spikelet) arising from back of lemma usually slightly above its middle, *basally* (on callus) *with rather long hairs* (½-⅔ as long as lemma). Perennial plant with subterranean stolons, frequently forming small tufts. ..................................................11. **VAHLODEA** FRIES.
— Spikelets 2.5–8 mm long; lemmas *oblong or oblong-lanceolate, more or less equal in length to glumes or no less than ⅔ as long as them, basally with shorter hairs* which are usually significantly shorter than half the length of the lemma. Perennial plant, usually forming more or less dense tufts. . .33.

33. Spikelets with 2–4 florets; lemmas laterally compressed, *keeled, with very short hairs on callus* (¹⁄₁₂-⅐ as long as lemmas, sometimes completely absent) but rachis of spikelet long-haired; awns more or less geniculate, *always far exserted* (by 2–6 mm) *from spikelet*, arising from back of lemmas always above their middle; panicles usually more or less condensed, often spikelike. .........................................13. **TRISETUM** PERS.
– Spikelets almost always with 2 florets; lemmas *rounded on back with relatively long* (⅕-⅓ as long as lemmas) *hairs on callus*; awns straight or weakly geniculate, *relatively short* (exserted from spikelet by no more than 2 mm), arising from back of lemmas below or above their middle (in latter case awns very short, not or scarcely exceeding tip of lemma, sometimes completely absent); panicles usually more or less diffuse, more rarely condensed, sometimes with spikelets modified into vegetative buds (vivipary). ................................................12. **DESCHAMPSIA** BEAUV.

34. Spikelets 4.5–7.5 mm long, aggregated in condensed (or more rarely more or less diffuse) paniculate inflorescence with smooth, often strongly abbreviated branchlets, *each with (1)2(3)* developed florets; glumes almost entirely hyaline, broadly lanceolate, *more or less equal to spikelet in length and usually significantly (often 1½ times) longer than lemmas of lower floret*; lemmas oblong-ovoid, with broad hyaline margin, bare or more or less pubescent mainly on nerves, on callus with rather numerous hairs up to 1.2 mm long. Perennial plant 10–40 cm high, with long subterranean

stolons (rhizomes), usually not forming tufts; leaf blades bare and smooth, rarely with scattered, very short hairs above; leaf sheaths usually closed from base to more than half their length . . . . . . . . . . . . . 23. **DUPONTIA** R. BR.
- Spikelets *with 2–8 developed florets; glumes significantly shorter than spikelet and always shorter than lemmas.* . . . . . . . . . . . . . . . . . . . . . . . . . . . .35.

35. Lemmas herbaceous, *almost without hyaline border,* bare or uniformly pubescent almost over whole surface, smooth or more or less scabrous, *gradually narrowing at tip and usually transforming into mucro or awn, more rarely awnless;* spikelets 4–15 mm long, sometimes modified into vegetative buds (vivipary). Caryopses usually *more or less fused with lemmas and paleas,* on ventral surface *with groove or furrow;* hilum oblong to linear, weakly coloured. Perennial plant, forming dense tufts or with subterranean stolons; leaf blades flat or folded lengthwise, frequently very narrow, setaceous. . . . . . . . . . . . . . . . . . . . . . . . . . . . . . . . . . . . 29. **FESTUCA** L.
- Lemmas on margin, at least distally, *with broad hyaline border,* bare or more or less pubescent basally (mainly on nerves), *usually blunt or obtuse at tip, rarely acutish but always without mucro or awn;* spikelets 4–10 mm long. Caryopses *free,* ellipsoid, *without groove or furrow* on ventral surface; hilum small, basal, usually broadly oval, rarely oblong, dark coloured. . .36.

36. Perennial plant *with long subterranean stolons (rhizomes),* not forming tufts; leaf blades 2–8 mm wide, usually flat; sheaths *always closed from base to more than half their length.* Panicles more or less diffuse or condensed, with completely smooth branchlets; lemmas and paleas bare and smooth, only callus of lemmas with a few short hairs. . . . . . . . . . . . . . . . . . . .
. . . . . . . . . . . . . . . . . . . . . . . . . . . . . . . . . . . 24. **ARCTOPHILA** (RUPR.) ANDERSS.
- Perennial (but often short-lived) plant, forming more or less dense tufts *without subterranean stolons,* but sometimes with epigeal prostrate vegetative shoots; leaf blades usually 1–3 mm wide, flat or folded lengthwise; sheaths *closed from base to no more than one-third of their length.* Panicles more or less diffuse or condensed with smooth or scabrous branchlets; lemmas usually more or less pubescent basally, rarely bare; paleas usually more or less scabrous or pubescent on keels, rarely bare and smooth. . . . . .
. . . . . . . . . . . . . . . . . . . . . . . . . . . . . . . . . . . . . . . . . . . . . .28. **PUCCINELLIA** PARL.

## GENUS 1 — Typhoides Moench (Digraphis Trin.) — REED CANARY GRASS

MONOTYPIC GENUS WHOSE sole species is widely distributed in extratropical latitudes of the Northern Hemisphere and introduced to the Southern Hemisphere.

1. ***Typhoides arundinacea*** (L.) Moench, Math. pl. (1794) 202.
   *Phalaris arundinacea* L., Sp. pl. (1753) 55; Hultén, Atlas, map 136.
   *Digraphis arundinacea* Trin., Fund. Agrost. (1820) 127; Grisebach in Ledebour, Fl. ross. IV, 454; Schmidt, Fl. jeniss. 129; Scheutz, Pl. jeniss. 192; Krylov, Fl. Zap. Sib. II, 153; Perfilev, Fl. Sev. I, 69; Rozhevits in Fl. SSSR II, 55; Leskov, Fl. Malozem. tundry 18; Hultén, Fl. Al. II, 131; Kuzeneva in Fl. Murm. I, 137; Karavayev, Konsp. fl. Yak. 49.
   Ill.: Fl. Murm. I, pl. XLII.

   Meadow plant of the temperate (especially northern temperate) zone, occurring with particular frequency in well-drained areas on river banks. May form pure stands or mixed stands with other grasses. Growing here and there in river valleys and in the southern fringe of the tundra zone.

   **Soviet Arctic.** Murman; Timanskaya Tundra (valleys of Pesha and Indiga); Eastern Bolshezemelskaya Tundra (upper course of Usa, upper reaches of Adzva); lower reaches of Yenisey (north to Zaostrovskoye settlement); Koryakia (on Bay of Korf).
   **Foreign Arctic.** Arctic Scandinavia.
   **Outside the Arctic.** Europe, except southern fringes; forest zone of European part of USSR, Siberia, the Soviet Far East and North America; in Asia distributed south to Northern Iran, Syria, Northern Mongolia, the Korean Peninsula and Japan. As an introduction in extratropical countries of the Southern Hemisphere.

## GENUS 2 — Anthoxanthum L. — VERNAL-GRASS

SMALL GENUS (about 5 species) distributed in temperate and subtropical latitudes of the Northern Hemisphere. One species, whose distribution is subarctic-alpine in general character, is rather widespread in the Arctic.

1. ***Anthoxanthum alpinum*** A. et D. Löve in Atvinnud. Haskol. Univ., Inst. appl. sc., Dept. Agricult., Rep. ser. 13, 3 (1948), 67; Golubtsova in Uch. zap. LGU, ser. biol. 23, 44; Shlyakova in Fl. Murm. I, 140; Jørgensen & al., Fl. pl. Greenl. 10.
   *A. odoratum* ssp. *alpinum* Löve — Böcher & al., Grønl. Fl. 299.
   *A. odoratum* var. *glabrescens* Celak.— Krylov, Fl. Zap. Sib. II, 157.
   *A. odoratum* L. — Ruprecht, Fl. samojed. cisur. 66; Lange, Consp. fl. groenl. 157; Gelert, Fl. arct. 1, 96; Andreyev, Mat. fl. Kanina 156; Tolmachev, Fl. Kolg. 13; Devold & Scholander, Fl. pl. SE Greenl. 133; Seidenfaden & Sørensen, Vasc. pl. NE Greenl. 179; Rozhevits in Fl. SSSR II, 56, pro parte; Perfilev, Fl. Sev. I, 69, pro maxima parte; Leskov, Fl. Malozem. tundry 18; Grøntved, Pterid. Spermatoph. Icel. 132; Hultén, Atlas, map 137; Polunin, Circump. arct. fl. 39, pro maxima parte.

   Subarctic-alpine species, particularly common in the European sector of the Arctic. Occurring most frequently in meadows, on slopes mainly of southern exposure, on fluvial deposits, and sometimes in grassy clearings within tall willow thickets. Common in most districts of the European Arctic, but nowhere an abundant plant playing an important role in the formation of plant communities.

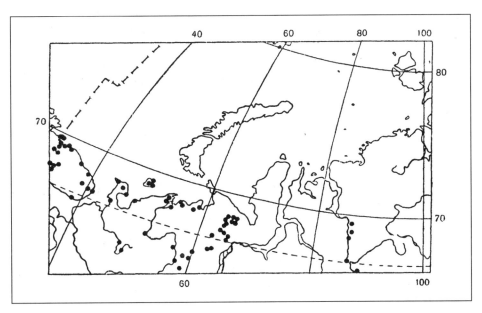

**Map II–1** Distribution of *Anthoxanthum alpinum* A. et D. Löve.

**Soviet Arctic.** Murman (more or less ubiquitous); Kanin; Timanskaya and Malozemelskaya Tundras; Kolguyev (rather common); Bolshezemelskaya Tundra (more or less ubiquitous, common); Polar Ural north to Mount Minisey; lower reaches of Yenisey near forest boundary (Dudinka, Khantayka). (Map II–1).

**Foreign Arctic.** SW and SE Greenland (not north of 64°); Iceland; Arctic Scandinavia.

**Outside the Arctic.** Widespread in northern and mountainous districts of Fennoscandia, and in the mountains of Central and parts of Southern Europe; occurring in the north of the European part of the USSR throughout the Kola Peninsula, on the shores of the White Sea, in the basins of the Mezen and the Pechora, and in the Northern Ural; apparently absent from the West Siberian Lowland; appearing again in the valley of the Yenisey, for whose basin there are records for districts extending from the forest-tundra to the southern taiga zone (mainly from the right bank and right tributaries); common in the Altay, Sayans and the mountains of Prebaikalia; becoming rarer east of Baykal, distributed east to Nerchinsk; the sole representative of the genus in the mountains of Siberia, as in the North; occurring in the subalpine zone of the Caucasus, the Tien Shan and the mountains of Eastern Kazakhstan.

The typical *A. odoratum* L., of which the species occurring in the Arctic was long considered a form (for the most part without any special designation), is well distinguished by the presence of more or less uniform, readily visible pile on the leaves, their sheaths, the rachis of the spikelets and the glumes. Due to the absence of pile, *A. alpinum* appears brighter green and its inflorescence more shining than in *A. odoratum*. The latter occurs nowhere in the Arctic. It is recorded from here and there in the northern forest zone, most often as an introduction.

## GENUS 3. **Hieröchloë** R. Br. — SWEET GRASS

GENUS CONTAINING ABOUT 30 species distributed in temperate and high latitudes of both hemispheres. Represented in the Arctic by three species, one of which (a plant whose general distribution has a boreal character) only penetrates parts of our region relatively close to the northern limit of forest but is rather characteristic of certain temperate arctic districts. Two other species, *H. alpina* and *H. pauciflora*, belong among the grasses most characteristic of the Arctic. The first of these possesses a circumpolar distribution, and is widespread in mountains of the temperate zone as well as in the Arctic. The second just exits arctic limits on the shores of the Pacific Ocean, and is characteristic of the Siberian and American Arctic.

1. Upper glumes *with conspicuous straight or geniculate awn exserted from spikelet.* Panicle more or less condensed or diffuse, about 3 cm long, *with few (9–15) spikelets* about 6 mm long. Culms 10–25 cm high, basally with shining lilac sheaths. Forming rather dense, small, usually somewhat asymmetrical tufts. Plant of dry, mainly alpine tundras. ................................. .............................................1. **H. ALPINA** (LILJEBL.) ROEM. ET SCHULT.
- Glumes *awnless* or with small mucro. Culms solitary or growing in groups, not forming tufts. ....................................................................2.

2. Small (10–20 cm high) plant *with narrow oblong one-sided panicle, about 2 cm long, with few (4–8) spikelets*; length of spikelets about 4 mm. Leaves narrow, more or less involute. Prostrate sterile shoots often formed at base of culm. Plant of moist lowland tundras. ..........2. **H. PAUCIFLORA** R. BR.
- Taller (20–40 cm), bright green, straight-stemmed plant with flat leaves. Panicle *more or less pyramidal, diffuse or somewhat condensed, 4–6(7) cm long, with numerous (25–50, sometimes more) spikelets* 4.5–5 mm long. Plant of mixed herb-grass meadows and tall shrub thickets. ............... ...............................................................3. **H. ODORATA** (L.) WAHLB.

   *1. Hieröchloë alpina* (Liljebl.) Roem. et Schult., Syst. veg. II (1817), 514; Grisebach in Ledebour, Fl. ross. IV, 408; Ruprecht, Fl. samojed. cisur. 66; Trautvetter, Pl. Sib. bor. 140; id., Fl. rip. Kolym. 77; Schmidt, Fl. jeniss. 128; Kjellman, Phanerog. sib. Nordk. 118; Scheutz, Pl. jeniss. 188; Lange, Consp. fl. groenl. 157; Gelert, Fl. arct. 1, 97; Simmons, Survey Phytogeogr. 42; M. Porsild, Fl. Disko 32; Lynge, Vasc. pl. N. Z. 109; Krylov, Fl. Zap. Sib. II, 160; Tolmachev, Fl. Kolg. 13; id., Fl. Taym. I, 91; id., Obz. fl. N. Z. 149; Petrov, Fl. Yak. I, 123; Perfilev, Fl. Sev. I, 70; Devold & Scholander, Fl. pl. SE Greenl. 142; Sørensen, Vasc. pl. E Greenl. 139; Seidenfaden & Sørensen, Vasc. pl. NE Greenl. 93, 179; Rozhevits in Fl. SSSR II, 60; id., Areal I, 23; Leskov, Fl. Malozem. tundry 18; Hultén, Fl. Al. II, 132; id., Atlas, map 138; Tikhomirov, Fl. Zap. Taym. 20; Kuzeneva in Fl. Murm. I, 142; A. E. Porsild, Ill. Fl. Arct. Arch. 22; Jørgensen & al., Fl. pl. Greenl. 12; Karavayev, Konsp. fl. Yak. 49; Böcher & al., Grønl. Fl. 299; Polunin, Circump. arct. fl. 55.
   *Aira alpina* Liljebl., Utk. Sv. Fl. (1792) 49.
   Ill.: Fl. arct. 1, 97, fig. 75; A. E. Porsild, l. c. 23, fig. 3,c.

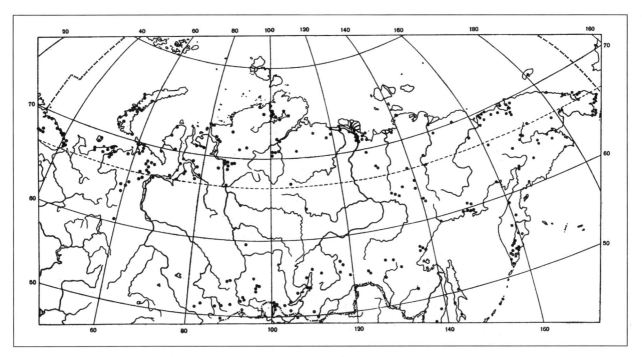

**Map II–2** Distribution of *Hierochloë alpina* (Liljebl.) Roem. et Schult.

Widely distributed arctic-alpine species. Characteristic plant of dry alpine tundras, growing in well-drained stony and sandy areas, sometimes in places almost devoid of snow cover during wintertime. Sometimes growing in large quantity and, for example in the East of Novaya Zemlya, included among the species of plants characterizing stony polygonal tundra. Here it is often associated with *Luzula confusa* and *L. nivalis*, sometimes forming together with these and certain other plants (*Potentilla emarginata, Oxytropis sordida*, etc.) patches of more or less closed vegetation alternating with more exposed areas of incompletely vegetated stony ground. Generally occurring in relative abundance under typically arctic conditions, becoming rare on the southern fringes of the Arctic Region. But also not very characteristic of districts of the true High Arctic. On the southern fringes of the Arctic most often found on sandy or sandy-stony crests of tundra ridges, sometimes in places with windblown soil. Becoming commoner on the fringe of the Arctic in typically alpine districts. Outside arctic limits a typical plant of many barrens in East Siberia and the Far East.

**Soviet Arctic.** Murman; Timanskaya and Malozemelskaya Tundras (relatively rare, not recorded for Kanin); Kolguyev; Bolshezemelskaya Tundra (common in northern districts); Pay-Khoy; Vaygach; Novaya Zemlya (north to Admiralteystvo Peninsula and Sedov Bay, very common at middle latitudes); Polar Ural; Yamal; Obsko-Tazovskiy Peninsula and Gydanskaya Tundra; lower reaches of Yenisey; Taymyr (north to Miller Bay and the lower reaches of the Lower Taymyra); lower reaches of Anabar and Olenek; lower reaches of Lena (but not observed in outer part of Lena Delta); lower reaches of Yana and Indigirka; lower reaches of Kolyma and the coast to the east of it (common); all Chukotka; Wrangel Island; Anadyr Basin; Koryakia. (Map II–2).

**Foreign Arctic.** Arctic Alaska and Canada (common); Canadian Arctic Archipelago (mainly in the east, reaching Ellesmere Island in the north); Greenland, except extreme south (in the north reaching the southern fringe of Peary Land); Spitsbergen (Western); Arctic Scandinavia.

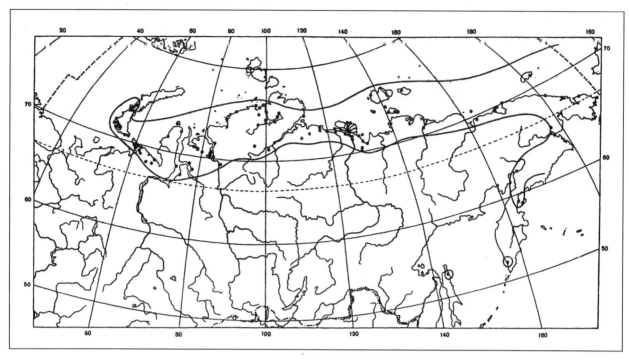

MAP II-3 Distribution of *Hierochloë pauciflora* R. Br.

**Outside the Arctic.** Mountains of Swedish Lappland and central part of Kola Peninsula; Northern Ural; elevated northern part of Central Siberian Plateau (in the south reaching the lower reaches of the Lower Tunguska and the upper reaches of the Vilyuy); summits of Yenisey Ridge; Verkhoyansk-Kolymsk mountain country; mountains of Kamchatka; Okhotsk coast; mountains of Lena-Amur watershed, Prebaikalia, Northern Mongolia; Sayans; Altay; isolated on barrens in the Sikhote-Alin, Sakhalin and Northern Japan; in America widespread in Alaska and subarctic districts of Canada but not extending very far to the south in the Rocky Mountains.

2. ***Hierochloë pauciflora*** R. Br. in Suppl. to App. Parry's Voyage, XI (1824), 293; Grisebach in Ledebour, Fl. ross. IV, 407; Trautvetter, Syll. pl. Sib. bor.-or. 63; Schmidt, Fl. jeniss. 128; Kjellman, Phanerog. sib. Nordk. 118; Scheutz, Pl. jeniss. 188; Gelert, Fl. arct. 1, 98; Simmons, Survey Phytogeogr. 43; Lynge, Vasc. pl. N. Z. 110; Krylov, Fl. Zap. Sib. II, 160; Petrov, Fl. Yak. I, 122; Tolmachev, Fl. Taym. I, 92; id., Obz. fl. N. Z. 149; Perfilev, Fl. Sev. I, 70; Rozhevits in Fl. SSSR II, 61; id., Areal I, 23; Hultén, Fl. Al. II, 135; Tikhomirov, Fl. Zap. Taym. 20; A. E. Porsild, Ill. Fl. Arct. Arch. 23; Karavayev, Konsp. fl. Yak. 49; Polunin, Circump. arct. fl. 57.

*H. racemosa* Trin. — Trautvetter, Fl. taim. 17.

Ill.: Fl. arct. 1, 97, fig. 76; Porsild, l. c. 23, fig. 3,b.

Plant characteristic of the Siberian and American Arctic and scarcely occurring beyond the limits of our region. Growing in wet (for the most part excessively wet) lowland sites, especially in polygonal tundra marshes where it grows in flat mossy areas, sometimes together with *Dupontia Fisheri, Carex stans* and other species of marshy tundra. Occurring locally in considerable quantity. Characteristic of districts with typically arctic floral composition. Not reaching parts of the region at higher latitude, possibly because of lack of development there of phytocenotic conditions suited to the needs of the species. The furthest northern localities are situated near the 74th to 76th parallels. Also becoming rare with approach to the northern limit of forest and corresponding change in the vegetation of marshy tundra.

**Soviet Arctic.** Novaya Zemlya (mainly in the southern half of the South Island, rare further north but reaching Krestovaya Bay on the west coast); Vaygach; Pay-Khoy; tundras at northern extremity of Polar Ural; Yamal; Obsko-Tazovskiy Peninsula; Gydanskaya Tundra; lower reaches of Yenisey; Taymyr (north to the "Zari" anchorage and the mouth of the Lower Taymyra); Preobrazheniye Island; lower reaches of Anabar and Olenek; fringes and islands of Lena Delta; coast between the Lena and the Yana; New Siberian Islands; coast east of the mouth of the Kolyma and islands adjacent to it; Wrangel Island; Chukotka (mainly in the north); lower reaches of Anadyr; district of mouth of Penzhina. (Map II–3).

**Foreign Arctic.** Arctic Alaska; arctic coast of Canada; northern extremity of Labrador; Canadian Arctic Archipelago north to Devon and Prince Patrick Islands.

**Outside the Arctic.** Reported as a rarity in the Karaginsk district, in extreme SW Kamchatka, and on the northern part of the east coast of Sakhalin. Records for Okhotsk are apparently erroneous.

3. ***Hierochloë odorata*** (L.) Wahlb., Fl. Ups. (1820) 32; Gelert, Fl. arct. 1, 98; Krylov, Fl. Zap. Sib. II, 158; Petrov, Fl. Yak. I, 128; Andreyev, Mat. fl. Kanina 159; Perfilev, Fl. Sev. I, 70; Rozhevits in Fl. SSSR II, 61; Leskov, Fl. Malozem. tundry 19; Gröntved, Pterid. Spermatoph. Icel. 140; Hultén, Fl. Al. II, 133; id., Atlas, map 140; Kuzeneva in Fl. Murm. I, 144; Löve & Löve, Consp. Icel. Fl. 83; A. E. Porsild, Ill. Fl. Arct. Arch. 23; Karavayev, Konsp. fl. Yak. 49; Polunin, Circump. arct. fl. 56; Böcher & al., Grønl. Fl. 299.

*H. borealis* Roem. et Schult. — Grisebach in Ledebour, Fl. ross. IV, 407; Ruprecht, Fl. samojed. cisur. 66; Trautvetter, Pl. Sib. bor. 139; id., Fl. rip. Kolym. 77; Scheutz, Pl. jeniss. 188.

*Holcus odoratus* L., Sp. pl. (1753) 1048.

Ill.: Porsild, l. c. 23, fig. 3, e.

Widely distributed boreal species, penetrating arctic limits (mainly in the European sector) sometimes to a considerable distance from the forest boundary, in this case retaining association with habitats supporting vegetational cover of a more temperate type. Growing in herbaceous openings within shrubbery, in meadows along the shores of rivers and streams, and on slopes of shorelines and ravines. Most often encountered in small quantity, but not a rarity where it occurs at all.

**Soviet Arctic.** Murmán; Kanin; coast of Cheshkaya Bay; Timanskaya and Malozemelskaya Tundras; Bolshezemelskaya Tundra (more or less ubiquitous); Pay-Khoy (north to River Oyu); coast of Baydaratskaya Bay; district of mouth of Ob; lower reaches of Yenisey, in forest-tundra and at short distance from the forest boundary (Tolstyy Nos, Dudinka, Khantayka, etc.); lower reaches of Kolyma; basin of River Velikaya, tributary of the Anadyr; Koryakia (district of Penzhina Bay and the Bay of Korf).

**Foreign Arctic.** Arctic Alaska; Labrador; west coast of Victoria Island (sole known locality within Canadian Arctic Archipelago); Iceland; Arctic Scandinavia.

**Outside the Arctic.** Widely distributed in the temperate zone of Europe, Asia and North America.

## GENUS 4    Milium L. — WOOD MILLET

SMALL GENUS DISTRIBUTED in temperate latitudes of the Northern Hemisphere. One species marginally penetrates the Arctic.

1. ***Milium effusum*** L., Sp. pl. (1753) 61; Grisebach in Ledebour, Fl. ross. IV, 444; Krylov, Fl. Zap. Sib. II, 183; Rozhevits in Fl. SSSR II, 119; Perfilev, Fl. Sev. I, 70; Leskov, Fl. Malozem. tundry 19; Gröntved, Pterid. Spermatoph. Icel. 141; Kuzeneva in Fl. Murm. I, 145; Hultén, Atlas, map 141; id., Amphi-atl. pl. 146.
   Ill.: Fl. Murm I, pl. XLV.

   Widely distributed boreal and nemoral forest plant, growing at the northern fringe of its range among shrub thickets. In the Arctic only near the southern fringe in the European sector, far from ubiquitous.

   **Soviet Arctic.** Murman; Kanin; Malozemelskaya Tundra; eastern Bolshezemelskaya Tundra (upper reaches of Usa); Polar Ural.

   **Foreign Arctic.** Iceland; Arctic Scandinavia.

   **Outside the Arctic.** Western Europe (except extreme south); greater part of European territory of USSR; Caucasus; southern West and Central Siberia; Preamuria, Korean Peninsula, Northern China; Japan, Sakhalin; Kamchatka; Northern Asia Minor; Tien Shan; Himalayas; NE North America.

## GENUS 5    Phleum L. — TIMOTHY

SMALL GENUS (no more than 20 species) distributed mainly in temperate and subtropical latitudes of the Northern Hemisphere, especially in the Old World (mainly in Europe and Western Asia). Penetrating the Southern Hemisphere in the mountains of Western America, ranging as far as subantartic districts. Represented in the Arctic by one native species of restricted distribution and one introduced species.

1. Inflorescence *oblong-ovoid or short-cylindrical, for the most part darkened (with lilac tint), with awned points of glumes protruding* and usually twice as long as glumes themselves. Sheath of uppermost culm leaf *strongly inflated.* Leaves *bare* or somewhat scabrous on margin. Plant 10–30 cm high. . . . . . . . . . . . . . . . . . . . . . . . . . . . . . . . . . . . . . . . . . . . . . . . . . . . .1. **PH. COMMUTATUM** GAUD.
– Inflorescence *long, narrowly cylindrical, green* (of same tone as whole plant). Glumes *with short awnlike mucro.* Sheath of uppermost leaf not inflated. Plant up to 50 cm or higher. (In Arctic only as introduction). . . . . . . . . . . . . . . . . . . . . . . . . . . . . . . . . . . . . . . . . . . . . . . . . . . . . . . . 2. **PH. PRATENSE** L.

   1. ***Phleum commutatum*** Gaud. in Alpina III (1808), 4; Jørgensen & al., Fl. pl. Greenl. 13; Böcher & al., Grønl. Fl. 297.
      *Ph. alpinum* ssp. *commutatum* (Gaud.) Hult., Atlas, map 142; id., Amphi-atl. pl. 234.
      *Ph. alpinum* var. *commutatum* (Gaud.) Mert. et Koch — Rozhevits in Fl. Az. Ross. 2, IV, 199; Krylov, Fl. Zap. Sib. II, 188; Leskov, Fl. Malozem. tundry 19.
      *Ph. alpinum* var. *americanum* Hult., Fl. Al. II, 136.
      *Ph. alpinum* L. — Grisebach in Ledebour, Fl. ross. IV, 458; Lange, Consp. fl. groenl. 155; Andreyev, Mat. fl. Kanina 159; id., Rast. vost. Bolshezem. tundry 72, fig. 15;

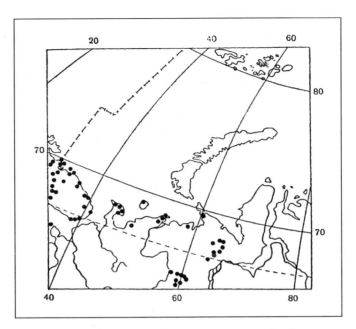

**Map II–4** Distribution of *Phleum commutatum* Gaud.

Devold & Scholander, Fl. pl. SE Greenl. 143; Seidenfaden, Vasc. pl. SE Greenl. 91; Perfilev, Fl. Sev. I, 71; Ovchinnikov in Fl. SSSR II, 135; Gröntved, Pterid. Spermatoph. Icel. 143; Kuzeneva in Fl. Murm. I, 146; Polunin, Circump. arct. fl. 60.
**Ill.:** Fl. Murm. I, pl. XLVI.

In recent special literature the name *Ph. commutatum* Gaud. (or *Ph. alpinum* ssp. *commutatum* Hult.) has been applied to the subarctic (and generally more widespread) form of *Ph. alpinum* L. s. l. We follow this usage and draw attention to the presence of rather distinct and in our opinion constant morphological differences between this plant and the Central European plant considered to be *Ph. alpinum* in the narrow sense. Recognition of *Ph. commutatum* as a separate species is supported by karyological data demonstrating its genetic distinctness, as well as by partial overlap of its range with that of *Ph. alpinum* s. str. although the two forms do not grow in immediately adjacent areas.

*Ph. commutatum* is a subarctic-alpine but far from circumpolar species, widespread in mountains of the temperate zone of the Northern Hemisphere and penetrating the Arctic in its Atlantic part. It grows on well-drained grassy slopes and in flat meadows, sometimes forming small almost pure stands. Within the European sector of the Soviet Arctic it is unevenly distributed, being absent from many districts adjacent to those where it has been recorded to occur. East of the Urals occurring nowhere in the Far North of the USSR.

**Soviet Arctic.** Murman; Northern Kanin; lower reaches of the Pesha; Timanskaya and Malozemelskaya Tundras; Kolguyev (rare, far from ubiquitous); Bolshezemelskaya Tundra (western part in district near Khaypudyrskaya Bay, eastern fringe on upper reaches of the Usa, Vorkuta); Polar Ural; Pay-Khoy (on coast of Yugorskiy Shar). (Map II–4).

**Foreign Arctic.** Labrador; South Greenland (from southern extremity to 70°N on both coasts); Iceland; Arctic Scandinavia.

**Outside the Arctic.** Northern Great Britain; central and northern zones of Fennoscandia; sporadically in NE of European part of USSR (in the Mezen and Pechora Basins); Northern Ural; higher alpine parts of the Alps (over 2500 m a.s.l.); Carpathians; mountains of Asia Minor; Caucasus; mountains of Southern Siberia,

Central Asia, Western China, Northern Korean Peninsula and Japan; Northern Kurile Islands; Kamchatka; Southern Alaska; mountains of Western North America; extreme NE Canada.

*Ph. alpinum* L. s. str. is distributed in the mountains of Central and Southern Europe (especially the Alps, but not at extremely high elevations). One of the races of *Ph. alpinum* s. l. occurs in the extreme south of South America, including Tierra del Fuego, and on the island of South Georgia.

2. **Phleum pratense** L. Sp. pl. (1753) 59; Grisebach in Ledebour, Fl. ross. IV, 457; Krylov, Fl. Zap. Sib. II, 187; Orchinnikov in Fl. SSSR II, 132; Perfilev, Fl. Sev. I, 71; Gröntved, Pterid. Spermatoph. Icel.; Hultén, Atlas, map 146; Kuzeneva in Fl. Murm. I, 145; Böcher & al., Grønl. Fl. 297.

Widespread European meadow plant, often used for hay and in many cases becoming naturalized. Sown locally and able to become naturalized also in far northern districts.

**Soviet Arctic.** Murman (Pechenga, district of Kola Bay, Teriberka, Kildin); River Pesha (within the forest-tundra).

**Foreign Arctic.** Extreme south and SW Greenland; Iceland; Arctic Scandinavia.

**Outside the Arctic.** Widespread throughout almost the whole of Europe, and distributed as an object of cultivation (and in many places naturalized) throughout the temperate and subtropical zones of the Northern Hemisphere, as well as in countries of the Southern Hemisphere.

## GENUS 6    Alopecurus L. — FOXTAIL

Relatively small (about 60–70 species), morphologically very distinctive genus, widely distributed in cold and temperate countries of both hemispheres, penetrating deeply into subtropical and tropical countries on mountain ranges. Of the 8 species reported in the Soviet Arctic, only one (*A. alpinus* Sm.) is a species with circumpolar distribution very characteristic of the Arctic. Approaching this species are certain very close but more southern species sometimes united with *A. alpinus* as subspecies or varieties. Species of two other groups represented in the Arctic, which may be united under the names *A. pratensis* L. s.l. and *A. geniculatus* L. s.l., also possess more southern ranges and only just penetrate the Arctic. The majority of species of the genus grow in various meadow or marsh-meadow communities and only certain species of alpine districts occur on screes, cliffs and stony or rocky slopes.

1. Plant 10–30 cm high, usually with culms *more or less ascending and rooting at lower nodes*. Spikelike panicles *narrowly long-cylindrical,* 2–6 cm long and 3–5 mm wide; spikelets *2–3 mm long;* glumes blunt at tip, pubescent mainly on nerves and on their basal part; anthers *0.4–0.6 mm long.* (*A. geniculatus* L. s. l.). . . . . . . . . . . . . . . . . . . . . . . . . . . . . . . . . . . . . . . . . .2.
- Culms usually *erect, not rooting at lower nodes*. Spikelike panicles *short-cylindrical or ellipsoid,* 1–8 cm long and 5–15 mm wide; spikelets *2.5–5.5 mm long;* glumes obtuse or acute at tip; anthers *1.6–2.5 mm long.* . . . . . . 4.

2. Awns of lemmas arising near their base, slightly geniculate at level of their tip and *exserted from spikelets by 1–2 mm;* anthers *1.2–1.7 mm long*, yellow but often with violet tinge. . . . . . . . . . . . . . . . . . . . . . . . . .6. **A. GENICULATUS** L.
– Awns of lemmas more or less reduced, *virtually not exserted from spikelet or exserted by 0.5–1.5 mm but then anthers 0.4–0.7 mm long*, initially orange-yellow, later bright yellow. . . . . . . . . . . . . . . . . . . . . . . . . . . . . . . . . . . . .3.

3. Awns exserted from spikelet *by 0.8–1.5 mm;* anthers *0.4–0.7 mm long*. . . . . . . . . . . . . . . . . . . . . . . . . . . . . . . . . . . . . . . . . . . . . . . . . . . . . .8. **A. AMURENSIS** KOM.
– Awns exserted from spikelet *by no more than 0.8 mm*, usually much reduced; anthers *0.7–1.2 mm long*. . . . . . . . . . . . . . . . . .7. **A. AEQUALIS** SOBOL.

4.  Glumes *with rather copious long pubescence over entire outer surface;* spikelike panicles 1–4 cm long, broadly ellipsoid or short-cylindrical; anthers 1.6–3 mm long. (*A. alpinus* Sm. s. l.). . . . . . . . . . . . . . . . . . . . . . . . .5.
– Glumes usually *with long pubescence only on keel and lateral nerves*, more rarely also with very short inconspicuous pubescence at their extreme base outside the lateral nerves and sometimes on the surface between the keel and the lateral nerves; spikelike panicles 2–8 cm long, cylindrical; anthers 2–3.5 mm long. (*A. pratensis* L. s. l.). . . . . . . . . . . . . . . . . . . . . . . . . .7.

5. Plant *30–80 cm high;* culms *with 3–4 nodes*, of which the uppermost is usually situated *above the middle of the culm (rarely near the middle)*. Spikelike panicles usually short-cylindrical; spikelets 2.5–3.8 mm long; glumes usually less copiously pubescent; awns of lemmas usually well developed and far exserted from spikelet, more rarely much reduced. . . . . . 
. . . . . . . . . . . . . . . . . . . . . . . . . . . . . . . . . . . . . . . . . . . . . . .3. **A. GLAUCUS** LESS.
– Plant *6–50 cm high;* culms *with 2–3 nodes*, of which the uppermost is situated *below their middle, sometimes even close to their base*. Spikelike panicles broadly ellipsoid or short-cylindrical; spikelets 3–6 mm long; glumes copiously pubescent. . . . . . . . . . . . . . . . . . . . . . . . . . . . . . . . . . . . . . .6.

6. Plant 6–40 cm high; blades of culm leaves *1.5–4 mm wide;* spikelike panicles *1–3 cm long and 5–10 mm wide;* spikelets 3–4.5 mm long; lemmas long-awned or with much reduced awn not exserted from spikelet. . . . . . . . 
. . . . . . . . . . . . . . . . . . . . . . . . . . . . . . . . . . . . . . . . . . . . . . . . .4. **A. ALPINUS** SM.
– Plant 20–50 cm high; blades of culm leaves *3–8 mm wide;* spikelike panicles *1.2–2.5 cm long and 10–15 mm wide;* spikelets 4–6 mm long; lemmas usually with well developed awn far exserted from spikelet. . . . . . . . . . . . . . . 
. . . . . . . . . . . . . . . . . . . . . . . . . . . . . . . . . . . . . . . . . . 5. **A. STEJNEGERI** VASEY.

7. *More or less glaucous green* plant 30–80 cm high, with long stolons; spikelets 4–5.5 mm long; tips of glumes *slightly bent sideways, appearing divergent* (spikelets "urn-shaped"); lemmas usually with awn *arising near*

*their middle and much reduced* (not exserted from spikelet), more rarely with well developed long awn. Saline meadows of sea shore.
............................................2. **A. ARUNDINACEUS** POIR.

– *Usually pure green* plant 20–70 cm high, with less developed, usually shorter subterranean stolons; spikelets 3.5–5 mm long, usually less pubescent than in preceding species; tips of glumes *not bent sideways, appearing convergent* (spikelets "elliptical"); lemmas usually with awn *arising near their base and well developed* (far exserted from spikelet), very rarely with weakly developed awn. .........................1. **A. PRATENSIS** L.

1. ***Alopecurus pratensis*** L., Sp. pl. (1753) 60; Grisebach in Ledebour, Fl. ross. IV, 462; Gelert, Fl. arct. 1, 99; Lynge, Vasc. pl. N. Z. 101; Krylov, Fl. Zap. Sib. II, 191; Petrov, Fl. Yak. I, 142; Perfilev, Fl. Sev. I, 72; Ovchinnikov in Fl. SSSR II, 150; Leskov, Fl. Malozem. tundry 20; Hultén, Atlas, map 151; id., Fl. Al. II, 143; Kuzeneva in Fl. Murm. I, 151; Hylander, Nord. Kärlväxtfl. I, 328; Karavayev, Konsp. fl. Yak. 50; Polunin, Circump. arct. fl. 37.

Ill.: Fl. SSSR II, pl. XI, fig. 2,3; Fl. Murm., pl. XLVII, 1; Hylander, l. c., fig. 45, A, B; Polunin, l. c. 37.

Species widely distributed over a considerable part of Eurasia, but predominantly boreal, introduced (often intentionally) in many other extratropical countries. Arctic and high alpine specimens of this species often have small overall size, a shorter spikelike panicle and also less scabrous panicle branchlets and leaf blades, forming the morphologically very weakly differentiated ecological-geographical race known under the name *A. pratensis* var. *alpestris* Wahlenb. (Flora Lapponica, 1812, 21). This scarcely merits higher taxonomic rank, since there is a very gradual transition to typical specimens of *A. pratensis* in the south. Among more northern specimens of this variety there occasionally occur specimens with much reduced awns scarcely exserted from the spikelets. In this respect these approach the next species, differing from it mainly in the form of the tips of the glumes and the green (not more or less glaucous) colour of the plant as a whole.

Usually growing in meadows and on gravelbars on the shores of rivers, lakes and streams, more rarely in various low shrub communities and on relatively dry grassy slopes, sometimes on sands near the sea shore.

**Soviet Arctic.** Murman; Kanin (to extreme north); Malozemelskaya and Bolshezemelskaya Tundras; Polar Ural (frequent); Pay-Khoy; Vaygach and Kolguyev; Novaya Zemlya (on the South Island in the district of Matochkin Shar and Belushya Bay, on the North Island only on rocks on the Bay of Glazov); Yamal (north to Tiutey); Obsko-Tazovskiy Peninsula; lower reaches of Yenisey (north to Dudinka); Anadyr Basin (only near Markovo settlement as an introduction). (Map II–5).

**Foreign Arctic.** Iceland (as introduction); Arctic Scandinavia.

**Outside the Arctic.** Almost all Europe; Caucasus; Central Asia; West Siberia; East Siberia (east to Baykal and the Vilyuy Basin). In the Far East only as an introduction; as an introduced or imported plant also in a considerable part of North America, North Africa and many other countries both of the Northern and Southern Hemispheres.

2. ***Alopecurus arundinaceus*** Poir. in Lam., Encycl. VIII (1808), 776; Hultén, Atlas, map 148; Hylander, Nord. Kärlväxtfl. I, 330.

*A. ventricosus* Pers., Syn. I (1805), 80; Krylov, Fl. Zap. Sib. II, 194; Petrov, Fl. Yak. I, 142; Perfilev, Fl. Sev. I, 73; Ovchinnikov in Fl. SSSR II, 149; Kuzeneva in Fl. Murm. I, 150; Karavayev, Konsp. fl. Yak. 50; non Huds., 1778.

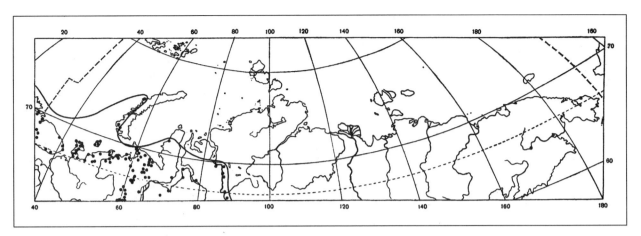

Map II-5 Distribution of *Alopecurus pratensis* L.

*A. ruthenicus* Weinm., Cat. Dorpat. (1810) 10; Grisebach in Ledebour, Fl. ross. IV, 463.
**Ill.:** Hylander, l. c., fig. 45, C, D.

Eurasian species, widely distributed in more or less saline meadows and marshes of the steppe and forest-steppe zones, also occurring in maritime saline meadows of Northern Europe and here penetrating arctic limits. The awns of the lemmas in specimens of this species are usually much reduced, but in many parts of its range specimens also occur with well developed awns far exserted from the spikelets. Such long-awned specimens, known under the name *A. arundinaceus* var. *exserens* (Griseb.) Marss. (Flora von Neu-Vorpommern und den Inseln Rügen und Usedom, 1869, 555) = *A. ruthenicus* var. *exserens* Griseb. (in Ledebour, l. c. 464), are almost just as common as short-awned specimens in the arctic part of the range and in many situations are possibly the result of hybridization between *A. arundinaceus* and *A. pratensis*. In the Arctic, as generally in Northern Europe, *A. arundinaceus* is a strictly littoral species, occurring away from the sea coast only as a rare introduction along roads and in settlements.

Growing in maritime saline meadows and marshes, sometimes also on sandbars and gravelbars on the sea shore, usually above high tide mark.

**Soviet Arctic.** Murman (sporadically along the whole shore); Kanin (only on the shore of Cheshkaya Bay and at the mouth of the Chizha); Malozemelskaya Tundra (a few sites on the sea shore).

**Foreign Arctic.** Arctic Scandinavia.

**Outside the Arctic.** Sea shores of Northern Europe (west to Denmark and northern Central Europe, further west only as an introduction); considerable part of the forest-steppe, steppe and semidesert zones of Eurasia from France to Prebaikalia and the Lena Basin (northern boundary of the species' distribution in Siberia reaching 65°N on the Ob, then shifting south to 60° in Krasnoyarsk Kray, but again reaching 66° in the Lena Basin); Algeria. Occurring as an introduction in many other countries both of Eurasia and North America.

**3. Alopecurus glaucus** Less. in Linnaea IX (1835), 206; Grisebach in Ledebour, Fl. ross. IV, 462; Petrov, Fl. Yak. I, 146; Ovchinnikov in Fl. SSSR II, 153; Hultén, Fl. Kamtch. I, 92; id., Fl. Al. II, 142.

*A. alpinus* var. *glaucus* (Less.) Kryl., Fl. Alt. (1914) 1580.

*A. tenuis* Kom. in Fedde, Repert. sp. nov. XIII (1914), 85; Ovchinnikov, l. c. 154.

*A. alpinus* var. *elatus* Roshev. in Fl. Az. Ross. 2, IV (1924), 221, pro parte.

*A. pseudobrachystachyus* Ovcz. in Fl. SSSR II (1934), 153.

*A. Roshevitzianus* Ovcz., l. c. 154.

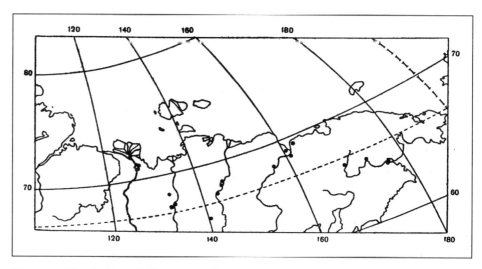

**Map II–6** Distribution of *Alopecurus glaucus* Less.

**Ill.**: Petrov, l. c., fig. 45; Fl. SSSR II, pl. X, fig. 6, 7, 12.

Species widely distributed in the Urals and mountainous districts of Siberia, here replacing the very close but more northern and alpine *A. alpinus*. Following Hultén we have chosen to retain *A. glaucus* in its wider sense, because the characters with which P.N. Ovchinnikov (Fl. SSSR II, 1934, 137-138) distinguished the species listed in synonymy (*A. tenuis, A. Roshevitzianus, A. pseudobrachystachyus* and *A. glaucus*) have proved far from constant in more extensive material and we have not been able to discover new, more stable characters. In the Urals, whence this species was described (from Mount Taganay), specimens with much reduced awns not exserted from the spikelets are rather common along with the typical long-awned form. Further east (through the Altay and Sayans) short-awned specimens become more and more rare, and in Yakutia and the Anadyr territory (where *A. glaucus* penetrates arctic limits) long-awned specimens occur almost exclusively. Despite the considerable overlap of the ranges of *A. alpinus* and *A. glaucus* in NE Siberia, the proportion of specimens transitional between them (probably of hybrid origin) is not that great there. Still rarer are specimens transitional between these species in the Urals, where *A. alpinus* occurs only at a few sites in the Northern Ural while *A. glaucus* is distributed almost exclusively in the mountains of the Middle and Southern Urals. Interrelations between *A. glaucus* and the closely related Prebaikalian species *A. brachystachyus* M.B. remain not entirely clear to us; the ranges of these species apparently still lack precise definition.

In the Arctic *A. glaucus* grows in meadowy areas of various kinds, commonly in marshy meadows on the shores of mountain streams and rivulets, more rarely in shrubbery or open forest.

**Soviet Arctic.** Lower reaches of Lena and northern part of Verkhoyansk Range; lower reaches of Kolyma (to mouth); Anadyr Basin. (Map II–6).

**Foreign Arctic.** Arctic Alaska (only in the district of the Bering Strait).

**Outside the Arctic.** Middle and Southern Urals (north to 60°); Altay and Tarbagatay; Sayan Mountains; mountains of East Siberia (SE to Bureinskiy Range); Kamchatka; Alaska; in the Rocky Mountains the very closely related species *A. occidentalis* Scribn. et Tweedy (according to Hultén not distinguishable from *A. glaucus*).

4. ***Alopecurus alpinus*** Sm., Fl. Brit. III (1804), 1386; Grisebach in Ledebour, Fl. ross. IV, 461; Lange, Consp. fl. groenl. 156; Gelert, Fl. arct. 1, 99; Lynge, Vasc. pl. N. Z. 100; Krylov, Fl. Zap. Sib. II, 123; Petrov, Fl. Yak. I, 147; Hanssen & Lid, Fl. pl. Franz Josef

L. 33; Scholander, Vasc. pl. Svalb. 64; Perfilev, Fl. Sev. I, 72; Ovchinnikov in Fl. SSSR II, 155; Tolmachev, Fl. Taym. I, 92; Hultén, Fl. Al. II, 140; A. E. Porsild, Ill. Fl. Arct. Arch. 24; Böcher & al., Grönl. Fl. 296; Polunin, Circump. arct. fl. 38.

*A. borealis* Trin., Fund. Agrost. (1820) 58; Ovchinnikov, l. c. 155; Karavayev, Konsp. fl. Yak. 50.

*A. alpinus* var. *borealis* (Trin.) Griseb. in Ledebour, l. c. 461.

*A. behringianus* Gand. in Bull. Soc. Bot. France LXVI (1920), 298.

?*A. altaicus* (Griseb.) Petr., Fl. Yak. I (1930), 146.

Ill.: Gelert, l. c., fig. 77; Petrov, l. c., fig. 46; Fl SSSR II, pl. X, fig. 10, 11; Porsild, l. c., fig. 3, g; Böcher & al., l. c., fig. 52, d; Polunin, l. c. 38.

Circumpolar high arctic species, displaying considerable polymorphism like the closely related *A. glaucus*. Typical Scottish specimens of this species are of relatively large size and possess short-cylindrical spikelike panicles almost always with short-awned spikelets (only occasionally do specimens occur with long awns far exserted from the spikelets). It is possible that these are not entirely identical with the specimens widespread in the Arctic and that with more careful study they will prove to belong to a separate species or subspecies, for which the name *A. alpinus* Sm. s. str. should be established. It is not fortuitous that A. Grisebach distinguished arctic specimens of *A. alpinus* under the name "*A. alpinus* var. *borealis* (Trin.) Griseb." from typical Scottish specimens (called by him "*A. alpinus* var. *scoticus* Griseb.") with the annotation that the latter variety "in Rossia non occurrit." The fact that *A. alpinus*, while present in Scotland, is completely absent from such relatively more northern countries as Iceland and Scandinavia, also attests to a certain distinctness of the typical Scottish race. High arctic specimens of *A. alpinus* are of small size and often possess more or less abbreviated spikelike panicles with much reduced awns on the lemmas. Long-awned specimens turn up almost everywhere, but their abundance noticeably increases in an eastward direction. Thus, among the material of *A. alpinus* from Novaya Zemlya present in the herbarium of the Botanical Institute of the USSR Academy of Sciences only a few of the specimens from Krestovaya Bay possess awns far exserted from the spikelets. In the New Siberian and Wrangel Islands long-awned specimens are more common, but here too specimens with short awns predominate. At the same time, in mainland districts of East Siberia long-awned specimens begin to noticeably predominate, while the short-awned almost completely disappear. On account of the great variation in awn length even within single populations, it is scarcely possible (following P. I. Ovchinnikov) to refer long-awned specimens to a separate species (*A. borealis* Trin.) and in our opinion it would be more correct in this case to distinguish only a variety, *A. alpinus* var. *borealis* (Trin.) Griseb. As we have already noted above, long- and short-awned specimens are almost equally common among Uralian material of *A. glaucus*, and likewise there is no basis for referring them to separate species. However, type specimens of *A. borealis* Trin. from St. Paul Island (like specimens of *A. behringianus* Gand. described from the same island) are not entirely identical with long-awned specimens of *A. alpinus* from the Arctic. With respect to the shape of the spikelike panicle and other characters, they somewhat deviate towards *A. Stejnegeri*. Specimens from more southern parts of the range of this species (from the Altay and the mountains of East Siberia) differ, like the Scottish, in the rather large size of the plant as a whole (up to 40-50 cm high), but almost always possess well developed long awns. Apparently they also form a weakly differentiated ecological-geographical race, appearing intermediate between *A. alpinus* and *A. glaucus*, with prior name at species rank [*A. altaicus* (Griseb.) Petr., based on *A. glaucus* var. *altaicus* Griseb.].

The role of *A. alpinus* in the vegetational cover of the Arctic noticeably increases from south to north (as may be said for *Deschampsia brevifolia, Phippsia algida*, etc.). Everywhere the species avoids oligotrophic communities, but colonizes com-

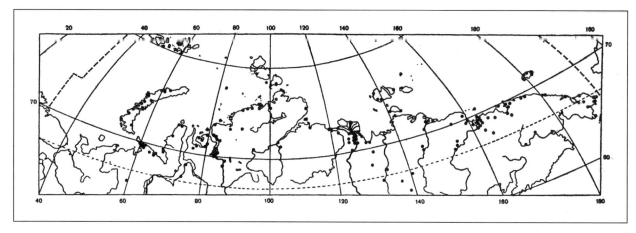

MAP II–7  Distribution of *Alopecurus alpinus* Sm.

paratively more fertile, rather moist and at the same time sufficiently aerated substrates of fine soil or fine soil and stones, especially in places protected by snow in winter. By forming subterranean stolons it normally grows as more or less extensive diffuse clones, only sometimes (for example, on moist fluvial alluvia) forming closed sward.

In the mainland tundras of East Siberia (where var. *borealis* generally predominates), the species is characteristic of nival meadows, moist gravelbars, moist shoreline slumps and shoreline mounds ("baydzharakhs"), more rarely occurring in patchy mossy tundras on well moistened slopes and in riverine herb-willow carrs (on sandy alluvium). It penetrates the most northern fringe of the forest zone in the littoral zone of the larger rivers, but further south is replaced there by *A. glaucus*.

In more northern districts of the polar coast of Siberia and on islands at high latitude (where var. *alpinus* with lemmas without or with non-exserted awns predominates), *A. alpinus* spreads out onto the watersheds and is locally one of the basic dominants of placorn patchy herb-moss high arctic tundras (for example, on islands of the New Siberian Archipelago); with its subteranean stolons and roots, the plant densely penetrates patches of incompletely vegetated loamy soil; it grows on the floodplains of rivers and streams, locally forming closed low sward. The anthocyanic tint of the plant as a whole much increases to the north.

**Soviet Arctic.** Karsk-Baydaratsk coast; Pay-Khoy; Vaygach; Novaya Zemlya (mainly on the North Island, south to Gusinaya Zemlya); Franz Josef Land; Yamal (south to Nakhodka Bay); Obsko-Tazovskiy Peninsula; Gydanskaya Tundra; lower reaches of Yenisey and islands in the Bay of Yenisey; Taymyr (south to the spurs of the Central Siberian Plateau); Severnaya Zemlya; Begichev Island; basins of the Anabar and Olenek; lower reaches of Lena north of Kyusyur and neighbouring mountains (Chekanovskiy Ridge, Kharaulakh Mountains); lower reaches of the Indigirka, Alazeya and Kolyma and neighbouring districts; New Siberian Islands; Chukotka north of the Anadyr watershed; Wrangel Island; district of Bering Strait; Anadyr Basin (rare, on higher areas of mountains). (Map II–7).

**Foreign Arctic.** Arctic Alaska (with St. Lawrence Island); arctic coast of Canada west of Hudson Bay (also Mackenzie Mountains); shores of Hudson Bay north of 60°; arctic part of Labrador (south to 56°N); Canadian Arctic Archipelago; almost all Greenland except its southeastern part (between 60° and 68°N); Spitsbergen and Bear Island.

**Outside the Arctic.** Scotland; Northern Ural (south to Muraveynyy Rock at about 61°30′N); Altay and Sayans (higher areas of mountains; specimens for the most part transitional to *A. glaucus*); northern fringe of Central Siberian Plateau; Verkhoyansk and Cherskiy Ranges (almost over their whole extent); Kamchatka (rare, in the Tigil Basin); Alaska. The very closely related *A. antarcticus* Vahl is found in Antarctic America.

5. ***Alopecurus Stejnegeri*** Vasey in Proc. U. S. Nat. Mus. X (1887), 153; Ovchinnikov in Fl. SSSR II, 156.
*A. alpinus* var. *Stejnegeri* (Vasey) Hult., Fl. Kamtch. I (1927), 90; id., Fl. Al. II, 141.
Ill.: Fl. SSSR II, pl. X, fig. 9.

Beringian species just penetrating the Arctic, very closely approaching *A. alpinus* and connected to it by rather numerous transitional specimens of hybrid origin. However in the Commander Islands and a large part of the Aleutians, where only this species occurs, it is very constant with respect to morphological characters and well distinguished from *A. alpinus*, as Hultén already noted. The differentiation of *A. Stejnegeri*, like that of *A. glaucus*, is not great, and both these species could be united with *A. alpinus* as subspecies.

Growing in marshy meadows on the shores of springs and streams, on gravelbars and clayey banks, sometimes near the sea coast.

**Soviet Arctic.** Chukotka Peninsula (Lawrence Bay — specimen of doubtful origin in the Mertens collections); Anadyr Basin (rather frequent); districts of Olyutorskiy and Karaginskiy Bays.

**Foreign Arctic.** St. Lawrence Island.

**Outside the Arctic.** Mountains of Okhotsk Coast between Okhotsk and the Bay of Gizhiga (mainly specimens transitional to related species); Kamchatka; Commander and Aleutian Islands.

6. ***Alopecurus geniculatus*** L., Sp. pl. (1753) 60; Grisebach in Ledebour, Fl. ross. IV, 464; Lange, Consp. fl. groenl. 156; Krylov, Fl. Zap. Sib. II, 197; Perfilev, Fl. Sev. I, 72; Ovchinnikov in Fl. SSSR II, 157; Hultén, Atlas, map 149; id., Fl. Al. X, Suppl. 1704; Kuzeneva in Fl. Murm. I, 151; Hylander, Nord. Kärlväxtfl. I, 331; Böcher & al., Grønl. Fl. 296; Polunin, Circump. arct. fl. 38.
Ill.: Fl. SSSR II, pl. X, fig. 2; Hylander, l. c., fig. 45.

European boreal species, only just penetrating the Arctic. As an importation or introduction this species is also distributed in many other countries both of the Northern and Southern Hemispheres.

Growing in marshy meadows, on more or less moistened sandy or clayey sites on the shores of waterbodies, sometimes as an introduction along roads and in settlements.

**Soviet Arctic.** Recorded for Murman (Pechenga district).

**Foreign Arctic.** SW Greenland (as introduction); Iceland; Arctic Scandinavia.

**Outside the Arctic.** Almost all Europe, except the more southern and southeastern part. As an imported or introduced plant in many districts of North America (including Southern Alaska), the Middle East, Japan, Australia and New Zealand.

7. ***Alopecurus aequalis*** Sobol., Fl. Petrop. (1799) 16; Petrov, Fl. Yak. I, 140; Ovchinnikov in Fl. SSSR II, 158; Hultén, Fl. Al. II, 139; id., Atlas, map 147; Kuzeneva in Fl. Murm. I, 152; Hylander, Nord. Kärlväxtfl. I, 331; A. E. Porsild, Ill. Fl. Arct. Arch. 24; Böcher & al., Grønl. Fl. 296; Karavayev, Konsp. fl. Yak. 50; Polunin, Circump. arct. fl. 38.
*A. aristulatus* Michx., Fl. Bor. Amer. I (1803), 43; Gelert, Fl. arct. 1, 100.
*A. fulvus* Sm. in Smith et Sowerby, Engl. Bot. XXI (1805), tab. 1467; Grisebach in Ledebour, Fl. ross. IV, 464; Krylov, Fl. Zap. Sib. II, 195; Perfilev, Fl. Sev. I, 72.
Ill.: Smith & Sowerby, l. c., tab. 1467; Petrov, l. c., fig. 43; Fl. Murm. I, pl. XLVII, 2; Hylander, l. c., fig. 45, f; Polunin, l. c. 38.

Boreal species widespread in cold and temperate countries of the Northern Hemisphere, almost entirely (excluding introductions) replacing *A. geniculatus* in Asia and North America. Specimens of this species from the eastern part of the Soviet Arctic differ somewhat from typical European specimens in having shorter (on average) anthers (0.7–0.9 mm long, not 0.9–1.2 mm as in typical specimens) and awns usually more or less exserted (by 0.2–0.8 mm) from the spikelets. Similar specimens are in general rather widespread both in the Siberian part of the species range (whence they are known under the name *A. fulvus* var. *sibiricus* Kryl.) and in North America, whence they have been described as a separate species (*A. aristulatus* Michx.) but can be distinguished only as a subspecies (*A. aequalis* ssp. *aristulatus* (Michx.) Tzvel. comb. nova). This subspecies occupies an apparently intermediate position between *A. aequalis* s. str. and the next species (*A. amurensis*), which could also be united with *A. aequalis* as a third subspecies.

Growing in wet sandy or silty sites on the shores of waterbodies, on gravelbars, on the margins of marshes and in marshy meadows, sometimes as an introduction along roads and in settlements.

**Soviet Arctic.** Murman (sporadically along the whole coast); lower reaches of Pechora (to its mouth); Bolshezemelskaya Tundra (eastern part); Polar Ural; Karsk-Baydaratsk Coast (at Kara); lower reaches of Ob (Salekhard district and on the Poluy); Anadyr and Penzhina Basins (ssp. *aristulatus*).

**Foreign Arctic.** Arctic Alaska (only Yukon Basin and Pribilof Islands); NW Canada (?); Labrador (to extreme north); Baffin Island (south of 71°N); Iceland; Arctic Scandinavia.

**Outside the Arctic.** Almost all Europe; Caucasus; Middle East; Central Asia; considerable part of Siberia (north to the lower reaches of the Ob and the Lower Tunguska, the upper reaches of the Olenek, the Vilyuy and Aldan Basins, and the upper reaches of the Indigirka and Kolyma); Kamchatka; Japan; North America (south to the states of California, New Mexico and Florida). As an introduction in many other countries of both hemispheres.

8. ***Alopecurus amurensis*** Kom. in Izv. Peterb. bot. sada XVI (1916), 151; Ovchinnikov in Fl. SSSR II, 158; Karavayev, Konsp. fl. Yak. 50.

Ill.: Fl. SSSR II, pl. X, fig. 3.

Eastern Asian species very close to *A. aequalis* ssp. *aristulatus* and connected to it by rather numerous transitional specimens and populations. Just penetrating the Arctic.

Like the preceding species, growing in wet sandy or gravelly sites on the shores of waterbodies, on the margins of marshes and in marshy meadows, sometimes occurring as a weed along roads and in settlements.

**Soviet Arctic.** Lower reaches of Anadyr (rare) and near the mouth of the Penzhina (Kamenskoye settlement).

**Foreign Arctic.** Not found.

**Outside the Arctic.** Transbaikalia and Preamuria; Primorskiy Kray; Yakutia (only east of the Lena, north to the upper reaches of the Yana and Indigirka); Okhotsk Coast and Kamchatka (sporadically); Sakhalin; Japan; NE China; Korean Peninsula.

## GENUS 7 — **Arctagrostis** Griseb. — ARCTAGROSTIS

NOTWITHSTANDING THAT THE number of described species in this small arctic-alpine genus has now reached 17, only two of them (both occurring in the territory of the USSR) are undoubtedly distinct although closely related species. One further species, *A. poaeoides* Nash, so far known only from Alaska and proposed by Hultén (Fl. Al. II, 148) as a distinct species, is in our opinion quite inadequately distinguished from *A. arundinacea* (Trin.) Beal and represents a variety of that species rather than a distinct species. The genus *Arctagrostis* was formerly referred to the tribe *Agrostideae* Kunth, but has recently been referred by Pilger (Das System der Gramineae, 1954) to the tribe *Festuceae* Nees and indeed shows an undoubted and rather close relationship to more "primitive" species of the genus *Poa* (for example, *P. eminens*), differing from these only in the complete reduction of the upper florets of the spikelet.

1. Plant 15–80 cm high; leaf blades 3–9 mm wide, on upper culm leaves usually shortened, *often several times shorter than their sheaths*. Panicles 3–20 cm long, usually condensed and dense; their branchlets scabrous, rarely almost smooth, often much abbreviated, *the longest of them up to 2–3 cm long, with relatively few (up to 10–15) spikelets*; spikelets *(3.8)4–6(7) mm long*; glumes almost always significantly shorter (often by more than 1 mm) than spikelets; anthers 2–3 mm long. . .1. **A. LATIFOLIA** (R. BR.) GRISEB.
– Plant 20–100 cm high; leaf blades 2–8 mm wide, on upper culm leaves *usually only slightly shorter than their sheaths*. Panicles 5–25 cm long, usually condensed and rather dense, more rarely diffuse; their branchlets scabrous or smooth, *the longest of them often up to 5–7 cm long, with numerous (over 15) spikelets* which are on average more laxly arranged than in the preceding species; spikelets *(2.2)2.5–3.8(4) mm long*; glumes usually narrower than in preceding species and with greater difference in size, the upper often almost equal to spikelet; anthers 1–2.4 mm long. . . . . . . . . . . . . . . . . . . . . . . . . . . . . . . . . . . . . . . . . . . . . . . . . . .2. **A. ARUNDINACEA** (TRIN.) BEAL.

   1. ***Arctagrostis latifolia*** (R. Br.) Griseb. in Ledebour, Fl. ross. IV (1853), 434; Gelert, Fl. arct. 1, 107; Lynge, Vasc. pl. N. Z. 101; Krylov, Fl. Zap. Sib. II, 201; Sørensen, Vasc. pl. E Greenl. 134; Perfilev, Fl Sev. I, 73; Rozhevits in Fl. SSSR II, 167; Hultén, Fl. Al. II, 145, excl. var.; Kuzeneva in Fl. Murm. I, 154; Hylander, Nord. Kärlväxtfl. I, 273; A. E. Porsild, Ill. Fl. Arct. Arch. 24; Karavayev, Konsp. fl. Yak. 50; Jørgensen & al., Fl. pl. Greenl. 13; Polunin, Circump. arct. fl. 40.
   *Colpodium latifolium* R. Br. in Suppl. to App. Parry's Voyage XI (1824), 286; Lange, Consp. fl. groenl. 166.
   *Cinna Brownii* Rupr., Fl. samojed. cisur. (1846) 66.
   *Vilfa gigantea* Turcz. ex Griseb., l. c. 435, in syn.
   *Arctagrostis stricta* Petr., Fl. Yak. I (1930), 156.
   *A. glauca* Petr., l. c. 159, pro parte.
   *A. aristulata* Petr., l. c. 161.
   *A. Tilesii* (Griseb.) Petr., l. c. 163, quoad pl.
   *A. arundinacea* auct. non Beal — Rozhevits, l. c. 168, pro parte.
   *A. anadyrensis* V. Vassil. in Bot. mat. Gerb. Bot. inst. AN SSSR XVII (1955), 52.

Ill.: Gelert, l. c., fig. 82; Petrov, Fl. Yak., fig. 50, 52; Fl. SSSR II, pl. XII, fig. 8; Fl. Murm. I, pl. XLVIII; V. Vasilev in Bot. mat. Gerb. Bot. inst. AN SSSR XVII, fig. 1, b; Porsild, l. c., fig. 4, c, d; Polunin, l. c. 40.

Circumpolar arctic species, penetrating far to the south in the mountain ranges of East Siberia. Displaying relatively little polymorphism; only in lower montane districts of non-arctic Siberia, apparently, has a very weakly differentiated ecological-geographical race been reported, in which the plant as a whole and the panicles are of larger size and the leaf sheaths usually strongly scabrous. However this race, first named "*Vilfa gigantea* Turcz." by N. S. Turchaninov and later labelled by V. N. Vasilev in the herbarium as a distinct species "*A. Turczaninovii* V. Vassil.," shows a rather gradual transition to typical *A. latifolia* in the north and in the barren zone of the mountains. At best it can be considered only as a subspecies, *A. latifolia* ssp. *gigantea* (Turcz. ex Griseb.) Tzvel., ssp. nova (A subspecie typica caulibus robustioribus, 50–80 cm alt., paniculis 10–20 cm lg. et vaginis vulgo scabris differt. Typus: Chamos dicto, inter cust. Urgudiense et Kluczeviense, 1829, Turczaninov). Particularly distinct from typical specimens of *A. latifolia* are specimens of this race with relatively lax greenish panicles collected in the upper part of the forest zone (usually in larch or cedar-pine forest). As for the other species synonymized with *A. latifolia*, they are in our opinion only variants with respect to particular characters. V. A. Petrov (l.c.) did not even consider it necessary to compare the new species described by him with the well known *A. latifolia* and *A. arundinacea*, restricting himself to noting that the latter two species occurred only in America. According to this interpretation the majority of specimens of *A. latifolia* from Yakutia were divided by him between 3 species: *A. stricta*, *A. glauca* (here were referred specimens both of *A. latifolia* and *A. arundinacea* with the often abortive panicles little exserted from the sheaths of the upper stem leaves) and *A. Tilesii*. Strangely V. A. Petrov referred to the last of these species specimens with larger spikelets ("up to 7.5–8 mm long"), although the type specimens of *Colpodium Tilesii* Griseb. present in the herbarium of the Botanical Institute of the USSR Academy of Sciences have spikelets in total 2.6–3.8 mm long, which attests to the undoubted synonymy of this species with *A. arundinacea*. As in *A. arundinacea*, specimens of *A. latifolia* from different parts of the range (from the Kola Peninsula, the Eastern Sayan, the lower reaches of the Yenisey and the Lena, the Dzhugdzhur Range, etc.) can possess lemmas with a sharp mucro at the tip, namely f. *aristulata* (Petr.) Roshev. (Fl. SSSR II, 168).

In the tundra zone *A. latifolia* is a plant of adequately drained habitats sufficiently covered by snow in winter. It usually occurs on fine-soil substrates; on stony sites and among rocks it colonizes areas enriched by fine soil. It is more characteristic of sites with closed tundra turf. By forming rhizomes it normally grows in rather extensive diffuse clones and is capable of forming more or less dense aggregations only under special conditions (in rock crevices and on rocky hillocks, on dry sandy bluffs).

The range of habitats of *A. latifolia* is very wide. In the Arctic it enters the tundras of marshy lowlands (tundras with flat hummocks, polygonal marshes, etc), among which *A. latifolia* grows on relatively well drained elevations of the microrelief, as well as mossy tundras with *Eriophorum* tussocks, moist patchy moss-sedge tundras (with *Carex hyperborea* s. l.), *Cassiope-Cladonia* and *Cladonia*-moss tundras. In *Dryas* tundras with little snow cover, *A. latifolia* mainly occurs only along the edge of sandy terraces of the larger rivers; sometimes it also grows luxuriantly on crumbling slumps of shoreline bluffs composed of alluvial or peaty-alluvial deposits. In the western part of the Soviet Arctic it also occurs in certain types of shrubby tundras. In East Siberia it often grows in the northern zone of open Dahurian larch forest (in tussocky, shrubby or shrubby-mossy open forest), and advances still further south in marshes.

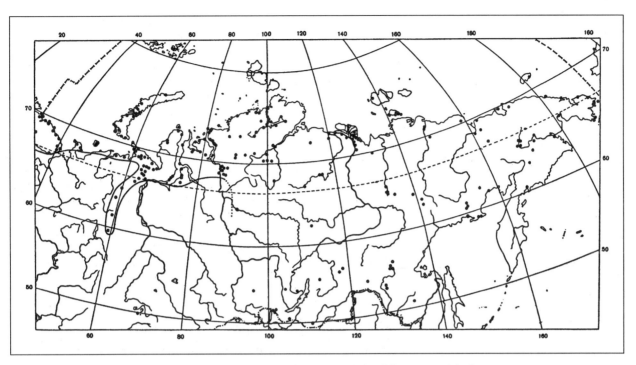

Map II–8 Distribution of *Arctagrostis latifolia* (R. Br.) Griseb.

In the High Arctic occurring infrequently and only in its more southern districts. High Arctic specimens apparently rarely bear fruit and often possess abnormally developed panicles. Also common in the Arctic are specimens with spikelets infected by a fungal disease, sometimes erroneously assumed to be a viviparous variety of the species.

**Soviet Arctic.** Murman; Northern Kanin; Kolguyev; Bolshezemelskaya Tundra; Polar Ural; Pay-Khoy; Novaya Zemlya (north to 76°); Yamal; Obsko-Tazovskiy Peninsula; Gydanskaya Tundra; lower reaches of Yenisey; Taymyr; Arctic Yakutia (frequent, along the whole coast); Wrangel Island; Chukotka Peninsula; Anadyr Basin (rare); Bay of Korf. (Map II-8).

**Foreign Arctic.** Arctic part of Alaska; arctic coast of Canada west of Hudson Bay (south to 60°N); arctic part of Labrador (south to 57°N); Canadian Arctic Archipelago; Greenland (on the whole coast north of 67°); Spitsbergen; Arctic Scandinavia.

**Outside the Arctic.** Kola Peninsula (Khibins Mountains); Urals (south to 60°N); the north of the Central Siberian Plateau (south to the Lower Tunguska Basin); barrens of Eastern Sayan; Prebaikalia; mountains of East Siberia from the Lena and Baykal to the Sea of Okhotsk and the Amur; Alaska and Yukon (upper reaches of the Yukon and its tributaries).

2. ***Arctagrostis arundinacea*** (Trin.) Beal, Grass. N. Amer. II (1896), 317; Scribner & Merrill, Grass. Alaska (1910) 55; Rozhevits in Fl. SSSR II, 168; Karavayev, Konsp. fl. Yak. 51.

*A. latifolia* var. *arundinacea* (Trin.) Griseb. in Ledebour, Fl. ross. IV, 435; Krylov, Fl. Zap. Sib. II, 202; Hultén, Fl. Al. II, 145.

*A. angustifolia* Nash in Britton & Rydberg in Bull. N.Y. Bot. Gard. II (1901–1903), 151.

*A. macrophylla* Nash, l. c. 151.

*A. latifolia* var. *arundinacea* f. *parviflora* Reverd. ex Kryl., Fl. Zap. Sib. II, 202.

*A. festucacea* Petr., Fl. Yak. I (1930), 155; Rozhevits, l. c. 167; Karavayev, l. c. 50.

*A. glauca* Petr., l. c. 159, pro parte (cum typo speciei).

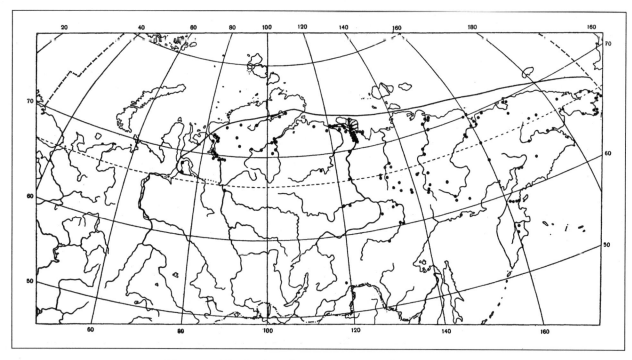

MAP II-9 Distribution of *Arctagrostis arundinacea* (Trin.) Beal.

*A. parviflora* (Reverd. ex Kryl.) Petr., l. c. 162.
*A. Tilesii* (Griseb.) Petr., l. c. 163, quoad nomen.
*A. ursorum* (Kom.) Kom. ex Roshev. in Fl. SSSR II, 168.
*A. latifolia* var. *angustifolia* (Nash) Hultén, l. c. 146.
*A. tenuis* V. Vassil. in Bot. mat. Gerb. Bot. inst. AN SSSR XVII (1955), 48.
*A. calamagrostidiformis* V. Vassil., l. c. 51.
*A. caespitans* V. Vassil., l. c. 53.
*A. viridula* V. Vassil., l. c. 54.
*Vilfa arundinacea* Trin., Gram. unifl. (1824) 157.
*Colpodium Tilesii* Griseb., l. c. 385.
*Poa ursorum* Kom. in Fedde, Repert. sp. nov. XIII (1914), 161.
**Ill.:** Trinius, Ic. gram. 1 (1828), tab. 55; Petrov, Fl. Yak. I, fig. 49, 53; V. Vasilev in Bot. mat. Gerb. Bot. inst. AN SSSR XVII, fig. 1, a, b, g, 2.

The range of this rather more southern, mainly East Siberian, species is almost completely overlapped by the range of *A. latifolia*, although specimens with characters more or less intermediate between the two species (probable hybrids) only occur rather rarely. In distinction from *A. latifolia*, *A. arundinacea* is significantly more polymorphic, although it would be erroneous to suppose that the variability of this species is to any degree greater than the variability of many other grass species in which such characters as the size of the spikelets, the form of the panicles and the width of the leaf blades similarly show much variation. Careful review of the extensive material of this species present in the herbarium of the Botanical Institute of the USSR Academy of Sciences has readily convinced us that, despite the real differences between particular specimens and populations of *A. arundinacea*, there is no morphological, ecological or geographical basis for dividing this species into any smaller taxonomic units. The type specimens of *A. arundinacea* possess rather large (3.4–4 mm long) spikelets, while specimens with smaller (2.5–3.5 mm long) spikelets, upon which such species as *A. Tilesii* and *A. parviflora* are based, are more common. The type specimens of *A. ursorum* and *A. calama-*

*grostidiformis* also have such small spikelets, but more laxly arranged in a more or less diffuse panicle. Occasionally there occur pseudocaespitose specimens (types of *A. festucacea* and *A. caespitans*) which have developed on gravelbars among large stones more or less preventing the longitudinal growth of rhizomes. The type of *A. caespitans* possesses smaller spikelets (2.2–2.6 mm long), approaching in this respect the Alaskan species *A. poaeoides* Nash, but spikelets of such a length also occur in certain other non-tufted specimens of *A. arundinacea* from Siberia. An erroneous account of *A. arundinacea* is given in the "Flora of the USSR," where R. Yu. Rozhevits apparently took as the basis for this species specimens of the subspecies *A. latifolia* ssp. *gigantea* mentioned above. Consequently there are errors in the key for identification of species (Fl. SSSR II, 166–167), where *A. arundinacea* is distinguished from *A. latifolia* only by a height of "1–1.5 m" (instead of 20–100 cm) and from *A. ursorum* by spikelets "usually 5–7 mm long" (instead of 2.2–4 mm long).

In distinction from the preceding species, *A. arundinacea* usually occurs in habitats which are significantly better drained (in the tundra zone, better insolated) and often also drier: found in meadows, on sands and gravelbars in the valleys of rivers and streams, in shrub communities (usually willow thickets) in valleys, and in the more southern parts of its range also in larch forest. Not native to the High Arctic.

**Soviet Arctic.** Obsko-Tazovskiy Peninsula; lower reaches of Yenisey; Taymyr (Pyasina, basin of Lake Taymyr); lower reaches of Olenek, Lena, Indigirka and Kolyma; Chukotka Peninsula (rather rare); Anadyr Basin; Penzhina Basin. (Map II–9).

**Foreign Arctic.** Just penetrating the arctic part of Alaska (but common in the district of the Bering Strait).

**Outside the Arctic.** Northern part of Central Siberian Plateau (south to the Lower Tunguska Basin); almost all Yakutia (south to the Aldansk and Olekmo-Charsk Mountains); Kolymsk Range; Kamchatka; Aleutian Islands, Alaska and Yukon (south to upper reaches of Yukon River).

## GENUS 8    Agrostis L. — BENT

THE OVER 150 SPECIES of this genus are distributed in all cold and temperate countries of both hemispheres, as well as in mountainous districts of the tropics. The majority of them are plants of mesophytic communities of meadows, marshes, forest glades and riverine sands and gravelbars. Many species of the genus penetrate arctic limits, but there is not a single high arctic species among them and extremely few species more or less characteristic of the Arctic. Of the 10 species occurring in the Soviet Arctic, only the subarctic *A. borealis* is characteristic for many districts of the European Arctic and the Northern European littoral species *A. straminea* is found in many parts of the coast of the Arctic Ocean. The remaining species are boreal and only just penetrate the Arctic. Clearly distinguished among them are a group of predominantly European species (*A. stolonifera, A. gigantea, A. tenuis, A. canina*) and a group of species mainly distributed in Siberia and North America (*A. Trinii, A. anadyrensis, A. clavata, A. scabra*). Under arctic conditions hybrids between species of *Agrostis* are not too frequent, so that as a rule material of this genus is easily identified. As in many other genera of grasses, it is of great importance for identifying species of *Agrostis* to measure precisely the length of the anthers, which can usually be found even in fruiting panicles. Viviparous forms are completely lacking in the genus *Agrostis*.

1. Paleas ½ - ⅔ *as long as* lemmas; lemmas almost always without awns, very rarely (mainly in hybrid specimens) with short awns arising from their distal part; anthers 1–1.5 mm long. .................................... .2.
– Paleas ⅒ - ⅕ *as long as* lemmas, sometimes completely absent. ........ .5.

2. Ligules of upper culm leaves *0.5–1.8 mm long*, blunt; leaf blades more or less scabrous on margin and lower surface, *on upper surface with relatively few scattered acicules or smooth*; plant 10–60 cm high, usually forming small tufts with short subterranean stolons. Panicles usually widely diffuse at flowering time; their branchlets rather long and fine, *bare or with relatively few scattered acicules*; spikelets 1.5–2.7 mm long; paleas usually half as long as lemmas. ........................................ .4. **A. TENUIS** SIBTH.
– Ligules of upper culm leaves *1.5–5 mm long*, blunt or acute; leaf blades *usually* (with exception of *A. straminea*) *also on upper surface with rather numerous acicules*. Branchlets of panicle usually *strongly scabrous with numerous densely arranged acicules*, weakly scabrous or smooth only in *A. straminea*; paleas usually ⅗ - ⅔ as long as lemmas. .................... .3.

3. Plant *30–100 cm high, without elongate epigeal vegetative shoots but with subterranean stolons*; culms usually more or less erect; leaf blades 1.5–6 mm wide, on both surfaces rather strongly scabrous with numerous acicules. Panicles 6–20 cm long, at flowering time and subsequently *widely diffuse*, with rather long branchlets strongly scabrous along their whole length; spikelets 1.8–3.5 mm long. ................. 1. **A. GIGANTEA** ROTH.
– Plant *6–50 cm high, without subterranean stolons but usually with elongate epigeal vegetative shoots* ("stolons"); culms usually ascending at base; leaf blades 1–3 mm wide. Panicles 2–12 cm long, more or less diffuse at flowering time, *subsequently usually with upwardly directed*, more or less abbreviated *branchlets*. ............................................... .4.

4. Plant of sea coast, with numerous elongate vegetative shoots rooting at their nodes; leaf blades *relatively weakly scabrous*, sometimes almost smooth. Panicles 2–6 cm long, with much abbreviated *branchlets which are smooth or weakly scabrous with scattered acicules*; spikelets 2–3.5 mm long. ......................................... .3. **A. STRAMINEA** HARTM.
– Plant of river or lake valleys, usually with less numerous elongate vegetative shoots; leaf blades *usually strongly scabrous on both sides*. Panicles 5–12 cm long, usually with less abbreviated *branchlets which are strongly scabrous with numerous acicules*; spikelets 1.6–3 mm long. .............. ................................................. 2. **A. STOLONIFERA** L.

5. Panicles *very elongate, 10–30 cm long, often almost as long as culms or not less than half as long as them*, at flowering time and subsequently widely diffuse; their branchlets rather long and fine, strongly scabrous along their whole length; lemmas *awnless or very rarely with short straight awns* aris-

ing near their tip; anthers *0.3–0.5 mm long*. Short-lived (possibly biennial) plant 15–70 cm high, forming dense tufts, without subterranean stolons. . . . . . . . . . . . . . . . . . . . . . . . . . . . . . . . . . . . . . . . . . . . . . . . . . . . . . . . . . . . . . .6.

– Panicles *shorter in comparison with length of culms, usually ⅛ - ⅓ as long as them*; lemmas *usually with rather long awns* arising near their middle or lower, *very rarely awnless*; anthers *0.6–1.5 mm long*. Perennial plant. . .7.

6. Glumes 1.8–2.6 mm long, *very long-acuminate at tip*; lemmas *½ - ⅔ as long as lower glumes*. Leaf blades on average narrower, 1–3 mm wide, flat or folded lengthwise; rather numerous leaves with very narrow blades usually present at base of tufts. . . . . . . . . . . . . . . . . . . . . . . . . . . .10. **A. SCABRA** WILLD.

– Glumes 1.5–2.3 mm long, *more shortly acuminate at tip*; lemmas *not so short in relation to lower glumes (¾ - ⅞ as long)*. Leaf blades on average broader, 1.5–5 mm wide; only a few leaves with very narrow blades sometimes present at base of tufts. . . . . . . . . . . . . . . . . . . . . . . . .9. **A. CLAVATA** TRIN.

7. Anthers *1–1.5 mm long*; spikelets 1.6–3 mm long; lemmas with rather long awns arising near their middle or lower and exceeding their tip by 1–2.5 mm; panicles 4–12 cm long, more or less diffuse, with relatively long branchlets. . . . . . . . . . . . . . . . . . . . . . . . . . . . . . . . . . . . . . . . . . . . . . . . .8.

– Anthers *0.6–0.8 mm long*. Plant forming tufts, without subterranean stolons. . . . . . . . . . . . . . . . . . . . . . . . . . . . . . . . . . . . . . . . . . . . . . . . . . . . . . .9.

8. Branchlets of panicle *strongly scabrous* with numerous acicules *along their whole length*. Plant forming dense or loose tufts, *usually without subterranean stolons*; leaf blades 1–3.5 mm wide; ligules of upper culm leaves *1.5–4 mm long*. . . . . . . . . . . . . . . . . . . . . . . . . . . . . . . . . . . . . . . . . 5. **A. CANINA** L.

– Branchlets of panicle *weakly scabrous* (with scattered acicules) to almost smooth. Plant forming dense tufts *with short subterranean stolons*; leaf blades on average narrower, 0.5–2 mm wide; ligules of upper culm leaves *0.5–2 mm long*. . . . . . . . . . . . . . . . . . . . . . . . . . . . . . . . . . . . 6. **A. TRINII** TURCZ.

9. Panicles usually noticeably elongate in comparison with culm length, 6–20 cm long; their branchlets *strongly scabrous with rather long acicules along their whole length*; spikelets 2–3 mm long; awns of lemmas *usually arising above their middle* and exceeding their tip by 0.5–2.5 mm, sometimes absent. Plant 20–50 cm high. . . . . . . . . . . . . . . . . . . . 7. **A. ANADYRENSIS** SOCZ.

– Panicles relatively short, 2–9 cm long; branchlets *weakly scabrous with scattered acicules, often almost smooth*; spikelets 2–3.5 mm long; awns of lemmas *arising near their middle or lower* and exceeding their tip by 1.5–3 mm. Plant 8–40 cm high. . . . . . . . . . . . . . . . . . . . . . . . . .8. **A. BOREALIS** HARTM.

**1. Agrostis gigantea** Roth, Fl. Germ. 1 (1788), 31; Philipson in Journ. Linn. Soc. London LI (1937), 94, pl. 9, 10; Petrov, Fl. Yak. I, 181; Hylander, Nord. Kärlväxtfl. I, 321; Böcher & al., Grønl. Fl. 293; Polunin, Circump. arct. fl. 36.

A. *diffusa* Host ex Bess., Fl. Gal. austr. 1 (1809), 68; Petrov, l. c. 178.

A. *alba* var. *gigantea* (Roth) Griseb. in Ledebour, Fl. ross. IV (1853), 437.

*A. alba* auct., non L. — Lange, Consp. fl. groenl. 158; Krylov, Fl. Zap. Sib. II, 206, pro parte; Perfilev, Fl. Sev. I, 74, pro parte; Shishkin in Fl. SSSR II, 183; Leskov, Fl. Malozem. tundry 21; Kuzeneva in Fl. Murm. I, 157; Karavayev, Konsp. fl. Yak. 51.

**Ill.:** Petrov, l. c., fig. 59, 60; Philipson, l. c., pl. 9, 10; Polunin, l. c. 36.

Widely distributed, mainly boreal European species, just penetrating the Arctic. Formerly it was widely known under the name "*A. alba* L.," but the type of the latter species has proved to be a specimen of *Poa nemoralis* L. with single-flowered spikelets (Philipson, l. c. 91). Apparently this species is a rather polymorphic complex even within the restricted scope of this work, with the name *A. gigantea* s. str. belonging to larger specimens from relatively drier habitats unlikely to penetrate the Arctic. Arctic specimens belong to a more widely distributed lower-growing ecological form (or race?) often with somewhat abbreviated panicle branchlets which usually grows in moister meadowy areas. However, no one has yet succeeded in establishing a morphological distinction between these forms at all clearly.

Growing in meadowy areas, as well as on sands and gravelbars in river and lake valleys, on grassy slopes, sometimes also as an introduction along roads and near settlements.

**Soviet Arctic.** Murman; Kanin (south of the Nadtey River), Malozemelskaya and Bolshezemelskaya Tundras; Polar Ural (only southern part); Anadyr Basin (near Markovo settlement, introduced).

**Foreign Arctic.** Extreme SW Greenland (introduced); Iceland; Arctic Scandinavia.

**Outside the Arctic.** Almost all extratropical Eurasia except NE and SE portions, north to Arctic Scandinavia, the Polar Ural, the Ob at 65°N, the Yenisey near the mouth of the Kureyka, and Southern Yakutia; North Africa. As an imported or introduced plant in many districts of North and South America and other countries.

**2. Agrostis stolonifera** L., Sp. pl. (1753) 62; Gelert, Fl. arct. 1, 108, pro parte; Philipson in Journ. Linn. Soc. London LI (1937), 94; Hultén, Fl. Al. II, 157; id., Atlas, map 158, pro parte; Hylander, Nord. Kärlväxtfl. I, 320, pro parte; Böcher & al., Grønl. Fl. 293; Polunin, Circump. arct. fl. 35.

*A. alba* var. *stolonifera* (L.) Sm., Engl. Fl. 1 (1824), 93; Grisebach in Ledebour, Fl. ross. IV, 437.

?*A. stolonizans* Bess. in Roemer et Schultes, Mant. III (1827), 567, in adnot.; Shishkin in Fl. SSSR II, 184; Leskov, Fl. Malozem. tundry 21; Kuzeneva in Fl. Murm. I, 158.

*A. stolonifera* var. *prorepens* Koch, Synops., Ed. 2 (1844), 902.

*A. prorepens* (Koch) Golub. in Zhurn. Russk. bot. obshch. VIII (1924), 120; G. Meyer in Korresp. Bl. Naturf. Ver. Riga XXVI (1883), 56, nomen.

*A. alba* auct. non L. — Krylov, Fl. Zap. Sib. II, 206, pro parte; Perfilev, Fl. Sev. I, 74.

**Ill.:** Philipson, l. c., pl. 11, 12; Fl. Murm. II, pl. LI; Polunin, l. c. 35.

Widely distributed boreal species. Its range almost coincides with that of the preceding species. We adopt the name *A. stolonifera* L. for one of the most widespread ecological-geographical races of the inadequately studied polymorphic complex *A. stolonifera* s.l. The biology of this race was intensively studied by M. M. Golubeva, who proposed for it the new specific name *A. prorepens* (Koch) Golub. This race is common in Northern Europe and is evidently the plant upon which Linnaeus based the species *A. stolonifera*. In the "Flora of the USSR" the majority of specimens of this species were referred to *A. stolonizans* Bess., a species by no means clearly delimited from *A. stolonifera* and at most constituting a more southern ecological-geographical race, intermediate between *A. stolonifera* s. str. and races with still smaller spikelets of the type of *A. maritima* Lam. and *A. albida* Trin. *Agrostis stolonifera* just penetrates the Arctic.

Growing mainly on sandbars and gravelbars on the shores of rivers and lakes, in springfed marshes and marshy meadows, sometimes entering water in shallow areas of waterbodies. Preferring soils with abundant moisture but well drained.

**Soviet Arctic.** Murman; Malozemelskaya and Bolshezemelskaya Tundras (only in the valleys of the larger rivers); Polar Ural (southern part); lower reaches of Ob.

**Foreign Arctic.** Beringian coast of Alaska (near Nome, introduced); SW coast of Greenland (north to 62°N, introduced); Iceland; Arctic Scandinavia.

**Outside the Arctic.** Greater part of extratropical Eurasia (north-east to the lower reaches of the Ob, the mouth of the Kureyka on the Yenisey, and the SW part of Yakutia). As an imported or introduced plant in North America and in many other countries of both hemispheres. Range not precisely clarified because of confusion with closely related species.

3. ***Agrostis straminea*** Hartm., Gram. Scand. (1819) 4 and Handb. Scand. Fl. (1820) 45.

   *A. alba* var. *maritima* auct. non G. Mey. — Grisebach in Ledebour, Fl. ross. IV, 438; Perfilev, Fl. Sev. I, 75.

   *A. alba* var. *salina* Pohle ex Tolm., Fl. Kolg. (1930) 13.

   *A. maritima* auct. non Lam. — Shishkin in Fl. SSSR II, 185; Kuzeneva in Fl. Murm. I, 160.

   *A. stolonifera* var. *prorepens* auct. non Koch — Hylander, Nord. Kärlväxtfl. I, 320, pro parte.

   Littoral species whose range has not been precisely established up to the present time because of confusion with other ecological-geographical races of the complex *A. stolonifera* s.l. Apparently occurring only in Northern Europe east to the Karsk-Baydaratsk coast, but it cannot be excluded that its range will prove to be amphiatlantic and that many records of *A. stolonifera* from Eastern Canada and Greenland will in fact prove to refer to the present species. In the "Flora of the USSR" and the "Flora of Murmansk Oblast" specimens of this species were referred to *A. maritima* Lam., although the latter species described from the vicinity of Narbonne in Southern France (topotypical specimens present in the herbarium of the Botanical Institute of the USSR Academy of Sciences) differs from *A. straminea* in a whole series of substantial characters: namely, the strongly scabrous panicle branchlets, the smaller (1.4–2 mm long) pale-green spikelets, and the stiffer greyish-green leaf blades. *Agrostis straminea* is closer to *A. stolonifera* s. str., but is clearly enough distinguished also from that species by the smooth or weakly scabrous branchlets of the strongly condensed panicles and the especially great development of stolonlike vegetative shoots. As in a series of other littoral species whose individuals are sometimes flooded by tides (for example, *Puccinellia phryganodes* and *Calamagrostis deschampsioides*), *A. straminea* manifests a noticeable reduction in the number of generative shoots on account of a copious development of vegetative shoots which creep and root at the nodes and whose secondary lateral shoots can apparently break off and give rise to new individuals.

   Growing in marshy meadows and on banks on the sea coast and in the mouths of the larger rivers (estuarine saltmarshes), also on maritime rocks, sometimes descending below high tide mark.

**Soviet Arctic.** Murman; Kanin; Malozemelskaya and Bolshezemelskaya Tundras; Kolguyev; Karskaya Tundra (near mouth of Kara).

**Foreign Arctic.** Arctic Scandinavia; possibly occurring in Iceland and NE North America.

**Outside the Arctic.** Shores of the Baltic, Norwegian and White Seas.

4. ***Agrostis tenuis*** Sibth., Fl. Oxon. (1794) 36; Philipson in Journ. Linn. Soc. London LI (1937), 85; Hultén, Fl. Al. II, 158; id., Atlas, map 159; Hylander, Nord. Kärlväxtfl. I, 322; Böcher & al., Grønl. Fl. 293; Polunin, Circump. arct. fl. 34.

   *A. capillaris* auct. non L. — Hudson, Fl. Angl., Ed. 2 (1762), 27; Shishkin in Fl. SSSR II, 185; Kuzeneva in Fl. Murm. I, 158.

*A. vulgaris* With., Bot. Arrang. Veg. Brit., Ed. 3, 11 (1796), 132; Grisebach in Ledebour, Fl. ross. IV, 438; Gelert, Fl. arct. 1, 108; Krylov, Fl. Zap. Sib. II, 207; Perfilev, Fl. Sev. I, 75.

Ill.: Fl. SSSR II, pl. XIII, fig. 6; Fl. Murm. I, pl. LII.

Predominantly European boreal species, just penetrating the Arctic. Formerly this species was sometimes known under the name "*A. capillaris* L.," but this apparently should be applied to one of the Southern European annual species of the genus, *A. delicatula* Pourr. (Philipson, l. c. 86–87). Specimens with somewhat abbreviated scabrous panicle branchlets occasionally encountered on the Kola Peninsula are apparently the hybrid *A. tenuis* × *A. stolonifera* (= *A.* × *Murbeckii* Fouill.), and specimens with a short awn on the distal part of the lemmas the hybrid *A. tenuis* × *A. borealis* (= *A.* × *lapponica* Mont.).

Growing mainly in meadowy areas on the banks of rivers and streams, on sandbars and gravelbars, on the margins of marshes, and sometimes as an introduction along roads.

**Soviet Arctic.** Murman.

**Foreign Arctic.** Extreme southern part of Greenland (introduced); Iceland; Arctic Scandinavia.

**Outside the Arctic.** Almost all Europe; Caucasus; Asia Minor; West Siberia (north to the basin of the Severnaya Sosva); Sayan Mountains and Yenisey Ridge. As an imported or introduced plant in a considerable part of North America from Southern Alaska and Southern Labrador to Mexico, as well as in many other extratropical countries.

5. ***Agrostis canina*** L., Sp. pl. (1753) 62; Grisebach in Ledebour, Fl. ross. IV, 440; Lange, Consp. fl. groenl. 158; Gelert, Fl. arct. 1, 108; Krylov, Fl. Zap. Sib. II, 204; Perfilev, Fl. Sev. I, 74; Kuzeneva in Fl. Murm. I, 157; Hylander, Nord. Kärlväxtfl. I, 322; Böcher & al., Grønl. Fl. 93; Polunin, Circump. arct. fl. 37.

Ill.: Fl. Murm. I, pl. XLIX; Polunin, l. c. 36.

Predominantly European boreal species, just penetrating the Arctic. In Siberia apparently completely replaced by the closely related species *A. Syreitschikovii* Smirn. and *A. Trinii* Turcz.

In the Soviet Arctic found only as an introduction along roads and near settlements.

**Soviet Arctic.** Murman (introduced near Murmansk, also reported for the settlement of Kharlovka but probably in error for *A. borealis*).

**Foreign Arctic.** Southern Greenland (north to 67°N on the west coast, to 65°N on the east coast); Iceland; Arctic Scandinavia.

**Outside the Arctic.** Almost all Europe, except the extreme south and south-east; NE part of North America (south from Newfoundland). As an imported or introduced plant in many extratropical countries.

6. ***Agrostis Trinii*** Turcz., Fl. baic.-dahur. (1856) 18, in nota; Petrov, Fl. Yak. I (1930), 170; Shishkin in Fl. SSSR II, 175; Karavayev, Konsp. fl. Yak. 51.

*A. canina* var. *rubra* Trautv., Pl. Sib. bor. (1877) 144.

Ill.: Petrov, l. c.; Fl. SSSR II, pl. XIII, fig. 5.

Boreal species widespread in East Siberia, penetrating the Arctic as far as the shore of the East Siberian Sea. Anther measurements in this species 1–1.5 mm long, not 0.7–1 mm as given in the "Flora of the USSR."

Growing in moderately moist meadows and on grassy slopes, on sands and gravelbars in river and lake valleys, on shores of waterbodies.

**Soviet Arctic.** Lower reaches of Yenisey (north to Dudinka); between the mouth of the Kolyma and the Bay of Chaun; Anadyr and Penzhina Basins (rather frequent).

**Foreign Arctic.** Not occurring.

**Outside the Arctic.** East Siberia east of the Yenisey and the Sayan Mountains, north to the northern fringe of the Central Siberian Plateau and the Verkhoyansk district; Mongolia; NE China; Korean Peninsula.

7. ***Agrostis anadyrensis*** Socz. in Fl. SSSR (1934) 176, 746.
    ?*A. geminata* Trin., Gram. unifl. (1824) 207.
    **Ill.:** Fl. SSSR II, pl. XIII, fig. 8.

    Species widely distributed in NE Siberia, occupying an intermediate position in the structure of the panicles and spikelets between such morphologically strongly differentiated species as *A. Trinii* and *A. clavata*, with some specimens of *A. anadyrensis* tending to varying degree towards one of these species, others towards the other. Therefore it appears very probable that *A. anadyrensis* has arisen as a result of hybridization between *A. Trinii* and *A. clavata*. This cannot only be of contemporary occurrence, since in the Anadyr Basin, where *A. anadyrensis* is very common, *A. clavata* has so far not been found. In those districts where both parent species (*A. Trinii* and *A. clavata*) survive to the present time (for example, in the Penzhina Basin), there undoubtedly occur contemporary hybrids between them which are externally indistinguishable from typical specimens of *A. anadyrensis*. It cannot be excluded that the prior name for *A. anadyrensis* may be *A. geminata* Trin., which applies to a species described from Unalaska Island (Aleutian Islands) representing according to Hultén's (Fl. Al. II, 156) supposition the hybrid *A. borealis* × *A. scabra*. The majority of specimens of *A. anadyrensis* differ from the type specimen of *A. geminata* (retained in the herbarium of the Botanical Institute of the USSR Academy of Sciences) in the larger size of the whole plant and of the panicles, however there are some specimens indistinguishable from it. Nevertheless, if the type specimen of *A. geminata* is genuinely of hybrid origin, it is more probable that the parent species involved in its production were *A. scabra* and *A. borealis*, that is species absent from the range of distribution of *A. anadyrensis*. Therefore, despite the great external similarity of *A. anadyrensis* and *A. geminata*, there is (at least for the present) greater support for accepting them as distinct, although very similar, species of hybrid origin.

    Growing mainly on sandy and gravelly shores of rivers and lakes, in meadows, and in shrubby communities in valleys.

    **Soviet Arctic.** Lower reaches of Lena (north to the district of Bulun); lower reaches of Kolyma; district of Bay of Chaun; Anadyr and Penzhina Basins (rather frequent).
    **Foreign Arctic.** Absent.
    **Outside the Arctic.** Mountainous districts of Yakutia east of the Lena, south to the Okhotsk district; apparently also occurring in Southern Alaska ("*Agrostis* sp." Hultén, l. c. 158).

8. ***Agrostis borealis*** Hartm., Handb. Scand. Fl., Ed. 3 (1838), 17; Gelert, Fl. arct. 1, 109; Krylov, Fl. Zap. Sib. II, 205; Petrov, Fl. Yak. I, 173; Perfilev, Fl. Sev. I, 74; Shishkin in Fl. SSSR II, 174; Leskov, Fl. Malozem. tundry 21; Hultén, Fl. Al. II, 153; id., Atlas, map 155; Kuzeneva in Fl. Murm. I, 156; Hylander, Nord. Kärlväxtfl. I, 323; A. E. Porsild, Ill. Fl. Arct. Arch. 24; Böcher & al., Grønl. Fl. 293; Hultén, Amphi-atl. pl. 98; Polunin, Circump. arct. fl. 36.
    *A. rubra* auct. non L. — Grisebach in Ledebour, Fl. ross. IV, 440, pro parte; Lange, Consp. fl. groenl. 157.
    *A. rupestris* auct. non All. — Grisebach, l. c. 439.
    *A. alpina* auct. non Scop. — Grisebach, l. c. 439.
    *A. viridissima* Kom. in Fedde, Repert. sp. nov. XIII (1914), 85.
    **Ill.:** Gelert, l. c., fig. 83; Fl SSSR II, pl. XIII, fig. 9; Porsild, l. c., fig. 4, a, b; Böcher & al., l. c., fig. 51, b; Polunin, l. c. 36.

Widely distributed subarctic species whose range, figured in the cited work of Hultén (Amphi-atl. pl.), shows much peculiarity. While completely absent from a great part of the Asian mainland east of the Urals, it reaches Asia again from North America through the Aleutian Islands and Kamchatka and extends on the Pacific Coast south to the Bureinskiy Range and Northern Japan. Hultén's completely isolated record of this species from the lower reaches of the Yenisey is not confirmed by material available to us and apparently refers to *A. Trinii*. Despite its wide range, *A. borealis* is rather constant with respect to morphological characters and very clearly delimited from closely related species (*A. canina* and *A. Trinii*), but apparently occasionally forms hybrids with *A. tenuis* and *A. stolonifera*.

Growing on stony and sandy slopes, on banks and gravelbars in river and lake valleys, and on relatively dry meadowy portions of tundras.

**Soviet Arctic.** Murman (rather frequent); Malozemelskaya Tundra (mainly the northern part of Timanskiy Ridge); Bolshezemelskaya Tundra (rather frequent); Kolguyev (rare); Polar Ural (frequent); Yugorskiy Peninsula; Vaygach (recorded by Hultén); lower reaches of Yenisey (recorded by Hultén for the Dudinka district, probably erroneously).

**Foreign Arctic.** Alaska (district of Bering Strait); NW part of Canada west of Hudson Bay; Labrador (to extreme north); southern part of Baffin Island (north to the Arctic Circle); Greenland (from extreme south north to 73° on both coasts); Arctic Scandinavia.

**Outside the Arctic.** Fennoscandia; Kola Peninsula; Northern Ural (south to 60°N); Kamchatka; Kurile Islands; Sakhalin; Bureinskiy Range (Dusse-Alin Mountains); Japan; Aleutian Islands; Alaska; considerable part of Canada (sporadically), south to Vancouver Island, the Saskatchewan River and the mountains of West Virginia. In the Rocky Mountains the closely related species *A. Bakeri* Rydb. (with weakly developed awn or awnless).

9. ***Agrostis clavata*** Trin. in Spreng., Neue Entdeck. II (1821), 55; Krylov, Fl. Zap. Sib. II, 208; Petrov, Fl. Yak. I, 168; Perfilev, Fl. Sev. I, 74; Shishkin in Fl. SSSR II, 178; Hultén, Atlas, map 157; Hylander, Nord. Kärlväxtfl. I, 324; Karavayev, Konsp. fl. Yak. 51.

*A. abakanensis* Less. ex Trin. in Mém. Ac. Pétersb., sér. 6, VI, 2 (1845), 325.
*A. laxiflora* auct. non Poir. — Grisebach in Ledebour, Fl. ross. IV, 441, pro parte.
*A. bottnica* Murb. in Bot. Notis. (1898) 13.

Ill.: Petrov, l. c., fig. 55; Fl SSSR II, pl. XIII, fig. 14.

Boreal species widespread in Siberia, penetrating as far west as Scandinavia whence it was described under the name *A. bottnica* Murb. In North America apparently completely absent, replaced by the closely related species *A. scabra* and *A. hiemalis*. Like the next species, easily distinguished from all preceding species by the very small anthers and by the panicles being very large in comparison with the total culm length and having long fine branchlets. Just reaching the Arctic.

Growing in meadows and on sandbars and gravelbars in river valleys; further south also in forest openings and shrubbery.

**Soviet Arctic.** Lower reaches of Ob (near Salekhard and on the Poluy); district of the Bay of Chaun (Ostrovnoye settlement on River Muktun, 20 VII 1952, Vikulova); Penzhina Basin and the vicinity of the Bay of Korf (rather frequent).

**Foreign Arctic.** Not occurring.

**Outside the Arctic.** Sweden and Finland (sporadically in the district of the Gulf of Bothnia between the Arctic Circle and 61°N); upper part of basin of Severnaya Dvina; Pechora and Mezen Basins (north to Ust-Tsilma); Urals (north to the upper reaches of the River Lyapin); West Siberia; East Siberia (north to Igarka on the Yenisey and Zhigansk on the Lena); Kamchatka; Sakhalin; Mongolia; NE China; Korean Peninsula; Japan.

**10. Agrostis scabra** Willd., Sp. pl. 1 (1798), 370; Hultén, Fl. Al. II, 155; Hylander, Nord. Kärlväxtfl. I, 324; Polunin, Circump. arct. fl. 36.

*A. laxiflora* (Michx.) Richards. in Bot. App. Franklin Journ. (1823) 731; Grisebach in Ledebour, Fl. ross. IV, 441, pro parte; non Poir., 1810.

*A. Michauxii* Trin., Gram. unifl. (1824) 206.

*A. hiemalis* auct. non Britt., Sterns. et Pogg. — Petrov, Fl. Yak. I, 167; Shishkin in Fl. SSSR II, 186.

Ill.: Fl. SSSR II, pl. XIII, fig. 11; Polunin, l. c.

Boreal species widespread in North America, reaching the Pacific Coast of Asia from Chukotka to Japan and Korea. At the northern limit of its range (including the Chukotka Peninsula) apparently occurring only at hot springs. Formerly erroneously identified with the closely related species *A. hiemalis* (Walt.) Britt., Sterns. et Pogg. distributed only in the eastern part of the USA and Canada. The latter species is distinguished from *A. scabra* by its smaller spikelets (1.5–1.7 mm long) and smaller anthers (about 0.2 mm long). *Agrostis scabra* has only been discovered within the Soviet Arctic very recently.

Growing on sandbars and gravelbars in river valleys, in moist meadows on the shores of hot springs and at the outfall of groundwaters, sometimes also as an introduction along roads and near settlements.

**Soviet Arctic.** Chukotka Peninsula (in vicinity of Chaplino Hot Springs, VIII 1956, Tikhomirov and Gabrilyuk); Koryakia (Bay of Korf, bank of Olyutorka and near Kultushnoye settlement, VIII 1960, Katenin and Shamurin).

**Foreign Arctic.** Not occurring.

**Outside the Arctic.** Asian Pacific Coast from Kamchatka to the Korean Peninsula and Japan; Aleutian Islands; considerable part of North America from Southern Alaska, Great Bear Lake, the southern part of Hudson Bay and Newfoundland in the north to California, New Mexico, Nebraska and Pennsylvania in the south. Introduced to other countries.

---

## GENUS 9   Calamagrostis Adans. — REED GRASS

THE GENUS *Calamagrostis* contains over 100 species distributed in almost all cold and temperate countries of both hemispheres. Some species also penetrate the tropical zone, occurring there almost exclusively in mountainous districts. Of the 12 species occurring in the Soviet Arctic, the majority consists of boreal and subarctic species; certain of them (*C. neglecta* and *C. lapponica*) are rather widespread in the Arctic and penetrate far to the north. Perhaps only one species, *C. Holmii* which constitutes one of the ecological-geographical races of the polymorphic complex *C. neglecta* s. l., can be counted among those characteristic of the Arctic. One further strictly littoral species, *C. deschampsioides*, is very characteristic of maritime marshy meadows and banks, sometimes even where subject to tidal flooding. One of the peculiarities of the genus is the great polymorphism of many of its species along with relative constancy of morphological characters within single populations (and within clones in rhizomatous species) and the very wide possibilities of hybridization. It is usually not difficult to find differences between separate populations within the same species, but at the same time very difficult to profer clear morphological distinctions between the major ecological-geographical races in such intricate species complexes as *C. canescens* s. l. or *C. neglecta* s. l. It is not always easy to determine whether particular specimens are of hybrid origin. Although hybrids in *Calamagrostis* are normally sterile, in many rhizomatous species they can readily reproduce vegetatively and consequently a single hybrid individual can in time produce a large clone which can easily be assumed to be a narrowly endemic "new" species. In the present treatment we have preferred to accept many species (*C. neglecta, C. Langsdorffii, C. angustifolia, C. lapponica*) in a wide sense, although it cannot be excluded that they can be divided into several less polymorphic species in a narrower sense through more comprehensive investigation and through the availability of more abundant, well labelled material.

1. Lemmas *½ - ⅔ as long as glumes*, almost entirely hyaline, *with 3 nerves* of which the central continues as a straight or almost straight awn which arises from the back of the lemma near its middle and scarcely exceeds its tip; callus hairs *1½ - 2 times as long as lemma*; paleas ½ - ⅔ as long as lemmas; rudiment of rachis above first (and sole) floret *completely absent or very short and always bare*; spikelets 5–7 mm long, aggregated in very dense paniculate inflorescence with very abbreviated scabrous branchlets; glumes very long and finely acuminate from lanceolate base, subulate. Plant with long subterranean stolons, not forming tufts; culms with (1)2(3) nodes; leaf blades 3–10 mm wide, more or less scabrous beneath, smooth or sparsely scabrous on upper surface. . . . . . . . . . . .1. **C. EPIGEIOS** (L.) ROTH.
– Lemmas *not less than ⅔ as long as glumes*, at least on their basal half herbaceous or herbaceous-hyaline, *with 5 nerves* of which the central continues as an awn; callus hairs *usually equalling or shorter than lemma, rarely slightly longer*; rudiment of rachis *usually present* in form of more or less hairy rachilla appressed to surface of palea, *very rarely completely absent*. . . . . . . . . . . . . . . . . . . . . . . . . . . . . . . . . . . . . . . . . . . . . . . . . .2.

2. Panicles relatively small, at flowering time and later *more or less diffuse*; their branchlets 1–3 per node, *completely smooth or with sparse acicules only distally*, without spikelets for a considerable distance from their base and with few spikelets in total; longest branchlets usually ⅖ - ⅔ as long as whole panicle; spikelets 3.5–6.5 mm long; awn arising from below or above middle of lemma, usually more or less bent and not much (but sometimes by up to 1.5 mm) exceeding its tip; paleas equalling lemmas in length; callus hairs ½ - ¾ as long as lemma; glumes weakly scabrous to completely smooth. Relatively short (usually up to 25 cm high) plant of marshy places on the sea coast with prostrate and ascending epigeal shoots often rooting at the nodes. . . . . . . . . . .8. **C. DESCHAMPSIOIDES** TRIN.
– Panicle branchlets *copiously scabrous with acicules or aciculiform hairs*, rarely (in *C. Holmii*) sparsely scabrous or even smooth, but then panicles *more or less condensed* with strongly abbreviated branchlets and paleas distinctly shorter than lemmas. . . . . . . . . . . . . . . . . . . . . . . . . . . . . . . . . . . . . .3.

3. Awn arising from back of lemma always considerably below its middle, strongly twisted on its basal half, then (usually at level of distal third of lemma) *geniculately bent, exceeding tip of lemma by 1.2–4 mm and usually far exserted from spikelet*; callus hairs always less than half as long as lemma; paleas *equalling or almost equalling length of lemmas*; spikelets 4–8 mm long, aggregated in dense spikelike panicles with much abbreviated branchlets. Plant of stony or rocky habitats, forming rather dense tufts (but with extravaginal shoots); stems with (1)2(3) nodes. . . . . . . . . . .4.
– Awns of lemmas *straight or relatively slightly bent*, not exceeding tips of lemmas or *exceeding them by no more than 0.8 mm*; callus hairs *usually more than half as long as lemma*, very rarely (in *C. neglecta*) slightly shorter than lemma (but then awn straight or almost so); paleas *always distinctly shorter than lemmas*. . . . . . . . . . . . . . . . . . . . . . . . . . . . . . . . . . . . . . . . 7.

4. Leaf blades linear or lanceolate-linear, 3–7 mm wide, flat, on upper surface *completely bare and smooth*, on lower surface and on margin more or less scabrous; shoots clothed basally with *rather numerous short scalelike sheaths without blades*. Spikelets 6–8 mm long; lemmas with awns exceeding their tips by 2.5–4 mm. . . . . . . . . . . . . . . . . . . . . . . . . .9. **C. KOROTKYI** LITW.
– Leaf blades narrowly linear (1–4 mm wide), often loosely folded lengthwise, on upper surface *more or less scabrous or short-haired*; shoots basally usually *with few scalelike sheaths without blades or completely lacking them*, but always with numerous sheaths of dead leaves loosely wrapped around bases of shoots. Lemmas with awns exceeding their tips by 1.2–3 mm. . . . . . . . . . . . . . . . . . . . . . . . . . . . . . . . . . . . . . . . . . . . . . . . . . . . . . .5.

5. Leaf blades on upper surface *with short but copious hairs*; plant on average larger, up to 80–100 cm high. Glumes 4–7 mm long, *gradually acuminate but not subulate* at tip; lemmas at tip with 2–4 relatively weakly expressed denticles, with dorsal awn arising near their base. . . . . . . . . . . . . .
. . . . . . . . . . . . . . . . . . . . . . . . . . . . . . . . . . . . . . . . .10. **C. PURPURASCENS** R. BR.
– Leaf blades on upper surface *scabrous with more or less numerous acicules, but without hairs*. Glumes *long and almost subulately acuminate at tip*, sometimes with more or less long mucro; lemmas at tip with 2–4 sometimes aristate-acuminate denticles. Smaller plant, usually up to 30–40 cm high. . . . . . . . . . . . . . . . . . . . . . . . . . . . . . . . . . . . . . . . . . . . . .6.

6. Glumes 5.5–8 mm long, on distal part *gradually* (almost subulately) acuminate, with their tips distinctly bent sideways; lemmas with awns arising usually *at level of one-third from their base* (i. e. significantly above base). . . . . . . . . . . . . . . . . . . . . . . . . . . . . . . . .12. **C. SESQUIFLORA** (TRIN.) TZVEL.
– Glumes 4–6 mm long, straight, *more abruptly* (almost subulately) acuminate at tip; lemmas with awns arising usually *near their base*. . . . . . . . . . . . .
. . . . . . . . . . . . . . . . . . . . . . . . . . . . . . . . . . . . . . . . . . . . . . .11. **C. ARCTICA** VASEY.

7. Plant with long or more or less abbreviated subterranean stolons; culms above base *with 1–2(3) nodes*; upper node at flowering time (and later) situated *near or below the middle of the culm*, which usually protrudes far

beyond the sheath of the upper culm leaf; leaf blades relatively narrow (1–4 mm wide), usually loosely folded lengthwise, more rarely flat. Panicles *dense, often spikelike, with strongly abbreviated branchlets*, of which the longest is usually ⅙-⅓ as long as the whole panicle; glumes *relatively short-acuminate*; lemmas herbaceous to beyond middle, usually scabrous over almost their whole surface; callus hairs equalling or (more often) shorter than lemmas. (*C. neglecta* s. l.). . . . . . . . . . . . . . . . . . . . . . . . .8.

– Plant with long subterranean stolons; culms above base *with 3–6 nodes*, on lower nodes sometimes with axillary vegetative shoots; uppermost node at flowering time and later situated *above the middle of the culm*, which on average protrudes less far from the sheath of the uppermost culm leaf; leaf blades on average broader (1–10 mm wide), usually flat, more rarely loosely folded lengthwise. Panicles *loose, often more or less diffuse, with relatively long branchlets*, of which the longest is usually ⅓-⅔ as long as the whole panicle; glumes *often long-acuminate*; lemmas herbaceous-hyaline, usually with acicules only on nerves; callus hairs almost equalling or slightly longer than lemmas. (*C. canescens* s. l). . . . .10.

8. Spikelets *3.8–6 mm long*, aggregated in dense or lax (often with drooping tip) paniculate inflorescence with strongly abbreviated branchlets; glumes relatively thin, but more or less scabrous with very short acicules over almost their whole surface; callus hairs very numerous, *equalling lemma in length or more rarely slightly shorter*; awn straight or slightly bent, arising at basal third of lemma or near its middle. Leaf blades almost always more or less scabrous on lower surface, on upper surface normally sparsely scabrous with very short (often tuberculate) acicules of relatively scattered disposition, but sometimes with sparse, rather long hairs; ligules on average longer, 2–5 mm long on upper culm leaves; joint of sheath and blade bare or with beard of short hairs. . . . .5. **C. LAPPONICA** (WAHLB.) HARTM.

– Spikelets *2–4.5 mm long*, aggregated in rather dense paniculate inflorescence with erect (not drooping) axis; callus hairs less numerous, *usually ½-¾ as long as lemma*. Leaf blades usually smooth or weakly scabrous on lower surface, on upper surface with numerous long or short acicules which are sometimes transformed into short (but still numerous) hairs; ligules on average shorter, 0.5–3 mm long on upper culm leaves; joint of sheath and blade always bare. . . . . . . . . . . . . . . . . . . . . . . . . . . . . . . . . . . . . 9.

9. Branchlets of panicle more or less scabrous, often weakly so, sometimes completely smooth; glumes relatively thin, *completely smooth or with short acicules of scattered disposition only on keel*, until flowering *usually rosy-violet, later with more or less golden margin, rarely greenish*, rather gradually acuminate at tip; lemmas *broadly hyaline* distally; awn arising at basal third of lemma or beyond (sometimes even on distal half) and usually not much exceeding its tip, at first straight or almost so, later more or less bent. Ligules of leaves very short, usually up to 1.5–2 mm long. . . . . . . .
. . . . . . . . . . . . . . . . . . . . . . . . . . . . . . . . . . . . . . . . . . . . . . .7. **C. HOLMII** LGE.

– Branchlets of panicle always strongly scabrous; glumes on average thicker and less hyaline, *always covered with short acicules not only on keel but also on their surface* (at least distally), *usually* (but not always) *darker coloured*, short- or long-acuminate at tip; lemmas distally *with narrower hyaline border and usually thicker overall*; awn straight or more rarely slightly bent, arising near or before middle of lemma and usually not exceeding its tip. Ligules of leaves longer on average. . . . . . . . . . . . . . . . . . . . . .
. . . . . . . . . . . . . . . . . . . . . . .6. **C. NEGLECTA** (EHRH.) GAERTN., MEY. ET SCHERB.

10. Rachilla of spikelet *absent or very short* (up to 0.5 mm long), *bare or more rarely with relatively few hairs*; lemmas *with very weakly developed, often almost unnoticeable awn* arising on their distal quarter, often at the very tip; glumes usually with very short acicules only on keels, more rarely with sparse acicules (inconspicuous even at high magnification) also over their surface. Plant 40–120 cm high; leaf blades usually 1.5–3.5 mm wide, flat or loosely folded lengthwise; ligules on leaves of vegetative shoots 1–2 mm long, on uppermost culm leaf 3–6 mm long, more or less scabrous on back to completely bare and smooth. . . . . . . . . . . . . .2. **C. CANESCENS** (WEB.) ROTH.

– Rachilla of spikelet *almost always present, rather copiously hairy*, 0.2–0.8 mm long; lemmas usually *with better developed awn arising below or above their middle (sometimes on distal third)*; glumes covered over their whole surface with rather long acicules which are conspicuous at high magnification and sometimes transformed into very short hairs. . . . . . .11.

11. Leaf blades usually *3–8 mm wide*, flat, *rather strongly divergent from axis of culm;* ligules on leaves of vegetative shoots 1.5–4 mm long, on uppermost culm leaf 4–8 mm long, usually with very short hairs on back; plant on average taller (40–120 cm high) *with relatively thick* culms. Glumes on average *larger* (3–6 mm long) *and longer-acuminate*. . . . . . . . . . . . . . . . . . . . .
. . . . . . . . . . . . . . . . . . . . . . . . . . . . . . . . . . 3. **C. LANGSDORFFII** (LINK) TRIN.

– Leaf blades usually *1.5–3.5 mm wide*, flat or loosely folded lengthwise, *less divergent from axis of culm* and usually firmer; ligules on leaves of vegetative shoots 1.5–3 mm long, on uppermost culm leaf 2.5–6 mm long, scabrous or with very short hairs on back; plant on average shorter (25–80 cm high) *with relatively thin* culms. Glumes on average *smaller* (2.4–4 mm long) *and shorter-acuminate*. . . . . . . . . . . . . . . . . . . .4. **C. ANGUSTIFOLIA** KOM.

**1. *Calamgrostis epigeios*** (L.) Roth, Tent. Fl. Germ. I (1788), 34; Grisebach in Ledebour, Fl. ross. IV, 432; Krylov, Fl. Zap. Sib. II, 224; Petrov, Fl. Yak. I, 188; Rozhevits in Fl. SSSR II, 194; Perfilev, Sl. Sev. I, 77; Hultén, Atlas, map 163; Kuzeneva in Fl. Murm. I, 164; Hylander, Nord. Kärlväxtfl. I, 314; Karavayev, Konsp. fl. Yak. 51.

*Arundo epigeios* L., Sp. pl. (1753) 81.

Ill.: Fl. SSSR II, pl. XIV, fig. 1; Fl. Murm. I, pl. LIII, fig. 1.

Widely distributed Eurasian species, occurring almost equally abundantly in the forest and steppe zones. Recorded only for a few sites in the European and West Siberian Arctic, where it is sometimes an introduction. Arctic specimens often have imperfectly developed spikelets and apparently rarely set seed.

Occurring usually at sites with more or less broken grass cover; on sands and gravelbars in valleys of rivers and lakes, on embankments of various kinds, and along roads.

**Soviet Arctic.** Murman (near Murmansk; recorded also for Ponoy settlement); Bolshezemelskaya Tundra (on the River Shapkina); Ob Basin (on the Poluy).

**Foreign Arctic.** Arctic Scandinavia.

**Outside the Arctic.** Almost all Europe and a considerable part of Asia (except southeastern and southern tropical districts and NE Siberia). As an introduction in eastern states of the USA and certain other countries.

2. ***Calamagrostis canescens*** (Web.) Roth, Tent. Fl. Germ. II, 1 (1789), 93; Hylander, Nord. Kärlväxtfl. I, 315.

*C. lanceolata* Roth, Tent. Fl. Germ. I (1788), 34; Grisebach in Ledebour, Fl. ross. IV, 431; Krylov, Fl. Zap. Sib. II, 223; Rozhevits in Fl. SSSR II, 203; Perfilev, Fl. Sev. I, 76; Hultén, Atlas, map 161; Kuzeneva in Fl. Murm. I, 164.

*C. vilnensis* Bess. in Roemer et Schultes, Add. ad Mant. II (1817), 602; Perfilev, l. c. 77.

*Arundo calamagrostis* L., Sp. pl. (1753) 81.

*Agrostis canescens* Web. in Wiggers, Prim. Fl. Holsat. (1780) 10.

**Ill.:** Fl. SSSR II, pl. XIV, fig. 12, 15; Fl. Murm I, pl. LIII, fig. 2.

Predominantly European boreal species, reaching north to the Kola Peninsula and Arctic Scandinavia. In NE Europe the range of this species significantly overlaps that of *C. Langsdorffii*, as a result of which numerous hybrid populations and clones appear here; these advance on average further northwards than do typical populations of *C. canescens*.

Thus, fully typical specimens of this species are known on the Kola Peninsula only from the Khibins district and from the south coast, while specimens more or less deviating towards *C. Langsdorffii* occur here almost everywhere. In many districts where *C. canescens* and *C. Langsdorffii* grow together, hybrid populations occur much more often than populations of the "pure" species. But despite the wide distribution of populations more or less intermediate between them, it can be postulated that these species were formerly fully separated both geographically and morphologically. It is not fortuitous that *C. canescens* is associated with more southern pine and mixed forests, appearing in Northern Europe as an undoubted relict of a warmer time, while *C. Langsdorffii* is basically native to the northern forest (taiga) zone and apparently penetrated Europe only with the onset of cooling during the Quaternary Period (together with a whole series of other taiga species with similar ranges).

**Soviet Arctic.** Murman (only hybrid populations more or less deviating towards *C. Langsdorffii*).

**Foreign Arctic.** Arctic Scandinavia (also hybrid populations; according to Hultén, typical populations reach 69°30'N).

**Outside the Arctic.** Greater part of Europe north-east to the middle course of the Severnaya Dvina and the Pechora, south-east to the Southern Ukraine and Saratov Oblast; Southern West Siberia and Northern Kazakhstan; further east sporadically to Baykal (almost exclusively hybrid populations).

3. ***Calamagrostis Langsdorffii*** (Link) Trin., Gram. Unifl. (1824) 225; Grisebach in Ledebour, Fl. ross. IV, 430; Gelert, Fl. arct. 1, 106; Krylov, Fl. Zap. Sib. II, 219; Petrov, Fl. Yak. I, 218; Tolmachev, Fl. Kolg. 13; Rozhevits in Fl. SSSR II, 213; Perfilev, Fl. Sev. I, 77; Leskov, Fl. Malozem. tundry 22; Kuzeneva in Fl. Murm. I, 166; Böcher & al., Grønl. Fl. 295; Karavayev, Konsp. fl. Yak. 51.

*C. scabra* C. Presl, Rel. Haenk. 1 (1830), 234.

*C. phragmitoides* Hartm., Handb. Scand. Fl. (1832) 20; Lange, Consp. fl. groenl. 159, pro parte; Perfilev, Fl. Sev. I, 76.

*C. flexuosa* Rupr. in Beitr. Pflanzenk. Russ. Reich. IV (1845), 34; Rozhevits, l. c. 209.

*C. elata* Blytt, Norsk Fl. (1847) 148; Krylov, l. c. 222; Rozhevits, l. c. 210; Perfilev, l. c. 77; Kuzeneva, l. c. 166.

*C. unilateralis* Petr., Fl. Yak. I (1930), 201.

*C. grandis* Petr., l. c. 216.

*C. fusca* Kom. in Izv. Bot. sada AN SSSR XXX (1932), 197; Rozhevits, l. c. 215; Karavayev, l. c. 52.

*C. canadensis* ssp. *Langsdorffii* (Link) Hult., Fl. Al. II (1942), 161.

*C. confusa* V. Vassil. in Bot. mat. Gerb. Bot. inst. AN SSSR XIII (1950), 49.

*C. purpurea* auct. non Trin. — Hultén, Atlas, map 167; Hylander, Nord. Kärlväxtfl. I, 316.

*C. canadensis* var. *Langsdorffii* (Link) Inman in A. E. Porsild, Ill. Fl. Arct. Arch. (1957) 26.

*C. canadensis* auct. non Beauv. — Polunin, Circump. arct. fl. 44, pro parte.

*Arundo Langsdorffii* Link, Enum. Hort. Berol. II (1821), 74.

**Ill.**: Trinius, l. c., tab. 4, fig. 10; Petrov, l. c., fig. 67, 75; Fl. SSSR II, pl. XV, fig. 1, 2, 10, 14; Fl. Murm I, pl. LIV.

Predominantly Siberian boreal species, widely distributed in the taiga zone and extending south in mountain ranges to Mongolia and Northern China. Displaying considerable polymorphism, and joined by numerous transitional populations with *C. canescens* in the west and with *C. angustifolia* in the east. However, the further subdivision of this species into ecological-geographical races is beset with great difficulties, because of which we prefer for the present to retain it in a wide sense. Specimens of *C. Langsdorffii* from Scandinavia and the European part of the USSR have on average more weakly developed awns on the lemmas (arising usually above the middle and not exceeding the tip) in comparison with typical East Siberian specimens, and can be accepted as a distinct, predominantly European geographical race whose prior name at species rank is *Calamagrostis phragmitoides* (described from Western Scandinavia). However, the absence of any other differences whatsoever between *C. phragmitoides* and *C. Langsdorffii*, as well as the absence of any well defined boundary between their ranges, compels us to abstain for the present from distinguishing *C. phragmitoides* as a separate species. Furthermore, it is very probable that the weaker awn development in specimens from the European part of the range of *C. Langsdorffii* is explained by longstanding introgression (through frequent hybridization) of this species (a relatively recent arrival in Europe) with the closely related but more "ancient" European species *C. canescens*, in which the awns are usually almost completely reduced.

*Calamagrostis phragmitoides* is in turn very closely approached by two other European species also included by us in *C. Langsdorffii*, *C. flexuosa* Rupr. (described from the vicinity of Pavlovsk in Leningrad Oblast) and *C. elata* Blytt (described from Norway). Type specimens of these species differ from the type specimens of *C. phragmitoides* only in having slightly larger spikelets and looser panicles with slender, often flexuous branchlets. These two species do not possess independent ranges and are apparently only ecological forms of *C. phragmitoides* associated with growing under densely shaded conditions. The other species cited by us in the synonymy are also very little different from typical specimens of *C. Langsdorffii*. The herbarium of the Botanical Institute of the USSR Academy of Sciences possesses types or isotypes of all species which, while not of course showing complete morphological agreement with isotypes of *C. Langsdorffii*, are insufficiently distinguished for them to be acceptable as distinct species. Thus, the isotype of *C. scabra* Presl (from Vancouver Island in North America) possesses longer-acuminate glumes than in isotypes of *C. Langsdorffii*; however, long-acuminate

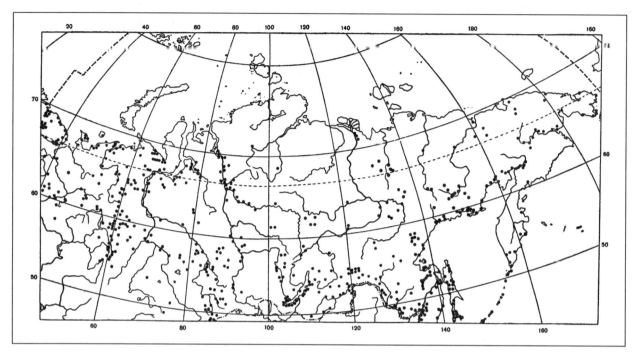

MAP II–10  Distribution of *Calamagrostis Langsdorffii* (Link) Trin.

glumes are in general more characteristic of *C. Langsdorffii*, and the short-acuminate glumes in the isotypes of that species are apparently the result of cultivation under more southern conditions (this species was described from specimens grown in the Berlin Botanical Garden). Somewhat better differentiated from the isotypes of *C. Langsdorffii* are the type specimens of *C. fusca* Kom. (described from Kamchatka) and *C. grandis* Petr. (described from the Dzhugdzhur Range), which noticeably deviate towards *C. angustifolia* and possibly belong to hybrid populations or clones. The type specimens of *C. unilateralis* Petr. and *C. confusa* V. Vassil. are in our opinion in full agreement with typical specimens of *C. Langsdorffii*, and can be placed in synonymy with the latter species without any reservation. However, *C. purpurea* Trin., described from the Baykal district, apparently represents a distinct though weakly differentiated Baykal-Sayan ecological-geographical race of the aggregate *C. canescens* s. l., possessing smaller, very numerous, densely arranged spikelets with very slightly scabrous glumes. We have found in the herbarium of the Botanical Institute of the USSR Academy of Sciences the type specimen of that species figured by Trinius which was unsuccessfully searched for by D. I. Litvinov (Tr. Bot. muzeya Ross. Ak. nauk, VIII, 56-62). Comparison of this with Scandinavian material of *C. Langsdorffii* clearly shows the error of attributing the name "*C. purpurea* Trin." (as prior) to *C. phragmitoides*, as done in recent works of Scandinavian authors (Hultén, l. c.; Hylander, l. c.).

In distinction from the previous species, *C. Langsdorffii* prefers well drained soils, occurring usually on the shores of waterbodies, in marshy meadows, and in shrub communities in valleys. The most northern specimens usually possess imperfectly developed sterile spikelets, and in the Anadyr-Penzhina territory, as well as in more southern districts of the Pacific Coast of Asia, specimens with the spikelets transformed into vegetative buds are occasionally found.

**Soviet Arctic.** Murman; Kanin; Malozemelskaya and Bolshezemelskaya Tundras; Kolguyev; Polar Ural; Yugorskiy Peninsula; Karskaya Tundra; lower reaches of Ob and Taz; Obsko-Tazovskiy Peninsula; southern part of Gydanskaya Tundra; lower reaches of Yenisey; lower reaches of Lena; lower reaches of Kolyma; Chukotka north of the Anadyr Basin (to the shore of the East Siberian Sea); Bering Strait district (Provideniye Bay, Arakamchechen Island, etc.); Anadyr and Penzhina Basins (rather frequent); Bay of Korf. (Map II–10).

**Foreign Arctic.** Beringian coast of Alaska (north to Norton Sound and St. Lawrence Island); arctic part of Canada west of Hudson Bay (rather frequent); Labrador (to extreme north); southern part of Baffin Island; Greenland (north to 70° on the west, to 65° on the east coast).

**Outside the Arctic.** NE part of Europe to Scandinavia and Poland in the west, to Moscow Oblast and the Southern Ural in the south; Siberia and the Far East; NE China; Korean Peninsula; Northern Japan; Alaska; Canada.

**4.** *Calamagrostis angustifolia* Kom. in Bot. mat. Gerb. Glavn. bot. sada AN SSSR VI (1926); Rozhevits in Fl. SSSR II, 214; Reverdatto in Sist. zam. mat. Gerb. Tomsk. univ. 1 (1941), 1.

?*C. czukczorum* Socz. in Fl. SSSR II (1934), 218, 749.

*C. tenuis* V. Vassil. in Bot. mat. Gerb. Bot. inst. AN SSSR VIII (1940), 66.

*C. magadanica* V. Vassil., l. c. XIII (1950), 49.

*C. hirsuta* V. Vassil., l. c. (1950) 50.

Ill.: Fl. SSSR II, pl. XV, fig. 13.

East Asian species similar to *C. canescens* in many respects and, like the latter, forming a large number of hybrid populations and clones transitional in characters to *C. Langsdorffii*. Very recently P. D. Yaroshenko (Hay fields and pastures of Primorskiy Kray, 1962, 25–29) on the basis of these "transitional forms" has proposed to reduce the rank of this species to that of a variety, *C. Langsdorffii* var. *angustifolia* (Kom.) Jarosch. However, in consideration of the no less abundant populations and clones transitional between *C. Langsdorffii* and two closely related species, the American *C. canadensis* (Michx.) Beauv. and the European *C. canescens* (Web.) Roth, this variety should be subordinated not to *C. Langsdorffii* but to one of these species, both of which were described earlier than *C. Langsdorffii*. Moreover, there are grounds for considering *C. angustifolia*, like *C. canescens*, a more "ancient" species than *C. Langsdorffii* which was in the past considerably better differentiated from the latter with respect to morphology. Therefore, the present existence of a considerable overlap of the ranges of the two species with formation of hybrid populations and clones may be considered secondary and a result of the onset of cooling during the Quaternary Period with associated advance of taiga and taigal elements southwards and to the sea coast. The type specimens of *C. angustifolia*, like many other specimens from the Amur Basin, possess a more or less distinct beard on the joints of the leaf sheaths and blades. But here too populations with completely bare joints are no less widely represented. The most northern specimens, as well as having bare sheath-blade joints, are usually of smaller size and possess more brightly coloured panicles, possibly forming a very weakly differentiated separate ecological-geographical race whose prior name at species rank is *C. czukczorum* Socz. (described from the Belaya River in the Anadyr Basin). The type of the latter species is at the present time, unfortunately, lost; however, indications in the original diagnosis of relatively long callus hairs and relatively long and slender panicle branchlets leave no doubt that this species belongs to the group of "*C. canescens* s. l." not to "*C. neglecta* s. l." to which it was referred by R. Yu. Rozhevits in the "Flora of the USSR." Evidently very similar to *C. czukczorum* are the species described by V. N. Vasilev from NE Siberia, *C. tenuis* V. Vassil. (Anadyr Basin), *C. magadanica* V. Vassil. (vicinity of Magadan) and *C. hirsuta* V. Vassil. (near the settlement of Kolyuchino on the Chukotka Peninsula). The insignificant differences existing between the type specimens of these species can scarcely provide grounds for considering them distinct.

Thus, the conversion of acicules on the glumes into very short hairs, which is a peculiarity of the type specimen of *C. hirsuta,* is sometimes found in certain populations and clones (usually under conditions of shade or abundant moisture) throughout the ranges of *C. Langsdorffii* and *C. angustifolia* and cannot possess substantial taxonomic importance.

*Calamagrostis angustifolia* grows mainly in marshy (especially peaty) meadows and marshes, on sandbars and gravelbars in river valleys, and sometimes in valley shrub communities (alder and willow carr), apparently preferring relatively poorly drained soils.

**Soviet Arctic.** District of Bay of Chaun (on the River Lyuleveyem); Anadyr and Penzhina Basins (rather frequent).

**Foreign Arctic.** Not occurring.

**Outside the Arctic.** Altay (eastern part), Yenisey Ridge and southern part of Central Siberian Plateau, Prebaikalia (sporadically); mountainous districts of Yakutia north to Verkhoyansk and Ozhogino settlement on the Indigirka; Okhotsk Coast and Preamuria; Kamchatka; Sakhalin; Kurile Islands; NE China; Korean Peninsula; Northern Japan.

5. ***Calamagrostis lapponica*** (Wahlb.) Hartm., Handb. Scand. Fl. (1820) 46; Trautvetter, Fl. taim. 18; Grisebach in Ledebour, Fl. ross. IV, 429; Krylov, Fl. Zap. Sib. II, 217; Rozhevits in Fl. SSSR II, 219; Perfilev, Fl. Sev. I, 78; Leskov, Fl. Malozem. tundry 22; Hultén, Fl. Al. II, 166; id., Atlas, map 164; Kuzeneva in Fl. Murm. I, 172; Hylander, Nord. Kärlväxtfl. I, 317; Böcher & al., Grønl. Fl. 295; A. E. Porsild, Ill. Fl. Arct. Arch. 26; Karavayev, Konsp. fl. Yak. 52; Polunin, Circump. arct. fl. 45.

*C. lapponica* var. *optima* Hartm., Handb. Scand. Fl., Ed. 5 (1849), 300.

*C. alascana* Kearney in U. S. Dept. Agricult. Div. Agrost. Bull. 11 (1898), 32.

*C. confinis* auct. non Beauv. — Gelert, Fl. arct. 1, 103; Tolmatchev, Contr. Fl. Vaig. 125; Perfilev, l. c. 78.

*C. sibirica* Petr., Fl. Yak. I (1930), 203; Karavayev, l. c. 52.

*C. Henriettae* Petr., l. c. 207.

*Arundo lapponica* Wahlb., Fl. lapp. (1812) 27.

**Ill.:** Wahlenberg, l. c., tab. 1; Petrov, Fl. Yak. I, fig. 68,69; Fl. SSSR II, pl. XVI, fig. 1; Fl. Murm. I, pl. LIII, fig. 8; Polunin, l. c. 44.

Almost circumpolar subarctic species, displaying relative constancy of morphological characters although not always clearly enough delimited from *C. neglecta*. Specimens from NE Asia have the awns of the lemmas on average attached more distally than in typical European specimens. Moreover, over almost the whole range of *C. lapponica* there occur two forms of this species joined by numerous intergrades but in many situations well differentiated: one (to which the type of the species belongs) with narrow panicles possessing more or less drooping tips and especially strongly abbreviated branchlets, almost always lacking a beard of hairs on the joints of the leaf sheaths and blades; the other ("var. *optima* Hartm.") with considerably looser panicles possessing longer branchlets, often with a conspicuous beard of hairs on the joints of the leaf sheaths and blades. These forms distinctly differ also in ecology. The first is a plant characteristic of stony and rocky tundras, while the second (in general a more southern and lower elevation form) usually occurs on gravel or sand in the valleys of rivers and lakes, occasionally descending there to below the upper boundary of forest. Lacking confidence in the constancy of the characters of these ecological forms (or races?), we prefer for the present to unite them under the single specific name *C. lapponica,* although V. A. Petrov (l.c.) has proposed the separate specific name *C. sibirica* Petr. (described from the upper reaches of the Tungir in Southern Yakutia) for the second of them and still earlier some authors identified this form with the American species *C. confinis* (Willd.) Beauv.; but according to reliably identified material pre-

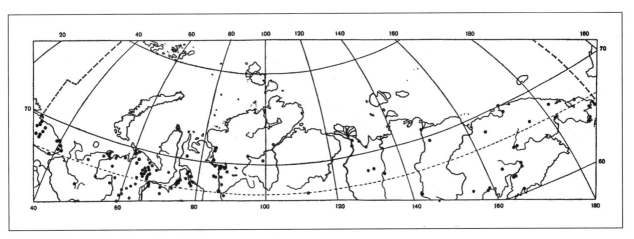

MAP II–11 Distribution of *Calamagrostis lapponica* (Wahlb.) Hartm.

sent in the Trinius herbarium, the latter species is a synonym (and prior name) for the very distinct American species *C. cinnoides* (Muhl.) Bart. Following R. Yu. Rozhevits, we also include among the synonyms of *C. lapponica* another of V. A. Petrov's species, *C. Henriettae* Petr., also described from the upper reaches of the Tungir from fully typical specimens of *C. lapponica* with somewhat looser interrupted panicle. Also scarcely different from *C. lapponica* is *C. alascana* Kearney described from Alaska in Kearney's monograph of the American species of *Calamagrostis*, and we follow Hultén (l. c. 1942) in including it among the synonyms of *C. lapponica*.

Like many other subarctic plants, this species is native to oligotrophic communities of dealkalinized acidic, usually adequately drained soils, sometimes of coarse mechanical composition (sandy, stony or rocky), and is often associated with a cover of fruticose lichens (reindeer moss); however, the range of habitats of *C. lapponica* in the more southern part of the tundra zone and the northern taiga subzone is rather wide.

The species is most characteristic of open reindeer-moss forest and glades on stony slopes and sandy terraces, and of dry tundras with rather little snow cover composed of dwarf shrubs and lichens; it also occurs on sandy and stony bluffs, stony summits, rocks and stony hillocks. It is rather common also in lichen-moss tundras and open forests (more often on elevations of microrelief, together with lichens), in dwarf shrub and shrub tundras, as well as in boggy (sphagnous) open forests, on peat hummocks and on bog margins; sometimes also recorded in meadowy areas as well as on sand and gravel deposits on floodplains.

In the zone below the arctic tundras the species is, as a rule, absent. In the basins of the Anadyr and Penzhina, as well as in the Verkhoyansk-Kolymsk mountain country, it often grows in openings among thickets of cedar-pine stlanik (on rocky or stony slopes). Occasionally growing profusely on ashes. Needs winter snow cover.

Beyond the limits of the Arctic the lower-elevational form of this species (*C. sibirica* Petr.) often penetrates sparse forests, especially of larch and birch.

**Soviet Arctic.** Murman; Malozemelskaya and Bolshezemelskaya Tundras; Polar Ural; Yugorskiy Peninsula; Vaygach Island; lower reaches of Ob; Obsko-Tazovskiy Peninsula; lower reaches of Yenisey (north to 71°43'); lower reaches of Lena; Tiksi; Chukotka Peninsula (Lawrence, Provideniye and Krest Bays); Anadyr and Penzhina Basins (rather common). (Map II–11).

**Foreign Arctic.** Alaska (district of Bering Strait); NW part of Canada west of Hudson Bay; Labrador (south to 55°N); southern part of Baffin Island; Greenland (west coast between 66° and 73°N, only var. *groenlandica* Lge.).

**Outside the Arctic.** Mountains of Scandinavia; Kola Peninsula and part of Karelia (occurring very sporadically, reaching Lake Ladoga); Timanskiy Ridge; Urals (to the Middle Ural inclusively); Altay; Sayan Mountains; Central Siberian Plateau; mountains of East Siberia south to Mongolia and NE China; Kamchatka; Sakhalin (Eastern Sakhalin mountains); Alaska; western part of Canada (Mackenzie Mountains, then sporadically south to the Saskatchewan River).

6. ***Calamagrostis neglecta*** (Ehrh.) Gaertn., Mey. et Scherb., Fl. Wett. I (1799), 94; Grisebach in Ledebour, Fl. ross. IV, 438; Gelert, Fl. arct. 1, 103; Lynge, Vasc. pl. N. Z. 103; Krylov, Fl. Zap. Sib. II, 218; Rozhevits in Fl. SSSR II, 215; Perfilev, Fl. Sev. I, 78; Hultén, Fl. Al. II, 167; id., Atlas, map 165; Kuzeneva in Fl. Murm. I, 168; Hylander, Nord. Kärlväxtfl. I, 317; Böcher & al., Grønl. Fl. 295; A. E. Porsild, Ill. Fl. Arct. Arch. 26; Karavayev, Konsp. fl. Yak. 52; Polunin, Circump. arct. fl. 44.

*C. stricta* (Timm) Koel., Descr. gram. (1802) 105; Petrov, Fl. Yak. I, 212; Leskov, Fl. Malozem. tundry 23.

*C. groenlandica* (Schrank) Kunth, Revis. Gram. I (1829), 79; Rozhevits, l. c. 216; Karavayev, l. c. 52.

? *C. inexpansa* A. Gray, N Amer. Gram. and Cyp. I (1834), No. 20; Hultén, Fl. Al. II, 164.

*C. borealis* Laest., Bidr. Vaextl. Torn. Lappm. (1860) 44.

*C. neglecta* var. *borealis* (Laest.) Trautv. in Tr. Peterb. bot. sada V (1877), 143; Kearney in U. S. Dept. Agricult. Div. Agrost. Bull. 11 (1898), 36; Tolmachev, Fl. Kolg. 13; Hultén, Fl. Al. II, 168.

*C. stricta* var. *borealis* (Laest.) Hartm., Scand. Fl., Ed. 2 (1879), 517; Lange, Consp. fl. groenl. 161.

*C. micrantha* Kearney, l. c. 36.

*C. kolgujewensis* Gand. in Bull. Soc. Bot. France LVI (1909), 533.

*C. jakutensis* Petr., Fl. Yak. I (1930), 214.

*C. neglecta* f. *arctica* Roshev. in Tolmachev & Pyatkov, Obz. rast. Diksona (1930) 153.

*C. neglecta* var. *micrantha* (Kearney) Stebbins in Rhodora XXXII (1930), 55.

*C. Reverdattoi* Golub. in Sist. zam. mat. Gerb. Tomsk univ. 4 (1936), 3.

*C. ochotensis* V. Vassil. in Bot. mat. Gerb. Bot. inst. AN SSSR VIII (1940), 214.

*Arundo neglecta* Ehrh., Beitr. VI (1791), 137.

*A. stricta* Timm in Siemss. Mecklenb. Mag. II (1795), 236.

*A. groenlandica* Schrank, Regensb. Denkschr. II (1818), 8.

**Ill.:** Petrov, Fl. Yak. I, fig. 72; Fl. SSSR II, pl. XV, fig. 15, 16; Fl. Murm. I, pl. LV; Porsild, l. c., fig. 41; Polunin, l. c. 44.

Circumpolar boreal species widespread in the forest zone of Eurasia and North America and penetrating rather far inside arctic limits. Displaying much polymorphism over the expanse of its wide range, and apparently constituting an intricate complex of ecological-geographical races linked by numerous transitional populations of hybrid origin. The heterogeneity of *C. neglecta* is also confirmed by cytological investigations, but we have not been able to divide this complex into separate smaller species at all satisfactorily. Three basic, more ecological than geographical, races are characterized by the size of the spikelets: *C. micrantha* Kearney, described from the Province of Saskatchewan in Canada, with spikelets 2–2.6 mm long; *C. neglecta* s. str., described from the vicinity of Uppsala in Sweden, with spikelets 2.6–3.2 mm long; and *C. stricta* (Timm) Koel., described from Germany, with spikelets 3.2–4 mm long. Spikelet size in these races is not correlated with other characters, including characters of the vegetative organs (thus, in all three races the leaf blades can either be very narrow and setiformly folded lengthwise or considerably wider and flat, and on their upperside they can either be scabrous or short-haired). The most common is the second of these races; the third occurs rather more rarely; the first occurs ocasionally in northern districts of the European part of the USSR and West Siberia, becoming more common in East

Siberia whence it has been described under two different specific names: *C. jakutensis* Petr. (from Yakutia) and *C. Reverdattoi* Golub. (near Nogayevo Bay on the Okhotsk Coast). However, both these species possess narrower (on average) leaf blades than does *C. micrantha* (judging from the original diagnosis of the latter). All more southern specimens of *C. neglecta* s.l. possess, as a rule, callus hairs ⅔-¾ as long as the lemmas, but more northern specimens commonly possess callus hairs not exceeding half the lemma length. Arctic specimens are normally also smaller and possess denser panicles with strongly abbreviated branchlets, and are sometimes distinguished as the separate variety *C. neglecta* var. *borealis* (Laest.) Trautv. based on the species *C. borealis* Laest. described from the northern part of Scandinavia. But both in Scandinavia and the Kola Peninsula this variety cannot be delimited at all clearly from typical specimens of *C. neglecta*. *Calamagrostis kolgujewensis* Gandoger, described from the collections of R. Pole on the dunes of Kolguyev, also possesses relatively small short-acuminate glumes and short callus hairs (about half as long as the lemmas) and is in our opinion fully included within the bounds of this more northern variety of *C. neglecta*. Somewhat more distinct (and perhaps transitional to *C. Holmii*) is the amphiatlantic ecological-geographical race described from Greenland and Labrador, *C. groenlandica* (Schrank) Kunth, which can be recognized as the subspecies *C. neglecta* ssp. *groenlandica* (Schrank) Matuszk. [Annal. Univ. Lublin, sect. C, 3 (1948), 242]. This is the most northern race of the complex *C. neglecta* s. l., possessing longer-acuminate glumes with brighter violet colouring (in *C. neglecta* ssp. *neglecta* the spikelets are usually paler coloured, brownish-violet or greenish). This subspecies differs from *C. Holmii* in having glumes scabrous not only on the keels, less hyaline lemmas and weaker awn development. Only this subspecies, widespread in Greenland and NE North America, occurs on Novaya Zemlya and Spitsbergen, while on Kolguyev and Vaygach apparently both subspecies occur and are linked by transitional populations although here somewhat isolated ecologically: *C. neglecta* ssp. *neglecta* occurs mainly on poorly vegetated sandy and gravelly areas, but *C. neglecta* ssp. *groenlandica* mainly in marshy areas of tundras with more abundant herb and shrub cover. In mainland districts of the European Arctic the morphological boundaries between these subspecies are completely obliterated, and east of Dikson Island *C. neglecta* ssp. *groenlandica* apparently completely disappears. Among Far Eastern material of *C. neglecta* s. l., in addition to specimens with small spikelets of the type of *C. jakutensis* and *C. Reverdattoi* mentioned above, there also occur specimens with relatively large spikelets (3.5–4 mm long) and with leaf blades more or less scabrous on the underside, significantly approaching the widespread North American species *C. inexpansa* A. Gray. An isotype of that species (Penn Yan, New York, Sartwell) present in the herbarium of the Botanical Institute of the USSR Academy of Sciences is very similar to these specimens, but has still larger spikelets (4–4.3 mm long), although according to the data of American authors the spikelet size in this species varies from 3 to 4.5 mm long. Thus it is very probable that *C. inexpansa* reaches NE Asia (including the Anadyr-Penzhina district). However, characters by which this species differs from specimens of *C. neglecta* with large spikelets are not entirely clear to us. With respect to the species *C. ochotensis* V. Vassil. described from the Ayan district on the Okhotsk Coast, we have not been able to find substantial differences between it and typical specimens of *C. neglecta*. According to the original diagnosis it differs from *C. neglecta* in having shorter leaf ligules and more acute lemmas with 4 denticles at their tip and a slightly bent awn; but the characters indicated could surely be considered characteristic of *C. neglecta*.

*Calamagrostis neglecta* grows in the Arctic mainly in more or less marshy meadowy areas of tundras, on the edges of marshes, on sands and gravelbars in

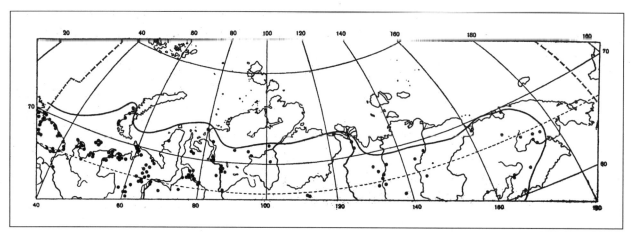

MAP II–12 Distribution of *Calamagrostis neglecta* (Ehrh.) Gaertn., Mey. et Scherb.

valleys of rivers and lakes (sometimes near the sea shore), and in shrub communities in valleys.

**Soviet Arctic.** Murman; Kanin; Malozemelskaya and Bolshezemelskaya Tundras; Kolguyev; Polar Ural; Yugorskiy Peninsula; Vaygach; Novaya Zemlya (mainly the South Island, on the North Island near Volchikha Bay; only ssp. *groenlandica*); Karsk-Baydaratsk Coast; lower reaches of Ob; southern part of Yamal (north to the Tiutey River); Obsko-Tazovskiy Peninsula; Gydanskaya Tundra; lower reaches of Yenisey (with Yenisey Bay); Dikson; northern slopes of Central Siberian Plateau; lower reaches of Lena and Kolyma; Anadyr and Penzhina Basins. (Map II–12).

**Foreign Arctic.** Alaska (district of Bering Strait); NW part of Canada west of Hudson Bay (rather frequent); Labrador (to extreme north), Canadian Arctic Archipelago (rather rare and only near the coast facing Greenland); Greenland (on both coasts north to 75°); Iceland; Spitsbergen; Arctic Scandinavia.

**Outside the Arctic.** Almost all Europe (except extreme south and south-east); northern part of Asia south to the Tien Shan, Mongolia and Northern Japan; disjunctly in the mountains of the Southern Transcaucasus, Turkey and Tadzhikistan; North America (south to the states of Oregon, Colorado, New York and New Hampshire).

7. **Calamagrostis Holmii** Lge. in Holm, Nov. Zem. Veget. (1885) 16; Gelert, Fl. arct. 1, 105; Tolmachev, Nov. dan. fl. Vayg. 125; Krylov, Fl. Zap. Sib. II, 217; Rozhevits in Fl. SSSR II, 217; Perfilev, Sl. Sev. I, 76; Hultén, Fl. Al. II, 164; ? Kuzeneva in Fl. Murm. I, 170; Karavayev, Konsp. fl. Yak. 52; Polunin, Circump. arct. fl. 42.

*C. neglecta* auct. non Gaertn., Mey. et Scherb. — Trautvetter, Pl. Sib. bor. 142, pro parte; id., Syll. pl. Sib. bor.-or. 64, pro parte; Tolmachev, Fl. Taym. I, 92; Tikhomirov, Fl. Zap. Taym. 21.

*C. kolymaensis* Kom. in Bot. mat. Gerb. Peterb. bot. sada II (1921), 129; Rozhevits, l. c. 219; Karavayev, l. c. 52.

*C. Bungeana* Petr., Fl. Yak. I, 209; Rozhevits, l. c. 218; Karavayev, l. c. 52.

?*C. Steinbergii* Roshev. in Izv. Bot. sada AN SSSR XXX (1932), 296 and in Fl. SSSR II (1934), 217.

*C. evenkiensis* Reverd. in Sist. zam. mat. Gerb. Tomsk. univ. 1 (1941), 3.

Ill.: Lange, l. c., tab. 1, fig. 2; Gelert, l. c., fig. 81; Petrov, Fl. Yak., fig. 70; Fl. SSSR II, pl. XV, fig. 17, 20, 21; Polunin, l. c. 43.

Almost exclusively Asian arctic species described from the mainland shore of Yugorskiy Shar. Very close to *C. neglecta* and not always very clearly distinguishable from the most northern race of that species, *C. neglecta* ssp. *groenlandica*, which replaces *C. Holmii* in many districts of the European and American Arctic.

Specimens of *C. Holmii* possessing looser panicles with relatively weakly scabrous or even smooth branchlets also significantly approach *C. deschampsioides,* sometimes only differing from that species in having shorter but usually more numerous branchlets on panicles which are always more or less condensed, and in the unequal length of the lemmas and paleas. Like *C. deschampsioides, C. Holmii* is highly variable with respect to spikelet size, length and place of attachment of awns, and length of callus hairs. In panicles of *C. Holmii* spikelets with two florets, as occasionally found in the Arctic and high mountains in other species of the genus, occur especially frequently and then the external resemblance of this species to some arctic species of the genus *Deschampsia* becomes still more complete. The appearance of spikelets with two florets in the genus *Calamagrostis* under extreme environmental conditions is an atavistic character, showing the existence of a rather close relationship between this genus and others of the tribe *Aveneae* which possess spikelets with many florets. When reviewing the abundant material of *C. Holmii* retained in the herbarium of the Botanical Institute of the USSR Academy of Sciences, we did not consider it possible to distinguish as separate units any of the subsequently described species listed above in synonymy. As in many other cases, the types of these species are more or less distinct from the isotype of *C. Holmii* present in the same herbarium, but with a large material the insubstantial nature of these differences becomes perfectly obvious. Thus, when entering coastal marshy areas, *C. Holmii* sometimes produces a peculiar form particularly approaching *C. deschampsioides* and apparently of hybrid origin. This form was described from the lower reaches of the Kolyma as the separate species *C. kolymaensis* Kom.; but specimens collected at the same time as the type specimens of this species but on more elevated portions of the shore show such a gradual transition to typical specimens of *C. Holmii* that the separation of *C. kolymaensis* is scarcely tenable. The type specimens of *C. Bungeana* Petr., described from the collections of A. Bunge from the Yana (near the settlement of Ulakhan-Sular), are larger and often have a greenish, only slightly tinted panicle; but these characters in our opinion have no substantial importance. In Yakutia *C. Holmii* penetrates furthest of all to the south on the barrens of mountain ranges, and it is not surprising that larger specimens occur here along with smaller typical specimens. Also little different from *C. Holmii* is *C. evenkiensis* Reverd., described from the district of Lake Khurinda in the Evenkiyskiy National Okrug, about whose differences from *C. Holmii* nothing is said in the original diagnosis. Another arctic species doubtfully included by us among the synonyms of *C. Holmii, C. Steinbergii* Roshev. (described from Litke Island on the coast of Yamal), remains inadequately clarified due to loss of the type specimen and the complete absence of other authentic specimens. In several characters in the original diagnosis this is closer to *C. Holmii,* but in the length of the callus hairs (⅓-½ as long as the lemmas) it approaches *C. neglecta* ssp. *groenlandica.* All the same, it may be more probable to suppose that it belongs to *C. Holmii,* and that the weak development of callus hairs, like the weak development of awns, was merely the result of general depauperation of the specimens (according to the original diagnosis they were in total 10-12 cm high).

Common plant of the more southern subzones of the tundra zone; absent from the highest arctic districts. Growing in places protected by snow in winter, more or less well drained; preferring fine soil substrates; avoiding both continuously moistened and very dry areas. Forming shoots within the soil or moss carpet, growing in loose (or in certain circumstances relatively dense) beds entwined in the tundra turf; sometimes accompanied by mosses.

Thus, in the lower reaches of the Lena this species often grows (though in relatively low abundance) in the tundras of marshy lowlands (polygonal, with flat hummocks, etc.), invariably colonizing their relatively well drained microeleva-

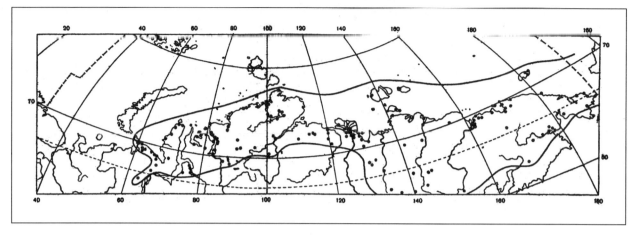

MAP II–13  Distribution of *Calamagrostis Holmii* Lge.

tions together with numerous mosses, prostrate species of *Salix*, etc.; it also occurs in *Eriophorum*-tussock tundras of watersheds. Wherever portions of marshy tundra with peaty gleyed soil dry out as a result of bank erosion, *C. Holmii* grows profusely, sometimes forming a closed sward; the species is also characteristic of bank slumps, riverine sand and gravel deposits, and shrub thickets of willow and dwarf birch in valleys; colonizing stony areas only when there is good snow cover and adequate moisture.

**Soviet Arctic.** Murman (recorded by O. I. Kuzeneva for the district of the settlement of Gremikha, but probably in error for the hybrid *C. neglecta* × *C. deschampsioides*); Bolshezemelskaya Tundra (eastern part); Polar Ural; Yugorskiy Peninsula; Vaygach; Yamal; Obsko-Tazovskiy Peninsula; Gydanskaya Tundra; coasts of Yenisey Bay and lower reaches of Yenisey (south to 69°N); Taymyr; northern slope of Central Siberian Plateau; lower reaches of Olenek and Anabar; lower reaches of Lena (to its mouth) and further east along the whole coast of the Arctic Ocean to the Chukotka Peninsula, including the New Siberian Islands and Wrangel Island; Anadyr Basin (rather frequent); Penzhina Basin (upper reaches of the River Palmatkina). (Map II–13).

**Foreign Arctic.** Distribution unclarified.

**Outside the Arctic.** Northern part of Central Siberian Plateau (southern boundary not established precisely, probably to 68°N); Verkhoyansk Range (south to the Tompo Basin at about 63°N) and mountainous districts of Yakutia to the east of this; St. Paul Island (according to Hultén).

8. ***Calamagrostis deschampsioides*** Trin., Icon. Gram. III (1836), tab. 354; Grisebach in Ledebour, Fl. ross. IV, 427; Gelert, Fl. arct. 1, 105; Rozhevits in Fl. SSSR II, 219; Perfilev, Fl. Sev. I, 76; Hultén, Fl. Al. II, 163; id., Atlas, map 162; Kuzeneva in Fl. Murm. I, 170; Hylander, Nord. Kärlväxtfl. I, 317; Karavayev, Konsp. fl. Yak. 52; Polunin, Circump. arct. fl. 42.

*C. inopia* Litw. in Bot. mat. Gerb. Glavn. bot. sada II (1921), 125.

*C. bracteolata* V. Vassil. in Bot. mat. Gerb. Bot. inst. AN SSSR XIII (1950), 52.

Ill.: Trinius, l. c., tab. 354; Gelert, l. c., fig. 80; Fl. SSSR II, pl. XV, fig. 22; Fl. Murm. I, pl. LIII, fig. 7.

Littoral arctic species of East Asian origin, extending south along the Pacific Coast of Asia to the mouth of the Amur and to Sakhalin. Easily distinguished from other species of the genus both by the panicle (with relatively few spikelets and rather long, smooth or slightly scabrous branchlets) and by the peculiar habitus, especially the presence of epigeal prostrate vegetative shoots which root at the

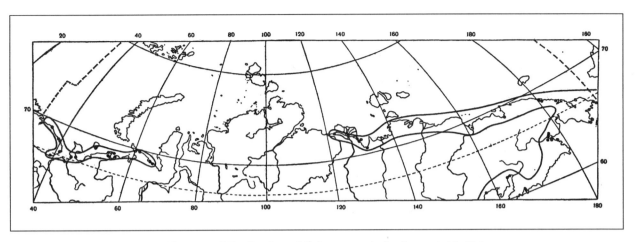

MAP II-14 Distribution of *Calamagrostis deschampsioides* Trin.

nodes. Apparently, like some other littoral species growing under similar conditions, for example *Puccinellia phryganodes* (Trin.) Scribn. et Merr., this species does not always produce reproductive culms and can easily be overlooked in the vegetative state. Perhaps this, as well as shortage of suitable habitats, is partly responsible for the gaps in the range of the species, which is in general rather sporadically distributed. Many characters of the reproductive organs (size of spikelets, degree of awn development and site of their origin from the back of the lemmas, length of callus hairs) show much variation in *C. deschampsioides*, as in *C. Holmii*, associated apparently with differences in the environmental circumstances of particular specimens and populations: namely, the structure and moisture content of the soil, the degree of vegetative cover, the duration of periods of flooding, and the salt content of the water. In particular, the degree of development of the awn (which may either arise near the base of the lemma and considerably exceed its tip, or arise on the distal part of the lemma and be much reduced) has no taxonomic importance in *C. deschampsioides*. European and West Siberian specimens somewhat differ from East Siberian by the (on average) more weakly developed awns and shorter callus hairs, forming a weakly characterized ecological-geographical race of more western range, but in both parts of the range identical specimens also occur. Among the synonyms listed above, *C. inopia* Litw. (described from N. S. Turchaninov's specimens from the district of the Bay of Uda) differs from the majority of other specimens by the greenish panicle and very weak development of the rachilla, and apparently merely represents a fortuitous individual deviation; *C. bracteolata* V. Vassil. from Bering Island differs according to the original diagnosis from *C. deschampsioides* by the larger spikelets and the presence of a scalelike bract at the base of the panicles, however the spikelet size given (5–5.5 mm long) falls fully within the range of typical *C. deschampsioides* and the scalelike bract (like in *Poa bracteosa* Kom. described from Kamchatka) is merely the consequence of abnormal development of a particular specimen or group of specimens (usually under conditions of high humidity) and possesses no taxonomic importance. On the Chukotka Peninsula specimens of *C. deschampsioides* are occasionally found with still larger spikelets (5.5–7 mm long); these have been described from Kodiak Island (Alaska) as the subspecies *C. deschampsioides* ssp. *macrantha* Piper, but according to Hultén (Fl. Al. II, 163) typical specimens of *C. deschampsioides* with shorter spikelets are also common on that island.

Growing in saltmarsh meadows more or less subject to tidal flooding on the sea coast and in the estuaries of larger rivers, also on maritime sandbars or gravelbars, and sometimes on maritime rocks.

**Soviet Arctic.** Murman (sporadically, mainly on eastern part of coast); Kanin; Malozemelskaya and Bolshezemelskaya Tundras (sporadically on the coast and in the estuaries of the larger rivers); Kolguyev; Karsk-Baydaratsk Coast; coasts of Olenek Bay and Buorkhaya Bay; district of Bay of Chaun and the coast east of it as far as the Bering Strait (rather frequent); coast near the mouth of the Anadyr (rather frequent, penetrating the lower reaches of the Anadyr); coast of Penzhina Bay; Bay of Korf. (Map II–14).

**Foreign Arctic.** Alaska (district of Bering Strait).

**Outside the Arctic.** Kamchatka; coast of Sea of Okhotsk south to the mouth of the Amur; northern part of Sakhalin; Aleutian Islands and Pribilof Islands; south coast of Alaska; western part of Hudson Bay (var. *Churchilliana* Polunin). Records for Japan, where *C. deschampsioides* is recorded as a high alpine plant, are doubtful.

9. ***Calamagrostis Korotkyi*** Litw. in Sp. rast. Gerb. russk. fl. VIII (1922), 182; Rozhevits in Fl. SSSR II, 228.

*C. sylvatica* auct. non DC. — Regel & Tiling, Fl. Ajan. 127.

**Ill.:** Fl. SSSR II, pl. XVI, fig. 10.

This species is a morphologically very weakly differentiated more northern and high alpine ecological-geographical race of the species *C. Turczaninowii* Litw. (Bot. mat. Gerb. Glavn. bot. sada II, 115), which was described almost simultaneously. Both these species, which perhaps it would be more correct to unite, are widespread in the Baykal district and thence extend very sporadically east to the Sikhote-Alin and north through Southern Yakutia and the mountains of the Okhotsk Coast to the Penzhina Basin and Northern Kamchatka. In addition, a closely related species, *C. Tweedyi* Scribn., occurs in the Rocky Mountains of North America. Specimens from Koryakia (Bay of Korf) differ in having rather looser panicles, approaching in habitus another closely related species, *C. monticola* Petr. ex Kom. (= *C. Sugawarae* Ohwi), distributed from the mountains of the Okhotsk Coast to the Sikhote-Alin and Sakhalin and differing from *C. Korotkyi* in having considerably shorter callus hairs.

Growing on stony and rocky slopes and on rocks, further south also in shrubbery and sparse forest (usually of larch or pine) but always near outcrops of bedrock.

**Soviet Arctic.** Penzhina Basin (Palmatkinskiy Range) and vicinity of Bay of Korf.

**Foreign Arctic.** Absent.

**Outside the Arctic.** Prebaikalia, Bureinskiy Range and Sikhote-Alin; mountains of Southern Yakutia; mountains along coast of Sea of Okhotsk (very sporadically).

10. ***Calamagrostis purpurascens*** R. Br. in Richardson, Bot. App. Franklin Journ. (1823) 731; Lange, Consp. fl. groenl. 160; Rozhevits in Fl. SSSR II, 22; Hultén, Fl. Al. II, 170, excl. var.; A. E. Porsild, Ill. Fl. Arct. Arch. 26; Böcher & al., Grønl. Fl. 294; Karavayev, Konsp. fl. Yak. 52; Polunin, Circump. arct. fl. 42.

*C. sylvatica* auct. non DC. — Trautvetter, Pl. Sib. bor.; id., Fl. rip. Kolym. 78.

*C. yukonensis* Nash in Bull. N. Y. Bot. Gard. II (1901), 154.

*C. arundinacea* auct. non Roth — Gelert, Fl. arct. 1, 102.

*C. Czekanowskiana* Litw. in Bot. mat. Gerb. Glavn. bot. sada II (1921), 117.

*C. wiluica* Litw. ex Petr., Fl. Yak. I, 193; Rozhevits, l. c. 229; Karavayev, l. c. 52.

*C. caespitosa* V. Vassil. in Bot. mat. Gerb. Bot. inst. AN SSSR XIII (1950), 54; non Steud., 1855.

**Ill.:** Petrov, Fl. Yak. I, fig. 63; Fl. SSSR II, pl. XVI, fig. 12, 13; V. Vasilev in Bot. mat. Gerb. Bot. inst. AN SSSR XIII, 55; Porsild, l. c., fig. 4, b; Böcher & al., l. c., fig. 51, a; Polunin, l. c. 43.

Rather widespread subarctic species, only penetrating the limits of the Soviet Arctic in a few districts. Specimens from NE Asia are very similar to North

American specimens of this species present in the herbarium of the Botanical Institute of the USSR Academy of Sciences, but nevertheless we are not entirely convinced of their complete identity with the type specimens (Mackenzie distr., between Point Lake and Arctic Sea) which we have not seen. Despite relative constancy of morphological characters and complete separation from closely related species, particular populations of *C. purpurascens* have several times been described as separate species. Among the synonymized species listed above, *C. yukonensis* Nash (Yukon distr., at Dawson) differs according to the original diagnosis only in the larger size of the plant, and was apparently compared by the author not with *C. purpurascens* but with the considerably more distant species *C. Tweedyi* Scribn. only by mistake.

*Calamagrostis Czekanowskiana* Litw. was described from the basin of the River Olenek from fully typical specimens of *C. purpurascens* but with abortive spikelets. *Calamagrostis wiluica* Litw. ex Petr. (from a sandy exposure on the bank of a tributary of the River Vilyuy) was, like *C. yukonensis*, described from larger specimens of *C. purpurascens* collected on the southern boundary of its range. It is no accident that to this day this species remains known only from the type specimens. *Calamagrostis caespitosa* V. Vassil. (Ola district of Magadan Oblast) was described from fully typical but late collected specimens of *C. purpurascens*. At the same time, right to the present day all authors (including even V. N. Vasilev who did much work on the flora of NE Siberia) have included in *C. purpurascens* such completely distinct (both morphologically and geographically) species as *C. arctica* and *C. sesquiflora*.

In its ecology and biocenotic relations *C. purpurascens* closely approaches the cryophilic steppe plants of the flora of Arctic Siberia; like many of these it shows an association with skeletal soils; it grows only in the most continental districts of subarctic East Siberia, practically nowhere reaching all the way to the sea coast; its furthest penetration into the territory of the tundra zone[2] occurs in the district of the Bay of Chaun and in the Anadyr Basin (where the northern forest boundary dips strongly to the South).

In the Verkhoyansk-Kolymsk mountain country the species is particularly characteristic of rock debris and (less unstable) talus screes formed from limestones, diabases, shales, argillites and sandstones; it also occurs on the lower rocky or stony summits; most common on slopes of southern exposure. Often growing on ledges and in rock crevices. Always common also in the more mesophilic variants of cryophytic steppe communities on slopes with stony soil in the zone below the barrens and in dry meadowy areas at the foot of slopes with stony soil. Very sparsely occurring also in larch woods with grass and bog cranberry on dry rocky slopes (more often south-facing), on dry vegetated gravelbars or in moister meadowy areas of mountain valleys, on dry stony lichen tundras, in openings among thickets of cedar-pine stlanik or on ashes of cedar-pine stlanik. The species penetrates the southern part of the tundra zone in similar habitats in the upper reaches of the Anabar and the district of the Kharaulakh Range, as well as in the basins of the Anadyr and Penzhina.

As in the next two species, the numerous loosely arranged sheaths of dead leaves wrapped around the shoots undoubtedly play a protective role both against deep frosts and against overheating of the plant.

**Soviet Arctic.** Lower reaches of Khatanga and Anabar (apparently not reaching their mouths); lower reaches of Lena (north to its delta); lower reaches of Kolyma (almost to the mouth) and further east to the district of the Bay of Chaun; Anadyr and Penzhina Basins (rather frequent).

**Foreign Arctic.** Alaska (district of Bering Strait); NW part of Canada west of Hudson Bay (rather frequent); Labrador (between 55° and 60°N); Canadian Arctic

---

[2] In the eastern part of the American North, especially on Ellesmere Island and Greenland, in districts of enhanced continentality (removed from the coast) *C. purpurascens* reaches the extreme polar limits of land (83°N).

Archipelago (sporadically); Greenland (on the west coast from 64° to 78°N, on the east from 67° to 76°N; replaced in the south by the closely related species *C. Poluninii* Sørens.).

**Outside the Arctic.** Northern part of Central Siberian Plateau (southern boundary inadequately clarified); lower reaches of Vilyuy; mountains east of the Lena south to the southern part of the Verkhoyansk Range; Alaska and almost all Canada (south to the NW states of the USA).

11. ***Calamagrostis arctica*** Vasey in U. S. Dept. Agricult. Div. Bot. Bull. XIII, 2 (1893), tab. 55.

*C. purpurascens* var. *arctica* (Vasey) Kearney in U. S. Dept. Agricult. Div. Agrost. Bull. XI (1898), 191.

*C. purpurascens* ssp. *arctica* (Vasey) Hult., Fl. Al. II (1942), 170, pro parte.

**Ill.**: Vasey, l. c., tab. 55.

Predominantly Pacific East Asian subarctic species, very close to *C. purpurascens* but easily distinguished from it by the more or less scabrous (not copiously hairy) upper leaf surfaces and by the glumes being longer-acuminate at their tips. We have not seen reliably determined specimens of this species from North America (it was described from St. Paul Island), but the good illustration of the type specimen and its detailed description in Vasey's work cited above leave no doubt that Asian specimens belong to the same species. Hultén (l. c.) apparently united under the name "*C. purpurascens* ssp. *arctica*" both this species and *C. sesquiflora*, species which are in fact very close but readily distinguishable (without any transitional specimens or populations) and completely isolated geographically. It is interesting to note that the ranges of all three Asian species of the *C. purpurascens* group spread out like a fan from the Chukotka Peninsula, corresponding to the three routes of their penetration of Asia from North America; while *C. purpurascens* has the most "continental" range in the northern part of East Siberia, *C. arctica* is to a considerable degree a maritime species with range extending along the continental coast of the Pacific Ocean from Chukotka through the Kolymsk and Dzhugdzhur Ranges, and the range of the third species, *C. sesquiflora*, is the most "marine" extending from Chukotka, Kamchatka and the Aleutian Islands through the Kurile Islands to Northern Japan. In extreme NE Asia the ranges of these three species become considerably closer, but even here they are apparently completely isolated from one another. There is a further species very close to *C. arctica* named by us *C. kalarica* Tzvel. sp. nova [A *C. arctica* foliorum laminis supra laevibus vel sublaevibus, ramis paniculae sublaevibus et arista in ⅓ parte inferiore lemmatis fixa differt. Typus speciei: "Transbaicalia, jugum Kalaricum inter ditionis fl. Kalar. et fl. Kuanda, tundra in reg. alp., 24 VII 1932, No. 502, Savicz" in Herb. Inst. Bot. Acad. Sc. URSS (Leningrad) conservatur]. This occurs in isolation on the barrens of the Kalarskiy Range in the northern part of Chitinskaya Oblast (collections of Savicz, 24 and 30 VIII 1932, Nos. 502 and 573). This species, whose range appears to be an extension of that of *C. arctica* into Transbaikalia, differs from *C. arctica* in having almost smooth panicle branchlets and a more distal origin of the awns (at the level of the basal third of the lemmas). Apparently, the species *C. Vaseyi* Beal from the Rocky Mountains of North America is also very close to *C. arctica*.

Like the preceding species, *C. arctica* grows on stony slopes and rocks, occurring exclusively in mountainous districts.

Thus, in the south-east of the Chukotka Peninsula (district of Chaplino settlement) the species grows on stony slopes of erosional mountain terraces, climbing to 600 m a. s. l. It mainly colonizes dry windswept areas with little snow, growing as isolated tufts among bare stones or among discontinuous cover of lichens (*Alectoria ochroleuca, Cetraria nivalis,* etc.) or of dwarf trailing shrubs (*Dryas, Loiseleuria, Diapensia, Empetrum*). According to the report of V. A. Gavrilyuk, *C.*

*arctica* forms heads and begins to flower later than all other grasses under the conditions of the SE Chukotka Peninsula.

**Soviet Arctic.** Chukotka Peninsula (district of Bay of Krest and near Chaplino settlement); Anadyr Basin (near its mouth).

**Foreign Arctic.** Not occurring.

**Outside the Arctic.** Mountains near the coast of the Sea of Okhotsk (Kolymsk, Pribrezhniy and Dzhugdzhur Ranges); Southern Alaska (Alaska Peninsula); Aleutian Islands (?) and Pribilof Islands.

**12.** *Calamagrostis sesquiflora* (Trin.) Tzvel. comb. nova

? *C. urelytra* Hack. in Tokyo Bot. Mag. XII (1897), 28 and in Bull. Herb. Boiss. VII (1899), 653.

*C. purpurascens* auct. non R. Br. — Rozhevits in Fl. SSSR II (1934), 228, pro parte.

*Trisetum sesquiflorum* Trin. in Bull. Acad. Sc. Petersb. (1836) 66.

*Avena sesquiflora* (Trin.) Griseb. in Ledebour, Fl. ross. IV (1853), 419.

This completely distinct species, described from Kamchatka and Unalaska Island (Aleutian Islands) was identified by R. Yu. Rozhevits with *C. purpurascens* due to some kind of misunderstanding. In distinction from the two preceding species, this species is exclusively maritime, distributed from Chukotka, Kamchatka and the Aleutian Islands to Northern Japan and scarcely penetrating the Arctic. Specimens from the southern part of its range (Northern Japan, part of the Kurile Islands, Moneron) possess leaf blades almost or completely smooth on the upperside, and possibly form a distinct ecological-geographical race described from Northern Japan under the name *C. urelytra* Hack. and linked to the more northern typical race by transitional populations. Another ecological-geographical race very close to *C. sesquiflora, C. foliosa* Kearney, is found in the mountains of California.

Growing on rocks, open stony or clayey slopes, and sometimes on gravelbars in the valleys of rivers and streams, always near the sea coast.

**Soviet Arctic.** Chukotka Peninsula (near Chaplino settlement); Koryakia (in the district of Capes Govena and Olyutorskiy).

**Foreign Arctic.** Not occurring.

**Outside the Arctic.** Commander and Aleutian Islands; Kamchatka; Kurile Islands; Moneron Island (off the southern part of Sakhalin); Northern Japan.

### HYBRIDS

In the Arctic, hybrids of the long-awned species (*C. purpurascens, C. arctica, C. sesquiflora* and *C. Korotkyi*), whether with each other or with other species, are completely absent. This is possibly the result of their converting to self-pollination or apogamy. Hybrids between closely related species in the complexes *C. neglecta* s. l. and *C. canescens* s. l. are, as already noted, rather common but due to the relatively insignificant differences between the species difficult to recognize although almost always possessing a high proportion of defective pollen. The more rarely occurring hybrids between species of more distant relationship, possessing characters more or less intermediate between the parent species, can usually be identified on external appearance. The following of these can be found in the Soviet Arctic: *C. lapponica* × *C. Langsdorffii, C. neglecta* × *C. Langsdorffii, C. Holmii* × *C. Langsdorffii, C. Holmii* × *C. angustifolia, C. deschampsioides* × *C. neglecta, C. deschampsioides* × *C. Holmii.* The first four hybrids most often inherit a more condensed panicle with more or less abbreviated branchlets from species of the complex *C. neglecta* s. l., and a more or less increased number of culm nodes and longer callus hairs from species of the complex *C. canescens* s. l. The last two hybrids are usually closer to *C. deschampsioides* in appearance, but have more scabrous panicle branchlets and lemmas and paleas of unequal length.

## GENUS 10 — **Apera** Adans. — SILKY BENT

GENUS CONTAINING THREE species native to Europe, extratropical Asia and North America. One species, a widespread weed, is sometimes introduced to the Arctic.

1. *Apera spica-venti* (L.) Beauv., Agrost. (1812) 151; Grisebach in Ledebour, Fl. ross. IV, 442; Krylov, Fl. Zap. Sib. II, 227; Rozhevits in Fll. SSSR II, 233; Perfilev, Fl. Sev. I, 78; Hultén, Atlas, map 171; Kuzeneva in Fl. Murm. I, 174.
   *Agrostis spica-venti* L., Sp. pl. (1753) 61.

   Widely distributed weed, often infesting winter crops. Sometimes occurring in the far north as an introduction near buildings and in waste places.
   **Soviet Arctic.** Reported for a few settlements in Murman.
   **Foreign Arctic.** Occasionally in Arctic Scandinavia.
   **Outside the Arctic.** Western and Central Europe; European part of USSR; Caucasus; central and southern districts of Siberia. As a rare introduction in southern districts of the Far East.

## GENUS 11 — **Vahlodea** Fries. — VAHLODEA

SMALL GENUS OF bipolar distribution, formerly sometimes united with *Deschampsia* Beauv. but well differentiated from it by the structure of the caryopses (possessing on their ventral side a well expressed longitudinal groove with oblong-linear hilum) and apparently considerably more closely related to the genus *Aira* L. Containing in total four closely related species which replace one another geographically, one of which, *V. atropurpurea* (Wahlb.) Fries, occurs within the Arctic. Another species, *V. flexuosa* (Honda) Ohwi [=*V. paramushirensis* (Kudo) Roshev.] is distributed along the Pacific Coast of Asia and North America from Hokkaido through the Kurile Islands, Kamchatka, the Commander and Aleutian Islands (also reaching St. Lawrence Island, the most northern point of its range!) as far as Vancouver Island. This species differs from *V. atropurpurea* in a whole series of substantive characters (completely free caryopses, shorter callus hairs, hairy leaf blades, panicle branchlets hairy along their whole length or only on the distal part, etc.), but is connected to it by a third species, *V. latifolia* (Hook.) Hult. growing in the Rocky Mountains of North America south to the states of California and Colorado. Hultén (Fl. Al. II, 182) unites *V. flexuosa* and *V. latifolia* as the single subspecies *V. atropurpurea* ssp. *paramushirensis* (Kudo) Hult., but in our opinion both these species, like *V. atropurpurea*, are sufficiently distinct to be recognized as separate species, both geographically and morphologically (in *V. latifolia* the leaves are completely bare and the panicle branchlets only have short acicules right to the spikelets). A fourth species, *V. magellanica* (Hook. f.) Tzvel. comb. nova (=*Aira magellanica* Hook. f., Fl. Antarct. II, 1847, 376), recognized by Hultén as a third subspecies of *V. atropurpurea*, is distributed in the district of the Strait of Magellan.

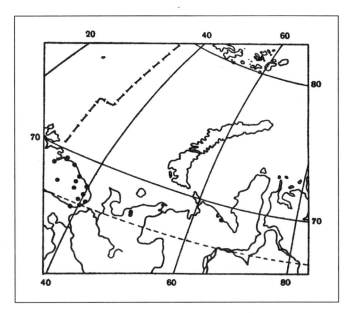

**MAP II-15** Distribution of *Vahlodea atropurpurea* (Wahlb.) Fries.

1. ***Vahlodea atropurpurea*** (Wahlb.) Fries in Bot. Notis. (1842) 141, 178; Lange, Consp. fl. groenl. 162; Gelert, Fl. arct. 1, 112; Rozhevits in Fl. SSSR II, 242; id. in Areal I, 25, map 10, a; Kuzeneva in Fl. Murm. I, 177; Hylander, Nord. Kärlväxtfl. I, 302; A. E. Porsild, Vasc. pl. W Can. Arch. 74; Hultén, Amphi-atl. pl. 204; Polunin, Circump. arct. fl. 75.

    *Aira atropurpurea* Wahlb., Fl. lapp. (1812) 37.

    *Deschampsia atropurpurea* (Wahlb.) Scheele in Flora XXVII (1844), 56; Grisebach in Ledebour, Fl. ross. IV, 423; Perfilev, Fl. Sev. I, 79.

    Ill.: Gelert, l. c., fig. 87; Fl. SSSR II, pl. XVIII, fig. 8; Fl. Murm. I, pl. LVIII; Polunin, l. c. 74.

    Amphiatlantic subarctic species, except for the Kola Peninsula known in the USSR only from a few widely separated localities. Despite its discontinuous range, it displays high constancy of morphological characters.

    Growing usually in more or less marshy meadows on the shores of rivers and streams and in sphagnum bogs with *Betula nana*, sometimes also in sparse willow or birch stands or on gravelbars.

    **Soviet Arctic.** Murman (frequent in whole district); northern part of the Timanskiy Ridge (on the Rivers Belaya and Velikaya); Karsk-Baydaratsk Coast; recorded by E. Hultén for the Polar Ural. (Map II-15).

    **Foreign Arctic.** NW Canada (at a single site on the coast of Hudson Bay near 61°N, and recorded for the district of the Mackenzie River); Labrador; South Greenland (south of 65°N); Arctic Scandinavia; recorded for the Canadian Arctic Archipelago (Somerset Island) but probably erroneously.

    **Outside the Arctic.** Mountains of Scandinavia (south to Southern Norway), the Khibins Mountains and the SE part of the Kola Peninsula; Newfoundland and the far eastern part of Canada; states of Maine and New Hampshire in the USA.

## GENUS 12    Deschampsia Beauv. — HAIR GRASS

GENUS INCLUDING UP TO 100 species, widely distributed in cold and temperate countries of both hemispheres. Represented in the Soviet Arctic by one species of the very distinct subgenus *Avenella* (Bl. et Fing.) Hyl. and 8 species of the larger subgenus *Deschampsia*. Three species of the genus, *D. borealis*, *D. brevifolia* and *D. alpina*, are high arctic, almost reaching the northern limit of the existence of vegetation, while the antarctic species *D. antarctica* Desv. is one of the two species of grasses so far known from Antarctica. Almost all the arctic species of subgenus *Deschampsia* are very closely related and readily hybridize with one another, apparently producing fertile hybrids. As a result of this, populations and specimens more or less intermediate between the different species are abundant and considerably impede identification. The complete lack of absolutely constant characters and insignificance of the differences between species possibly justifies the position taken by many authors right to the present time that many species of subgenus *Deschampsia* are units of lower taxonomic rank (subspecies or even varieties). Species of *Deschampsia* occur mainly in meadow or marsh-meadow communities. Some species grow on sands and gravelbars in river valleys, sometimes also on the sea coast although there are no strictly littoral species among them.

1. Plant 15–40 cm high, forming loose, rather sprawling tufts; leaf blades very narrow, *always setiformly folded lengthwise but not stiff*, 0.3–0.8 mm in diameter, *on upper (inner) surface without distinct ribs* and usually more or less covered with small papillae. Panicles 4–10 cm long, widely diffuse at flowering time and later; their branchlets slightly scabrous, with 1–3 nodes, usually more or less flexuous and with secondary branchlets strongly divergent from rachis; spikelets 4.5–5.5 mm long, always with 2 florets; segments of spikelet rachis *very short and thick*, about ⅕ as long as adjacent lemma; lemmas 4–5.5 mm long, on their basal part with a geniculate awn *which exceeds the tip of the lemma by 1.5–3 mm.* . . . . . . . . . . . . . . . . . . . . . . . . . . . . . . . . . . . . . . . . . . . . . . . . . . . . . . . . . . . . . 1. **D. FLEXUOSA** (L.) TRIN.

    – Plant forming dense tufts (except in cases where the tufts are continuously drifted over by sand or other fluvial deposits); leaf blades *flat or folded lengthwise*, sometimes very narrow but *always with distinct ribs on their upper surface*. Spikelets with 2–3 florets; segments of spikelet rachis *thin and rather long*, usually about ⅓ as long as adjacent lemma; lemmas with awns *exceeding their tips by no more than 1 mm*, frequently much reduced. . . . . . . . . . . . . . . . . . . . . . . . . . . . . . . . . . . . . . . . . . . . . . . . . . . . . . . . . . . . .2.

2. Plant 30–70 cm high, forming dense and rather large tufts; leaf blades flat or loosely folded lengthwise, on their upper surface with very prominent sharp ribs, *on margin and on ribs strongly scabrous with numerous, densely arranged acicules*. Panicles at flowering time widely diffuse, more or less pyramidal, *with branchlets scabrous along their whole, or almost their whole length;* spikelets 3–4.5 mm long, usually pinkish- or brownish-vio-

let, more rarely greeenish; lemmas 2.5–4 mm long, with straight or slightly bent awn *arising near base of lemma* and usually not exceeding its tip. . . . . . . . . . . . . . . . . . . . . . . . . . . . . . . . . . . . . . . . . . . . . . . . .2. **D. CAESPITOSA** (L.) BEAUV.

– Plant often smaller; leaf blades on upper surface with less prominent ribs, *on these and on margin with relatively few scattered acicules, occasionally completely smooth.* Panicles usually *with branchlets smooth or only scabrous on their distal half;* lemmas with awn *arising from their basal, middle or distal part* and often more or less reduced, *sometimes completely absent.* . . . . . . . . . . . . . . . . . . . . . . . . . . . . . . . . . . . . . . . . . . . . . . . . . . . . . . .3.

3. Panicles 2.5–10 cm long, usually more or less condensed and dense, more rarely diffuse; their branchlets usually completely smooth, usually more or less abbreviated and thickened; spikelets 4–6 mm long, usually with slight pinkish- or brownish-violet tinge, *in whole or part modified into vegetative buds* (vivipary); awns of lemmas usually much reduced, arising near their tip or completely absent. Plant 10–40 cm high; leaf blades 1–3 mm wide, usually loosely folded lengthwise and rather stiff. . . . . . . . . . . . . . . . . . . . . . .
. . . . . . . . . . . . . . . . . . . . . . . . . . . . . . . . . . . . . .8. **D. ALPINA** (L.) ROEM. ET SCHULT.

– Spikelets *never modified into vegetative buds;* lemmas and paleas normally developed (not elongated). . . . . . . . . . . . . . . . . . . . . . . . . . . . . . . . . . . . . . . . . 4.

4. Plant of poorly vegetated, usually shifting soils on the shores of rivers and coastal bays (sandy, gravelly or clayey alluvia), frequently forming rather loose tufts, 20–50 cm high; leaf blades flat or folded lengthwise, usually rather long and soft, those of upper culm leaf not much abbreviated. Panicles 7–20 cm long, *noticeably elongate in comparison with their width,* usually *more than ⅔ as long as rest of culm,* more or less diffuse; their branchlets often somewhat abbreviated, with scattered acicules distally or more rarely along almost their entire length; spikelets 4–6 mm long, usually more or less greenish, more rarely with slight pinkish-violet tinge; glumes narrowly lanceolate, elongate, the upper usually 4–6 mm long (*more or less equal in length to whole spikelet*), the lower *usually exceeding adjacent lemma in length* ; awns of lemmas usually arising near their middle (more rarely near their base) and not exceeding their tips. . . . . . . . . . . . .
. . . . . . . . . . . . . . . . . . . . . . . . . . . . . . . . . . . . . . . . . . . . .9. **D. OBENSIS** ROSHEV.

– Plant always densely tufted. Panicles diffuse or more or less condensed, frequently with more or less abbreviated branchlets, *never so long in comparison with the rest of the culm which exceeds the panicle by more than 1½ times;* spikelets more or less pinkish- or brownish-violet, more rarely greenish; glumes (except in *D. brevifolia*) broadly lanceolate, up to 4 mm long, the upper *usually shorter than spikelet,* the lower *usually not exceeding adjacent lemma in length.* . . . . . . . . . . . . . . . . . . . . . . . . . . . . . . . . . . . . . . . . . . . . .5.

5. High arctic plant 5–30 cm high, forming dense tufts; leaf blades usually rather stiff, loosely folded lengthwise, more rarely flat, those of upper culm leaf much abbreviated. Panicles 2–10 cm long, *condensed or diffuse, with branchlets directed obliquely upwards and usually noticeably abbreviated;*

awns of lemmas usually not exceeding their tips, sometimes much reduced. .................................................... 6.
- More southern plant, 10–60 cm high. Panicles 5–15 cm long, at flowering time and later *widely diffuse, with long branchlets spreading horizontally.* ............................................................... 7.

6. Spikelets *4–5.5 mm (usually about 4.5 mm) long;* glumes rather large, the upper *about 4 mm long,* the lower often slightly exceeding adjacent lemma. ........................................ 7. **D. BREVIFOLIA** R. BR.
- Spikelets *3–4 mm (usually about 3.5 mm) long;* glumes smaller, the upper *about 3 mm long,* the lower usually slightly shorter than adjacent lemma, more rarely equal to it in length. ......... 6. **D. BOREALIS** (TRAUTV.) ROSHEV.

7. Spikelets 3.4–4.5 mm long, more or less pinkish-violet with golden or silvery border; segments of spikelet rachis *bare or with a few hairs only on their distal half.* ........................... 4. **D. ANADYRENSIS** V. VASSIL.
- Segments of spikelet rachis *copiously hairy along their whole length or on their distal two-thirds.* ............................................... 8.

8. Panicles at flowering time and later *pyramidal,* with rather thickish branchlets; spikelets 3.5–4.5 mm long, *usually pinkish- or brownish-violet, more rarely greenish;* awns of lemmas usually *arising near their base and not exceeding their tip.* Leaf blades rather stiff, loosely folded lengthwise or flat, somewhat greyish-green. ...................... 3. **D. GLAUCA** HARTM.
- Panicles at flowering time and later *elongate-pyramidal,* with very thin, often flexuous branchlets; spikelets 3–4.5 mm long, *usually weakly tinted or greenish* ; awn of lemmas usually *arising near their middle and frequently exceeding their tip, often much reduced, sometimes completely absent.* Leaf blades softer, flat or folded lengthwise (sometimes almost setiform), green. ................... 5. **D. SUKATSCHEWII** (POPL.) ROSHEV.

   *1. Deschampsia flexuosa* (L.) Trin., Gram. Suppl. (1836) 9; Grisebach in Ledebour, Fl. ross. IV, 420; Perfilev, Fl. Sev. I, 79; Rozhevits in Fl. SSSR II, 244; Hultén, Fl. Al. II, 180; id., Atlas, map 181; Kuzeneva in Fl. Murm. I, 78; Hylander, Nord. Kärlväxtfl. I, 301; Hultén, Amphi-atl. pl. 250; Polunin, Circump. arct. fl. 50.
   *Aira flexuosa* L., Sp. pl. (1753) 65; Lange, Consp. fl. groenl. 162; Gelert, Fl arct. 1, 112. *A. montana* L., l. c. 65.
   *Avenella flexuosa* (L.) Parl., Fl. Ital. 1 (1848), 246.
   Ill.: Fl. SSSR II, pl. XVIII, fig. 9; Hylander, l. c., fig. 36d; Polunin, l. c. 50.
   Widely distributed, predominantly boreal species with disjunct range. Belonging to the morphologically very distinctive subgenus *Avenella* (Bl. et Fing.) Hyl., containing only a few closely related species and fully deserving to be separated as a distinct genus. Arctic and high alpine specimens of this species have on average larger (4.5–5.5 mm long, instead of 3.5–4.5 mm) and more intensively coloured spikelets, as well as smooth or almost smooth panicle branchlets, constituting a weakly differentiated ecological-geographical race known under the names *D. flexuosa* var. *montana* (L.) Gremli and *D. flexuosa* ssp. *montana* (L.) A. Löve. Although Linnaeus described this race as a separate species, *Aira montana* L., its differences from more southern specimens of *D. flexuosa* are so weakly expressed that it scarcely merits the rank of subspecies.

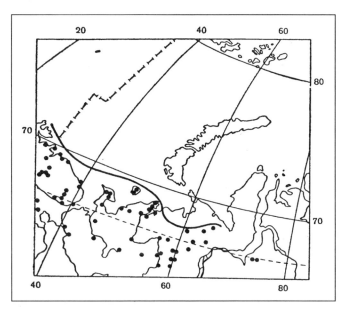

MAP II–16  Distribution of *Deschampsia flexuosa* (L.) Trin.

In the Arctic this species grows predominantly in more or less sparse forest and shrub communities, especially in birch stands, as well as on relatively dry, more or less stony or sandy areas of open tundras.

**Soviet Arctic.** Murman; Kanin; Kolguyev Island; Malozemelskaya and Bolshezemelskaya Tundras; Polar Ural; Obsko-Tazovskiy Peninsula (middle course of Nyda, 29 VII 1953, Norin). (Map II–16).

**Foreign Arctic.** Labrador; Greenland (south of 70°N); Iceland; Arctic Scandinavia.

**Outside the Arctic.** Almost all Europe except the south of the Balkan Peninsula and the southeast of the European part of the USSR; mountains of Caucasus and Turkey; Yenisey Ridge; Primorskiy and Khabarovsk Kray (south of the Dzhugdzhur Range); Sakhalin; Kamchatka (southern part); Kurile Islands; all Japan; NE China; Aleutian Islands; Pacific Coast of North America (Alexander Archipelago and near the City of Vancouver); eastern part of Canada and USA (east of the Mississippi Basin). Very closely related species or subspecies in the mountains of Northern and Equatorial Africa, and in Borneo, the Philippines, New Guinea, Taiwan and Southern South America. Found in New Zealand as an introduction.

2. ***Deschampsia caespitosa*** (L.) Beauv., Agrost. (1812) 91, 149, 160; Grisebach in Ledebour, Fl. ross. IV, 421; Krylov, Fl. Zap. Sib. II, 229, pro parte; Rozhevits in Fl. SSSR II, 421; Hultén, Fl. Al. II, 175; id., Atlas, map 180, pro parte; Kuzeneva in Fl. Murm. I, 180; Hylander, Nord. Kärlväxtfl. I, 397; A. E. Porsild, Ill. Fl. Arct. Arch. 27; Polunin, Circump. arct. fl. 48, pro parte.

*Aira caespitosa* L., Sp. pl. (1753) 64; Gelert, Fl. arct. 1, 113, pro parte.

Ill.: Fl. SSSR II, pl. XVIII, fig. 14; Fl. Murm. I, pl. LX; Polunin, l. c. 49.

Widely distributed, predominantly boreal species represented over the expanse of its wide range by a whole series of ecological-geographical races which are weakly differentiated morphologically and inadequately studied. Specimens from the Soviet Arctic belong to the most common and apparently typical race of this species with relatively larger, intensively coloured spikelets.

Growing usually in marshy meadows, on the shores of waterbodies, and on gravelbars, as well as in more or less marshy shrub communities (willow and birch carr).

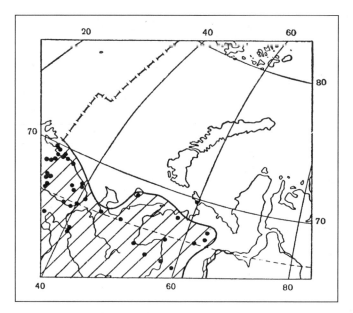

MAP II–17  Distribution of *Deschampsia caespitosa* (L.) Beauv.

**Soviet Arctic.** Murman; Kanin; Malozemelskaya and Bolshezemelskaya Tundras; Kolguyev (near Bugrino settlement, possibly of introduced origin); Polar Ural; Yugorskiy Peninsula (rare); Vaygach (Varnek Bay, probably introduced). (Map II–17).

**Foreign Arctic.** Arctic Alaska (records possibly referring to other species); NW Canada (records possibly referring to other species); Labrador; Canadian Arctic Archipelago (only Baffin Island south of the Arctic Circle); Iceland; Arctic Scandinavia.

**Outside the Arctic.** Almost all Europe; Middle East; West Siberia and Kazakhstan (north to the Severnaya Sosva, east to the Yenisey and Baykal, south to the Tien Shan); NW China; North America (except extreme south). Introduced to many extratropical countries of the Southern Hemisphere.

3. ***Deschampsia glauca*** Hartm., Handb. Scand. Fl. (1820) 448; Tsvelev in Bot. mat. Gerb. Bot. inst. AN SSSR XXI (1961), 43.

*D. caespitosa* var. *glauca* (Hartm.) Sam. in Svensk Bot. Tidskr. XIII (1919), 253; Hultén, Fl. Al. II, 176; Hylander, Nord. Kärlväxtfl. I, 298.

?*D. pumila* auct. — Ostenfeld in Meddl. om Grønl. LXIV, 6 (1923), 253; A. E. Porsild, Ill. Fl. Arct. Arch. 27; Polunin, Circump. arct. fl. 50; non *D. pumila* Fom. et Woron., 1907.

*D. borealis* auct. — Rozhevits in Fl. SSSR II, 246, pro parte; Kuzeneva in Fl. Murm. I, 180.

Ill.: Porsild, l. c., fig. 5, f; Polunin, l. c. 50.

Apparently almost circumpolar arctic-alpine species, occupying an intermediate position between the preceding species and such arctic species as *D. borealis* and *D. brevifolia*. It represents a more northern and alpine race of *D. caespitosa*, differing from that species in the smaller (on average) size of the plant as a whole, the smooth or almost smooth panicle branchlets, and the stiffer, somewhat greyish green leaf blades which are slightly scabrous on the upperside. At the base of tufts of *D. glauca*, as in other arctic species of the genus, there are numerous sheaths of dead leaves, considerably more conspicuous than in *D. caespitosa*. In East Siberia this species also rather closely approaches *D. Sukatschewii*, with which it forms transitional populations.

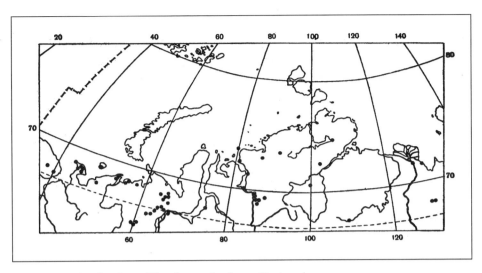

MAP II–18  Distribution of *Deschampsia glauca* Hartm.

The distribution of *D. glauca* outside the Soviet Union is still not entirely clear to us. Ostenfeld, on the basis of statistical measurements of the length of the callus hairs of the lemmas and some other characters, proposed the species *D. pumila* from Arctic America based on *D. brevifolia* γ *pumila* Griseb. (Ledebour, Fl. ross. IV, 422). The type of this species, which according to Ostenfeld is the specimen from Kamchatka illustrated by Trinius (Sp. Gram. III, tab. 254, c), is unlikely to belong to the species widespread in Arctic America, but is most likely a dwarf specimen of *D. Sukatschewii* with very narrow leaf blades. Moreover, the name "*D. pumila* Ostenf." is a subsequent homonym of the Caucasian species *D. pumila* (Stev.) Fom. et Woron. and cannot be used on this account. As for the North American specimens of the species described by Ostenfeld, they probably belong either to *D. borealis* or to *D. glauca*. In particular, the specimen illustrated by A.E. Porsild as "*D. pumila*" belongs more probably to *D. glauca* than to *D. borealis*, although the higher arctic range of *D. pumila* indicated in A. Porsild's work corresponds better with *D. borealis*. Ecological-geographical races very close to *D. glauca* apparently occur in the mountains of Central and Southern Europe. Such, for example, is *D. Biebersteiniana* Roem. et Schult. (=*Aira brevifolia* M. B.) described from the Great Caucasus Mountains ("circa acidulam Nartsana"), which possibly should be considered a subsequent synonym of *D. glauca*. *Deschampsia glauca* displays rather great variation with respect to spikelet size, usually rather closely approaching *D. borealis*. But in the high mountains of the Middle and Prepolar Urals specimens with very large spikelets are known, in this respect identical to *D. brevifolia* but possessing large diffuse panicles. It is possible that such specimens belong to a distinct North-Uralian ecological-geographical race (like, for instance, *Calamagrostis uralensis* Litw., which differs from *C. Langsdorffii* also in having larger spikelets), but specimens closely approaching them also occur in the Ponoy Basin on the Kola Peninsula where typical *D. glauca* is widespread.

Occurring usually on the shores of mountain streams, lakes and rivers on gravelbars or in more or less marshy meadows, sometimes also on the margins of sphagnum bogs.

**Soviet Arctic.** Murman (specimens available from the basins of the Ponoy and Iokanga); Kanin (northern mountainous part); Kolguyev; northern part of Timanskiy Ridge; Bolshezemelskaya Tundra; Polar Ural; Pay-Khoy; Obsko-Tazovskiy Peninsula; Taymyr (south from the Byrranga Mountains, rather rare); northern edge of Central Siberian Plateau (specimens partly transitional to *D.*

*Sukatschewii*); lower reaches of Lena; northern part of Verkhoyansk Range (including the district of Tiksi). (Map II–18).

**Foreign Arctic.** Arctic Scandinavia; distribution of species not precisely clarified due to confusion with *D. caespitosa* and other arctic species. Apparently occurring also in Iceland and the NE part of North America.

**Outside the Arctic.** Mountains of Northern Scandinavia; Urals (south to 60°N); Central Siberian Plateau (south to the basin of the Lower Tunguska); Verkhoyansk and Momskiy Ranges. Very closely related species in the mountains of Europe and the Rocky Mountains of North America.

4. ***Deschampsia anadyrensis*** V. Vassil. in Bot. mat. Gerb. Bot. inst. AN SSSR VIII, 5 (1940), 68, fig. 2.

In the herbarium of the Botanical Institute of the USSR Academy of Sciences there are five type specimens of this species (Anadyr Basin; sandy-silty floodplain on lower course of Anadyr opposite Lake Krasnoye, 10 VIII 1933, No. 1591, Vasilev). In appearance they are more reminiscent of *D. glauca* than *D. Sukatschewii*, but differ from both these species in almost or completely lacking hairs on the segments of the spikelet rachis and in having shorter (on average) ligules on the stem leaves. Similar specimens have not been found among the remaining material from the Anadyr Basin (for the most part collected later and therefore with fallen florets) referred by us to *D. Sukatschewii*, although some specimens of the latter species approach both *D. anadyrensis* and *D. glauca* in the structure of the panicles and in external appearance. Since *D. anadyrensis* is so far known only from a single locality, the distinctness of this species remains rather doubtful and in need of confirmation from more extensive material.

According to the label on the type, growing on sandy or gravelly river shores.

**Soviet Arctic.** Anadyr Basin (only type specimens known).

So far not found in the **Foreign Arctic** or **outside the Arctic**.

5. ***Deschampsia Sukatschewii*** (Popl.) Roshev. in Fl. SSSR II (1934), 246; Karavayev, Konsp. fl. Yak. 53.

*D. caespitosa* ssp. *orientalis* Hult., Fl. Kamtch. I (1927), 109; id., Fl. Al. II, 176, pro parte.
*D. Komarovii* V. Vassil. in Bot. mat. Gerb. Bot. inst. AN SSSR VIII, 12 (1940), 214.
*Aira Sukatschewii* Popl. in Och. po fitosots. i fitogeogr. (1929) 382.
**Ill.:** Fl. SSSR II, pl. XVIII, fig. 15.

Species widely distributed in East Siberia, displaying much polymorphism and divided into some ecological-geographical races of weak morphological differentiation. At least four of these apparently merit the rank of subspecies and may be characterized as follows.

*D. Sukatschewii* ssp. *Sukatschewii* (=*Aira Sukatschewii* Popl. s. str. Lectotypus: Transbaicalia inter fl. Nercza et fl. Kuenga; ad ripam fl. Olov prope pag. Olov, 13 VII 1911, No. 1790, Sukatschew et Poplavskaja). Rather tall (30–60 cm high) plant, with large panicles noticeably elongate in comparison with their width and with numerous, rather small (2.6–4 mm long), relatively weakly coloured spikelets. Lemmas with awns well developed and slightly exceeding their tips. Widespread in Prebaikalia and SE Siberia, scarcely penetrating the Arctic.

*D. Sukatschewii* ssp. *submutica* (Trautv.) Tzvel. comb. nova (=*Aira caespitosa* var. *submutica* Trautv. in Acta Hort. Petrop. V, 1877, 141. Lectotypus: Sibiria orient. ad fl. Olenek, haud procul ab ostio fl. Maigda mediae, 15 VII 1874, Chekanowski et Müller). Basically similar to the preceding subspecies, but plant on average shorter (15–40 cm high) with awns of lemmas much reduced and often completely absent. Here belong the majority of specimens from the Soviet Arctic, where this subspecies is connected by transitional populations with *D. glauca* and *D. obensis*.

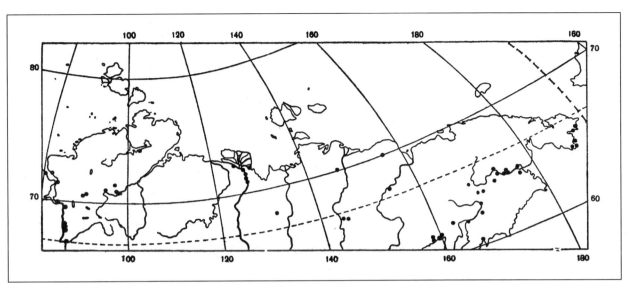

Map II-19 Distribution of *Deschampsia Sukatschewii* (Popl.) Roshev.

*D. Sukatschewii* ssp. *orientalis* (Hult.) Tzvel. comb. nova (=*D. caespitosa* ssp. *orientalis* Hult., l. c. Lectotypus: Kamtchatka australis, Akhomten Bay, 275 m s. m., 13 IX 1920, No. 1362, Hultén). Spikelets less numerous than in "ssp. *Sukatschewii*" and slightly larger (4–5 mm long). Lemmas with awns usually slightly exceeding their tips. Occurring in parts of the species' range adjacent to the Pacific Coast (especially in Kamchatka) and scarcely penetrating the Arctic. Hultén (l. c.) considers that "ssp. *orientalis*" has an almost circumpolar distribution and records it not only for Northern Siberia but also for Arctic Scandinavia, apparently attributing to it specimens of *D. obensis* with smaller spikelets and certain specimens of *D. glauca* from Scandinavia. Very close to this subspecies is the species *D. paramushirensis* Honda described from the Kurile Islands, distinguished by still larger spikelets and in this respect more closely approaching the North Pacific sublittoral species *D. beringensis*.

*D. Sukatschewii* ssp. *minor* (Kom.) Tzvel. comb. nova (=*D. caespitosa* f. *minor* Kom., Fl. Kamch. I, 1927, 152. Lectotypus: Kamtchatka, systema lac. Kronotzkoje, Krascheninnikov-mons, 18 VIII 1909, Komarov). Possessing panicles shorter in comparison with their width and more intensively coloured, in this respect approaching *D. glauca* and *D. borealis* but on the whole showing closer relationship with *D. Sukatschewii* s. str. Lemmas with awns usually much reduced and rarely exceeding their tips. Numerous specimens of this subspecies, constituting a higher-alpine race of *D. Sukatschewii* s. l., were collected by V. L. Komarov in the mountains of Kamchatka. It is very probable that this subspecies will also be discovered in mountainous districts of the Okhotsk Coast and of the Anadyr-Penzhina Basin.

It is possible that *D. Komarovii* V. Vassil. (l. c.), described from the Okhotsk Coast (Ayan district, valley of lower course of River Lantar, sometimes flooded by tides, 26 VIII 1935, No. 412, V. Vasilev), is a distinct maritime ecological-geographical race of *D. Sukatschewii*, differing from typical specimens by the very narrow leaf blades (which are setiformly folded lengthwise) along with the relatively small size of the plant as a whole. Specimens identical with the type of this species are also known from other sites on the Okhotsk Coast and from Kamchatka (including the district of the Bay of Korf within the area of the present Flora), but it is very probable that they are merely extreme variants of *D. Sukatschewii* with respect to leaf width. *Deschampsia Sukatschewii* in the sense we have accepted, like many species of mountainous districts of East Siberia belonging to other genera, shows a closer approach to arctic and subarctic species of its genus (*D. borealis, D. glauca, D. obensis*) than to the boreal *D. caespitosa*. *Deschampsia Sukatschewii* is characterized both by relatively weak development of acicules on the leaf blades (which are usually pure green and rather soft) and by much variation with respect to the degree of awn development. As in the majority of arctic species of the genus, in *D. Sukatschewii* the awn can be attached to the basal, middle or distal part of the lemma.

Growing mainly on sandy or gravelly shores of rivers and streams, more rarely in better vegetated meadowy areas of stream valleys.

**Soviet Arctic.** Lower reaches of Yenisey (north to 72°30'N); Taymyr (south of the Byrranga Mountains); lower reaches of Lena (including Tiksi district); basins of the Yana, Indigirka, Alazeya and Kolyma almost to their mouths; Anadyr and Penzhina Basins; SE Chukotka Peninsula (vicinity of Chaymenskiy Hot Spring, VI 1935, F. Golovachev; also untypical specimens transitional to *D. borealis* from Arakamchechen Island and Provideniye Bay). (Map II–19).

**Foreign Arctic.** Recorded by Hultén for Arctic Alaska; possibly occurring also in other districts of the American Arctic.

**Outside the Arctic.** East Siberia (western range boundary advancing approximately to the Yenisey); Mongolia; NE China; Korean Peninsula; Japan; Alaska. Probably reaching Canada.

6. ***Deschampsia borealis*** (Trautv.) Roshev. in Fl. SSSR II (1934), 246, 750; Karavayev, Konsp. fl. Yak. 53, pro parte.

*D. caespitosa* var. *minor* auct. non Kunth — Trautvetter, Fl. taim. 19.

*D. arctica* var. *borealis* (Trautv.) Kryl., Fl. Zap. Sib. II (1928), 233.

*D. borealis* var. *glacialis* Roshev. in Tolmachev, Fl. Taym. I (1932), 93.

*D. brevifolia* auct. non R. Br. — Tsvelev in Bot. mat. Gerb. Bot. inst. AN SSSR XXI (1961), 42, pro parte.

*Aira caespitosa* var. *borealis* Trautv. in Acta Hort. Petrop. (1871) 86.

High arctic, apparently almost circumpolar species, although the very limited material from the American Arctic at our disposal does not allow us to establish its range completely. Previously we (l. c.) united this species with *D. brevifolia* on the basis of the almost complete coincidence of their ranges and the existence of transitional populations and specimens. But under extreme arctic conditions (for example, on Novaya Zemlya) these two species are relatively well distinguished and apparently native to habitats with different ecological conditions. It is interesting that on the North Island of Novaya Zemlya the species *D. borealis* and *D. alpina* predominate, but on the South Island *D. brevifolia*. Moreover, the most typical specimens of *D. brevifolia* with very elongate glumes distinctly approach such species of riverine or maritime dunes as *D. obensis* and *D. beringensis*, while *D. borealis* shows a very clear affinity with *D. caespitosa*, being connected to that species through small-spikelet specimens of *D. glauca*. Specimens of *D. borealis* from more southern arctic districts (for example, the northern part of the Verkhoyansk Range) have larger, often more or less diffuse panicles, and are connected by a whole series of transitional populations both with *D. glauca* in the west and with *D. Sukatschewii* in the east. Apparently even among high arctic specimens of this species, the eastern specimens (for example, from Wrangel Island) more closely approach *D. Sukatschewii* than do the western, although the differences are extremely insignificant. The specimens from Taymyr (ad fl. Taimyr, 1843, Middendorff) cited by R. Yu. Rozhevits in the Latin diagnosis of *D. borealis* should be considered the type specimens, although in his Russian diagnosis he states that the species was described from Novaya Zemlya.

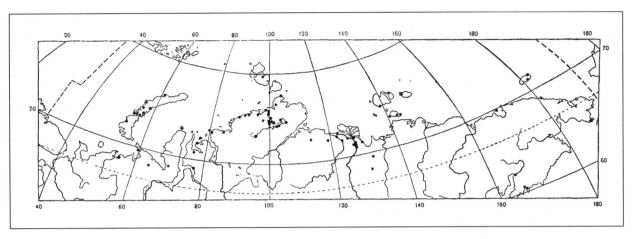

MAP II–20 Distribution of *Deschampsia borealis* (Trautv.) Roshev.

Growing mainly in more or less marshy areas of stony-mossy or moss-lichen tundras, sometimes forming moss-hairgrass communities, more rarely on gravel-bars and moist grassy slopes.

**Soviet Arctic.** Bolshezemelskaya Tundra (Vangurey Range, specimens more or less transitional to *D. glauca*); Vaygach (rare); Novaya Zemlya (mainly the North Island and the district of Matochkin Shar); Yamal; Belyy Island; Gydanskaya Tundra; Taymyr (mainly north from the Pyasina Basin and Lake Taymyr); Severnaya Zemlya; lower reaches of the Lena, Olenek and Anabar (mainly specimens more or less transitional to *D. glauca* or *D. Sukatschewii*); Verkhoyansk Range (north of 67°N, mainly specimens more or less transitional to other species); New Siberian Islands (frequent on all islands); Ayon Island; North and SE Chukotka; Wrangel Island; mouth of Penzhina, Bay of Korf. (Map II–20).

**Foreign Arctic.** Distribution not precisely clarified. Apparently occurring on islands of the Canadian Arctic Archipelago and Greenland (north of 68°N), as well as on Spitsbergen.

**Outside the Arctic.** Occurring only on the highest peaks of the Prepolar Ural and the Verkhoyansk Range (north of 67°N).

7. ***Deschampsia brevifolia*** R. Br. in Suppl. to App. Parry's Voyage XI (1824), 291; Grisebach in Ledebour, Fl. ross. IV, 422; A. E. Porsild, Ill. Fl. Arct. Arch. 27; Polunin, Circump. arct. fl. 49.

*D. arctica* (Spreng.) Ostenf. in Meddl. om Grønl. LXIV (1923), 167; Shishkin & Krylov, Fl. Zap. Sib. II, 232; Rozhevits in Fl. SSSR II, 248; Perfilev, Fl. Sev. I, 81; Karavayev, Konsp. fl. Yak. 53.

*Aira arctica* Spreng., Syst. Veg. IV, 2 (1827), 32.

*A. caespitosa* var. *brevifolia* (R. Br.) Trautv. in Acta Hort. Petrop. 1 (1871), 86.

*A. brevifolia* (R. Br.) Lge., Consp. fl. groenl. (1880) 163; non M. B., 1819.

*A. caespitosa* auct. non L. — Gelert, Fl. arct. 1, 113, pro parte.

**Ill.:** Trinius, Sp. Gram. III (1836), tab. 256a; Porsild, l. c., fig. 5, a; Polunin, l. c. 49.

High arctic circumpolar (or almost circumpolar) species, differing from *D. borealis* in having larger (4–5.5 mm long) spikelets with more elongate glumes but likewise possessing condensed or slightly diffuse panicles with more or less abbreviated branchlets. The isotype of this species illustrated by Trinius (l. c.) retained in the herbarium of the Botanical Institute of the USSR Academy of Sciences has much abbreviated panicle branchlets and large, weakly coloured spikelets resembling the spikelets of *D. obensis* in appearance. The only specimens from the Soviet Arctic identical with it are those from Sverdrup and Belyy Islands in the Kara Sea;

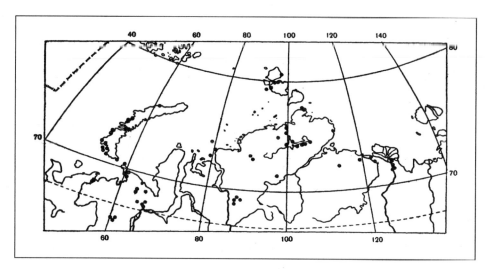

MAP II–21  Distribution of *Deschampsia brevifolia* R. Br.

but these are closely connected with the majority of the other specimens referred by us to this species, which possess more diffuse panicles with more intensively coloured spikelets. Specimens from more southern arctic districts with still more diffuse panicles occupy a more or less intermediate position between *D. brevifolia* and large-spikelet specimens of *D. glauca*. The position of the lemma awns in *D. brevifolia* is just as inconstant as in *D. borealis*; they can arise near the base of the lemma, from its middle or from its distal part (which actually happens quite frequently despite R. Yu. Rozhevits' statement in the "Flora of the USSR" that the awns in this species arise from the basal quarter of the lemma in all florets).

Growing on gravelbars and in stony, clayey or sandy areas of tundra near the shores of waterbodies, apparently almost always in communities with very sparse herb cover.

**Soviet Arctic.** Bolshezemelskaya Tundra (eastern part); Polar Ural (partly specimens transitional to *D. glauca*); Vaygach; Novaya Zemlya (mainly the South Island and the district of Matochkin Shar); islands of Kara Sea; Taymyr (mainly north from the Pyasina Basin and Lake Taymyr); Severnaya Zemlya (only vegetative specimens, but apparently belonging to this species); lower reaches of the Anabar, Lena and Olenek (mainly specimens more or less deviating towards *D. glauca* or *D. obensis*). (Map II–21).

**Foreign Arctic.** Arctic Alaska; NW Canada (only northern coast and Mackenzie Mountains); Canadian Arctic Archipelago (from extreme north to the entrance of Hudson Bay, on Baffin Island only north of 70°N); Spitsbergen (?).

**Outside the Arctic.** Highest peaks of Urals north of 50°N and NW part of Central Siberian Plateau (Putorana Mountains); only specimens more or less transitional to *D. glauca*.

8. ***Deschampsia alpina*** (L.) Roem. et Schult., Syst. veg. II (1917), 686; Grisebach in Ledebour, Fl. ross. IV, 422; Lynge, Vasc. pl. N. Z. 106; Devold & Scholander, Fl. pl. SE Greenl. 136; Scholander, Vasc. pl. Svalb. 65; Perfilev, Fl. Sev. I, 81, pro parte; Rozhevits in Fl. SSSR II, 248, pro parte; Kuzeneva in Fl. Murm. I, 182; Hylander, Nord. Kärlväxtfl. I, 298; A. E. Porsild, Ill. Fl. Arct. Arch. 27; Hultén, Amphi-atl. pl. 222; Polunin, Circump. arct. fl. 49; Tsvelev in Bot. mat. Gerb. Bot. inst. AN SSSR XXI (1961), 43.

*D. laevigata* (Sm.) Roem. et Schult., l. c. 686.

*Aira alpina* L., Sp. pl. (1753) 65; Lange, Consp. fl. groenl. 163.

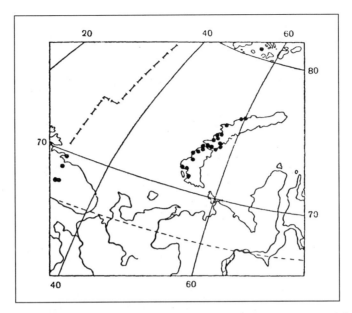

MAP II-22  Distribution of *Deschampsia alpina* (L.) Roem. et Schult.

*A. laevigata* Sm. in Trans. Linn. Soc. X (1810), 137.
*A. caespitosa* auct. non L. — Gelert, Fl. arct. 1, 113, pro parte.
Ill.:Trinius, Sp. Gram. III (1836), tab. 254; Porsild, l. c., fig. 5, d, e; Polunin, l. c. 49.

Arctic-alpine amphiatlantic species, very close to *D. brevifolia* and having the appearance of a viviparous variety of that species. However, the ranges of these species not only do not coincide but scarcely even overlap, which confirms their distinctness. Very interesting is the behaviour of these species in the European and American Arctic: in North America and Greenland *D. alpina* represents a more southern race in comparison with *D. brevifolia* which is there one of the highest-arctic species; but in the European Arctic it is *D. alpina* which reaches the northern limit of existence of flowering plants (although certainly also penetrating rather far to the south), while *D. brevifolia* has a somewhat more southern distribution at least in the eastern part of the European Arctic. Thus, on the North Island of Novaya Zemlya *D. alpina* and *D. borealis* occur almost exclusively, but on the South Island *D. brevifolia* predominates while *D. alpina* is almost completely absent. East of Novaya Zemlya *D. alpina* does not occur at all, and all previous records of this species for Arctic Siberia refer either to *D. obensis* or *D. brevifolia*. Apart from (more or less) conversion of the spikelets into vegetative buds, the characteristic features of *D. alpina* are the completely smooth panicle branchlets and almost complete reduction of the lemma awns as a consequence of vivipary. As a result of almost complete transition to vegetative reproduction, this species shows only insignificant variation and is well delimited from other species.

Growing usually in meadows and on gravelbars on the shores of rivers and streams, on very moist portions of stony slopes, and in moderately moist loamy polygonal tundras.

**Soviet Arctic.** Murman (recorded for the whole district, but in the herbarium of the Botanical Institute there are only a few specimens from the Rybachiy Peninsula and the vicinity of Murmansk); Novaya Zemlya (North Island and district of Matochkin Shar, more rarely on the South Island); Franz Josef Land. (Map II-22).

**Foreign Arctic.** NW part of Labrador (south to 55°N); southern part of Baffin Island; Greenland (only south of 70°N); Iceland; Jan Mayen; Spitsbergen; Bear Island; Arctic Scandinavia.

**Outside the Arctic.** Scotland; mountains of Scandinavia (south to Southern Norway); Khibins Mountains.

9. ***Deschampsia obensis*** Roshev. in Izv. Bot. sada AN SSSR XXX (1932), 771 and in Fl. SSSR II (1934), 247; Perfilev, Fl. Sev. I, 80; Tsvelev in Bot. mat. Gerb. Bot. inst. AN SSSR XXI, 44.

*D. caespitosa* var. *grandiflora* Trautv., Fl. taim. (1847) 19, pro parte.

Eurasian subarctic species, characterized by some degree of elongation of many parts of the plant, including the panicles (which frequently become almost as long as the remainder of the culm), leaf blades and glumes. The spikelets of *D. obensis* are usually pale greenish or very slightly coloured, and the tufts of this species sometimes become very loose as a result of having been drifted over by sand or other fluvial deposits. The type specimens (Lectotypus: Sinus fl. Obj, prope pag. Nachodka, VII 1912, No. 39, Buschevitch) have relatively few, rather large spikelets and very loose tufts. There are not too many similar specimens in the herbarium of the Botanical Institute of the USSR Academy of Sciences, but they are very closely connected with the considerably more numerous specimens also referred by us to *D. obensis* but possessing smaller spikelets and denser tufts. Specimens of *D. obensis* with small spikelets are in their turn connected by transitional specimens and populations with *D. Sukatschewii* and to a lesser degree with *D. glauca*. Therefore it can be postulated that elongation of the panicles and correlated elongation of the glumes may be the result of longstanding adaptation of an ancestral species of the *D. Sukatschewii* -type to growing under conditions of shifting soil, and that populations of *D. obensis* may have arisen polytopically in different parts of the range of the source species. But there is another and perhaps more acceptable explanation of the origin of *D. obensis*. The fact is that, besides this species, there is a whole series of species closely related to it distributed on riverine and maritime dunes along the northern coasts of Eurasia and North America. These are *D. Wibeliana* (Sond.) Parl. endemic to the lower reaches of the Elbe, *D. bottnica* (Wahlb.) Trin. endemic to the Gulf of Bothnia, *D. mezensis* Senjan.-Korcz. et Korcz. endemic to the Mezen, *D. beringensis* Hult. a species of the northern coasts of the Pacific Ocean, and *D. Mackenzieana* Raup endemic to the district of the Mackenzie River and Lake Athabasca. As rightly stated by Hultén (Fl. Al. II, 174-175), all these species share an undoubted affinity and are more closely related to one another than to species of the group *D. caespitosa* L. s.l. (including *D. glauca, D. Sukatschewii*, etc.), to which they are connected by numerous transitional specimens. The hybrid origin of all specimens transitional between *D. bottnica* and *D. caespitosa* has already been firmly established, and consequently the same may be postulated with respect to specimens transitional between *D. obensis* and *D. Sukatschewii*. Among the species of the series listed above, *D. bottnica* is the most differentiated from the *D. caespitosa* s. l. group. Contrary to that species, *D. obensis* most closely approaches the group *D. caespitosa* s. l.; however, specimens of *D. obensis* (and to a still greater degree specimens of *D. Wibeliana* and *D. mezensis*) are very similar to specimens of the hybrid *D. bottnica* × *D. caespitosa*, known under the species name *D.* × *Neumaniana* (Dörfl.) Hyl. Hence it may be postulated that an ancestral species close to *D. bottnica* was formerly widespread on alluvial soils in the lower reaches of rivers and on the sea coast, shifting its range according to changes in the outline of the continents. In later times the range of this species was divided into several separate parts, and the species itself, as a result of frequent hybridization with species of the group *D. caespitosa* s. l. which had invaded from the south, in many cases became closer to those species and produced forms similar to *D. obensis*. The closely related North Pacific species *D. beringensis* Hult. described from Bering Island, differing (on average) from *D. obensis* in having larger spikelets, more scabrous panicle branchlets, broader leaf

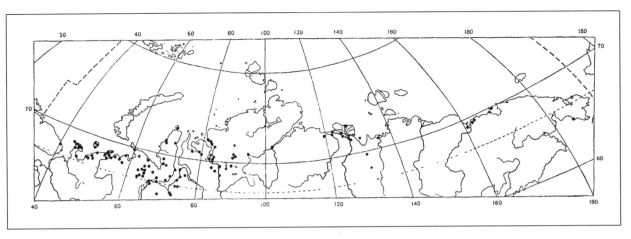

**MAP II–23** Distribution of *Deschampsia obensis* Roshev.

blades, and awns usually slightly exceeding the tips of the lemmas, was also recorded for the Arctic (Chukotka Peninsula) in the "Flora of the USSR" (II, 1934, 251); but specimens so identified from the north coast of the Chukotka Peninsula are in our opinion more correctly referred to *D. obensis,* and specimens from the Beringian coast of that peninsula belong rather to *D. Sukatschewii* although showing some points of resemblance to *D. beringensis.* Apparently the range of that species, as shown on Hultén's (Fl. Al. II, map 120) map, is bounded on the north by the Pribilof and Aleutian Islands and nowhere enters the Arctic.

Growing usually on poorly stabilized sands of valleys on the lower courses of rivers or near the sea coast, more rarely on clayey or silty river deposits.

**Soviet Arctic.** Murman [only one doubtful specimen with much abbreviated panicle branchlets (near mouth of River Lumbovka, 27 VIII 1928, No. 920, Zinserling)]; Kolguyev (frequent); Kanin; Malozemelskaya and Bolshezemelskaya Tundras; Polar Ural (rarely, on the larger rivers); Pay-Khoy; Vaygach; lower reaches of Ob and its tributaries; Yamal and Obsko-Tazovskiy Peninsula; lower reaches of Yenisey (north of Khantayka) and islands of Yenisey Bay; Taymyr (south of the Byrranga Mountains, rather rarely); Khatanga Basin (north of the Central Siberian Plateau); lower reaches of Lena (north of Kyusyur); northern part of Verkhoyansk Range (specimens transitional to *D. Sukatschewii);* near mouths of Kolyma and Medvezhya; Ayon Island and Cape Bering (some specimens transitional to *D. beringensis).* (Map II–23).

**Foreign Arctic.** Apparently replaced by other species.

**Outside the Arctic.** On the Ob and its tributaries south to 63°N.

## GENUS 13    **Trisetum** Pers. — TRISETUM

Genus containing about 60 species distributed at extratropical latitudes in both hemispheres. Four species occur in the Arctic. One of them, *T. spicatum* s. l., is the most widespread representative of the genus in the World and one of the common tundra plants. *Trisetum sibiricum* ssp. *litoralis* (Rupr.) Roshev., a race originating in arctic districts of Eurasia, occurs outside the Arctic only in the Verkhoyansk Mountains. *Trisetum sibiricum* s. str. and *T. molle* are boreal species with limited penetration of the Arctic. *Trisetum subalpestre* is sporadically distributed in East Siberia, including arctic districts, and occurs disjunctly in the mountains of Fennoscandia.

All species except *T. spicatum* are associated in the Arctic with intrazonal (meadowy) assemblages and nowhere play a role of any importance in the formation of communities. *Trisetum spicatum* is a common plant of patchy, stony or sometimes mossy tundras.

1. Culm below panicle and panicle branchlets *bare,* smooth or more or less scabrous. . . . . . . . . . . . . . . . . . . . . . . . . . . . . . . . . . . . . . . . . . . . . . . . . . . . . . . . . . . . .2.
– Culm below panicle and panicle branchlets *short-haired.* . . . . . . . . . . . . . .4.

2. Leaves *3–7 mm wide,* flat, scabrous on both surfaces, with rather long scattered hairs, *not ciliate* on margin. Inflorescence a more or less *diffuse* panicle, with *branchlets 0.5–4 cm long.* Spikelets 5–7 mm long. Glumes greenish. Lemma from greenish yellow to golden brown with long *awn 5–10 mm long.* (*T. sibiricum* s. l.). . . . . . . . . . . . . . . . . . . . . . . . . . . . . . . . . . . . . . .3.
– Leaves *0.5–2 mm wide,* flat or involute, scabrous on both surfaces, without long hairs, *short-ciliate* on margin. Inflorescence a lax *spikelike* panicle with *branchlets 0.3–0.5 cm long.* Spikelets 4–5 mm long. Glumes and lemmas green or more rarely with violet tinge. Lemma with shorter *awn 3–5 mm long.* . . . . . . . . . . . . . . . . . . . . . . . . . . . . .2. **T. SUBALPESTRE** (HARTM.) NEUM.

3. Plant *30–65 cm high.* Inflorescence a relatively large diffuse panicle 7–12 cm long, 3–5 cm wide, with *branchlets up to 4 cm long.* Spikelets 2–3 flowered, 5–7 mm long; glumes yellowish green, lemma *yellow* or rarely brownish. Awn of lemma always considerably longer than spikelet, up to *10 mm long.* Leaves 4–7 mm wide. . . . . . . . . . . .1. **T. SIBIRICUM** RUPR. S. STR.
– Plant *15–30 cm high.* Inflorescence a smaller panicle 3–5 cm long, 2–3 cm wide, with *branchlets 0.5–2 cm long.* Spikelets usually 2-flowered, 5–6 mm long; glumes yellow with violet back, lemma *golden-brown* or more rarely with violet tinge. Awn of lemma equal to or slightly longer than spikelet, *5–8 mm long.* Leaves 3–5 mm wide. . . . . . . . . . . . . . . . . . . . . . . . . . . . . . . . . . . . . . . . . . . . . . . . . . . . . . . . . .1a. **T. SIBIRICUM** SSP. **LITORALIS** (RUPR.) ROSHEV.

4. Plant low, 10–30 cm high, strongly downy. Inflorescence a *dense spikelike* panicle with *rounded* tip. Spikelets *3.5–4.5 mm long, violet* or more rarely greenish violet. *Upper glume broad, considerably shorter than lemma.* Lemma *bare.* . . . . . . . . . . . . . . . . . . . . . . . . . . . . . . . . . .3. **T. SPICATUM** (L.) RICHT.

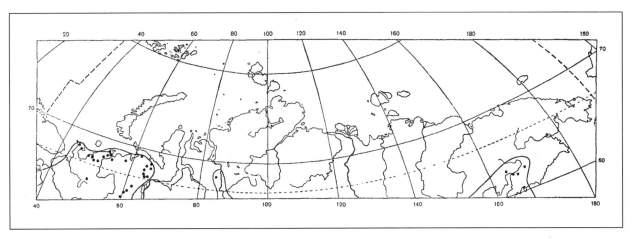

MAP II-24 Distribution of *Trisetum sibiricum* Rupr.

– Plant taller, 25–45 cm high. Inflorescence a *lax spikelike* panicle with *acuminate* tip. Spikelets *5–8 mm long, green. Upper glume narrow, long,* almost completely concealing lemma. Lemma *slightly downy*. [*T. molle* (Michx.) Kunth s. l.]. .................................................5.

5. Plant tall and slender, 25–40 cm high. Inflorescence a *narrow spikelike* panicle, 4–6 cm long, 0.5–1 cm wide. *Spikelets 5–6 mm long.* ............. ...................................4. **T. MOLLE** (MICHX.) KUNTH S. STR.

– Plant 30–45 cm high. Inflorescence a *large spikelike* panicle, up to 10 cm long, 1–2.5 cm wide. *Spikelets 6–8 mm long.* ............................ .........................4a. **T. MOLLE** SSP. **ALASCANUM** (NASH) REBR.

1. ***Trisetum sibiricum*** Rupr. in Beitr. Pflanzenk. Russ. Reich. 2 (1845), 65; Krylov, Fl. Zap. Sib. II, 234; Rozhevits in Fl. SSSR II, 253; Perfilev, Fl. Sev. I, 81; Leskov, Fl. Malozem. tundry 26; Hultén, Fl. Al. II, 183; Karavayev, Konsp. fl. Yak. 53; Polunin, Circump. arct. fl. 75.

*T. flavescens* auct. non Beauv. — Gelert, Fl. arct. 1, 111.

*Avena flavescens* L. — Grisebach in Ledebour, Fl. ross. IV, 417, pro parte.

Species widely distributed in the forest and northern steppe zones of Eastern Europe and Asia. Penetrating the Arctic only in the European North and in the Penzhina Basin. Occurring in euthermal areas with deep thawing of permafrost (or its absence) on meadowy slopes in valleys of rivers and streams, in meadows among willow thickets (*Salix phylicifolia, S. glauca*), and in sparse spruce or spruce-birch woods. Growing in small quantity, nowhere playing an important role in the herb cover within arctic limits.

Northern plants differ from typical Siberian plants only in having shorter stems and a less diffuse (but not condensed!) panicle. Certain plants from the Malozemelskaya Tundra and the Polar Ural have brownish (not green as normally) colouration of the glumes and lemmas.

**Soviet Arctic.** Kanin; Timanskaya and Malozemelskaya Tundras (frequent); Bolshezemelskaya Tundra (common in the east on the Usa, Vorkuta and Kara); Polar Ural (to 67°20′N); Pay-Khoy (very rare); lower reaches of Yenisey (River Solenaya); Penzhina Basin (frequent); Bay of Korf. (Map II-24).

**Foreign Arctic.** Not occurring. A.E. Porsild's records for Alaska refer to *T. sibiricum* ssp. *litoralis* (Rupr.) Roshev.

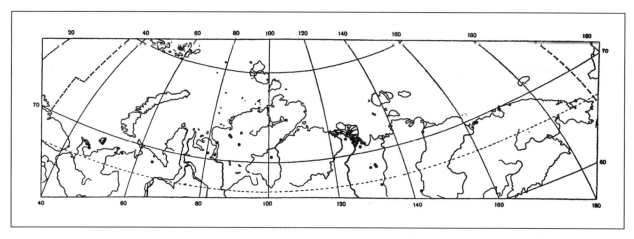

**MAP II–25** Distribution of *Trisetum sibiricum* ssp. *litoralis* (Rupr.) Roshev.

**Outside the Arctic.** Forest zone of Europe east of the Gulf of Finland and the Wisla; Siberia south of 60°N; Far East, including Sakhalin, Kamchatka and Commander Islands. Recorded for North America (Central Yukon, Sitka).

*1a.* ***Trisetum sibiricum*** ssp. ***litoralis*** (Rupr.) Roshev. in Izv. Gl. Bot. sada 21 (1922), 90.
*T. sibiricum* var. *litorale* Rupr. in Beitr. Pflanzenk. Russ. Reich. 1 (1845), 65; Andreyev, Mat. fl. Kanina 60; Tolmachev, Fl. Kolg. 14; id., Fl. Taym. I, 94; Rozhevits in Fl. SSSR II, 253; Perfilev, Sl. Sev. I, 81.
*T. sibiricum* auct. non Rupr. — Gelert, Fl. arct. 1, 112; Karavayev, Konsp. fl. Yak. 53, pro parte; Polunin, Circump. arct. fl. 75, pro parte.
*Avena Ruprechtii* Griseb. in Ledebour, Fl. ross. IV (1853), 418.
*A. flavescens* auct. non L. — Trautvetter, Pl. Sib. bor. 141, pro parte.
*A. flavescens* var. *agrostidea* Trautv., Syll. pl. Sib. bor.–or. 63.

The arctic, predominantly Siberian race *T. sibiricum* ssp. *litoralis* (Rupr.) Roshev. is well distinguished from *T. sibiricum* s. str. This is a significantly shorter plant (height not exceeding 40 cm, usually 15–20 cm) with compact, often almost spikelike inflorescence and golden brown colouration of the glumes and lemmas.

Occurring mainly on the shores of rivers, on sandbars and gravelbars and on meadowy slopes, more rarely among willow thickets and on mossy tundra at the foot of sloping banks. Over the extent of its range *T. sibiricum* ssp. *litoralis* varies to a considerable degree with respect to the size and shape of the inflorescence. Specimens described by Ruprecht from Kanin have a rather large panicle (6 cm long and 2.5 cm wide) with branchlets up to 10 mm long. Plants from the Siberian Arctic, where this race is considerably more widespread, have a very dense short panicle of almost oval shape with short (up to 5 mm) branchlets.

On the lower reaches of the Lena plants have been found with a small-flowered lax inflorescence reminiscent in shape of the inflorescence of *T. subalpestre*, which led Trautvetter (Syll. pl. Sib. bor.–or.) to separate them under the name *Avena flavescens* var. *agrostidea*. In the shape and colouration of the glumes, leaf width and the nature of their downiness, these plants undoubtedly belong to *T. sibiricum* ssp. *litoralis*. The name *Avena flavescens* var. *agrostidea* should be considered a synonym of this subspecies, not a synonym of *T. subalpestre* (Gelert, Fl. arct.).

**Soviet Arctic.** Kanin (northern part); Kolguyev (common); Eastern Bolshezemelskaya Tundra (Khalmer–Yu); lower reaches of Yenisey (north to 72°N); Taymyr (in the basin of the Pyasina and Lake Taymyr); lower reaches of Khatanga; Arctic Yakutia (lower reaches of the Olenek, Lena, Indigirka and Kolyma); Rautan Island in the Bay of Chaun. (Map II–25).

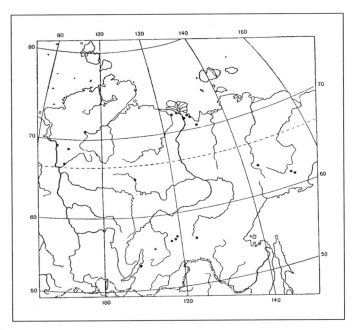

MAP II–26  Distribution of *Trisetum subalpestre* (Hartm.) Neum.

**Foreign Arctic.** Arctic Alaska.

**Outside the Arctic.** Northern fringe of Central Siberian Plateau; Verkhoyansk-Kolymsk mountain country.

2. ***Trisetum subalpestre*** (Hartm.) Neum., Sver. fl. (1901) 755; Rozhevits in Fl. SSSR II, 254; Hultén, Atlas, map 191; Karavayev, Konsp. fl. Yak. 53.

   *T. agrostideum* (Laest.) Fries — Gelert, Fl. arct. 1, 111.

   *Avena agrostidea* Fries — Grisebach in Ledebour, Fl. ross. IV (1853), 418; Scheutz, Pl. jeniss. 183.

   *A. flavescens* auct. non L. — Trautvetter, Pl. Sib. bor. 141, pro parte (respecting plants from Kolung–bas).

   Peculiar species, in appearance approaching *T. molle* from which it differs in completely lacking downiness and in having a laxer inflorescence. Distributed sporadically in East Siberia.

   Penetrating the Arctic only on the Yenisey and the Lena. Occurring very rarely near snowfields, on stream banks where snow long persists, on gravelbars and in mixed herb meadows. Further south more common in (larch or spruce) forest, as well as in alpine meadows in the barren zone.

**Soviet Arctic.** Lower reaches of Yenisey; River Kheta (vicinity of Volochanka); lower reaches of Olenek and Lena. (Map II–26).

**Foreign Arctic.** Arctic Scandinavia (south of 70°N).

**Outside the Arctic.** Northern and northwestern fringe of Central Siberian Plateau; upper reaches of Vilyuy; Transbaikalia; Southern Yakutia; upper reaches of Indigirka and Kolyma.

3. ***Trisetum spicatum*** (L.) Richt., Pl. Eur. (1890) 59; Lynge, Vasc. pl. N. Z. 127; Tolmachev, Fl. Kolg. 14; Andreyev, Mat. fl. Kanina 160; Tolmachev, Fl. Taym. I, 94; Rozhevits in Fl. SSSR II, 255; Perfilev, Fl. Sev. I, 81; Tolmachev, Obz. fl. N. Z. 149; Leskov, Fl. Malozem. tundry 26; Hultén, Fl. Al. II, 184; id., Atlas, map 190; Kuzeneva in Fl. Murm. I, 83; Böcher & al., Grønl. Fl. 290; A. E. Porsild, Ill. Fl. Arct. Arch. 28; Karavayev, Konsp. fl. Yak. 53, pro parte; Polunin, Circump. arct. fl. 74.

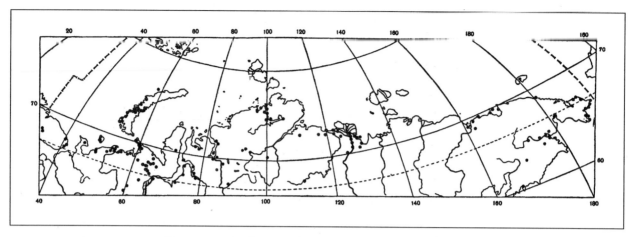

MAP II–27  Distribution of *Trisetum spicatum* (L.) Richt.

*T. spicatum* ssp. *spicatum* Hultén in Sv. Bot. Tidskr. 53, 2 (1959), 204.
*T. subspicatum* (L.) Beauv. — Gelert, Fl. arct. 1, 110.
*Avena subspicata* Clairv. — Grisebach in Ledebour, Fl. ross. IV, 418; Schmidt, Fl. jeniss. 128; Trautvetter, Pl. Sib. bor. 140; Scheutz, Pl. jeniss. 189.

One of the most widespread species of the genus and probably the richest in the degree of variation of particular characters. The existence of a large number of different forms and the presence of transitional specimens to a high degree impedes the taxonomic definition and identification of this species. Botanists have made repeated attempts to treat this species systematically. For North America 14 varieties were recognized [Louis-Marie, The Genus Trisetum in America, Rhodora 30–31 (1928–1929)] and their number continues to increase although there has so far been no agreement regarding their taxonomic evaluation. Hultén (l. c., 1959) considers many races originally described as species to be subspecies of *T. spicatum*, including *T. molle* (Michx.) Kunth, *T. alascanum* Nash, *T. groenlandicum* Steud., *T. labradoricum* Steud., etc.

We basically accept this interpretation, but consider the race *T. spicatum* ssp. *molle* (together with *T. spicatum* ssp. *alascanum*) with long narrow glumes to be a separate species.

In the Soviet Arctic *T. spicatum* is represented by two forms. The typical Lapland race of *T. spicatum*, identified by Hultén as *T. spicatum* ssp. *spicatum*, is characterized by a dense spikelike inflorescence with rounded tip, small spikelets (3.5–4.5 mm long) and short broad violet–coloured glumes. This subspecies is found from the Kola Peninsula to the lower reaches of the Yenisey, with only an insignificant degree of variation. East of the Yenisey, together with typical *T. spicatum* distributed in the more northern districts (Taymyr, lower reaches of Lena, Wrangel Island), there are also found plants with lax inflorescence, sometimes interrupted on its lower part. Their glumes are usually narrower and longer, but the actual spikelets are small and violet–coloured.

This race is known as f. *elatior* Kryl. and is apparently the same as var. *Maydenii* (Gand.) Fernald in North America.

In East Siberia, where *T. molle* grows together with *T. spicatum*, a large number of intermediate forms occur.

*Trisetum spicatum* is distributed throughout the Arctic. It is one of the obligatory components of stony or rocky tundras on Novaya Zemlya, and occurs in patchy tundras, on rocks, and sometimes on sandbars on the shores of rivers and lakes throughout Siberia. Sometimes in meadowy, mossy or dwarf-willow tundras.

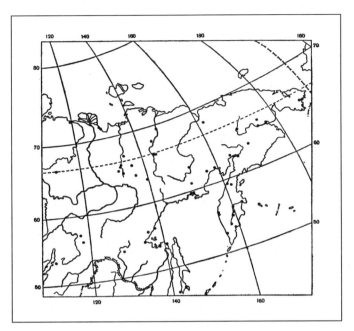

MAP II–28 Distribution of *Trisetum molle* (Michx.) Kunth.

**Soviet Arctic.** Murman; Kanin; Kolguyev; Malozemelskaya and Bolshezemelskaya Tundras; Polar Ural; Pay-Khoy; Vaygach; Novaya Zemlya (to 76°N); Yamal; Gydanskiy Peninsula; lower reaches of Yenisey; Taymyr (north to the mouth of the Lower Taymyra); Arctic Yakutia (lower reaches of the Olenek, Lena and Kolyma); Chukotka; Wrangel Island; Anadyr Basin; Penzhina Basin (rare); Bay of Korf. (Map II–27).

**Foreign Arctic.** Arctic Alaska; Canadian Arctic Archipelago; Northern Canada; Labrador; Greenland; Spitsbergen; Arctic Scandinavia.

**Outside the Arctic.** Prepolar Ural; mountains of East Siberia (untypical form); Kamchatka; Canada; mountains of Fennoscandia. Plants from mountainous districts of the Altay, Sayans and Khamar-Daban belong to another subspecies, *T. spicatum* ssp. *ovatipaniculatum* Hultén, also distributed in the mountains of Central Europe.

4. ***Trisetum molle*** (Michx.) Kunth, Rev. Gram. I (1829), 101; Trinius in Mem. Ac. Petrop. VI, 1 (1830), 64.

*T. spicatum* ssp. *molle* (Michx.) Hult. in Sv. Bot. Tidskr. 53, 2 (1959), 216.
*T. spicatum* var. *molle* (Michx.) Piper in Contr. U. S. Nat. Herb. 11 (1906), 125.
*Avena mollis* Michx., Fl. Bor.–Amer. (1803) 72.

Species closely related to *T. spicatum*, widespread in the forest zone of North America. In Eurasia reaching west to the Lena, occurring almost everywhere in Yakutia but very rare in the north, penetrating the Arctic in the lower reaches of the Olenek, Indigirka and Kolyma.

Boreal species characteristic of dry forest and meadow communities in valleys. Characterized by slender stems, a lax narrow inflorescence and narrow green glumes concealing the lemmas. In districts where *T. spicatum* also occurs, various forms transitional between these species are produced, which impedes identification. The majority of botanists, although drawing a sharp distinction between the *T. molle* and *T. spicatum* types, have nevertheless considered *T. molle* as a variety or subspecies of the latter due to the large number of transitional forms.

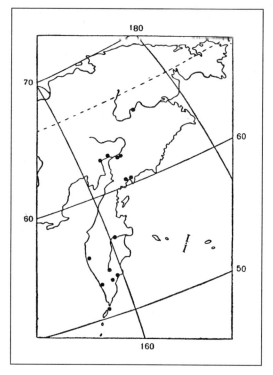

**Map II-29** Distribution of *Trisetum molle* ssp. *alascanum* (Nash) Rebr.

In the Arctic *T. molle* occurs in willow thickets, on gravelbars and in mixed herb meadows in stream valleys.

**Soviet Arctic.** Lower reaches of Olenek; Buorkhaya Bay (Tiksi, Kharaulakh); lower reaches of Kolyma; Beringian coast of Chukotka Peninsula; Anadyr Basin (very rare); lower reaches of Penzhina. (Map II-28).

**Foreign Arctic.** Arctic Alaska; Labrador.

**Outside the Arctic.** East Siberia east of the Lena (including Transbaikalia); Kamchatka; forest region of Canada.

**4a. *Trisetum molle* ssp. *alascanum* (Nash) Rebr. comb. nova.**

*T. alascanum* Nash in Bull. N. Y. Bot. Gard. 3, 6 (1901), 155; Scribner & Merrill in Contr. U. S. Nat. Herb. 13, 3 (1910), 65.

*T. spicatum* ssp. *alascanum* (Nash) Hult. in Sv. Bot. Tidskr. 53, 2 (1959), 210.

*T. spicatum* var. *alascanum* (Nash) Malte ex Louis-Marie in Rhodora 30 (1928), 239; Hultén, Fl. Aleut. Isl. (1960) 90.

This race is distributed along the Pacific coasts of Kamchatka, the Japanese Islands and North America. It penetrates the Arctic only in Koryakia, occurring on meadowy slopes and in willow thickets in river valleys.

Differing from *T. molle* only in its larger size (spikelets 7 mm or more, inflorescence large and dense with acuminate tip, plant tall) and greater downiness.

The majority of authors consider *T. alascanum* as a subspecies or variety of *T. spicatum*, since they reject the specific distinctness of *T. molle*. In such cases the undoubted closeness of *T. spicatum* var. *alascanum* and *T. spicatum* var. *molle* is noted.

These two closely related races can be contrasted with *T. spicatum* s. str. with respect to the structure of the glumes and the shape and colour of the whole inflorescence.

**Soviet Arctic.** Middle course of the Anadyr; lower reaches of the Penzhina; Bay of Korf. (Map II–29).
**Foreign Arctic.** Not occurring.
**Outside the Arctic.** Kamchatka, Kurile and Aleutian Islands; Hokkaido Island; Pacific Coast of North America (south to 40°N).

---

## GENUS 14    Helictotrichon Bess. ex Roem. et Schult. — OAT GRASS

LARGE GENUS WIDELY distributed in temperate latitudes of Eurasia, Africa and America and in the upper zones of mountains in the tropics (Asia, Africa). Only two species, belonging to different sections, penetrate the Arctic.

1. *New growth extravaginal;* shoots of new growth spreading horizontally, forming in the first year a rosette of radical leaves which are rather long and narrow and often folded lengthwise. Leaves (with sheaths) *bare; leaf blades* of flowering culms *broad,* (2)3–7(11) mm wide, flat or more rarely folded lengthwise, thick, glaucescent, the lower subrotund distally (normally abruptly constricted to short point), more or less acute; *ligule large* (up to 5–6 mm long), broad. Spikelets usually large, (11)13–18(26) mm long, often of variegated colour, *with 3–4(–6) florets; glumes broadly lanceolate,* acute or subobtuse, *much shorter than whole spikelet*. Lemmas (9)12–14(17) mm long, like the glumes with broad translucent scarious margin above; their awns *flattened at base.* ...........................
 ................................1. **H. DAHURICUM** (KOM.) KITAGAWA.
– *New growth only intravaginal* (true tussock plant). *Radical leaves with setiformly involute blade,* stiff, sometimes arcuately curved, *squarrosely pubescent* (hairs later partly or wholly caducous); *leaves of flowering culms* with narrow involute blades, many times shorter than the *squarrosely pubescent sheath and with very short truncate ligule.* Spikelets smaller, 11–14(16) mm long, *with 2(–3) florets,* stramineous (sometimes with lilac tint); *glumes narrowly lanceolate, long-acuminate,* the upper slightly shorter than length of whole spikelet. Lemmas (8)11–12 mm long, *with their awns cylindrical at base.* ..........2. **H. KRYLOVII** (N. PAVL.) HENRARD.

1. **Helictotrichon dahuricum** (Kom.) Kitagawa in Rep. Inst. Sc. Research Manchoukuo 3, 1 (1939), 77; Karavayev, Konsp. fl. Yak. 53.
 *Avenastrum dahuricum* Roshev. in Fl. SSSR II (1934), 275.
 *Avena planiculmis* ssp. *dahurica* Kom. in Fl. Kamch. I (1927), 159 (incl. var. *kamtschatica*).
 *A. planiculmis* Turcz., Fl. baic.–dahur. 3 (1856), 322; Hultén, Fl. Kamtch. I (1927), 117; non Schrad.

 Characteristic plant of river valleys in the mountains of NE Siberia, penetrating the tundra zone only near the two northern extremities of the arc of mountains running from the Lena to the Chaun (northern extremity of Kharaulakh Mountains; mountains east of Bay of Chaun, Anadyr and Penzhina Basins). The majority of collections of this species have been made in the Amur Basin, where it is extremely widely distributed all the way to Transbaikalia. Further north, in the Aldan Basin, the species is extremely rare and confined to mountainous districts.

In the Verkhoyansk-Kolymsk mountain country *H. dahuricum* is especially common in the zone below the barrens (with abundance of open unforested spaces); it grows on vegetated gravel deposits, in valley meadows together with *Zerna Pumpelliana*, *Festuca altaica* and *Leymus interior*, in grassy willow thickets, and more rarely in groves of poplar or Chosenia; sometimes it can also be found on the higher accumulative terraces in lichenaceous dwarf–birch carr and low willow carr (*Salix Krylovii*), which there replace meadow and grassy willow thickets. *Helictotrichon dahuricum* is also characteristic of dry meadows at the foot of southfacing slopes (composed of stones and fine soil), of dry glades and of grassy larch woods on southfacing slopes.

The habitats of this species in the tundra zone are very similar to those in more southern portions of the arc of mountains from the Lena to the Chaun.

The tundra-dwelling and very similar subarctic populations of *H. dahuricum* differ somewhat from populations of the Amur Basin; in particular, the Kamchatka plant was distinguished by V. L. Komarov as the separate variety var. *kamtschatica* Kom. (on the basis of the dark bronzy-lilac colour of the lemmas). Dark colouration of the lemmas is generally characteristic of subarctic and tundra populations, usually growing in open areas; but some specimens from willow or Chosenia groves etc. possess bright green lemmas, like the typical form, and the form with dark coloured lemmas can also be found occasionally in the Amur Basin. Northern populations are usually characterized by shorter leaves, less leafy fertile culms, fewer florets in the spikelets, and rather smaller size of the lemmas, but these differences are inconstant and provide no basis even for distinguishing a separate variety.

We note that in the mountains of Europe (Sudeten, Carpathians, Eastern Alps, Northern Balkans) and Asia Minor there is found a plant morphologically very similar to ours, *H. planiculme* (Schrad.) Pilger.

**Soviet Arctic.** Lower reaches of Lena (Bulkur); Northern Kharaulakh (near SW coast of Buorkhaya Bay); district of Bay of Chaun (60 km west of Ust-Chaun settlement); Chukotskiy Range (middle course of River Ekiatam); central and southern part of Amur Basin; lower part of Penzhina Basin.

**Foreign Arctic.** Absent.

**Outside the Arctic.** Verkhoyansk-Kolymsk mountain country (Verkhoyansk and Cherskiy Ranges); mountains in east and SE part of the Aldan Basin; northern part of Kamchatka; Amur Basin west to the Nercha, Onon and Chita; Northern Sakhalin.

2. ***Helictotrichon Krylovii*** (N. Pavl.) Henrard in Blumen, 3 (1940), 431; Karavayev, Konsp. fl. Yak. 54; id., Bot. zhurn. XLIII, 4, 481.

*Avena Krylovii* N. Pavl. in Animadv. syst. Herb. Univ. Tomsk. 5–6 (1933), 1.

*Avenastrum Krylovii* Roshev. in Fl. SSSR II (1934), 279.

Plant of subarctic mountain steppes of the Verkhoyansk-Kolymsk mountain country; additionally found in the vicinity of Yakutsk (whence it was described).

Closely related to Asian mountain steppe species of the complex *H. desertorum* s. l.

In the tundra zone collected only at one site, 60 km south of Ust-Chaun settlement together with a series of other steppe species (*Carex duriuscula*, *Potentilla* spp., etc.).

**Soviet Arctic.** District of Bay of Chaun (60 km from Ust-Chaun settlement).

**Foreign Arctic.** Absent.

**Outside the Arctic.** Middle and upper course of Yana; upper course of Indigirka; mainstream left bank of Lena in vicinity of Yakutsk.

## ♣ GENUS 15  Beckmannia Host — SLOUGH GRASS

OLIGOTYPIC GENUS (2 species) distributed in temperate latitudes of the Northern Hemisphere. Both species just penetrate marginal districts of the Arctic.

1. Stems at base *bulbously thickened*; spikelets 2-flowered, more or less *strongly inflated*; lemma without beak at tip or with very small beak. . . . . . . . . . . . . . . . . . . . . . . . . . . . . . . . . . . . . . . . . . . . . . . . . . . . .1.B. ERUCIFORMIS (L.) HOST.
- Stems at base *without bulbous thickening*; spikelets 1-flowered (sometimes with vestige of second floret), *slightly inflated*; lemma extended as sharp beak at tip. . . . . . . . . . . . . . . . . . . . . . . . .2. B. SYZIGACHNE (STEUD.) FERN.

   1. *Beckmannia eruciformis* (L.) Host, Gram. Austr. III (1805), 5; Grisebach in Ledebour, Fl. ross. IV, 453; Rozhevits in Fl. SSSR II, 288; Perfilev, Fl. Sev. I, 83; Kuzeneva in Fl. Murm. I, 187.
   *Phalaris erucaeformis* L., Sp. pl. (1753) 55.
   Plant of the temperate zone, mainly of Eastern and Central Europe, generally foreign to the far north but sometimes introduced there by man.
   **Soviet Arctic.** In Murman (district of Kola Bay) as an introduction.
   **Foreign Arctic.** Not reported.
   **Outside the Arctic.** Widely distributed in the European part of the USSR, including the Caucasus and Crimea, and in European countries adjacent to the USSR.

   2. *Beckmannia syzigachne* (Steud.) Fern. in Contrib. Gray Herb. LXXIX (1928), 27; Rozhevits in Fl. SSSR II, 288; Hultén, Fl. Al. II, 188; Karavayev, Konsp. fl. Yak. 54; Böcher & al., Grønl. Fl. 297.
   *B. eruciformis* var. *baicalensis* Kusn. — Krylov, Fl. Zap. Sib. II, 249.
   *B. eruciformis* auct. non Host — Trautvetter, Fl. rip. Kolym. 78; Gelert, Fl. arct. 1, 114.
   *Panicum syzigachne* Steud. in Flora XXIX (1846), 19.
   Boreal Asian-American species, just penetrating the Arctic. Growing in valleys of rivers and streams, usually on incompletely vegetated alluvia.
   **Soviet Arctic.** Lower reaches of Ob (at Salekhard); lower reaches of Lena (at northern limit of forest); lower course of Anadyr; Penzhina Basin.
   **Foreign Arctic.** West coast of Alaska; extreme south of Greenland (introduced).
   **Outside the Arctic.** Widespread in the northern forest zone, reaching in the northwest the middle course of the Usa and the lower reaches of the Ob; in the Yenisey Basin only in the south; widespread in Yakutia, approaching treeline on the Yana, Indigirka and Kolyma and reaching it on the Lena; reaching south to the Dzhungarskiy Alatau, Northern China, the Korean Peninsula and Central Japan; in North America in Alaska and western parts of Canada and the USA, south to California. Rarely in the St. Lawrence Basin (as an introduction?).

## GENUS 16 — **Phragmites** Adans. — REED

GENUS CONTAINING A few species distributed mainly in tropical countries. One of them, the almost cosmopolitan *Ph. communis* Trin., is reported in Northern Europe from certain districts on the fringe of the Arctic.

1. ***Phragmites communis*** Trin., Fund. Agrost. (1820) 134; Krylov, Fl. Zap. Sib. II, 280; Perfilev, Fl. Sev. I, 83; Lavrenko & Komarov in Fl. SSSR II, 304; Hultén, Atlas, map 134; Kuzeneva in Fl. Murm. I, 188.
   *Arundo phragmites* L., Sp. pl. (1753) 81; Grisebach in Ledebour, Fl. ross. IV, 393.
   **Ill.:** Fl. Murm. I, pl. LXII.

   Plant of abundantly watered riverine marshes and shallow waterbodies, forming there dense continuous beds. Occurring in quantity in the north of the forest zone but rarely reaching the polar limit of forest, near which it is often reported in a sterile state.

   **Soviet Arctic.** Murman (districts of Pechenga and Kola Bay); South Kanin.
   **Foreign Arctic.** Arctic Scandinavia.
   **Outside the Arctic.** From the northern edge of the forest zone in Scandinavia and the European part of the USSR, the middle courses of the Ob, Yenisey and Lena, and Central Kamchatka south to North Africa, Syria, Iraq, Northern India, Northern China, the island of Taiwan, and Japan; also in Southern Africa, Australia, North and South America.

   American plants are usually contrasted with plants from the Old World as *Ph. communis* var. *Berlandieri* (Fourn.) Fern.

## GENUS 17 — **Molinia** Schrank — MOOR GRASS

SMALL GENUS (with few species) distributed in the Old World in the temperate zone of the Northern Hemisphere. One species penetrates arctic limits.

1. ***Molinia coerulea*** (L.) Moench, Meth. (1794) 183; Grisebach in Ledebour, Fl. ross. IV, 395; Krylov, Fl. Zap. Sib. II, 254; Rozhevits in Fl. SSSR II, 312; Perfilev, Fl. Sev. I, 84; Kuzeneva in Fl. Murm. I, 190; Hultén, Atlas, map 200; id., Amphi-atl. pl. 160.
   *Aira coerulea* L., Sp. pl. (1753) 63.
   *Sesleria coerulea* Ard. — Gröntved, Pterid. Spermatoph. Icel. 151.

   Plant of the northern temperate zone, almost exclusively European, just penetrating arctic limits in Northern Scandinavia and Iceland.

   **Soviet Arctic.** Murman (Pechenga, district of Kola Bay, Teriberka).
   **Foreign Arctic.** Iceland; Arctic Scandinavia.
   **Outside the Arctic.** Greater part of Europe, including the NW part of the forest zone (east of the White Sea only in more southern districts); Southern West Siberia. As an introduction in North America.

## GENUS 18    Koeleria Pers. — JUNE GRASS

GENUS OF ABOUT 80 species distributed in almost all extratropical countries of both hemispheres, occurring mainly in more or less mesophilic communities of various kinds: meadows, meadowy steppes and sparse forests. Of the three species occurring in the Soviet Arctic, one (*K. seminuda*) just penetrates its limits, but the others *(K. asiatica* and *K. Pohleana)* are rather characteristic arctic plants, growing almost exclusively on rather dry sandy or stony areas of tundras. In the American Arctic and Greenland the genus *Koeleria* is not represented except in Alaska and the Yukon Basin.

1. Plant 8–30 cm high, forming dense or rather loose tufts; vegetative shoots *appearing thickened at base due to the numerous leaf sheaths, usually with 4–6 developed leaves* (excluding dead leaves); sheaths of culm leaves *bare and smooth*. Panicles 2–4 cm long; glumes bare, *more or less subobtuse at tip;* lemmas acutish or subobtuse at tip, only on lower part more or less hairy to almost bare. . . . . . . . . . . . . . . . . .3. **K. POHLEANA** (DOMIN) GONTSCH.
– Vegetative shoots *not thickened at base, usually with 2–3 developed leaves* (excluding dead leaves); sheaths of culm leaves *hairy*. Glumes bare or hairy, *with acute tip, usually long-acuminate;* lemmas usually hairy over almost their whole surface, more rarely bare or almost bare, usually long-acuminate at tip. . . . . . . . . . . . . . . . . . . . . . . . . . . . . . . . . . . . . . . . . . .2.

2. Plant *20–60 cm high,* forming rather dense tufts; culms *on lower part usually sparsely hairy to almost bare;* sheaths often also relatively weakly hairy. Panicles 3–7 cm long, greenish or with faint rosy-violet tinge; glumes relatively short-acuminate at tip, the lower *3–4 mm long,* the upper *4–6 mm long;* lemmas with scattered hairs over almost their whole surface or only on their lower part; anthers 1.6–2.6 mm long. . . . . . . . . . . . . . . . . . . . . . . . . . . . . . . . . . . . . . . . . . . . . . .1. **K. SEMINUDA** (TRAUTV.) GONTSCH.
– Plant *5–35 cm high,* forming dense or loose tufts; culms *usually densely hairy along their entire length;* sheaths usually densely hairy. Panicles 1–4 cm long, usually more or less brownish-violet or rosy-violet, rarely greenish; glumes with long-acuminate, sometimes awnlike tip, on average narrower in comparison with their length, the lower *2.5–3.5 mm long,* the upper *3–4.5 mm long;* lemmas usually more or less hairy over almost their whole surface, rarely almost bare; anthers 1.2–2 mm long. . . . . . . . . . . . . . . . . . . . . . . . . . . . . . . . . . . . . . . . . . . . . . . . . . . . .2. **K. ASIATICA** DOMIN.

    *1. Koeleria seminuda* (Trautv.) Gontsch. in Fl. SSSR II (1934), 331; Karavayev, Konsp. fl. Yak. 54.
    *K. cristata* var. *seminuda* Trautv., Pl. Sib. bor. (1877) 138.
    *K. cristata* auct. non Pers. — Gelert, Fl. arct. 1, 115, pro parte.
    *K. gracilis* ssp. *seminuda* (Trautv.) Domin, Monogr. Koeleria (1907) 227.
    ? *K. gracilis* ssp. *sibirica* Domin, l. c. 227.
    *K. gracilis* ssp. *gracilis* var. *arctica* Domin, l. c. 211.
    ? *K. sibirica* (Domin) Gontsch., l. c. 331.
    *K. janaensis* Petr. ex Karav., Konsp. fl. Yak. (1958) 54, nomen nudum.

Ill.: Domin, l. c., tab. XIII, fig. 18; Fl. SSSR II, pl. XXV, fig. 11.

East Siberian boreal species, very closely approaching *K. macrantha* (Ledeb.) Spreng. (= *K. gracilis* Pers., nom. illegit.) which is widespread in meadowy steppes of Eurasia, and occupying a position somewhat intermediate between that species and *K. asiatica*. Both *K. macrantha* and *K. asiatica* are connected with *K. seminuda* by rather numerous populations and specimens with transitional characters, apparently of hybrid origin. However, it cannot be excluded that *K. seminuda* is entirely of hybrid origin and arose in the area of secondary overlap of the previously separate ranges of *K. macrantha* and *K. asiatica*. This hypothesis is partially supported by the existence of specimens of undoubtedly hybrid origin very similar to *K. seminuda* at sites of contact between *K. macrantha* and a species very close to *K. asiatica*, *K. caucasica* (Trin. ex Domin) Gontsch., in the Caucasus and the Urals. Judging from type specimens present in the herbarium of the Botanical Institute of the USSR Academy of Sciences, *K. sibirica* (Domin) Gontsch. (described by Domin from the mouth of the Lower Tunguska) and *K. janaensis* Petr. from the vicinity of Verkhoyansk (treated by V. A. Petrov as a separate species) are both scarcely distinguishable from *K. seminuda*. We have selected as lectotype of *K. seminuda* a specimen with the following label: "ad Lena inf., prope pag. Ajakit, 28 VII 1875, Czekanovski."

Growing on sand and gravel deposits in river valleys, in valley shrub communities, and outside arctic limits in meadowy steppes and sparse forests.

**Soviet Arctic.** Lower reaches of Lena (district of Ayakit settlement).

**Foreign Arctic.** Not occurring.

**Outside the Arctic.** Whole of East Siberia north to the basin of the Lower Tunguska, the lower reaches of the Lena and the district of Verkhoyansk.

2. **Koeleria asiatica** Domin in Bull. Herb. Boiss., Ser. 2, V (1905), 947; id., Monogr. Koeleria (1907) 250; Krylov, Fl. Zap. Sib. II, 262; Tolmachev, Fl. Taym. I, 94; Perfilev, Fl. Sev. I, 84; Goncharov in Fl. SSSR II, 335; Tikhomirov, Fl. Zap. Taym. 22; Karavayev, Konsp. fl. Yak. 54; Polunin, Circump. arct. fl. 57; Wiggins in Contr. Dudley Herb. V, 5 (1959), 129.

*K. hirsuta* auct. non Gaud. — Trautvetter, Fl. taim. 20; id., Pl. Sib. bor. 139; Kjellman, Phanerog. sib. Nordk. 118; Gelert, Fl. arct. 1, 115.

*K. Mariae* V. Vassil. in Bot. mat. Gerb. Bot. inst. AN SSSR VII, 5 (1940), 70, fig. 3.

*K. Cairnesiana* Hult., Fl. Al. II (1942), 190.

Ill.: Domin, Monogr. Koeleria, tab. XVI, fig. 1–2; Fl. SSSR II, pl. XXV, fig. 15; V. Vasilev in Bot. mat. Gerb. Bot. inst. AN SSSR VII, 5, fig. 3; Hultén, l. c., fig. 1, b.

Almost exclusively Asian arctic species, in the relatively narrow sense adopted by Domin and N. F. Goncharov scarcely reaching beyond arctic limits. However, there is a very close species, *K. caucasica* (Trin. ex Domin) Gontsch., widespread in the alpine zone of the Caucasus mountains, the Urals, Altay and Tien Shan, and another no less close species (scarcely distinguishable from specimens of *K. asiatica* with bare glumes), *K. atroviolacea* Domin, occurring in the Altay and Sayan Mountains. In all parts of the range of *K. asiatica* from which sufficient herbarium material is available, this species shows very great variation with respect to the degree of pubescence of the glumes. The lectotype of the species (ad fl. Taimyr, 74°15', VII 1843) and its duplicates possess densely hairy lemmas but bare or slightly hairy glumes. Rarer but found throughout the range of the species are specimens with densely hairy glumes, "var. *lanuginosa* Domin" (Monogr. Koeleria 251) described from the Lena Delta, from which "var. *sublanuginosa* Kryl." (l. c. 263) from the Ob Sound scarcely differs. More distinct and most closely approaching *K. atroviolacea* is "var. *leiantha* Domin" (Monogr. Koeleria 252) with bare or almost bare glumes, occurring mainly in Northern Yakutia; however, this variety apparently does not form pure populations but occurs intermixed with other vari-

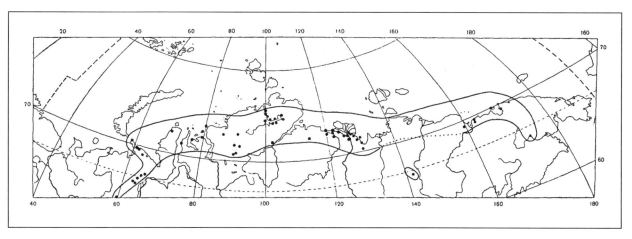

MAP II-30   Distribution of *Koeleria asiatica* Domin.

eties of the species. Also common in Yakutia are specimens more or less deviating towards *K. seminuda*, which are frequently hybrids with that species. From the Gydanskaya Tundra (basin of Gyda-Yam Bay, Lake Yesena-To, 11 VII 1927, Gorodkov) an undoubtedly hybrid specimen is known with scabrous (not hairy) panicle branchlets and with the lemmas short-awned below their tips. R. Yu. Rozhevits described a separate species, *K. Gorodkovii* Roshev. (Izv. Bot. sada AN SSSR XXX, 296), on the basis of this specimen, but in the opinion of S. A. Nevskiy (with whom we agree) it is a hybrid between *K. asiatica* and one of the species of the genus *Trisetum* (most probably *T. sibiricum* Rupr.). We think it possible to unite with *K. asiatica* the species *K. Mariae* V. Vassil. and *K. Cairnesiana* Hult. described by V. N. Vasilev from the Anadyr Basin and by Hultén from the Yukon Basin. Apart from the rather larger than average size of the plant as a whole (25–35 cm high), we have been unable to find any other (more or less) substantial differences between them and *K. asiatica*. In the diagnosis and key to species identification in Domin's monograph, loose tufts with decumbent bases of the shoots is shown as an important distinguishing feature of *K. asiatica*. But from the large material available to us at the present time, it is perfectly obvious that dense tufts without any "rhizomes" are more characteristic of this species and that a certain loosening of the tufts with formation of basally elongated (decumbent) shoots is merely the result of growing on riverine alluvia or shifting sands which continually drift over the tufts. Domin, having only very little material of *K. asiatica* available to him, attached too much significance to his fortuitous finding of elongated shoot bases in certain specimens, and on this basis referred *K. asiatica* together with such completely different species as *K. Delavignei* Czern. ex Domin and *K. polonica* Domin to his subsection "*Pseudorepentes* Domin," while *K. caucasica*, a species extremely close to *K. asiatica*, was referred by him to the subsection "*Caespitosae verae* Domin," Later this same mistake was repeated by N. F. Goncharov (Fl. SSSR II, 321, 335).

*Koeleria asiatica* is normally associated with dry tundras with little snow cover and dry areas in valleys, shoreline bluffs, etc.; rare outside the Arctic.

The considerable unevenness observable in the distribution of this species within the Arctic is linked to its special edaphic requirements. The species grows mainly on poorly consolidated sandstones which release sandy or sandy–stony eluvia during weathering, and on dry alluvial deposits of sand or sand and gravel.

Thus, in the lower reaches of the Lena the species grows very commonly on the Chekanovskiy Ridge, composed mainly of Lower Cretaceous sandstones of just the right type, in patchy or continuous *Dryas*-tundras. In the northern part of the

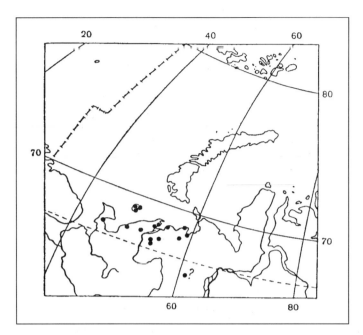

MAP II–31  Distribution of *Koeleria Pohleana* (Domin) Gontsch.

Kharaulakh Mountains, where outcrops of clayey shales and denser sandstones predominate, *K. asiatica* is rarer and concentrated mainly on sand and gravel deposits of mountain rivers. Similar behaviour is shown by *Salix nummularia* and *Papaver Czekanovskii,* common companions of the present species.

**Soviet Arctic.** Polar Ural; Pay-Khoy; Vaygach (in insula Waigatsch versus Novaja Semlja locis pluribus, 1897, Feilden, according to Domin); Yamal (River Olega-Yaga); Gydanskaya Tundra; islands of Yenisey Bay; Taymyr (rather frequent, north to the mouths of the Pyasina and Lower Taymyra); lower reaches of the Khatanga, Anabar and Olenek (reaching south almost to 72°N); lower reaches of Lena (including delta); Kharaulakh Mountains and district of Buorkhaya Bay; lower reaches of Kolyma and further east to the district of Chaun Bay (rather frequent); Ayon Island; Wrangel Island; Anadyr Basin (rather rare). (Map II–30).

**Foreign Arctic.** Arctic Alaska (Chip River); Yukon Basin (rare).

**Ouside the Arctic.** Northern Ural (south to 60°N); Cherskiy Ridge (single locality on left bank of Indigirka).

3. **Koeleria Pohleana** (Domin) Gontsch. in Fl. SSSR II (1934), 323; Leskov, Fl. Malozem. tundry 26; Polunin, Circump. arct. fl. 58.

*K. glauca* var. *Pohleana* Domin, Monogr. Koeleria (1907) 64; Tolmachev, Fl. Kolg. 14; Perfilev, Fl. Sev. I, 84.

*K. gracilis* auct. non Pers. — Perfilev, l. c. 85, pro parte; Leskov, l. c. 26.

*K. asiatica* auct. non Domin — Leskov, l. c. 26.

**Ill.:** Fl. SSSR II, pl. XXV, fig. 1.

Eastern European arctic species, rather close to the widespread but considerably more southern species *K. glauca* DC., whose range extends only as far north as Lake Onega and the Vychegda. Displaying relative constancy of morphological characters, which is to a considerable degree the result of the complete separation of its range from the ranges of other species of the genus. Like *K. glauca,* one of the characteristic plants of sand dunes. The intergeneric hybrid *Koeleria Pohleana* × *Trisetum spicatum,* possessing lemmas with small awns below their tips, was collected from the mouth of the Peschanka on Kolguyev (4 VIII 1905, Zhuravskiy).

Growing on dry sandy slopes and poorly vegetated windswept dunes, constituting a characteristic plant of sandy tundras.

**Soviet Arctic.** Kanin; Kolguyev; Malozemelskaya and Bolshezemelskaya Tundras; Polar Ural (stated at about 66°N, but possibly an error in labelling). (Map II–31).

Not occurring in the **Foreign Arctic** or **outside the Arctic**.

---

## GENUS 19    Melica L. — MELIC

EXTENSIVE GENUS DISTRIBUTED in temperate latitudes of both hemispheres and in mountainous districts of the tropical zone. One species just penetrates the Arctic.

1. ***Melica nutans*** L., Sp. pl. (1753) 66; Grisebach in Ledebour, Fl. ross. IV, 399; Krylov, Fl. Zap. Sib. II, 274; Rozhevits in Fl. SSSR II, 351; Perfilev, Fl. Sev. I, 86; Hultén, Atlas, map 198; Kuzeneva in Fl. Murm. I, 193.

    Forest plant of the temperate zone, in Northern Europe here and there entering the tundra zone where it grows among shrubs.

    **Soviet Arctic.** A few sites in Murman.

    **Foreign Arctic.** Arctic Scandinavia.

    **Outside the Arctic.** Widely distributed in Western Europe and in the temperate northern, central and southern zones of the USSR; east of the White Sea not reaching latitudes close to the northern limit of forest; southeastern part of range extending into China, the Korean Peninsula and Japan.

---

## GENUS 20    Pleuropogon R. Br. — SEMAPHORE GRASS

SMALL GENUS WITH interesting distribution: one species in the Arctic (circumpolar high–arctic), the remaining five in cordilleran North America (four species in the state of California, one in Oregon). The arctic species, *P. Sabinii* R. Br. (type of the genus!), is very different in morphology from the cordilleran. The chromosome number has been determined for *P. Sabinii* as 2n = 40, but in the cordilleran species the number of chromosomes fluctuates between 2n = 16 and 2n = 32 (Jørgensen & al., Fl. pl. Greenl. 18). The cordilleran species are plants of moist meadows, marshes, and the margins of waterbodies and mountain streams, predominantly native to the forest zone. Their ranges are situated south of the region where total glaciation developed during the Pleistocene; the range of the arctic species is centered on the unglaciated region of the Arctic; its ecology is close to the ecology of the cordilleran species. It is interesting that the distribution of the arctic species extends further southwards in Siberia than in America.

1. ***Pleuropogon Sabinii*** R. Br. in Suppl. to App. Parry's Voyage XI (1824), 289; Trautvetter, Syl. pl. Sib. bor.–or. 83; Gelert, Fl. arct. 1, 116; Simmons, Survey Phytogeogr. 48; Lynge, Vasc. pl. N. Z. 110; Krylov, Fl. Zap. Sib. II, 275; Tolmachev & Pyatkov, Obz. rast. Dikona 154; Tolmachev, Fl. Taym. I, 95; Hanssen & Lid, Fl. pl. Franz Josef L. 35; Perfilev, Fl. Sev. I, 86; Nekrasova in Fl. SSSR II, 353; Tolmachev,

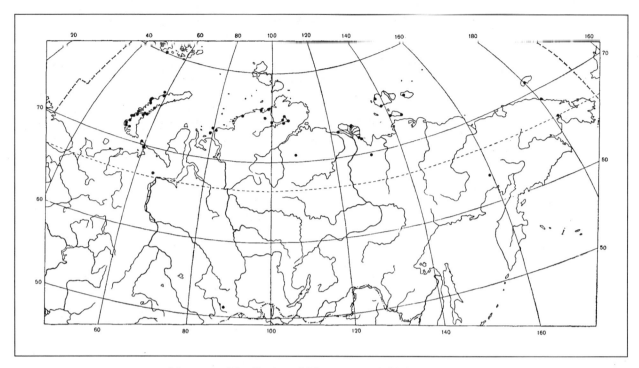

**Map II–32** Distribution of *Pleuropogon Sabinii* R. Br.

Obz. fl. N. Z. 149; Seidenfaden & Sørensen, Vasc. pl. NE Greenl. 94; Tikhomirov, Fl. Zap. Taym. 23; Rozhevits in Areal, I, 22; A. E. Porsild, Ill. Fl. Arct. Arch. 32; Böcher & al., Grønl. Fl. 286; Karavayev, Konsp. fl. Yak. 55; Jørgensen & al., Flow. pl. Greenl. 18; Polunin, Circump. arct. fl. 61.

**Ill.:** R. Brown, l. c., tab. D; Fedchenko & Flerov, Fl. Yevr. Ross., fig. 106; Fl. SSSR II, pl. XXVII, fig. 5, a; Porsild, l. c., fig. 13, c; Böcher & al., l. c., fig. 52.

Morphologically sharply delimited species, with constant retention of basic specific characters over the whole extent of its range. Significant variation from south to north has been observed only in the size of the plant (from 20–30 to 5–7 cm), the leafiness of the culm, the length of the leaves, the number and size of the spikelets, and the number of florets per spikelet (general reduction northwards); in some more southern specimens the spikelets are lilaceous bright green (contrary to the majority of other populations, for which dark purple lemmas and paleas are characteristic), whitish scarious only near their tips.

We should also mention a form growing at sites with deeper water (f. *aquatica* Roshev. in herb.), sterile, with much extended linear-filiform culms and leaves.

*Pleuropogon Sabinii* is a tundra hygrophyte, which also does not avoid more strongly hydric habitats; it can be found together with *Alopecurus alpinus* (in districts at higher latitude), *Carex stans* or rhizomatous cottongrasses of marshy lowlands, *Dupontia Fisheri* and *Arctophila fulva*; growing in minerotrophic or slightly peaty soil. On Novaya Zemlya often growing along mountain streams; in many other districts (for example, in Arctic Yakutia) it is more characteristic of lowland terrain, at the foot of mountains etc. The habitats of *P. Sabinii* include the shores of streams and tundra rivers, oxbows on floodplains, pools in thermokarst sinkholes, silty or silt-and-pebble lakeshores, wet mossy areas near melting snowfields, strongly hydric old cryoturbation fissures (for example, in polygonal marshes), and marshy sedge-moss or cottongrass-moss tundras. On the north coast of Taymyr (Vostochnaya Bay) the species was found in patchy tundra on a slope. Reproducing by rhizomes, so normally occurring in beds.

The distribution of this species is almost entirely restricted to more northern portions of the Arctic; it is absent from the mainland tundras of Eastern Europe (west of the Polar Ural). In the lower reaches of the Lena, *P. Sabinii* is rare in the typical tundra subzone where it occurs only locally on the coast; it becomes more common in the arctic tundra subzone.

So much the more interesting are the few points where the species has been identified outside the tundra zone; all are located in mountainous areas. Thus, high on the Anabar-Khatanga watershed on the NE part of the Central Siberian Plateau, *P. Sabinii* has been found in the alpine tundra zone. In the northern part of the Verkhoyansk Range it is found on a silt-and-gravel deposit of the Altan (basin of the middle course of the Omoloy) at the edge of a gigantic ice-sheet within the forest zone; also found there were *Phippsia algida, Salix reptans, Saxifraga hyperborea*, etc. A still more southern locality for *P. Sabinii* is found in the Kolyma Mountains [left bank of Tik near Kedon settlement (Omolon Basin), grassy willow carr in floodplain on sandbar-gravelbar, frequent!]. Finally, a locality for the species highly disjunct from the main range has long been known in the high alpine region of the SE Altay (wet stony pass between the Rivers Dzhyumala and Ak-Kol at elevation of over 2800 m a. s. l., abundant). The spread of the species to the Altay is enigmatic.

**Soviet Arctic.** Polar Ural; Vaygach; North and South Islands of Novaya Zemlya; Franz Josef Land; Sibiryakov and Dikson Islands; western part of north coast of Taymyr; Central Taymyr; district of Lena Delta; Tiksi; New Siberian Islands; Wrangel Island; north coast of Chukotka (Cape Schmidt); district of Bay of Krest. (Map II–32).

**Foreign Arctic.** Canadian Arctic Archipelago; eastern part of arctic coast of Canada (Melville Peninsula, northern extremity of Labrador Peninsula); NW, North and East Greenland; Spitsbergen.

**Outside the Arctic.** Anabar-Khatanga watershed (barrens); northern part of Verkhoyansk Range (on the Altan, a tributary of the Omoloy); Kolyma Mountains (near Kedon settlement); SE Altay.

---

## GENUS 21    Dactylis L. — COCKSFOOT

SMALL GENUS NATIVE to temperate and southern Europe and neighbouring parts of the Old World. One species occurs here and there in the European Arctic as an introduction.

1. ***Dactylis glomerata*** L., Sp. pl. (1753) 71; Grisebach in Ledebour, Fl. ross. IV, 368; Krylov, Fl. Zap. Sib. II, 277; Ovchinnikov in Fl. SSSR II, 361; Perfilev, Fl. Sev. I, 87; Hultén, Fl. Al. II, 194; id., Atlas, map 204; Kuzeneva in Fl. Murm. I, 194.

    Widely distributed European plant, long cultivated and introduced to many countries outside its natural range. Sown (and partly naturalized) here and there in the far north of Europe.

    **Soviet Arctic.** Murman (Pechenga).
    **Foreign Arctic.** Arctic Scandinavia.
    **Outside the Arctic.** Greater part of Europe (including Scandinavia and northern regions of the USSR); SW Siberia; Middle East and Central Asia; North Africa. Introduced in East Siberia and the Far East.

## GENUS 22    Poa L. — BLUEGRASS

THE GENUS *Poa* contains about 500 species, distributed almost throughout the World but in tropical countries almost exclusively in mountainous districts. On account of such a large number of species which are sometimes weakly differentiated morphologically and connected to one another by specimens with transitional characters (especially in the largest sections of the genus, *Poa* and *Stenopoa*), *Poa* is one of the most difficult genera of grasses with respect to systematics. The treatment of *Poa* by R. Yu. Rozhevits in the "Flora of the USSR" (II, 1934, 366–426) was undoubtedly a great achievement in the study of the numerous species of this genus, but it is very out-of-date at the present time. This particularly applies to the system of the genus, since the "series" into which R. Yu. Rozhevits divided the species of *Poa* in the "Flora of the USSR" in many cases contain very dissimilar species. Thus he referred to the "series" *Alpinae* Roshev. (Fl. SSSR II, 411), along with *P. alpina* L. belonging to the section *Bolbophorum* Aschers. et Graebn., such completely isolated species as *P. abbreviata* R. Br. and *P. pseudoabbreviata* Roshev., as well as *P. Albertii* Rgl., *P. udensis* Trautv. et Mey. and *P. Tanfiljewii* Roshev. belonging to the section *Stenopoa* Dum. (the last of these species being scarcely distinguishable from *P. nemoralis* L.). It is not surprising that the key for species identification constructed on the basis of such a system in the "Flora of the USSR" is such that it is completely impossible to identify many species with it. However, it should be recognized that, as the number of species included in an identification key increases (and there are none too few *Poa* species in the "Flora of the USSR"), difficulties in composing such a key also much increase. Even for the present "Arctic Flora," in which we include 32 species of the genus (including some which are doubtful in our opinion), giving a good key for identification is a very difficult task. In all cases when using our key, attention should be paid not only to the few fundamental characters (italicized in opposing couplets), but also to all the remaining characters indicated for each species. Many species of *Poa* are characteristic of the Arctic and widely distributed there (*P. alpigena, P. arctica, P. alpina, P. glauca, P. abbreviata, P. paucispicula* and *P. pseudoabbreviata*), and indeed occupy a very prominent and important position in vegetative communities. Other species only just penetrate the limits of the Soviet Arctic, usually either in the extreme north-west (species of European origin, *P. compressa, P. trivialis* and *P. subcaerulea*) or in the extreme north-east (Asian or Beringian species, *P. stepposa, P. botryoides, P. malacantha* and *P. eminens*). Hybrids between species of *Poa* are not rare, but they occur far from as frequently as, for example, in the genus *Calamagrostis* and only between species of certain sections. Moreover, apomictic species are known in many sections of *Poa*, and many arctic species (not only from the sections *Poa* and *Bolbophorum*) possess varieties with spikelets modified into vegetative buds. Such viviparous varieties usually behave as separate ecological-geographical races with a quite distinct range which far from coincides with the range of the basic species. It can be postulated that the origin of such viviparous races belongs to the coldest time of the Quaternary Period, when the transition to vivipary was for some species the sole means of surviving glaciation in refugia of restricted area.

1. Lemmas including callus *completely bare* (but often with short acicules on keel and nerves); paleas scabrous on keels with very short acicules; anthers 1.5–2.5 mm long; panicles 4–10 cm long, with rather long scabrous branchlets arranged in twos to fives on lower nodes. Plant 20–80 cm high, without long subterranean stolons, usually forming small tufts; sheaths of culm leaves closed for half to two-thirds of their length, often more or less scabrous; ligule of uppermost culm leaf 0.5–1.2 mm long; leaf blades 1.5–4 mm wide, usually flat. ........................30. **P. SIBIRICA** ROSHEV.
– Lemmas *more or less hairy*, at least on callus or at base of keel and marginal nerves. ............................................................2.

2. Culms and leaf sheaths *strongly compressed along their whole (or almost their whole) length*, often with sharp ribs. Plant 10–30 cm high, with long subterranean stolons. Panicles 2–8 cm long, usually rather dense and narrow, often secund, with more or less scabrous branchlets; spikelets 3–6 mm long; lemmas 2–3 mm long, appressed hairy only on keel and marginal nerves on their basal part, on callus with small cluster of long tangled hairs; paleas scabrous on keels; anthers 1–1.6 mm long. .................
................................................18. **P. COMPRESSA** L.
– Culms *cylindrical*, rarely slightly compressed only near base (but sheaths may be more or less compressed). ......................................3.

3. Anthers *0.4–1 mm long, oblong;* spikelets 3–7 mm long, usually more or less rosy-violet; panicle branchlets always arranged in ones or twos per node, *usually smooth, more rarely scabrous for whole length (then anthers 0.4–0.6 mm long)*. Relatively low, often dwarf plant 3–25 cm high, usually forming dense or loose tufts, always without subterranean stolons. .....4.
– Anthers *1.2–2.5 mm long, oblong-linear or linear, rarely about 1 mm long but then panicle branchlets scabrous for whole length* with numerous acicules. ................................................................8.

4. Lemmas *on lower half rather copiously short-haired over whole surface,* hairless on callus or with a few crinkled hairs; paleas on keels more or less hairy basally and scabrous distally, or entirely scabrous; anthers 0.6–1 mm long; panicles small (0.5–2.5 cm long) and very dense, *with much abbreviated* smooth or more rarely slightly scabrous *branchlets.* Plant 3–15 cm high, forming very dense tufts; leaf blades *relatively stiff,* usually involute, up to 1 mm in diameter. ......................14. **P. ABBREVIATA** R. BR.
– Lemmas *more or less hairy only on keel and marginal nerves, sometimes also on intermediate nerves;* panicles *usually larger, with longer branchlets.* Leaf blades usually *softer,* often flat. ...................................5.

5. Panicles 2–8 cm long, more or less diffuse, with long slender *branchlets scabrous with short acicules along their whole length;* lemmas slightly hairy on keel and marginal nerves, without cluster of long tangled hairs on callus; paleas scabrous on keels with very short acicules; anthers *0.4–0.6 mm*

*long*. Plant 4–20 cm high, forming very dense tufts; leaf blades flat or involute, 0.3–1.2 mm wide.................17. **P. PSEUDOABBREVIATA** ROSHEV.

– Panicles *with completely smooth branchlets;* anthers *0.5–1 mm long*. Plant 5–25 cm high, usually forming rather loose tufts; leaf blades 0.7–3 mm wide, flat and soft. ...............................................................6.

6. Lemmas hairy basally on keel and marginal nerves and usually also *on the conspicuous intermediate nerves, without differentiated cluster of long tangled hairs* on callus but often with a few crinkled hairs; paleas *more or less appressed hairy* on keels, but *without acicules;* panicle branchlets usually *with rather numerous spikelets*. Annual or perennial plant. ..............
................................................................32. **P. ANNUA** L.

– Lemmas *with inconspicuous intermediate nerves,* hairy basally on keel and marginal nerves, *with small cluster of long tangled hairs* on callus; paleas *scabrous* on keels *with very short acicules;* panicle branchlets very long and slender, *usually with only 1–3 spikelets*. Perennial plant. (*P. leptocoma* Trin. s. l.). ..........................................................................7.

7. Plant *15–25(30) cm high,* forming *very loose tufts;* leaf blades *usually more or less scabrous* on margin and on upper surface with very short acicules or bristles. Spikelets *greenish,* almost always without violet tinge. ...........
.............................................................15. **P. LEPTOCOMA** TRIN.

– Plant *5–20 cm high,* usually forming *denser tufts;* leaf blades *smooth or almost smooth* (with isolated acicules). Spikelets *almost always rosy-violet*.
..........................................16. **P. PAUCISPICULA** SCRIBN. ET MERR.

8. Lemmas *without cluster of long tangled hairs* on callus, but occasionally with a few crinkled hairs; intermediate nerves usually inconspicuous; paleas more or less appressed hairy on keels; anthers 1.2–2 mm long; panicles relatively small (2–6 cm long), more or less diffuse; their branchlets *smooth or more rarely slightly scabrous* (with a few scattered acicules), arranged always in ones or twos per node, without spikelets only on basal quarter to half of their length and generally with rather numerous spikelets. Plant 5–35 cm high, forming more or less dense tufts, always without subterranean stolons. .......................................9.

– Lemmas *with well expressed (fully differentiated) cluster of long tangled hairs* on callus, *more rarely without cluster of hairs but then panicle branchlets with numerous acicules along their whole length;* paleas usually only with acicules (occasionally elongated) on keels, more rarely hairy on basal part. ..................................................................10.

9. Plant *forming dense tufts* composed of fertile culms and abbreviated vegetative shoots whose bases are usually enclosed by the sheaths of dead leaves; leaf blades *somewhat greyish-green, thickish,* usually flat, 2–5 mm wide. Spikelets 4–8 mm long; glumes *broadly ovoid, often of almost equal size;* lemmas rather copiously hairy on keel and marginal nerves, occasionally also with hairs between nerves; keels of paleas *with acicules dis-*

*tally and more or less hairy basally and medially.* Including viviparous variety. . . . . . . . . . . . . . . . . . . . . . . . . . . . . . . . . . . . . . . . . . . . . .13. **P. ALPINA** L.

– Plant *forming rather loose tufts,* usually without abbreviated vegetative shoots; leaf blades *bright green,* rather soft and flat, 1–4 mm wide. Spikelets 3–5 mm long; glumes *lanceolate-ovoid, usually of very unequal size;* lemmas always without hairs between nerves; paleas *entirely hairy on keels, without acicules.* No viviparous variety. . . . . . . .31. **P. SUPINA** SCHRAD.

10. Spikelets 2.4–4.5 mm long; glumes *very different in size, the lower almost always with 1 nerve, the upper with 3 nerves;* lemmas *with conspicuous intermediate nerves, appressed hairy* on keel usually to half their length but on marginal nerves *only at their very base, sometimes almost bare* but always with small cluster of long tangled hairs on callus; paleas on keels *with very short, often tuberculate acicules inconspicuous even at high magnification;* anthers 1–1.8 mm long; panicles more or less diffuse, with branchlets scabrous along their whole length, arranged in twos to fives on lower nodes. Plant 20–80 cm high, without subterranean stolons, usually forming small loose tufts; ligule of uppermost culm leaf long and pointed, 2.5–5 mm long; leaf blades 1.5–4 mm wide; sheaths of culm leaves *closed for basal ⅓-½ of their length, usually more or less scabrous with retrorse acicules.* . . . . . . . . . . . . . . . . . . . . . . . . . . . . . . . . . . . . . . . . . . . . .29. **P. TRIVIALIS** L.

– Glumes *normally less different in size, usually both with 3 nerves;* lemmas *more or less hairy on marginal nerves for basal ¼-½ of their length,* occasionally also hairy between nerves; paleas on keels *always with longer acicules, occasionally transformed into hairs basally.* Sheaths of culm leaves *closed for basal ⅙-⅓ of their length, usually smooth, more rarely scabrous* but then *intermediate nerves of lemmas inconspicuous* and panicle branchlets usually much abbreviated. . . . . . . . . . . . . . . . . . . . . . . . . . . . . . . . . . . . . .11.

11. Plant *always without subterranean stolons, forming dense or more rarely looser tufts,* with more or less numerous flowering culms (smooth or more or less scabrous below panicle) *but usually without abbreviated vegetative shoots;* leaf sheaths *closed for basal ⅙-¼ of their length,* occasionally scabrous; leaf blades usually rather narrow, 0.6–3 mm wide, long-acuminate. Panicles with branchlets *more or less ribbed and always strongly scabrous along their whole length* (with numerous short acicules), often more or less abbreviated; spikelets 3–6 mm long; lemmas *with conspicuous or inconspicuous intermediate nerves, with relatively small cluster of long tangled hairs on callus* or sometimes completely without this; anthers 1–2 mm long. Viviparous varieties non-existent. (Section *Stenopoa* Dum.).
. . . . . . . . . . . . . . . . . . . . . . . . . . . . . . . . . . . . . . . . . . . . . . . . . . . . . . . . . . . .12.

– Plant *almost always with more or less long subterranean stolons, not forming tufts or more rarely forming rather dense tufts which consist of* one or a few flowering culms (always smooth below the panicle) and *one or a few abbreviated vegetative shoots;* leaf sheaths *closed for basal ¼-⅓ of their length,* always smooth; leaf blades 0.5–5 mm wide, on uppermost culm leaves almost always several times shorter than their sheaths, usually short-acuminate. Panicles with branchlets *cylindrical or weakly ribbed, smooth or weakly scabrous* (with isolated or more or less sparse, usually rather long acicules); lemmas *usually with more or less conspicuous intermediate nerves, on callus always with well developed cluster of long tangled hairs* which are occasionally very numerous (but in *P. eminens* callus more or less uniformly covered with hairs over almost its whole surface); anthers 1.3–2.5 mm long. Many species with viviparous varieties. . . . . . .21.

12. Usually pure green plant, 20–50 cm high, forming relatively loose tufts; culms smooth, *with 3–5 nodes of which the uppermost is often situated near their middle, relatively more leafy and rising not very far above sheath of uppermost leaf*; leaf sheaths smooth, or rarely slightly scabrous; blades of uppermost culm leaves *usually only slightly shorter than their sheaths.* Panicles *usually with rather long branchlets,* at flowering time more or less diffuse, rarely compact; lemmas always bare between nerves. (*P. nemoralis* L. s. l.). . . . . . . . . . . . . . . . . . . . . . . . . . . . . . . . . . . . . . . . . . .13.
– Usually greyish-green or glaucous green plant, forming dense tufts; culms smooth or scabrous below panicle, *with 2–3 nodes situated close together near their base, relatively less leafy and rising far above sheath of uppermost leaf*; leaf sheaths smooth or scabrous; blades of uppermost culm leaves *usually more or less abbreviated*. Panicles usually *with much abbreviated branchlets* (in comparison with total panicle length); lemmas bare or hairy between nerves. . . . . . . . . . . . . . . . . . . . . . . . . . . . . . . . . . . . . . . .16.

13. Ligule of uppermost culm leaf *1–3 mm long, usually longer than wide,* often rather pointed. Segments of spikelet rachis bare, but usually more or less scabrous with acicules or tubercles; callus of lemmas with small cluster of long tangled hairs. . . . . . . . . . . . . . . . . . . . . . . . . . .22. **P. PALUSTRIS** L.
– Ligule of uppermost culm leaf *0.2–1 mm long, always shorter than wide,* broadly rounded. . . . . . . . . . . . . . . . . . . . . . . . . . . . . . . . . . . . . . . . . . . . .14.

14. Callus of lemmas *without cluster of long tangled hairs*; segments of spikelet rachis more or less hairy. . . . . . . . . . . . . . .20. **P. TANFILJEWII** ROSHEV.
– Callus of lemmas *always with small cluster of long tangled hairs*. . . . . . . .15.

15. Segments of spikelet rachis *bare,* but usually more or less scabrous with acicules or tubercles. . . . . . . . . . . . . . . . . . . . . . . . . . .21. **P. LAPPONICA** PROKUD.
– Segments of spikelet rachis more or less *hairy*. . . . . . . . .19. **P. NEMORALIS** L.

16. More or less *glaucous green* plant of rocks and stony slopes, 10–35 cm high; culms below panicle *smooth or slightly scabrous;* sheaths and lower surface of leaf blades (at least on their basal half) *smooth*; ligule of uppermost culm leaf 1–2.5 mm long, or very rarely (in some specimens of *P. glauca*) 0.3–1 mm long. Panicle branchlets usually with spikelets relatively few (1–4), on average larger (usually 4–6 mm long), more or less rosy-violet. (*P. glauca* Vahl s. l.). . . . . . . . . . . . . . . . . . . . . . . . . . . . . . . . . . . . . . . . .17.
– More or less *greyish green or green* plant of stony slopes and other more or less steppelike habitats, 20–40 cm high; culms below panicle *always*

*scabrous;* sheaths and lower surface of leaf blades *more or less scabrous;* ligule of uppermost culm leaf 1–3.5 mm long. Panicle branchlets, when not too abbreviated, with spikelets relatively numerous (3–10), on average smaller (usually 3–5 mm long), greenish or more or less rosy-violet; callus of lemmas almost always with small cluster of long tangled hairs, occasionally consisting of only a few hairs. (*P. ochotensis* Trin. s. l.). . . . . . . . .19.

17. Lemmas on basal part *with short hairs between nerves,* almost always with small cluster of long tangled hairs on callus. . . . . . . . .28. **P. BRYOPHILA** TRIN.
— Lemmas hairy *only on keel and marginal nerves, often also on intermediate nerves.* . . . . . . . . . . . . . . . . . . . . . . . . . . . . . . . . . . . . . . . . . . . . . . . . . . . .18.

18. Callus of lemmas *completely bare;* pubescence on keel and marginal nerves relatively short. . . . . . . . . . . . . . . . . . . . . . . .27. **P. ANADYRICA** ROSHEV.
— Callus of lemmas *with small cluster of long tangled hairs* (sometimes consisting of only a few hairs); lemmas on keel and nerves with relatively more copious and longer pubescence. . . . . . . . . . . . . . .26. **P. GLAUCA** VAHL.

19. Lemmas on basal part *with short hairs between nerves;* panicles usually with rather strongly abbreviated branchlets. . . . . .24. **P. FILICULMIS** ROSHEV.
— Lemmas hairy *only on keel and marginal nerves,* often also on intermediate nerves. . . . . . . . . . . . . . . . . . . . . . . . . . . . . . . . . . . . . . . . . . . . . . . . . . . .20.

20. Branchlets of *dense, more or less spikelike panicle much abbreviated, up to 1 cm long;* spikelets usually more or less rosy-violet. Uppermost culm leaf (usually situated near base of culms) *with much abbreviated blade,* which is always several times shorter than its sheath. . . . . . . . . . . . . . . . . . . . . . . . . . . . . . . . . . . . . . . . . . . . . . .25. **P. BOTRYOIDES** (TRIN. EX GRISEB.) ROSHEV.
— At least some of the panicle branchlets longer (*over 1 cm long*), more or less diffuse at flowering time; spikelets greenish or with faint rosy-violet tinge. Uppermost culm leaf usually situated higher, *with less abbreviated blade,* sometimes almost equal in length to its sheath. . . . . . . . . . . . . . . . . . . . . . . . . . . . . . . . . . . . . . . . . . . . . . . . .23. **P. STEPPOSA** (KRYL.) ROSHEV.

21. Panicles 8–20 cm long, usually long and narrow with much abbreviated branchlets arranged in twos to fives per node; spikelets 6–8 mm long; glumes usually almost equal to lemmas in length, the upper with 3–5 nerves; callus of lemmas *more or less uniformly covered with rather long crinkled hairs, which seem to form a wrap around the base of each floret;* lemmas 4–7 mm long, with 5–9 not always conspicuous nerves, *hairy* on keel and marginal nerves *only near base* but more or less scabrous with acicules over almost all their remaining surface. Plant of meadows and banks of sea coast, 20–50 cm high, with rather thick, more or less branching rhizomes; ligule of uppermost culm leaf 0.5–2 mm long; leaf blades 2–8 mm wide, very stiff. . . . . . . . . . . . . . . . . . . . . . . . . .1. **P. EMINENS** C. PRESL.
— Callus of lemmas *with fully differentiated cluster of long tangled hairs on its back* (at base of keel), *sometimes also with shorter hairs at base of marginal nerves but never with continuous wrap of hairs of more or less uni-*

*form length surrounding base of each floret*; lemmas rather copiously hairy *on basal ⅓-⅔ of their length* on keel, *on basal ¼-⅔* on marginal nerves. (Section *Poa*)....................................................22.

22. Lemmas on their basal part *rather uniformly copiously or sparsely hairy between nerves,* on their remaining part *usually more or less scabrous with short acicules,* more rarely smooth; paleas scabrous on keels, on basal part with rather long acicules often transformed into appressed or more or less erect hairs, *between nerves with scattered* (sometimes only isolated) *hairs or much elongated acicules*; panicles at flowering time diffuse, usually more or less pyramidal, with long and slender, often flexuous branchlets which usually bear 1–3 (more rarely up to 5) spikelets and are arranged in twos to threes, more rarely fours, per node. Ligule of upper culm leaf 1.5–4 mm long. (*P. arctica* R. Br. s. l.)........................................23.
– Lemmas on their basal part *hairy only on keel and marginal nerves, more rarely also on intermediate nerves, sometimes* (in *P. sublanata*) also with hairs *between marginal and intermediate nerves but always bare to very base between keel and intermediate nerves*; remaining surface of lemma (excluding keel) *usually smooth* or with a few very short acicules; paleas scabrous on keels, on basal part with elongated acicules, *smooth and bare between keels* or more rarely with a few very short and inconspicuous acicules; branchlets of panicle often with more numerous spikelets.......27.

23. Plant *35–70 cm high,* not forming tufts but with short subterranean stolons; leaf blades 2–5 mm wide, flat, *that of uppermost culm leaf almost equal in length to its sheath.* Panicles 8–15 cm long, pyramidal, broadly diffuse at flowering time; spikelets 6–8 mm long, usually with faint rosy-violet tinge, often greenish; lemmas relatively weakly hairy between nerves (or sometimes bare), but always more or less scabrous with acicules.............................................3. **P. PLATYANTHA** KOM.
– Plant *more lowly, usually 10–35 cm high;* leaf blades on average narrower, *that of uppermost culm leaf considerably (usually several times) shorter than its sheath.* Panicles 3–9 cm long; spikelets usually rosy-violet, more rarely greenish. ...............................................24.

24. Spikelets 6–8 mm long; lemmas *copiously and rather long hairy* on their basal third, on keel and marginal nerves *for almost ⅔-¾ of their length with very numerous crinkled hairs over 1 mm long;* panicles small in comparison with overall size of plant (usually 4–6 cm long), with relatively short (usually up to 3 cm long) branchlets. Plant 20–35 cm high, usually not forming tufts but with short subterranean stolons. .................
............................................5. **P. LANATA** SCRIBN. ET MERR.
– Lemmas *less copiously hairy or, if copiously hairy, then panicles with smaller (4–6 mm long) spikelets;* keel and marginal nerves *with shorter and less numerous hairs;* panicles always larger in comparison with overall size of plant........................................................25.

25. Spikelets *6–8 mm long;* glumes almost equalling lemmas, *the upper usually more than 4 mm long;* lemmas *4–6 mm long.* Plant with short subterranean stolons, sometimes forming rather dense tufts; shoots clothed basally *with numerous, loosely arranged sheaths of dead leaves.* . . . . . . . . . . . . . . . . . . . . . . . . . . . . . . . . . . . . . . . . . . . . . . . . . .4. **P. MALACANTHA** KOM.
- Spikelets *4–6.5 mm long;* glumes shorter than lemmas, *the upper usually less than 4 mm long;* lemmas *3–4.5 mm long.* Shoots clothed basally *with relatively numerous, tightly appressed sheaths of dead leaves.* . . . . . . . . . .26.

26. Plant *forming dense tufts, with a few short subterranean stolons or sometimes completely lacking them.* . . . . . . . . . . . . . .7. **P. TOLMATCHEWII** ROSHEV.
- Plant *with more or less elongate subterranean stolons, not forming dense tuft.* . . . . . . . . . . . . . . . . . . . . . . . . . . . . . . . . . . . . . . . . . .6. **P. ARCTICA** R. BR.

27. Spikelets 6–8 mm long, greenish or with faint rosy-violet tinge; glumes almost equal to lemmas in length, *the lower with 3–5, the upper with 5 nerves;* lemmas 4–6 mm long, *with 7–9 not always conspicuous nerves,* rather copiously and long hairy on keel for ⅔ of their length, on marginal nerves for ⅕ - ⅓ of their length, on remaining surface (except keel) smooth or almost smooth; panicles 6–15 cm long, more or less compact and dense but with rather long smooth branchlets arranged in twos to fives per node. Plant of streamside meadows and gravelbars, 20–50 cm high, with relatively thick, more or less branching rhizomes; ligule 1–2 mm long; leaf blades 1.5–4 mm wide, very stiff, flat or loosely folded lengthwise. . . . . . . . . . . . . . . . . . . . . . . . . . . . . . . . . . . . . . . . . . . . . . . . . . . . . . . . . .2. **P. TRAUTVETTERI** TZVEL.
- Glumes *almost always with 3 nerves;* lemmas *with 5 usually conspicuous nerves.* Plant with rather more slender rhizomes. . . . . . . . . . . . . . . . . . . . .28.

28. Spikelets *6–8 mm long;* lemmas *with scattered short acicules over almost their whole surface;* paleas between keels *with scattered rather long acicules which are sometimes transformed into hairs;* panicles at flowering time usually pyramidal with long, more or less flexuous branchlets bearing only 1–4 spikelets. Plant with short subterranean stolons, occasionally forming tufts; ligule of uppermost culm leaf 2–3.5 mm long (see couplet 23 above, since in *P. platyantha* and *P. malacantha* specimens with the lemmas completely bare between the nerves occasionally occur). . . . . . . . . .23.
- Spikelets usually *3–5.5 mm long,* rarely (in *P. sublanata)* longer; lemmas *usually with acicules only on keel, smooth or almost smooth over their surface;* paleas *bare and smooth, or more rarely with very short acicules* (inconspicuous even at high magnification). (*P. pratensis* L. s. l.). . . . . . .29.

29. Spikelets *4–7 mm long;* lemmas *with very large cluster of long tangled hairs* on callus, *very copiously and rather long hairy for ⅔ - ¾ of their length on keel, for half their length on marginal nerves, and for ⅓ of their length on intermediate nerves,* sometimes with hairs spreading over their surface between marginal and intermediate nerves; panicles 6–15 cm long; their branchlets slightly scabrous (to completely smooth), usually more or less abbreviated (in comparison with overall length of panicle), arranged in

twos to fives per node and *usually bearing only 1–4 spikelets.* Plant of riverine sands and gravelbars, 20–50 cm high, with long subterranean stolons; ligule of uppermost culm leaf *1.5–3.5 mm long;* leaf blades 1.5–3.5 mm wide, usually flat. .................................9. **P. SUBLANATA** REVERD.

– Spikelets *(3)3.5–5.5(6) mm long;* lemmas *less copiously hairy, usually bare on intermediate nerves;* branchlets of panicle (at least the longer of them) *usually bearing numerous spikelets* (4–15). Ligule of uppermost culm leaf *0.8–2.5 mm long,* usually rather obtuse. ............................30.

30. Plant 8–20 cm high, with short subterranean stolons (but usually not forming tufts), *more or less glaucous green with mealy bloom* which in the dry state is most conspicuous on the glumes; leaf blades relatively short and broad, 1.5–3 mm wide. Panicles 2.5–6 cm long, at flowering time and later more or less pyramidal, with horizontally spreading, rather thickish branchlets arranged *usually in ones or twos on lower nodes, rarely in threes.* ...................................11. **P. SUBCAERULEA** SM.

– Often taller *green* plant (without glaucescent mealy bloom). Branchlets of panicle *usually arranged in threes to fives on lower nodes,* more rarely (in small specimens of *P. alpigena)* in ones to twos but then more slender and directed more or less obliquely upwards. ...........................31.

31. Plant 20–50 cm high, *forming small but dense tufts (clusters of shoots united by a rhizome and consisting of one or a few flowering culms and one or a few abbreviated vegetative shoots which are closely approximated to the base of the culms and enclosed by rather numerous sheaths of dead leaves);* leaf blades of vegetative shoots usually rather stiff, setiformly folded lengthwise and directed upwards, up to 1 mm in diameter, more rarely flat and 0.5–1.5 mm wide. Panicles 4–12 cm long, with a few abbreviated, slightly scabrous branchlets which at flowering time and later are usually directed more or less obliquely upwards; glumes usually up to 3–3.5 mm long. .................................................12. **P. ANGUSTIFOLIA** L.

– Plant *not forming dense clusters of shoots; shoots usually solitary, more rarely approximated but then loosely arranged and with only a small number of sheaths of dead leaves at their bases;* leaf blades of vegetative shoots often broader and always softer. Glumes up to 4 mm long. ............32.

32. Plant 10–40 cm high; leaf blades of vegetative shoots *usually very narrow,* 0.6–2 (more rarely up to 3) mm wide, often folded lengthwise. Panicles at flowering time *usually oblong or elongate-pramidal;* their branchlets usually somewhat abbreviated, *smooth or almost smooth* (with a few acicules on their terminal part), at flowering time and later usually *relatively slightly divergent;* spikelets 3–5 mm long, usually more or less rosy-violet, rarely greenish, often modified into vegetative buds (vivipary). ................
..........................................10. **P. ALPIGENA** (FR.) LINDM.

– Plant 20–60 cm high; leaf blades of vegetative shoots *on average broader,* 1–4 mm wide, usually flat. Panicles at flowering time *usually more or less pyramidal;* their branchlets *usually slightly scabrous* (with acicules scattered over their entire or almost their entire length), rarely almost smooth,

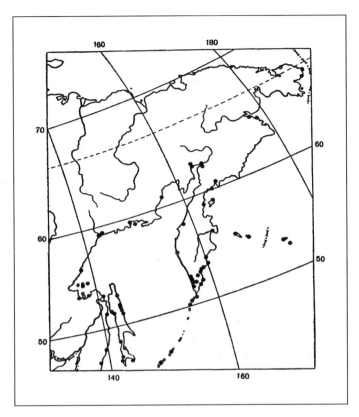

MAP II–33   Distribution of *Poa eminens* Presl.

at flowering time and later *usually strongly divergent from rachis of panicle, often spreading horizontally;* spikelets on average larger (3.5–5.5 mm long), usually with faint rosy-violet tinge or greenish, not modified into vegetative buds. . . . . . . . . . . . . . . . . . . . . . . . . . . . . . . . . .8. **P. PRATENSIS** L.

1. ***Poa eminens*** C. Presl, Reliq. Haenk. 1 (1830), 273; Rozhevits in Fl. SSSR II, 425; Hultén, Fl. Al. II, 205; Polunin, Circump. arct. fl. 66.
   *P. glumaris* Trin. in Mém. Ac. Pétersb. sér. 6, 1 (1831), 379; Gelert, Fl. arct. 1, 121.
   *Glyceria glumaris* (Trin.) Griseb. in Ledebour, Fl. ross. IV (1853), 392.
   *Poa kurilensis* Hack. in Bull. Herb. Boiss., ser. 2, IV (1904), 524.
   *P. Trinii* Scribn. et Merr., Grass. Alaska (1910) 73.
   **Ill.:** Fl. SSSR II, pl. XXXII, fig. 6.

   Littoral, predominantly Beringian species belonging to Section *Arctopoa* (Griseb.) Tzvel. comb. nova (*Glyceria* sect. *Arctopoa* Griseb. in Ledebour, l. c. 392). This is apparently one of the most primitive sections of the genus, as indicated by such morphological characters as the well developed thickened rhizomes, large spikelets, frequent increase of the number of nerves on the glumes to 5 and on the lemmas to 7–9, relatively weak development of a keel on the lemmas, and especially the nature of the pubescence of the callus of the lemmas (copiously and more or less uniformly covered with rather long crinkled hairs). The last of these characters in our opinion indicates a certain remote connection between the genus *Poa* and such very isolated and primitive hydrophilic genera of grasses as *Scolochloa* and *Arctophila* and with certain genera of the tribe *Aveneae* Dum., which is in general very closely related to the tribe *Festuceae* Dum. Moreover, the species and species-groups of *Poa* related to *P. eminens* which sometimes possess dioecious spikelets, are distributed mainly in temperate districts of North America rich in relics, while in the territory of Eurasia (excluding *P. Trautvetteri* Tzvel.) only two morphologically very isolated species occur: *P. tibetica* Munro and *P. subfastigata* Trin., which show a certain affinity with *P. eminens* although belonging to other sections of the genus. The variation of the species in the arctic part of its range, as in other parts, is not great and usually amounts to partial or complete abortion of spikelets accompanied sometimes by strong elongation of the glumes (whence the name "*P. glumaris* Trin."). It should also be noted that the spikelets of *P. eminens* at various stages of their development (from their leaving the sheath of the uppermost culm leaf to fruiting) seem externally very different, as already indicated by Hultén (l. c. 206). It is probable that *P. eminens* is distributed rather sporadically in NE Asia and considerably rarer than on the American coast of the Bering Sea due to a lack of habitats with suitable ecological conditions.

Growing on maritime sandbars or gravelbars, sometimes also in maritime marshy meadows, usually in the estuaries of the larger rivers and on the shores of small bays.

**Soviet Arctic.** Beringian coast of Chukotka Peninsula (Uelen, Lawrence Bay); Anadyr Basin (Anadyr, Geka district); lower reaches of Penzhina (common); Bay of Korf. (Map II–33).

**Foreign Arctic.** West coast of Alaska north to 68°N; east coast of Labrador north to 56°N.

**Outside the Arctic.** Pacific coast of Asia south to Olga Bay (Primorskiy Kray) and the island of Hokkaido; west coast of North America from Alaska to Vancouver Island; southern part of Hudson Bay (isolated locality); eastern part of Canada from Labrador to the Magdalen Islands (47°N).

2. **Poa Trautvetteri** Tzvel. sp. nova.
*P. glumaris* var. *laevigata* Trautv. in Acta Hort. Petrop. V, 1 (1877), 137; non *P. laevigata* Scribn., 1897.

The variety *P. glumaris* var. *laevigata* Trautv., described from rather numerous, well collected specimens of A. L. Chekanovskiy (ad Lena inferiores, prope pag. Goworowo, 25 VII 1875, Czekanowski), is in our opinion sufficiently distinct from *P. eminens* C. Presl (= *P. glumaris* Trin.) to be recognized as a separate species, showing in certain respects even greater similarity to another littoral Beringian species of *Poa* (but from the section *Poa*), *P. macrocalyx* Trautv. et Mey.

*Poa Trautvetteri* more closely resembles *P. eminens* in the structure of the rhizomes, in the increased number of nerves on the glumes and lemmas in comparison with the norm for the genus *Poa*, and in having smooth panicle branchlets and a short ligule; but resembles *P. macrocalyx* in having rather long branchlets of a relatively short panicle, and in the pubescence of the lemmas (long-haired on the keel and marginal nerves to a considerable extent, and with the callus possessing a fully differentiated cluster of long hairs on its back as characteristic of the genus *Poa*). In consideration of the very restricted range of *P. Trautvetteri* (after A. L. Chekanovskiy it has been collected only once more in 1958 by B. N. Ovchinnikov 3 km below Govorovo settlement on the Lena), it might be assumed to be a hybrid between the above two species; but *P. eminens* at the present time only just ranges north of the Bering Strait on the Alaskan coast and *P. macrocalyx* does not occur in the Arctic at all (reaching only as far north as the Commander and Aleutian Islands). Moreover, *P. Trautvetteri* has the lemmas almost completely smooth between the nerves, a character not belonging to either *P. eminens* or *P. macrocalyx*. The remaining hypothesis (which seems to us highly probable) is that the hybrid origin of *P. Trautvetteri* should be referred to a very distant time of milder climatic conditions, when the ranges of the two littoral species extended considerably further to the north and west. We possess no information with respect to the ability of *P. Trautvetteri* to reproduce by seed (it is quite probable that the caryopses in this species do not develop, although the anthers are normally developed), but it can readily reproduce vegetatively (with the aid of rhizomes) and therefore must be considered a legitimate species although possibly represented only by a few relict populations.

Growing on the shores of rivers in marshy meadows, on gravelbars and on riverine alluvia.

**Soviet Arctic.** Lena below Govorovo settlement. Endemic to this district.

3. **Poa platyantha** Kom. in Bot. mat. Gerb. Glav. bot. sada RSFSR V, 10 (1924), 148; Rozhevits in Fl. SSSR II, 423.

*P. penicillata* Kom., l. c. 148; Rozhevits, l. c. 396.

*P. occidentalis* auct. non Vasey — Hultén, Fl. Kamtch. I, 129.

? *P. Eduardii* Golub. in Sist. zam. po mat. Gerb. Tomsk. gos. univ. 4 (1936), 2.

? *P. hispidula* var. *aleutica* Hult., Fl. Aleut. Isl. (1937) 88; id., Fl. Al. II, 211.

**Ill.:** Fl. SSSR II, pl. XXXI, fig. 21; pl. XXIX, fig. 12.

East Asian species whose range is not entirely clear to us because of an abundance of very close and possibly partly identical species described from North America and Eastern Asia (*P. stenantha* Trin., *P. flavicans* Ledeb., *P. Turneri* Scribn., *P. hispidula* Vasey, *P. scabriflora* Hack., etc.). All these species possess rather large spikelets (6–9 mm long) as well as lemmas scabrous or hairy between the nerves, and are apparently more "primitive" than the other large species–group of section *Poa*, *P. pratensis* L. s. l., as is also confirmed by their geographical distribution. The type specimen of *P. penicillata* Kom. (Kamchatka, Avacha Basin on shore of River Koryatskaya, 24 VII 1908, No. 3496, Komarov) differs from the type specimen of *P. platyantha* Kom. (Kamchatka, Bolshaya River Basin in district of Nachika village, 30 VII 1908, Komarov), apart from somewhat abnormal panicle development, only in having the lemmas bare between the nerves. Inspection of much herbarium material from Kamchatka shows that such specimens are only extreme variants with respect to a single character (the lemma being generally rather weakly hairy between the nerves) and cannot be attributed to a separate species. However, in some cases such specimens may be hybrids between *P. platyantha* and the very closely related littoral species *P. macrocalyx* Trautv. et Mey. From the Arctic only one specimen of *P. platyantha* of doubtful origin, possibly collected on the Aleutian Islands, is so far known.

Outside arctic limits growing in sparse forest and shrubbery, in forest glades and on riverine gravelbars, ascending in the mountains to the upper limit of forest. On Kamchatka specimens with viviparous spikelets (f. *vivipara* Kom.) occur occasionally.

**Soviet Arctic.** Beringian coast of Chukotka Peninsula (Senyavin Strait, Mertens).

**Foreign Arctic.** Absent.

**Outside the Arctic.** Coast of Sea of Okhotsk in district of Magadan; Kamchatka (north to 58°N); Kurile, Commander and Aleutian Islands. Possibly also the south coast of Alaska and the island of Hokkaido.

4. **Poa malacantha** Kom. in Bot. mat. Gerb. Glavn. bot. sada RSFSR V, 10 (1924), 148; Rozhevits in Fl. SSSR II, 422.

*P. bracteosa* Kom., l. c. 147; Rozhevits, l. c. 422.

*P. Komarovii* Roshev. in Izv. Glavn. bot. sada SSSR XXVIII (1927), 286 and in Fl. SSSR II, 422; Hultén, Fl. Al. II, 212; Polunin, Circump. arct. fl. 66.

**Ill.:** Fl. SSSR II, pl. XXXI, fig. 18, 20.

Arctic–alpine, predominantly Beringian species, replacing the preceding species in the tundra zone and in the barren zone of mountains and connected to it by specimens and populations with transitional characters. In comparison with

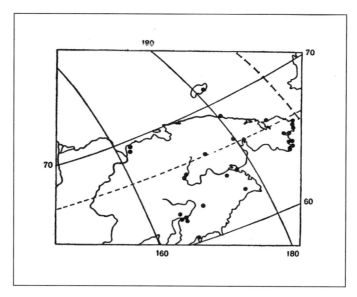

MAP II–34  Distribution of *Poa malacantha* Kom.

*P. platyantha* this plant is smaller, with not so large panicles and more strongly coloured spikelets, often forming rather dense tufts. From another closely related species, *P. arctica* R. Br., this species is usually easily distinguished (apart from the larger spikelets) by the numerous loosely arranged sheaths of dead leaves which clothe the base of the shoots like a veil. There occasionally occur in the far east of the Chukotka Peninsula specimens with greenish spikelets which significantly approach the species *P. stenantha* Trin. (widespread in Alaska and the Aleutian Islands), differing from it only in having smooth or very slightly scabrous panicle branchlets and a better developed cluster of long tangled hairs on the callus of the lemmas. As in many other cases, there are differences between the type specimens of the three species united by us as a single species under the name "*P. malacantha* Kom." (we have preferred not to use the name "*P. bracteosa* Kom." published one page earlier because that species was separated mainly due to the presence of a membranous bract at the base of the panicles, a character associated with abnormal panicle development and posessing no systematic importance); but it is perfectly obvious from the large herbarium material from Kamchatka that these differences are purely individual or populational. Thus, the type specimen of *P. bracteosa* Kom. (Kamchatka, Avacha Basin, stony tundra near foot of Koryatskaya volcano, 17 VIII 1908, Komarov), erroneously identified with *P. nivicola* Kom. by Hultén (Fl. Kamtch. I, 125), is densely tufted and possesses lemmas almost bare between the nerves and smooth panicle branchlets; the type specimen of *P. malacantha* Kom. (Kamchatka, subalpine meadows on River Kashkan, 25 VI 1909, No. 2832, Komarov) is loosely tufted, with lemmas slightly hairy between the nerves and slightly scabrous panicle branchlets; the lectotype of *P. Komarovii* Roshev. (Kamchatka, Paratunka Basin, near Klyuchi village, 28 VI 1908, No. 1080, Komarov) is relatively densely tufted, with lemmas rather copiously hairy and almost smooth panicle branchlets. It should also be noted that Kamchatka has proved to be not at all the "centre of endemism for bluegrasses" that R. Yu. Rozhevits supposed it to be (Izv. Glavn. bot. sada SSSR XXVI, 285–287). At the present time only the species *P. nivicola* Kom. can be considered endemic to Kamchatka, and it is very likely that it too will be discovered somewhere or other in neighbouring districts of Eastern Asia.

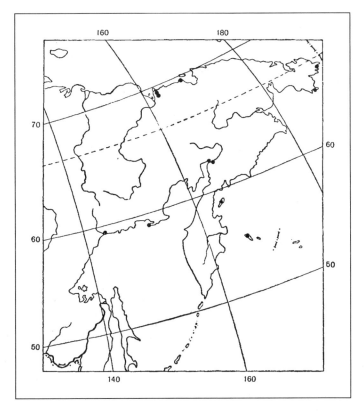

MAP II-35  Distribution of *Poa lanata* Scrib. et Merr.

There is a viviparous variety of this species: *P. malacantha* var. *vivipara* (Roshev.) Tzvel. comb. nova (= *P. Komarovii* var. *vivipara* Roshev., l. c., 1927, 286), known in the Soviet Arctic from the Anadyr Basin (Rarytkin Range) and from the vicinity of Provideniye Bay. It is rather common on Kamchatka and the Commander Islands.

In the tundra zone *P. malacantha* grows on stony and clayey slopes, screes and gravelbars; further south it occurs on barrens, sometimes penetrating the upper part of the forest zone (larch forest or thickets of cedar-pine stlanik). It colonizes well drained, slightly or moderately moist sites devoid of closed vegetational cover, usually exposed to winds. Of ubiquitous occurrence in far eastern districts of the Soviet Arctic, but in low abundance.

Thus, on the SE coast of the Chukotka Peninsula *P. malacantha* can be found everywhere on poorly vegetated rocky mountain slopes all the way to the summits, as well as in patchy dwarf shrub-moss or dwarf shrub-lichen tundras (with *Dryas* and *Empetrum)* at the edge of bare patches, on slumping portions of clayey or stony slopes, on dry riverine gravelbars, etc. On the southern part of the Koryak Coast the plant occurs near the summits of volcanic cones (above the zone of closed thickets of cedar-pine stlanik) on talus screes and on open windswept areas with cushions of *Diapensia obovata, Loiseleuria procumbens,* etc.

**Soviet Arctic.** Lower reaches of Kolyma (River Medvezhya); Ayon Island; Wrangel Island; northern and Beringian coasts of Chukotka Peninsula (frequent, Toygunen Peninsula, Cape Dezhnev, Lawrence Bay, Provideniye Bay); Bay of Krest; Anadyr Basin; Penzhina Basin; Bay of Korf. (Map II–34).

**Foreign Arctic.** Northern and western Alaska (near Point Barrow and in the district of the Bering Strait).

**Outside the Arctic.** Northern coast of Sea of Okhotsk (district of Okhotsk and Magadan); Kamchatka; Kurile Islands; Commander and Aleutian Islands; southern part of Alaska.

5. **Poa lanata** Scribn. et Merr., Grass. Alaska (1910) 72; Hultén, Fl. Al. II (1942), 213; Polunin, Circump. arct. fl. (1959) 66.

*P. petraea* Trin. ex Kom. in Fl. Kamch. I (1927), 173; Hultén, Fl. Kamtch. I, 72; Rozhevits in Fl. SSSR II, 424.

**Ill.:** Scribner & Merrill, l. c., pl. 16.

Northern Beringian species very close to *P. malacantha* Kom., partly occupying an intermediate position between that species and *P. platyantha* Kom. but differing from them in having relatively small panicles and very copiously long-haired lemmas. Judging from the original diagnosis and figure, the type specimen of *P. lanata* differs from the type specimens of *P. petraea* Trin. ex Kom. (Kamchatka, 1831, Peters) only in the slightly larger size of the plant as a whole, which in our opinion should not be considered a sufficient basis for recognizing two separate species. The viviparous variety described by Hultén, *P. lanata* var. *vivipara* Hult. (Fl. Aleut. Isl., 1937, 90), has not so far been found in the USSR.

Both inside and outside the Arctic *P. lanata* is very sporadically distributed, occurring mainly on sandy and gravelly shores of rivers and streams, usually near the sea coast.

**Soviet Arctic.** Lower reaches of Kolyma (River Medvezhya); Karchik Peninsula in the district of the Bay of Chaun; Beringian coast of Chukotka Peninsula (Puoten Bay, Provideniye Bay); lower reaches of Penzhina. (Map II–35).

**Foreign Arctic.** West coast of Alaska north to Cape Lisburne.

**Outside the Arctic.** Coast of Sea of Okhotsk (near Okhotsk); Karaginskiy Island; Commander and Aleutian Islands; southern part of Alaska; a very closely related species (*P. Smirnowii* Roshev.) in the Sayan Mountains (Tunkinskiye Barrens).

6. **Poa arctica** R. Br. in Suppl. to App. Parry's Voyage XI (1824), 288; Grisebach in Ledebour, Fl. ross. IV, 373; Lindman in Lynge, Vasc. pl. N. Z. 118; Krylov, Fl. Zap. Sib. II, 289; Perfilev, Fl. Sev. I, 91; Rozhevits in Fl. SSSR II, 410; Hultén, Fl. Al. II, 201; Hylander, Nord. Kärlväxtfl. I, 258; Kuzeneva in Fl. Murm. I, 212; A. E. Porsild, Ill. Fl. Arct. Arch. 29; Hultén, Amphi–atl. pl. 24; Karavayev, Konsp. fl. Yak. 56; Polunin, Circump. arct. fl. 66.

*P. flexuosa* Wahlb., Fl. suec. (1824) 108; Lange, Consp. fl. groenl. 178; non Sm., 1800.

*P. cenisia* auct. non All. — Gelert, Fl. arct. 1, 122.

? *P. Williamsii* Nash in Bull. N. Y. Bot. Gard. II, 6 (1901), 151.

*P. rigens* auct. non Hartm. — Hanssen & Lid, Fl. pl. Franz Josef L. 35; Scholander, Vasc. pl. Svalb. 93.

*P. petschorica* Roshev. in Izv. Bot. sada AN SSSR XXX (1932), 775 and in Fl. SSSR II, 410.

**Ill.:** Lindman, l. c., tab. 46, fig. 2; Fl. SSSR II, pl. XXX, fig. 18, 19; Fl. Murm., pl. LXXI; Porsild, l. c., fig. 6, e.

Circumpolar arctic species, one of the most characteristic and most widespread plants in the Arctic. Displaying much polymorphism and forming numerous more or less distinct populations differing mainly in the overall size of the plant, the degree of development of subterranean stolons, the shape of the panicle, the size and colour of the spikelets, the degree of pubescence of the lemmas, etc. In a very interesting work on polymorphism in *P. arctica*, Nannfeldt (On the Polymorphy of *Poa arctica* R. Br., Symb. Bot. Ups. IV, 4, 1940) recognizes for Scandinavia alone six ecological–geographical races of this species as subspecies, the rank being subsequently changed to varieties by Hylander (l. c.). Two of these subspecies, characterized by dense tufts and almost lacking subterranean stolons, are more correctly referred to the next species; the remaining four, "ssp. *elongata* (Bl.) Nannf.," "ssp. *microglumis* Nannf.," "ssp. *depauperata* (Fr.) Nannf." and "ssp. *tromsensis* Nannf.," belong to *P. arctica* in a narrower sense and apparently also occur in the Soviet Arctic. However, we have so far not succeeded in dividing the extensive herbarium material available into these four subspecies, both because of blurred boundaries between them and because of the existence of a large number of local populations not falling within their definitions. This also applies to the subspecies from Alaska distinguished by Hultén (l. c., 1942, 202) as "ssp. *Williamsii* (Nash) Hult." and "ssp. *longiculmis* Hult.," the first of which he also recorded for Kamchatka and the Chukotka Peninsula. In more southern districts of the Far East, starting from Kamchatka and the Okhotsk Coast, there do indeed occur specimens of *P. arctica* differing from more northern specimens in the (on average) larger size of the plant, more or less abbreviated subterranean stolons, and faintly coloured spikelets (usually specimens from larch forests or shrubby valley communities); it is possible that these belong to "ssp. *Williamsii*," but they are morphologically very weakly differentiated. Moreover, in Hultén's subsequent work (l. c., 1958) this subspecies is recorded by him only for America. In NE Asia some increase in the variation of *P. arctica* is generally observed (for example, specimens with rather strongly scabrous panicle branchlets and specimens with characters more or less transitional to *P. malacantha* Kom. and *P. lanata* Scribn. et Merr. occur there); this is doubtless associated with the circumstance that precisely there and in North America this species is accompanied by a whole group of closely related species of more southern distribution whose high arctic derivative it apparently represents. The East Asian and North American connections of *P. arctica* are also reflected in its present range: this species is completely absent from the mountains of Central and Southern Europe (where *P. cenisia* All. occurs, a species similar to *P. arctica* in appearance but quite different from it) and penetrates deeply into the forest and forest-steppe zones of Asia only on the barrens of its northeastern mountain ranges, reaching as far west as Baykal. *Poa arctica* shows only a very remote affinity to the group *P. pratensis* L. s. l. As for *P. petschorica* Roshev., this species was described from a single herbarium sheet (Pechora Bay, near Bolvannoye setlement, 25 VIII 1924, Popov) containing specimens of not entirely normal development possessing broader leaves, denser panicles and brightly coloured spikelets apparently as a result of growing under conditions of excessive moisture or temporary flooding. At any rate it is not a separate species.

The viviparous variety *P. arctica* var. *vivipara* Hook. (Hooker, Fl. Bor. Amer. II, 1840, 246) occurs relatively rarely and usually gives the impression more of the hybrid *P. arctica* × *P. alpigena*, being connected to the viviparous variety of the latter species by specimens with transitional characters. In the Soviet Arctic it is known only from Novaya Zemlya (specimens assumed by Lindman to be of hybrid origin, but without firm grounds) and from the Chukotka Peninsula (Provideniye Bay). In North America viviparous specimens of *P. arctica* apparently occur more often than in Eurasia.

A characteristic plant of mesic portions of tundras (placorn, valley, lowland or alpine) which are more or less well-drained and snow-covered in winter. Spreading by forming slender subterranean stolons, it usually occurs as diffuse beds in the turf of mossy, lichenaceous, herb-moss or dwarf shrub-moss tundras; in extensive lowland marshes it colonizes elevations of the microrelief (for example, the hummocks of polygonal marshes). On riverine gravelbars and loose shoreline banks it grows luxuriantly and sometimes forms small lawns. In alpine habitats it is common in *Cassiope*-tundras, moist patchy moss-sedge tundras, and in meadows at the edges of snowfields; sometimes it spreads as a luxuriant carpet over accumulations of fine soil in the crevices of rocky hillocks, also near marmot colonies (Arctic Kharaulakh); rarer in closed *Dryas*-tundras; avoiding open com-

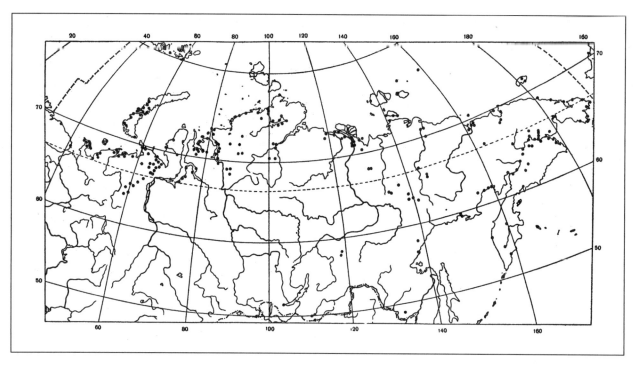

MAP II–36   Distribution of *Poa arctica* R. Br.

munities of strongly windswept stony areas almost devoid of snow cover in winter (as well as tundra marshes). It readily colonizes fertilized sites near settlements and around old camps of reindeer-herders, where it acquires an appearance not proper to the tundra plant (large carpets, taller stems, green leaves, etc.). In the High Arctic it gradually (from south to north) loses its importance in the vegetative cover. In East Siberia it penetrates the northern part of the forest zone, occurring in shrub communities in valleys (dwarf birch and willow carr), also in mossy larch forest in valleys, open larch forest on mountainsides, etc.

**Soviet Arctic.** Murman (Iokanga, Ponoy); Kanin (common); Kolguyev; Malozemelskaya and Bolshezemelskaya Tundras; Polar Ural; Pay-Khoy; Vaygach; Novaya Zemlya; Franz Josef Land; Yamal; Obsko-Tazovskiy Peninsula; Gydanskaya Tundra; lower reaches of Yenisey (common); Taymyr; lower reaches of Olenek; lower reaches and delta of Lena; Tiksi; New Siberian Islands; lower reaches of Indigirka and Kolyma; district of Bay of Chaun; Ayon Island; Wrangel Island; Chukotka; Anadyr Basin; lower reaches of Penzhina; Bay of Korf. (Map II–36).

**Foreign Arctic.** Alaska (west and north coasts); Northern Canada; Labrador; Canadian Arctic Archipelago; Greenland (from extreme south to 80°N); Spitsbergen; Arctic Scandinavia.

**Outside the Arctic.** Mountains of Scandinavian Peninsula (disjunct portion of range in Southern Norway); Khibins Mountains; Urals (south approximately to 62°N); northern part of Central Siberian Plateau; Prebaikalia and mountains east of Baykal; mountains of NE Yakutia (Verkhoyansk and Cherskiy Ranges, etc.); mountains of the north and west coasts of the Sea of Okhotsk (Pribrezhnyy, Dzhugdzhur and Kolyma Ranges, etc.); Bureinskiy Range; Sikhote-Alin; Kamchatka; Kurile Islands; reported for Sakhalin and Hokkaido; Commander and Aleutian Islands; Alaska; Greenland; northern part of Canada (south to Lake Superior and in the Rocky Mountains to the states of Oregon and New Mexico).

7. ***Poa Tolmatchewii*** Roshev. in Izv. Bot. sada AN SSSR XXX (1932), 299 and in Tolmachev, Fl. Taym. I, 97; id. in Fl. SSSR II, 411.
? *P. trichopoda* Lge., Fl. Dan. XVII (1877–1883), tab. 2885; non Heldr. et Sart., 1859.
? *P. filipes* Lge., Consp. fl. groenl. (1890) 175.
*P. arctica* ssp. *caespitans* (Simm. ex Nannf.) Nannf. in Symb. Bot. Upsal. IV, 4 (1940), 71; A. E. Porsild, Ill. Fl. Arct. Arch. 30; Hultén, Amphi-atl. pl. 24.
*P. cenisia* f. *caespitans* Simm. ex Nannf., l. c., in syn.
*P. arctica* var. *caespitans* (Simm. ex Nannf.) Hyl., Nord. Kärlväxtfl. I (1953), 260.
**Ill.:** Lange, l. c. (1877–1883), tab. 2885; Tolmachev, Fl. Taym. I, fig. 20; Fl. SSSR II, pl. XXX, fig. 20; Porsild, l. c., fig. 6, d.

This very sporadically distributed species, described from a few specimens from Taymyr (lower reaches of River Yamu-Tarida, 13 IX 1928, No. 834, Tolmachev), was subsequently also discovered in the Tiksi district and very recently a fully typical specimen (identified as "*P. Ganeschinii* Roshev.") was found by us among herbarium material from the Khibins Mountains (vicinity of Khibinogorsk, Yukspoora Plateau, 2 IX 1934, A. Semenova-Tyan-Shanskaya). The range has thereby proved to be considerably expanded and there is consequently every reason for identifying this species with the subspecies *P. arctica* ssp. *caespitans* Nannf. described in detail by Nannfeldt (l. c.), which also differs from *P. arctica* by its densely tufted growth form. We also think it very probable that this species can be identified with the Greenlandic species *P. filipes* Lge. (= *P. trichopoda* Lge.), judging from the original description and figure which are identical with *P. Tolmatchewii*. However Nannfeldt, after seeing the type of *P. filipes*, did not consider it possible to identify "ssp. *caespitans*" with that species because of the structure of the anthers, which are strongly abortive in "ssp. *caespitans*" (since caryopses are nevertheless formed in this subspecies, Nannfeldt considered it to be an apomictic subspecies) but in *P. filipes* normally developed with good pollen. Specimens of *P. Tolmatchewii* usually also have abortive anthers (but often larger than in ssp. *caespitans)*, on account of which we have also not yet decided to completely identify *P. Tolmatchewii* with *P. filipes*. The very sporadic distribution of *P. Tolmatchewii* and its abortive anthers suggest a possible hybrid origin of this species, something which seems to us highly probable. If this is the case, *P. arctica* R. Br. and *P. glauca* Vahl may be considered the parent species, as is fully supported by the fact that the specimens of *P. Tolmatchewii* known to us always occur at localities from which the two stated species are also known (though considerably more rarely). Furthermore, there are specimens of *P. Tolmatchewii* more closely approaching *P. glauca* from the vicinity of Tiksi (whence an especially large herbarium material is presently available): these have strongly scabrous, more or less ribbed branchlets of a panicle which has the general appearance of *P. arctica;* collected there were two closely associated tufts, one of which belongs to *P. glauca*, the other to *P. Tolmatchewii*. Of course, this hypothesis needs to be tested by observations and experiment.

The viviparous variety of this species occurs extremely rarely and until most recent time was known only from the small alpine district of Dovre in Southern Norway; but it has recently also been found in the territory of the USSR in the lower reaches of the Lena (Tuora-Sis Range, Sokuydakh Mountain, moist talus scree, 11 VIII 1957, Norin and Yurtsev). The following is the pertinent synonymy.

*Poa Tolmatchewii* var. *stricta* (Lindeb.) Tzvel. comb. nova. — *P. stricta* Lindeb. in Bot. Notis. (1855) 10; non Roth, 1821; nec D. Don, 1821. — *P. arctica* ssp. *stricta* (Lindeb.) Nannf. in Symb. Bot. Upsal. IV, 4 (1940), 67. — *P. arctica* var. *stricta* (Lindeb.) Hyl., Nord. Kärlväxtfl. I (1953), 259. — *P. arctica* var. *vivipara* auct. non Hook.; Yurtsev in Bot. zhurn. XLIV, 8, 1174.

*Poa Tolmatchewii* grows on open stony or rocky slopes, screes and gravelbars.

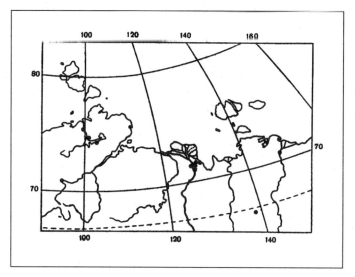

MAP II–37  Distribution of *Poa Tolmatchewii* Roshev.

**Soviet Arctic.** Taymyr (basin of Lake Taymyr); lower reaches of Lena; Tiksi; lower reaches of Indigirka (Chokurdakh). Recorded by Nannfeldt for Novaya Zemlya (Belushya Bay, Malyye Karmakuly, Matochkin Shar, Bezymyannyy Bay). (Map II–37).

**Foreign Arctic.** Canadian Arctic Archipelago; Labrador (northern part); Greenland (north of the Arctic Circle); Spitsbergen.

**Outside the Arctic.** Mountains of Scandinavian Peninsula (two isolated localities in Norway); Khibins Mountains (Yukspoora Plateau); northern part of Central Siberian Plateau (in the Kheta Basin).

**8. *Poa pratensis*** L., Sp. pl. (1753) 67; Grisebach in Ledebour, Fl. ross. IV (1853), 378; Krylov, Fl. Zap. Sib. II, 297, pro parte; Perfilev, Fl. Sev. I, 89; Rozhevits in Fl. SSSR II, 388; Hultén, Fl. Al. II, 219; Kuzeneva in Fl. Murm. I, 203; Karavayev, Konsp. fl. Yak. 55; Polunin, Circump. arct. fl. 67.

? *P. rigens* Hartm., Scand. Fl. (1820) 448.

*P. turfosa* Litw. in Sp. Gerb. Fl. SSSR VIII (1922), 135; Rozhevits, l. c. 389; Kuzeneva, l. c. 204.

*P. pinegensis* Roshev. in Izv. Bot. sada AN SSSR XXX (1932), 773 and l. c. (1934) 393; Perfilev, l. c. 88.

*P. pratensis* ssp. *eupratensis* Hiit., Suom. Kasvio (1933) 204; Hylander, Nord. Kärlväxtfl. I, 257.

**Ill.:** Fl. SSSR II, pl. XXIX, fig. 1, 6.

Widely distributed boreal species replaced in the Arctic by the very closely related species *P. alpigena* (Fr.) Lindm. Occurring far from universally in the Soviet Union, already becoming rather rare in the forest-steppe zone and restricted to more northern vegetational communities (sphagnum bogs, marshy meadows, etc.). Among the species included by us as synonyms of *P. pratensis*, *P. turfosa* Litw.was described from numerous specimens from a mossy bog in the valley of the River Unzha in Vladimir Oblast; these are quite different from flat–leaved specimens of *P. angustifolia* L. which are sometimes confused with *P. pratensis*, but agree fully with typical northern specimens of *P. pratensis* except that the stems and sheaths of the radical leaves are actually very slightly complanate basally (something to which substantial importance can scarcely be attributed). *Poa pinegensis* Roshev. was described from specimens of *P. pratensis* with relatively numer-

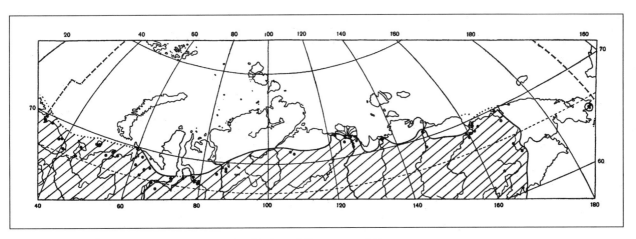

MAP II–38  Distribution of *Poa pratensis* L.

ous leaves, evidently collected in a very moist habitat (Pinega Basin, Yula Valley, at springs, VII 1899, Pole). Specimens of *P. pratensis* identical with the type specimens of that species both in habitus and in leaf pubescence also occur in many other districts of the Soviet Union. The isotype of *P. rigens* Hartm. present in the herbarium of the Botanical Institute of the USSR Academy of Sciences (Kvikkjokk, L. Laestadius) has almost smooth, horizontally spreading branchlets in a relatively small panicle, comparatively large spikelets and very copious pubescence on all five nerves of the lemmas. The last character brings it significantly close to *P. sublanata* Reverd., from which it differs only in having a small pyramidal panicle with more numerous spikelets. On the basis of its almost smooth panicle branchlets, *P. rigens* could also be referred to *P. alpigena* (Fr.) Lindm. (then the name of that species should be changed to "*P. rigens* Hartm."), but its larger spikelets and horizontally spreading branchlets in a short-pyramidal panicle speak rather for a closer relationship with *P. pratensis* s. str., to which we refer it provisionally. It is quite probable that *P. rigens* Hartm. represents a separate apomictic race of the group *P. pratensis* s. l., but at present it remains inadequately studied. No viviparous variety of *P. pratensis* s. str. is so far known.

In the Soviet Arctic *P. pratensis* occurs considerably more rarely than *P. alpigena*, and is associated with warmer and more protected areas, occurring usually in shrub thickets, on southfacing sandy or stony slopes, on shores of waterbodies, and sometimes also as an introduction near settlements.

**Soviet Arctic.** Murman; Kanin; Kolguyev; lower reaches of Pechora (Rosvinskoye); eastern part of Bolshezemelskaya Tundra (near the River More-Yu, etc.); Polar Ural; Vaygach; southern part of Ob Sound and Tazovskaya Bay; lower reaches of Yenisey (Dudinka, River Chokoto etc., also in the vicinity of Norilsk on the Boganidka River and at Boganidskoye Lake); southern extremity of Taymyr (River Kheta near the Volochanka); River Popigay; lower reaches of the Olenek, Lena, Yana, Indigirka and Kolyma; slopes of the Northern Anyuyskiy Range and the adjacent part of the coastal plain (Cape Medvezhiy, Cape Laptev); Anadyr Basin; Penzhina Basin; Bay of Korf (Kultushnoye). (Map II–38).

**Foreign Arctic.** Distribution not precisely clarified on account of confusion with *P. alpigena*, but apparently absent from Spitsbergen, the arctic part of Alaska, the Canadian Arctic Archipelago and the more northern part of Greenland.

**Outside the Arctic.** Almost all Europe and the more northern part of Asia (absent from the steppe and desert zones, but reappearing in mountainous districts); North America; southern part of South America. As an introduction in many other coun-

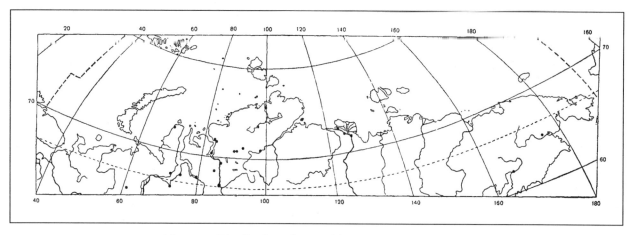

MAP II–39 Distribution of *Poa sublanata* Reverd.

tries of the World. Confused with other closely related species of the group *P. pratensis* L. s. l.

9. ***Poa sublanata*** Reverd. in Sist. zam. mat. Gerb. Tomsk. univ. No. 2–3 (1934), 1.
? *P. pratensis* var. *laxiflora* Lge., Consp. fl. groenl. (1890) 177.

The most typical specimens of this species from the lower reaches of the Ob, Yenisey and Taz (it was described from the lower reaches of the Yenisey, sandbars of islands in the Bolshaya Kheta near its mouth, 16 VIII 1914, Reverdatto) are well distinguished from all species of the group *P. pratensis* L. s. l. by having panicles conspicuously elongated in comparison with their width and rather large (5–7 mm long) spikelets arranged in ones to threes at the tips of the branchlets, also by the (on average) longer ligules of the culm leaves. But this form is connected by gradual transitions with more numerous plants (also from sands in river valleys) which possess smaller (4–5 mm long) and more numerous spikelets on the branchlets and approach either *P. pratensis* L. s. str. or *P. alpigena* (Fr.) Lindm. It is not entirely clear to us whether the most typical populations of *P. sublanata* actually represent the remnants of some relict species subsequently blended with other forms of *P. pratensis* s. l. through frequent hybridization, or whether to the contrary they are hybrid populations arising from hybridization of more widespread forms of *P. pratensis* of the type of *P. rigens* Hartm. with *P. arctica* R. Br. Despite the rather large anthers (2–2.5 mm), the most typical populations of *P. sublanata* are apparently sterile, but readily reproduce with the aid of long rhizomes. In the lower reaches of the Ob and Yenisey, as well as of the Lena (Tit-Ary Island), there occasionally also occurs a viviparous variety of this species, *P. sublanata* var. *vivipara* Tzvel. var. nova (a typo spiculis viviparis differt).

*Poa sublanata* grows on shifting or more or less stabilized sands in river valleys, occasionally also spreading to gravelbars.

**Soviet Arctic.** Yamal; coast of Ob Sound; lower reaches of Taz; lower reaches of Yenisey (common, north to Sopochnaya Korga); Taymyr (not entirely typical specimens; Lower Taymyra, Dudypta); Olenekskaya Channel of Lena (not entirely typical specimens); Tit–Ary in the lower reaches of the Lena; Anadyr (not entirely typical specimens); mouth of Penzhina. (Map II–39).

**Foreign Arctic.** Distribution unclarified. Similar forms apparently occur in Canada and Greenland (for example, *P. pratensis* var. *laxiflora* Lge.).

**Outside the Arctic.** Not occurring.

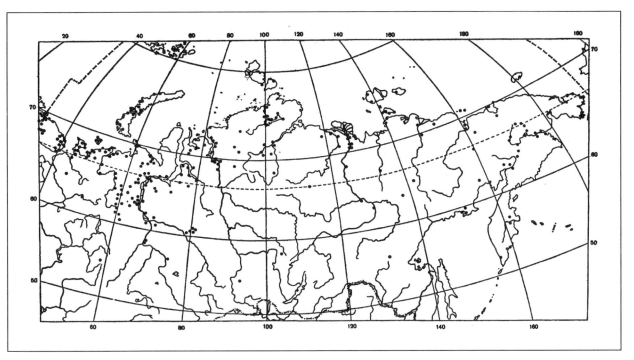

MAP II–40  Distribution of *Poa alpigena* (Fries) Lindm.

10. ***Poa alpigena*** (Fries) Lindm., Svensk Fanerogamfl. (1918) 91; id. in Lynge, Vasc. pl. N. Z. 114; Tolmachev, Fl. Taym. I, 95; Perfilev, Fl. Sev. I, 90; Rozhevits in Fl. SSSR II, 390; Scholander, Vasc. pl. Svalb. 89; Hultén, Fl. Al. II, 197; Kuzeneva in Fl. Murm. I, 205; A. E. Porsild, Ill. Fl. Arct. Arch. 28; Karavayev, Konsp. fl. Yak. 55.

*P. pratensis* var. *alpigena* Fries, Herb. Norm. Fasc. 9 (1842), 93; id., Summ. veg. (1846) 76, nom. nud.; Blytt, Norg. Fl. 1, 130; Lange, Consp. fl. groenl. 176; Krylov, Fl. Zap. Sib. II, 298.

*P. pratensis* auct. non L. — Gelert, Fl. arct. 1,121, pro parte; Polunin, Circump. arct. fl. 67, pro parte.

*P. pratensis* ssp. *alpigena* (Fries) Hiit., Suom. Kasvio (1933) 205; Hylander, Nord. Kärlväxtfl. I, 258.

Ill.: Fl. SSSR II, pl. XXIX, fig. 5; Porsild, l. c., fig. 6, a.

Widespread species very characteristic of the Arctic whose most typical specimens possess elongated, slightly diffuse panicles with smooth branchlets and rather small (3–5 mm long) rosy-violet spikelets, as well as very narrow leaf blades; but in more southern parts of its range, *P. alpigena* shows such a gradual transition to *P. pratensis* L. s. str. that a morphological boundary between these two species can only be drawn very approximately. Perhaps the most stable character of *P. alpigena*, which can usually be applied in doubtful cases, is possession of completely or virtually smooth panicle branchlets; in *P. pratensis* these are usually covered with scattered acicules over their entire or almost their entire length. But occasional specimens in which the branchlets are smooth but spread horizontally at flowering time in a relatively short panicle with larger spikelets, are apparently better referred to *P. pratensis*. The boundaries between *P. alpigena* and *P. sublanata* are also not very distinct; moreover, specimens very closely approaching the latter species (and *P. rigens*), with a very large tuft of long tangled hairs on the callus of the lemmas and pubescence on their intermediate nerves, perhaps constitute a

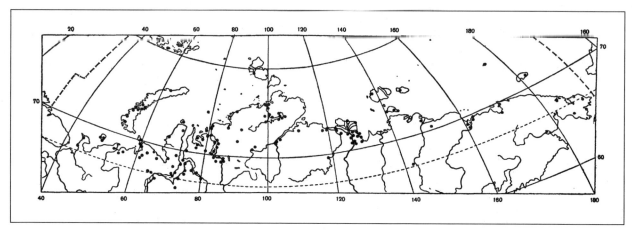

**Map II-41** Distribution of *Poa alpigena* var. *colpodea* (Th. Fries) Scholand.

separate but very weakly differentiated ecological race usually associated with sands and gravelbars.

Unlike *P. pratensis* L. s. str., *P. alpigena* has a viviparous variety which is very widespread in the Arctic, but whose range is distinctly smaller than that of non–viviparous *P. alpigena* and nowhere passes beyond its limits; in many more northern parts of the range (on islands of the Arctic Ocean) this variety occurs considerably more frequently than the typical form. The viviparous variety has the following synonymy.

*Poa alpigena* var. *colpodea* (Th. Fries) Scholand., Vasc. pl. Svalb. (1934) 89; A. E. Porsild, Ill. Fl. Arct. Arch. 28. — *P. stricta* ssp. *colpodea* Th. Fries in Öfvers. Vet. Ak. Förhandl. XXVI (1869), 138. — *P. stricta* auct. non Lindeb.: Perfilev, Mat. fl. N. Z. Kolg. 14. — *P. alpigena* f. *vivipara* Roshev. in Fl. SSSR II, 390. — *P. alpigena* var. *vivipara* Hult., Fl. Al. II (1942), 198; non Scholand., 1934.

Scholander (l. c.), in addition to the viviparous variety indicated above characterized by very small spikelets and a more or less condensed panicle, distinguishes another, also viviparous variety: *P. alpigena* var. *vivipara* (Malmgr.) Scholand. (l. c. 88), corresponding to *P. flexuosa* var. *vivipara* Malmgr. (in Öfvers. Vet. Ak. Förhandl., 1862, 253) and possessing larger spikelets and more diffuse panicles. But it is hardly possible to draw any at all clear morphological boundary between these two varieties, and very probable that at least many specimens of "var. *vivipara* (Malmgr.) Scholand." are of hybrid origin resulting from hybridization of *P. alpigena* with *P. arctica*.

*Poa alpigena* is in many respects similar to *P. arctica* (see above) in its ecology: just like that species it participates in the formation of tundra turf along with green mosses, dwarf shrubs and other herbs; it requires adequate (but not excessive!) soil moisture and winter snow cover; very often it grows on the streets of settlements and near the encampments of reindeer-herders; usually growing in carpets. In comparison with *P. arctica*, *P. alpigena* is less associated with alpine tundras but is very characteristic of lowland tundras and extends northwards to the extreme polar limits of land; it occurs more frequently than *P. arctica* in more poorly drained areas. *Poa alpigena* only just penetrates the northern part of the forest zone, where it is associated mainly with landforms of valleys and extensive lowlands, namely moist gravelbars, willow carr, low silt-and-pebble lakeshores, hummocks of polygonal marshes, and shoreline slopes formed from loose substrates; it grows profusely around lemming burrows and on shoreline mounds (baydzharakhs), forming communities of meadowlike character along with species of the mixed-herb tundra.

In the high arctic tundras of the polar islands of Siberia, *P. alpigena* is a constant component in the patchy and polygonal herb–moss tundras of watersheds; it grows in depressions of the microrelief together with species of mosses and lichens, polar willow (*Salix polaris*) and certain herbs; also common on shoreline baydzharakhs; not characteristic here of extensive, poorly drained lowlands.

**Soviet Arctic.** Murman (frequent); Kanin; Kolguyev; Malozemelskaya and Bolshezemelskaya Tundras; lower reaches of Kara; Yugorskiy Shar; Vaygach; Novaya Zemlya (north to 76°N); Franz Josef Land (Hooker, Rudolf and other islands); shores of Ob Sound and Tazovskaya Bay; Gydanskaya Tundra; lower reaches of Yenisey; Taymyr; Severnaya Zemlya; lower reaches of Olenek; lower reaches and deltaic part of Lena; Tiksi; New Siberian Islands; lower reaches of Indigirka and Kolyma; district of Bay of Chaun; Wrangel Island; Chukotka Peninsula; Anadyr Basin; Penzhina Basin. (Map II–40).

*Poa alpigena* var. *colpodea* (Th. Fries) Scholand. occurs sympatrically with *P. alpigena*, but is absent from the Kola Peninsula and the basins of the Anadyr and Penzhina. (Map II–41).

**Foreign Arctic.** Alaska; northern part of Canada; Labrador; Canadian Arctic Archipelago; Greenland (except SE part); Spitsbergen; Arctic Scandinavia.

**Outside the Arctic.** Scandinavia (north of 60°N), Kola Peninsula; Urals; on the Ob; northern part of Central Siberian Plateau; Yenisey Ridge; mountains of NE Siberia (on barrens) south to the Verkhoyansk Range, the Shantar Islands and Kamchatka inclusively; Commander and Aleutian Islands; Alaska and northern part of Canada (south to the southern extremity of Hudson Bay and Newfoundland). The distribution of this species is still inadequately studied due to confusion with closely related species.

**11. *Poa subcaerulea*** Sm., Engl. Bot. XIV (1802), tab. 1004.
*P. humilis* Ehrh., Beitr. VI (1791), 84, nom. nud.; non Lejeune, 1811.
*P. pratensis* var. *humilis* (Ehrh.) Griseb. in Ledebour, Fl. ross. IV (1853), 379.
*P. irrigata* Lindm. in Bot. Notis. (1905) 88; Rozhevits in Fl. SSSR II, 390; Hultén, Fl. Al. II, 212; Kuzeneva in Fl. Murm I, 204.
**Ill.:** Smith, l. c., tab. 1004; Fl. SSSR II, pl. XXIX, fig. 4.

*Poa subcaerulea* Sm. is apparently a North Atlantic species, sufficiently distinguished from other species of the group *P. pratensis* L. s. l. by the glaucous green colour and small size of the plant as a whole, as well as by panicle branchlets which are horizontally spreading or even deflexed after flowering and arranged in ones to threes per node. The range of *P. subcaerulea* was shown largely erroneously in the "Flora of the USSR" and "Flora of Murmansk Oblast," since it was based on specimens of *P. pratensis* incorrectly identified as "*P. irrigata* Lindm." This species is not characteristic of the Arctic, occurring here apparently only as an introduction.

Growing in moist sandy sites and peaty meadows usually near the sea coast and on the shores of larger waterbodies, sometimes also occurring as an introduction along roads and in settlements.

**Soviet Arctic.** Murman (sporadically); on the River Tanyurer in the Anadyr Basin (probably as an introduction).

**Foreign Arctic.** Arctic Scandinavia.

**Outside the Arctic.** Great Britain; Fennoscandia; NW European part of USSR; south coast of Alaska (as introduction). Distribution of species still inadequately clarified, probably also occurring in Denmark and the north of Central Europe.

**12. *Poa angustifolia*** L., Sp. pl. (1753) 67; Rozhevits in Fl. SSSR II, 388; Hultén, Fl. Al. II, 200; Kuzeneva in Fl. Murm. I, 204.
*P. pratensis* var. *angustifolia* (L.) Sm., Fl. Brit. (1800) 105; Grisebach in Ledebour, Fl. ross. IV, 379; Krylov, Fl. Zap. Sib. II, 298.

*P. setacea* Hoffm., Deutsch. Fl., Ed. 2, 1 (1800), 44; Rozhevits, l. c. 389.
*P. strigosa* Hoffm., l. c.; Rozhevits, l. c. 389.
*P. pratensis* ssp. *angustifolia* (L.) Lindb. fil. apud Hylander, Nord. Kärlväxtfl. I (1953), 257.

**Ill.:** Fl. SSSR II, pl. XXIX, fig. 2, 3.

This species is distributed mainly in the steppe and forest-steppe zones, but penetrates far into the forest zone in very dry habitats (open sandy slopes, limestone outcrops, etc.). In the "Flora of Murmansk Oblast" it is recorded by O. I. Kuzeneva for a few places in Arctic Murman, but apparently erroneously since all specimens from there which we have seen have proved to belong to other species of the group *P. pratensis* L. s. l. It should be noted that the fundamental difference between *P. angustifolia* and other closely related species is not possession of narrow setiform longitudinally folded leaves (which is very often the case also in *P. alpigena*), but the ability to grow in the form of small but very dense clusters of shoots united by a long rhizome.

Possibly occurring in the Arctic only on very dry sandy or stony slopes of southern exposure or as an introduction near settlements.

**Soviet Arctic.** Recorded for Murman (Pechenga, mouth of Tuloma, Fl. Murm. I, 204), but probably erroneously.

**Foreign Arctic.** Not occurring.

**Outside the Arctic.** Almost all Eurasia [except the southeastern tropical part and the far north; in the European part of the USSR reaching north to the Khibins Mountains (on the slopes above Lake Imandra) and Pinega (that is to 65°N), and as far east as Berezovo on the Ob; completely absent from NE Siberia as far as the southern part of Yakutia]. Occurring as an introduction in Kamchatka. The North American range of this species is not precisely clarified, but it is recorded by Hultén for southern Alaska (as an introduction).

13. **Poa alpina** L., Sp. pl. (1753) 67; Grisebach in Ledebour, Fl. ross. IV, 370; Lange, Consp. fl. groenl. 1, 176; Gelert, Fl. arct. 1, 123; Lindman in Lynge, Vasc. pl. N. Z. 116; Krylov, Fl. Zap. Sib. II, 287; Perfilev, Fl. Sev. I, 91; Rozhevits in Fl. SSSR II, 411; Scholander, Vasc. pl. Svalb. 92; Hultén, Fl. Al. II, 198; Hylander, Nord. Kärlväxtfl. I, 254; Kuzeneva in Fl. Murm. I, 212; A. E. Porsild, Ill. Fl. Arct. Arch. 30; Hultén, Amphi-atl. pl. 230; Karavayev, Konsp. fl. Yak. 56; Polunin, Circump. arct. fl. 66.

**Ill.:** Lindman, l. c., tab. XLVI, 1; Fl. SSSR II, pl. XXXI, fig. 1; Fl. Murm. I, pl. LXXII; Porsild, l. c., fig. 6, e; Polunin, l. c. 65.

Widely distributed arctic-alpine species of the section *Bolbophorum* Aschers. et Graebn., which contains (besides *P. alpina* and a few very closely related species) also a large group of steppe and semidesert species with bulblike swelling at the base of the vegetative shoots (*P. bulbosa* L. s. l.). Despite its wide range, *P. alpina* is almost completely absent from NE Siberia and therefore cannot be considered a circumpolar species. The polymorphism of this species is relatively slight in the Arctic and scarcely merits special discussion.

The viviparous variety of *P. alpina* has its own range, far from coincident with the range of the basic species. The following synonymy refers to it.

*P. alpina* var. *vivipara* L., Sp. pl. (1753) 67; Rozhevits, l. c. 411. — *P. vivipara* Willd., Enum. hort. berol. (1809) 103. — *P. bulbosa* auct. non L.: Gelert, l. c. 123.

*Poa alpina* grows in poorly vegetated well drained areas which are adequately covered by snow in winter. It is common in the Eastern European Arctic but nowhere grows in masses, only as isolated tufts. Thus on Novaya Zemlya (in the district of Matochkin Shar Strait), where the viviparous variety occurs exclusively, this grows in moderately snow-covered portions of clayey-stony tundra and of rocky slopes, always as small tufts. On Gusiniy Island the species is characteristic of moist gravelbars along streams. On Vaygach the typical (non-viviparous) form is

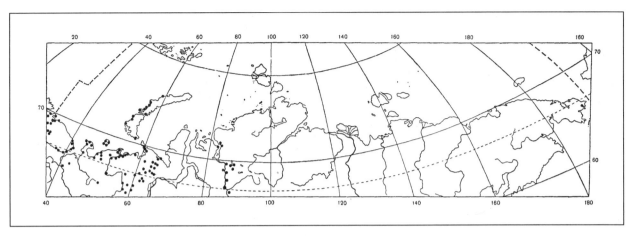

MAP II–42   Distribution of *Poa alpina* L.

common on rocky slopes and rock ledges; further south (in the Bolshezemelskaya Tundra) it is associated with well snow-covered sites on slopes cooled by the waters of melting snowfields, also on riverine gravelbars and sandy shoreline slopes; penetrating the forest-tundra on sandy embankments along the railroad.

**Soviet Arctic.** Murman (common); Kanin; Kolguyev (viviparous variety); Malozemelskaya and Bolshezemelskaya Tundras; Polar Ural (viviparous variety reported from the upper reaches of the Kara); Pay-Khoy; Vaygach; Novaya Zemlya (viviparous variety on the North Island); lower reaches of Yenisey (north to the Chayka River, 71°53′N); Dudypta River in Taymyr; Chukotka Peninsula (Lawrence Bay). (Map II–42).

**Foreign Arctic.** Arctic part of Alaska (sporadically, including the district of the Bering Strait); arctic coast of Canada (west of Hudson Bay, sporadically); southern part of Baffin Island; Labrador; Greenland (except NW part); Iceland; Spitsbergen; Arctic Scandinavia.

**Outside the Arctic.** Ireland and northern part of Great Britain; Scandinavian Peninsula; Northern European part of the USSR south to Northern Estonia, Lake Onega, Vychegda and Kosvinskiy Rock in the Urals; mountains of Central Europe from the Pyrenees to the Carpathians; Caucasus; mountains of Central Asia; Altay; Central Siberian Plateau; western part of Sayan Mountains; Alaska; the greater part of Northern Canada (but very sporadically) south to Vancouver Island, Lake Superior and Newfoundland, and in the Rocky Mountains south to the states of Utah and Colorado. In more southern parts of the range the viviparous variety normally does not occur.

14. **Poa abbreviata** R. Br. in Suppl. to App. Parry's Voyage (1824) 287; Grisebach in Ledebour, Fl. ross. IV, 377; Lange, Consp. fl. groenl. 172; Gelert, Fl. arct. 1, 124; Lindman in Lynge, Vasc. pl. N. Z. 113; Hanssen & Lid, Fl. pl. Franz Josef L. 35; Sørensen, Vasc. pl. E Greenl. 141, 149; Perfilev, Fl. Sev. I, 92; Rozhevits in Fl. SSSR II, 412; Scholander, Vasc. pl. Svalb. 80; Hultén, Fl. Al. II, 222; A. E. Porsild, Ill. Fl. Arct. Arch. 30; Hultén, Amphi-atl. pl. 20; Polunin, Circump. arct. fl. 66; Yurtsev in Bot. zhurn. XLIV, 8, 1175.

*P. Malmgrenii* Gand. in Bull. Soc. Bot. France LXVI (1919), 302.

**Ill.:** Lindman, l. c., tab. XLV, fig. 1; Fl. SSSR II, pl. XXXI, fig. 2; Porsild, l. c., fig. 7, a; Polunin, l. c. 65.

Morphologically very distinctive high-arctic circumpolar species, referred by Nannfeldt (in Symb. Bot. Ups. 1, 5, 1935, 25) to the separate monotypic section *Abbreviatae* Nannf. but in our opinion also showing many characters in common

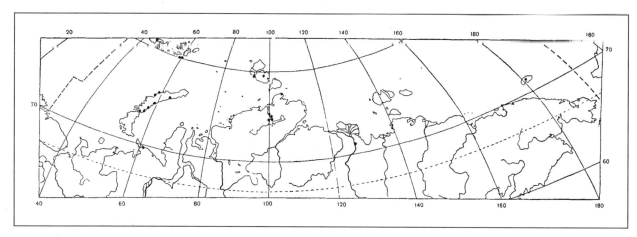

MAP II–43  Distribution of *Poa abbreviata* R. Br.

(especially the small anthers, absence of extravaginal shoots, etc.) with the species of section *Oreinos* Aschers. et Graebn. (*P. laxa* Haenke, *P. flexuosa* Sm., etc.). The large gap in the range of *P. abbreviata* between Taymyr and Wrangel Island which existed until very recently has now been much reduced by the unexpected discovery of this species on Mount Sokuydakh in the lower reaches of the Lena (Yurtsev, l. c.). After the Amderma district (whence we have not, however, seen specimens), this new locality is the most southern in the territory of the USSR. Despite its distribution in areas of extreme arctic conditions, this species does not produce a viviparous variety and displays only very insignificant polymorphism. According to Sørensen (l. c. 146) *P. Malmgrenii* Gand. described from Spitsbergen also belongs to this species.

Growing in open windswept areas under conditions of minimal winter snow cover. Colonizing sites devoid of vegetational turf, growing in discrete though sometimes numerous tufts. Especially characteristic of mountainous districts of the High Arctic, where it prefers strongly weathered (stones and fine soil) eluvia of mountain rocks (especially limestones, also basalts, etc.). Habituated here both to areas of relatively surplus moisture (but always well drained) and to areas which are (on the surface) highly desiccated.

Thus, on Alexander Island (Franz Josef Archipelago) the species grows in profusion in open communities (without moss cover) on high and relatively dry areas on the steep slope above Dezhnev Bay. Here it is common on small-stone eluvia of basalts; isolated tufts are also found on the ancient beach ridge composed of large shingle.

On the east coast of the North Island of Novaya Zemlya (at the entrance to Matochkin Shar Strait), the species is characteristic of open clayey-stony areas with little snow cover, in association with *Papaver radicatum* ssp. *polare*, *Potentilla emarginata*, etc.; it becomes rarer at a distance from the shore.

On the Island of the October Revolution (Severnaya Zemlya), the species is comparatively common in the central part of the island on the high and dry stony plateaux formed by limestones, as well as on the steep slopes of ancient erosional terraces.

Found in the lower reaches of the Lena only on the highest summit of the Tuora-Sis Range (Sokuydakh) at elevations from 800 to 950 m a. s. l.; here it grows massively in open communities of loamy-stony areas (on limestone eluvium), but is virtually absent from rock debris formed from diabases.

**Soviet Arctic.** North Island of Novaya Zemlya; Franz Josef Land; Yugorskiy Shar (near Amderma); Taymyr (Lower Taymyra, Mod Bay); Island of the October Revolution;

Tuora-Sis Range in the lower reaches of the Lena (Mount Sokuydakh); Wrangel Island. (Map II–43).

**Foreign Arctic.** Northern part of Alaska; arctic coast of Canada; Canadian Arctic Archipelago (common); Greenland (north of the Arctic Circle, common); Spitsbergen.

**Outside the Arctic.** Northern cordillera of North America (a few sites).

15. ***Poa leptocoma*** Trin. in Mém. Ac. Pétersb., sér. 6, 1 (1831), 374, excl. var. β; Hultén, Fl. Al. II, 215, pro parte; Polunin, Circump. arct. fl. 66, pro parte.

*P. stenantha* var. *leptocoma* (Trin.) Griseb. in Ledebour, Fl. ross. IV (1853), 373.

*P. flavidula* Kom. in Bot. mat. Gerb. Glavn. bot. sada RSFSR 10 (1924), 146.

Predominantly South Beringian species, occupying together with the next species a very isolated position in the genus but apparently belonging to the arctic-alpine section *Oreinos* Aschers. et Graebn., characterized mainly by the absence of extravaginal shoots and the small anthers. The species *P. flavidula* Kom. described by V. L. Komarov from Kamchatka should be referred to this species, although occupying a somewhat intermediate position between *P. leptocoma* and *P. paucispicula*. *Poa leptocoma* var. β Trin. (= *P. patens* Trin. in herb.) has significantly larger anthers and belongs to *P. nivicola* Kom. (l. c. 147), a separate species of the alpine zone of Kamchatka which shows a closer relationship to *P. arctica* than to the group *P. leptocoma* Trin. s. l. *Poa leptocoma* apparently does not occur at all in the Arctic, but is completely replaced by the next species; we record it only on the basis of a single specimen of doubtful origin (which may have been collected on the Aleutian Islands).

Growing in moist meadows and on gravelbars on the shores of rivers and streams, sometimes also at hot springs.

**Soviet Arctic.** Beringian coast of Chukotka Peninsula (Sinus St. Laurentii, Chamisso).

**Foreign Arctic.** Not occurring.

**Outside the Arctic.** Kamchatka (southern part); Kurile Islands; Aleutian Islands; Alaska; northern cordillera from Alaska to the states of California, Nevada, Utah and New Mexico.

16. ***Poa paucispicula*** Scribn. et Merr., Grass. Alaska (1910), 69; Hultén, Fl. Kamtch. I, 131; Raup in Sargentia VI, 114.

*P. nivicola* Kom. in Bot. mat. Gerb. Glavn. bot. sada RSFSR V, 10 (1924), 147, pro parte (excluding type specimen); Rozhevits in Fl. SSSR II, 379, pro parte.

*P. taimyrensis* Roshev. in Izv. Bot. sada AN SSSR XXX (1932), 229 and in Fl. SSSR II (1934), 409; Tolmachev, Fl. Taym. I, 97; Tikhomirov, Fl. Zap. Taym. 24; Karavayev, Konsp. fl. Yak. 56.

*P. leptocoma* auct. non Trin. — Hultén, Fl. Al. II, 215, pro parte; Polunin, Circump. arct. fl. 66, pro parte.

Ill.: Scribner & Merrill, l. c., pl. 15; Fl. SSSR II, pl. XXVIII, fig. 10; Tolmachev, l. c., fig. 19; Polunin, l. c. 65.

Predominantly East Siberian arctic-alpine species, replacing the preceding species in the alpine zone of mountains and in the Arctic and connected with it by transitional specimens. Following Raup (l. c.) and Hitchcock (Manual of Grasses, Ed. II, 1951, 121) we retain it at the rank of full species, although some authors (Hultén, Fl. Al. II; Polunin, l. c.) consider it a synonym of *P. leptocoma* without any reservations. However, in the Soviet Arctic *P. paucispicula* is a species with very stable morphological characters and does not show any forms transitional to *P. leptocoma*. As already noted by Hultén (Fl. Kamtch. I), the specimens listed by V. L. Komarov in the description of *P. nivicola* Kom. have proved to belong to two completely different species, one of which is *P. paucispicula* (to which about half the specimens belong), while the other is the species endemic to Kamchatka erro-

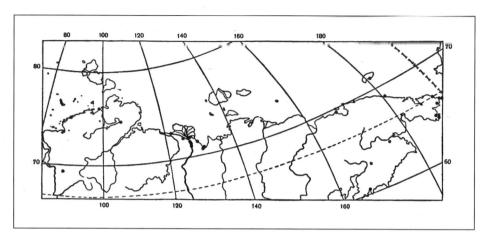

MAP II–44  Distribution of *Poa paucispicula* Scribn. et Merr.

neously identified by Hultén with *P. bracteosa* Kom. We think it possible to retain the name "*P. nivicola* Kom." for the second of these species by selecting one of V. L. Komarov's specimens as lectotype (Kamchatka, upper reaches of River Kashkan, near Pushchinoy village, 24 VI 1909, No. 2802, Komarov). Three herbarium sheets have such a label, on one of which in V. L. Komarov's hand is written "*Poa nivicola* Kom. sp. nova." Also proven to belong to *P. paucispicula* is the species *P. taimyrensis* Roshev., described from a few small specimens (East Taymyr, lower reaches of River Yamu-Nera, 21 VIII 1928, No. 825, Tolmachev) which in no other way differ from typical specimens of *P. paucispicula*. Strangely *P. taimyrensis* was referred by R. Yu. Rozhevits (Fl. SSSR II, 409) to the series "*Lanatiflorae* Roshev." together with such completely different species as *P. lanatiflora* Roshev. [= *Colpodium lanatiflorum* (Roshev.) Tzvel.] and *P. Soczawae* Roshev., which in our opinion is a synonym of *P. glauca* Vahl. Thus, the range of *P. paucispicula* has proved to extend west to Taymyr.

Growing in late thawing sites of winter snow accumulations, with good or surplus (continuously flowing) moisture, normally on marshy (sometimes alluvial) soils.

Thus, in Arctic Yakutia (Chekanovskiy Ridge, northern extremity of Kharaulakh Mountains) the species occurs normally only near snowfields (in the valleys of mountain rivulets, in sinkholes and at the foot of northfacing slopes, in rocky mossy beds of small streams flowing out from beneath large snow patches). Here *P. paucispicula* grows both in relatively closed nival meadows (with *Taraxacum arcticum, Ranunculus nivalis, Salix polaris*, etc.) and at later thawing sites in the open communities of silted gravelbars (together with *Phippsia* spp. and *Saxifraga hyperborea*) or on loamy-stony slopes covered with small mosses (with *Ranunculus pygmaeus* etc.).

The species grows under similar conditions also on the SE extremity of the Chukotka Peninsula, where it is particularly common; locally forming small beds; preferring incompletely vegetated (silty or loamy) sites; very rarely also encountered at the edge of bare loamy patches in dwarf shrub-moss tundras.

**Soviet Arctic.** NW edge of Central Siberian Plateau (Norilsk); Taymyr (lower reaches of the Yamu-Nera and Lower Taymyra, Cape Sterlegov); coast of Olenek Bay (Stannakh-Khocho); lower reaches of Lena (Ayakit, Tigiya, Chaytumus); Buorkhaya Bay (Tiksi, Nyuayba settlement); Wrangel Island; Cape Schmidt; Beringian coast of Chukotka Peninsula (Lawrence Bay, Arakamchechen Island, Chaplino); Anadyr Basin; Penzhina Basin; Bay of Korf. (Map II–44).

**Foreign Arctic.** St. Lawrence Island and Alaska in the district of the Bering Strait.

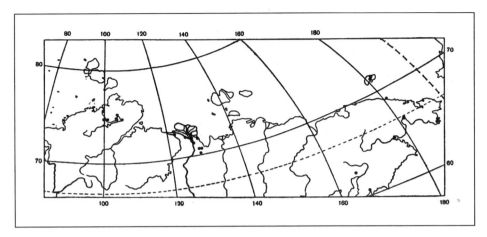

MAP II–45  Distribution of *Poa pseudoabbreviata* Roshev.

**Outside the Arctic.** Kamchatka[3]; northern cordillera from Alaska to Washington State.

**17. *Poa pseudoabbreviata*** Roshev. in Bot. mat. Gerb. Glavn. bot. sada RSFSR III (1922), 91; id. in Fl. SSSR II (1934), 413.

*P. laxa* auct. non Haenke — Gelert, Fl. arct. 1, 123, pro parte.

*P. subabbreviata* Roshev. ex Tolmatch., Fl. Taym. I, 96, nom. nud.

*P. brachyanthera* Hult., Fl. Aleut. Isl. (1937) 86; id., Fl. Al. II, 204; Karavayev, Konsp. fl. Yak. 56.

**Ill.:** Fl. SSSR II, pl. XXXI, fig. 6.

Arctic-alpine species of section *Oreinos* Aschers. et Graebn., occupying a rather isolated position in the section and displaying great stability of morphological characters. Readily distinguished from all other species of the genus due to its very small anthers, very weakly pubescent lemmas and rather strongly scabrous long thin panicle branchlets, but apparently sometimes overlooked on account of the small size of the plant as a whole. The type specimens of this species (Garganskiy Pass, 30–31 VII 1902, Komarov) are identical to arctic specimens of *P. brachyanthera* Hult., and it may be postulated that the presently existing large gap in its range will be significantly reduced after further studies of the floras of the barrens of East Siberia.

Characteristic plant of dry stony tundras of mountainous districts of the Siberian Arctic, especially the most continental (Arctic Yakutia). It preferentially colonizes open windswept sites (sometimes devoid of snow cover in winter), and selects incompletely vegetated areas (stony tundras with sparse herbs; dry convex patches in stony polygonal *Dryas*-tundras).

Thus, at the northern extremity of the Kharaulakh Mountains (on the Tiksi coast) *P. pseudoabbreviata* is a common (almost massively occurring) plant of summits and of the southwestern slopes (facing the winds off the continent which prevail in winter) of low hills composed of clayey shales; it also colonizes overgrown shale screes; also found on dry gravelbars. Further south (upper course of the Kharaulakh) the species occurs only near mountain summits (over 800 m a. s. l.) on steep stony (shale) southfacing slopes.

On Chekanovskiy Ridge the species occurs more rarely and normally on dry bare patches in polygonal or terraced alpine *Dryas*-tundras developed on the sand-and-gravel eluvium of sandstones; it is also characteristic of the (here less frequent) open sparse-herb communities of the most windswept portions of summits and convexities of slopes, where the loose sandstone eluvium eroded by strong winds acquires an almost dusty character.

[3] In 1962 this species was found by V. N. Siplivinskiy on the barrens of the Barguzinskiy Range.

**Soviet Arctic.** Taymyr (the Lower Taymyra and the north shore of Lake Taymyr); coast of Olenek Bay (Stannakh-Khocho); lower reaches of Lena (Sietachan, Arangastakh, upper reaches of Kharaulakh); Tiksi; Wrangel Island; Cape Schmidt; Beringian coast of Chukotka Peninsula (Bay of Krest, Providenye Bay); River Oklan (town of Stadukhino); Bay of Korf. (Map II–45).

**Foreign Arctic.** Not occurring.

**Outside the Arctic.** Prebaikalia (Tunkinskiye Barrens) and Northern Transbaikalia (Kodar Range); Aleutian Islands; Alaska (southern part).

*18.* ***Poa compressa*** L., Sp. pl. (1753) 69; Grisebach in Ledebour, Fl. ross. IV (1853), 371; Krylov, Fl. Zap. Sib. II, 224; Perfilev, Fl. Sev. I, 90; Rozhevits in Fl. SSSR II, 408; Hultén, Fl. Al. II, 205; Kuzeneva in Fl. Murm. I, 210; Hylander, Nord. Kärlväxtfl. I, 263.

**Ill.:** Fl. SSSR II, pl. XXX, fig. 14; Fl. Murm. I, pl. LXVIII, fig. 4.

Predominantly European species of the forest and forest-steppe zones, introduced to many other countries of the World. It occupies a completely isolated position in the genus and is sometimes referred to a separate section *Tichopoa* Aschers. et Graebn., but it significantly approaches species of the section *Stenopoa* Dum. in the structure of the spikelets and panicles. Occurring in the Arctic only as an introduction, apparently reproducing there only by vegetative means with the aid of subterranean stolons.

Growing on stony and clayey slopes, dry gravelbars, along roads and near settlements.

**Soviet Arctic.** Murman (as an introduction).

**Foreign Arctic.** West coast of Alaska (near Nome, introduced).

**Outside the Arctic.** Almost all Europe; Caucasus; Asia Minor. As an introduction in the southern part of Alaska and Canada, many states of the USA, West Siberia and Kamchatka, India, and many other countries.

*19.* ***Poa nemoralis*** L., Sp. pl. (1753) 69; Grisebach in Ledebour, Fl. ross. IV, 375; Lange, Consp. fl. groenl. 174; Gelert, Fl. arct. 1, 125; Krylov, Fl. Zap. Sib. II, 291; Perfilev, Fl. Sev. I, 92; Rozhevits in Fl. SSSR II, 400; Hultén, Fl. Al. II, 217; Kuzeneva in Fl. Murm. I, 208; Hylander, Nord. Kärlväxtfl. I, 264.

**Ill.:** Fl. SSSR II, pl. XXX, fig. 5; Fl. Murm. I, pl. LXVIII, fig. 1.

Species widespread in the forest-steppe and forest zones of the Northern Hemisphere, usually associated with broadleaved or mixed forests and almost completely absent from a considerable part of the Siberian taiga. Like the nine following species it belongs to the very distinctive section *Stenopoa* Dum., whose most characteristic features are the presence of only extravaginal shoots (which however are never creeping), the absence of persistent abbreviated vegetative shoots (that is, all vegetative shoots formed in autumn produce flowering stems the following summer), the very inconspicuous or completely absent intermediate nerves on the lemmas, and the always strongly scabrous panicle branchlets. The numerous species of this section are rather closely tied to one another by a whole series of transitional populations and specimens possibly of hybrid origin, so that under a broader species concept they might all be considered subspecies of a single gigantic polytypic species (*P. nemoralis* L. s. l.). Only with some strain can one succeed in dividing the section *Stenopoa* Dum. into a few smaller groups of closely related species, that is series characterized by a certain homogeneity with respect to morphology, ecology and geography although the morphological boundaries between them are perhaps no more distinct than those between their included species. Among the species of this section, *P. nemoralis* L. s. str. is apparently one of the most ancient species, which was already widespread in the deciduous broadleaved forests of the Tertiary Period. In the zone of the oak forest-steppe of

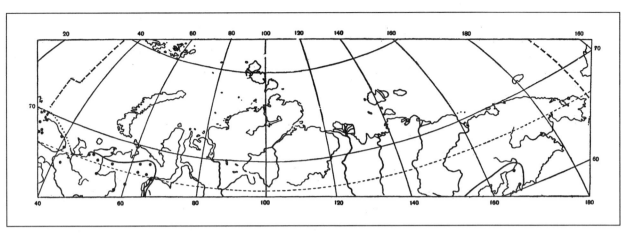

Map II-46  Distribution of *Poa nemoralis* L.

Europe this species is characterized by relatively stable morphology, but becomes very polymorphic in NE Europe and Siberia where real differences between it and other species of section *Stenopoa* Dum. sometimes completely disappear.

Despite its relatively southern range, *P. nemoralis* penetrates many districts of the Arctic, occurring here in the habitats most protected from deep frosts and winds, usually in valley shrub communities, sparse forests and forest openings, more rarely in meadows and on gravelbars.

**Soviet Arctic.** Murman (rather rare); Timanskaya Tundra; Eastern Bolshezemelskaya Tundra; Polar Ural (rare); lower reaches of Ob (Salekhard); Penzhina Basin (River Palmatkina); Bay of Korf. (Map II–46).

**Foreign Arctic.** Southern part of Greenland; Iceland; probably penetrating Arctic Scandinavia.

**Outside the Arctic.** Almost all Europe; considerable part of Asia from Berezovo on the Ob, the Lower Tunguska, Southern Yakutia, the Okhotsk Coast and Kamchatka in the north to Iran, Northern India and Northern China in the south; southern part of Alaska and the Aleutian Islands; a few places on the west coast of North America; NE states of the USA from the Great Lakes to Newfoundland and Delaware.

20. **Poa Tanfiljewii** Roshev. in Fl. SSSR II (1934), 413; id. in Tr. Bot. inst. AN SSSR, ser. I, II (1936), 96.

Ill.: Fl. SSSR II, pl. XXXI, fig. 7.

Species not entirely clear to us, differing from *P. nemoralis* only in the absence or very weak development of the cluster of long tangled hairs on the callus of the lemmas. In almost all species of section *Stenopoa* Dum. this cluster is relatively weakly developed and sometimes has a tendency to completely disappear, so that it is hardly possible to ascribe substantial importance to this character. However, in *P. nemoralis* the cluster of long tangled hairs on the callus of the lemmas is almost always well developed, and specimens with very weak development of (or completely without) this cluster occur only in one part of that species range (Northern Uralia); this provides some argument in favour of retaining *P. Tanfiljewii* as a separate species.

Growing mainly on gravelbars and sandy shores of rivers and streams, in sparse forests and sometimes on rocks.

**Soviet Arctic.** Timanskaya Tundra (sporadically); lower reaches of Pechora.
**Foreign Arctic.** Not occurring.
**Outside the Arctic.** Prepolar Ural (upper reaches of Pechora and Severnaya Sosva).

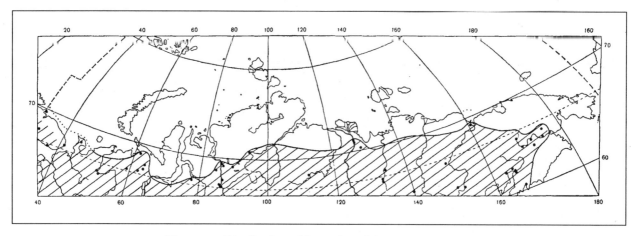

MAP II–47  Distribution of *Poa palustris* L.

**21. Poa lapponica** Prokud. in Zhurn. Inst. Bot. AN USSR XX (1939), 198; Kuzeneva in Fl. Murm. I, 210.

Ill.: Fl. Murm. I, pl. LXVIII, fig. 3.

Like the preceding, this species differs from *P. nemoralis* in only a single character, the bare rachis of the spikelet, something to which substantial importance can hardly be ascribed. Moreover, in *P. glauca* Vahl, *P. stepposa* (Kryl.) Roshev. and certain other species of section *Stenopoa* the spikelet rachis may be either bare or hairy, and the pubescence of the spikelet rachis in *P. nemoralis* is very variable and sometimes consists of only a few hairs. The possibility of a hybrid origin of *P. lapponica* also cannot be excluded (*P. nemoralis* × *P. palustris* or *P. nemoralis* × *P. glauca*). Thus, the distinctness of this species requires confirmation, especially through observations in nature and experiments in cultivation.

Growing in sparse forest and shrubbery, in meadows, sometimes on stony or sandy slopes, also on gravelbars.

**Soviet Arctic.** Murman (rare); Eastern Bolshezemelskaya Tundra (rare).

**Foreign Arctic.** Probably occurring in Arctic Scandinavia.

**Outside the Arctic.** Kola Peninsula (Khibins); NE European part of USSR (on the Pinega and in the upper reaches of the Pechora); Sayan Mountains and Yenisey Ridge (doubtful specimens).

**22. Poa palustris** L., Syst., Ed. 10 (1759), 874, emend. Roth, 1789; Krylov, Fl. Zap. Sib. II, 293; Perfilev, Fl. Sev. I, 92; Rozhevits in Fl. SSSR II, 397; Hultén, Fl. Al. II, 218; Kuzeneva in Fl. Murm. I, 205; Hylander, Nord. Kärlväxtfl. I, 265.

*P. serotina* Ehrh., Beitr. VI (1791), 86, nom. nud.; Grisebach in Ledebour, Fl. ross. IV, 375.

*P. fertilis* Host, Gram. Austr. III (1805), 10.

? *P. rotundata* Trin. in Mém. Ac. Pétersb., sér. 6, 1 (1831), 378; Grisebach, l. c. 374.

Ill.: Host, l. c., tab. 15; Fl. SSSR II, pl. XXIX, fig. 14; Fl. Murm. I, pl. LXVIII, fig. 2.

Species widespread in the forest and forest-steppe zones of the Northern Hemisphere, displaying very great polymorphism but so far not subject to satisfactory subdivision into smaller taxonomic units with more stable morphological characters.

Despite universal recognition *P. palustris* is only very poorly delimited from *P. nemoralis* (especially in Siberia and NE Europe), constituting with respect to that species not only an ecological race of more open and usually moister habitats, but also a geographical race native basically to the zone of the taiga and other forests of more northern type. Thus, in Siberia (except for Primorskiy Kray) it almost com-

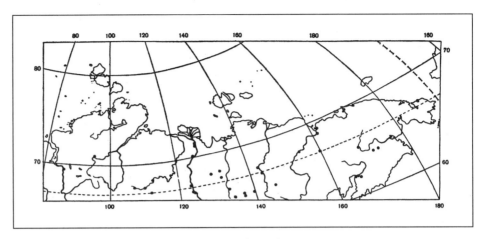

MAP II–48 Distribution of *Poa stepposa* (Kryl.) Roshev.

pletely replaces *P. nemoralis,* occurring here not only in unforested areas but also in various forest and shrub communities from spruce-fir forest to birch forest-islands. In NE Siberia *P. palustris* becomes considerably rarer and apparently intergrades with inadequately studied closely related species (for example, *P. rotundata* Trin.). A series of transitional populations and specimens connects *P. palustris* with *P. stepposa* (Kryl.) Roshev.

In the Arctic *P. palustris* occurs in meadows, on shores of waterbodies, on moist gravelbars, and sometimes also in willow and alder carr in river valleys.

**Soviet Arctic.** Murman (frequent as an introduction); Kanin (rare); Bolshezemelskaya Tundra; basin of Tazovskaya Bay (lower reaches of Pur); lower reaches of Yenisey (north to Dudinka, common); Kheta Basin; lower reaches of Lena and Kolyma; Anadyr Basin (common); Penzhina Basin; Bay of Korf. (Map II–47).

**Foreign Arctic.** Probably occurring in the arctic part of Scandinavia.

**Outside the Arctic.** Northern and Central Europe south to Great Britain, Northern Italy, Greece and the Caucasus; considerable part of Asia from the Arctic to Iran, Northern China and Japan; North America from southern Alaska, the southern part of Hudson Bay and Newfoundland to the states of California, New Mexico, Nebraska, Missouri and Virginia. Introduced to certain other countries.

23. **Poa stepposa** (Kryl.) Roshev. in Fl. SSSR II (1934), 401 and 754.
*P. sterilis* auct. non M. B. — Trautvetter in Acta Hort. Petrop. V, 1 (1877), 137.
*P. attenuata* var. *stepposa* Kryl., Fl. Alt. i Tomsk. gub. VII (1914), 1656; id., Fl. Zap. Sib. II, 285.
*P. attenuata* auct. non Trin. — Karavayev, Konsp. fl. Yak. 55, pro parte.
Ill.: Fl. SSSR II, pl. XXX, fig. 6.

*Poa stepposa* (Kryl.) Roshev. is one of many ill-defined species of section *Stenopoa* characteristic of steppe or more or less steppelike vegetational communities. In comparison with such species as *P. nemoralis* and *P. palustris*, these have a considerably more xeromorphic habit: strongly scabrous culms and leaves, a conspicuously reduced number of nodes of a culm which rises far above the sheath of the uppermost culm leaf, and more or less abbreviated branchlets of a panicle which in *P. botryoides* and closely related species becomes spikelike and strongly reminiscent of the panicle of the genus *Koeleria* Pers. Apparently the "oldest" of the described species of this group is *P. ochotensis* Trin. (Mém. Ac. Pétersb., sér. 6, 1, 1831, 377), which was included by R. Yu. Rozhevits (Fl. SSSR II, 395) in the series "*Pratenses* Roshev." due to some kind of misunderstanding; but inadequate herbarium material from the Okhotsk Coast does not yet allow us to determine the

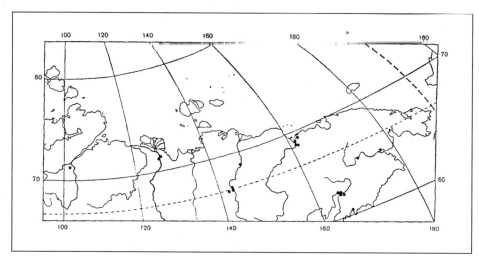

MAP II–49  Distribution of *Poa filiculmis* Roshev.

connections between this species and such extremely closely related species as *P. sphondyloides* Trin., *P. attenuata* Trin. etc. with sufficient clarity. Among all these species *P. stepposa* (possibly the prior name for this species will prove to be *P. transbaicalica* Roshev.) differs clearly enough from *P. attenuata* Trin. and *P. botryoides* (Trin. ex Griseb.) Roshev. both in range and morphology; but it is sometimes united with these species and shows no less affinity with *P. palustris* L., appearing to form a connecting link between the group of steppe species *P. ochotensis* Trin. s. l. and more mesophilic species of the type of *P. nemoralis* L. and *P. palustris* L. This typical steppe species is widespread in steppes of the plains and lower hills of Southern Siberia and Kazakhstan, and in association with the sharply increased continentality of climate in NE Asia it here advances far to the north, possibly even forming here a distinct ecological-geographical race.

In the Arctic this species occurs only in the most southern districts (bordering forest-tundra) and here in the driest (more or less steppelike) vegetational communities, usually on dry rocky or stony slopes and in dry open larch forests, sometimes also on gravelbars and sands in river valleys.

**Soviet Arctic.** Lower reaches of Yenisey (north to Dudinka); Kheta Basin (Volochanka); lower reaches of Lena and Kolyma; district of Bay of Chaun; Anadyr Basin; Penzhina Basin. (Map II–48).

**Foreign Arctic.** Not reported.

**Outside the Arctic.** Widespread in the forest-steppe and steppe zones of Eurasia, from the right bank of the Volga in the west to the Okhotsk Coast in the east.

**24.** ***Poa filiculmis*** Roshev. in Bot. mat. Gerb. Bot. inst. AN SSSR XI (1949), 29.

This species differs from the preceding only in having the lemmas short-haired between the nerves; in panicle structure it occupies a somewhat intermediate position between *P. stepposa* and *P. botryoides*. Such parallel pairs of species, one without and one with pubescence between the nerves, occur rather frequently in section *Stenopoa* (for example, *P. glauca* and *P. bryophila*, *P. botryoides* and *P. argunensis* Roshev., etc.), with the range of the second species (with pubescence between the nerves) always completely enclosed within the range of the first and occupying a considerably smaller area. Differences in ecology between these species have usually not been found, and it would possibly be more correct to consider the species with pubescence between the nerves not as distinct species but as varieties of the species not possessing such pubescence. However, the question of species criteria in section *Stenopoa* is so complicated that we still prefer to retain doubtful species of the type of *P. filiculmis, P. anadyrica* and *P. lapponica* as distinct. The fact is that, as we have already remarked, the boundaries between all species of this section are very blurred, but it is hardly possible to unite them all into one gigantic complex. Thus *P. filiculmis,* apart from its very close connection with *P. stepposa,* shows no less close an affinity to *P. botryoides* and in the Arctic is also very closely connected to *P. bryophila* Trin. of the group *P. glauca* Vahl s. l., appearing to constitute a more southern (almost steppe) race of it.

Growing on dry stony slopes, in steppelike meadows, sometimes also on sands and gravelbars in river valleys.

**Soviet Arctic.** Lower reaches of Lena and Kolyma; Anadyr Basin; Penzhina Basin; Bay of Korf. (Map II-49).

**Foreign Arctic.** Not reported.

**Outside the Arctic.** Northern edge of Central Siberian Plateau in the Khatanga Basin; middle course of Indigirka.

**25. Poa botryoides** (Trin. ex Griseb.) Roshev. in Fl. Zabayk. I (1929), 83; id. in Fl. SSSR II (1934), 403; Trinius ex Besser in Flora XVII, 1, Beibl. (1834), 28, nomen nudum.

*P. serotina* var. *botryoides* Trin. ex Griseb. in Ledebour, Fl. ross. IV (1853), 375.

Ill.: Fl. SSSR II, pl. XXX, fig. 9.

Predominantly a Sayan-Dahurian species, very close to *P. stepposa* but differing from it in having very dense panicles with branchlets always much abbreviated, as well as much abbreviated leaf blades. In distinction from *P. stepposa,* this is an exclusively mountain species although preferring areas with more or less distinctly solonetzic soil. Already of rather sporadic occurrence in Yakutia, it scarcely penetrates the Arctic.

**Soviet Arctic.** Anadyr Basin (River Tanyurer, 20 VIII 1941, Avramchik).

**Foreign Arctic.** Not occurring.

**Outside the Arctic.** Eastern Altay; Sayans; Transbaikalia; Verkhoyansk-Kolymsk mountain country (north to 68°20'N); Preamuria; Mongolia; Northern China.

**26. Poa glauca** Vahl, Fl. Dan., Fasc. 17 (1790), 3; Lange, Consp. fl. groenl. 172; Gelert, Fl. arct. 1, 124; Tolmachev, Fl. Taym. I, 96; Sørensen, Vasc. pl. E Greenl. 144, 149; Rozhevits in Fl. SSSR II, 398; Hultén, Fl. Al. II, 208; Kuzeneva in Fl. Murm. I, 206; Hylander, Nord. Kärlväxtfl. I, 264; A. E. Porsild, Ill. Fl. Arct. Arch. 30; Karavayev, Konsp. fl. Yak. 55; Polunin, Circump. arct. fl. 66.

*P. caesia* Sm., Fl. Brit. (1800) 103; Grisebach in Ledebour, Fl. ross. IV, 374; Perfilev, Fl. Sev. I, 91.

*P. conferta* Blytt, Norg. Fl. 1 (1861), 123; non Elliott, 1821.

*P. Soczawae* Roshev. in Izv. Bot. sada AN SSSR XXX, 1–2 (1932), 298; id. in Fl. SSSR II, 410.

*P. Ganeschinii* Roshev. in Fl. SSSR II (1934), 398; id. in Tr. Bot. inst. AN SSSR, ser. I, II (1936), 97; Kuzeneva, l. c. 206.

*P. evenkiensis* Reverd. in Sist. zam. mat. Gerb. Tomsk. univ. 8 (1936), 2.

*P. glauca* var. *conferta* (Blytt) Hyl., l. c. 263.

Ill.: Vahl, l. c., tab. 964; Fl. SSSR II, pl. XXIX, fig. 15, 16; Fl. Murm. I, pl. LXIX, LXX.

Circumpolar (but with a series of disjunctions) arctic-alpine species of section *Stenopoa* Dum., displaying very great polymorphism and repeatedly described under different specific names whose list is far from exhausted by the synonymy stated above. Connected by transitional specimens and populations (possibly of hybrid origin) both with the mesophilic species *P. nemoralis* and *P. palustris* and with the more xerophilic species *P. botryoides* and *P. stepposa,* but significantly closer to the latter two species. Apparently it was the xerophilic mountain-steppe species of section *Stenopoa,* not forest species of the *P. nemoralis* type, which were

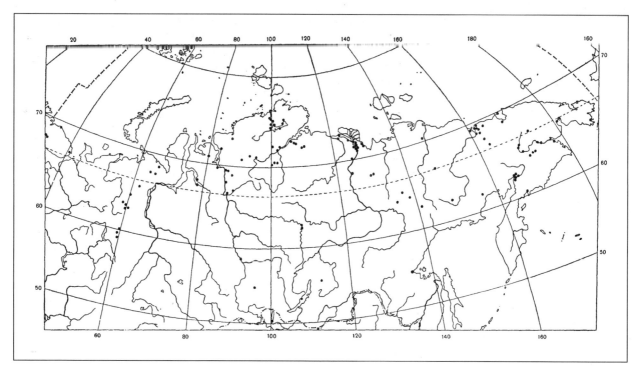

MAP II–50 Distribution of *Poa glauca* Vahl.

transformed under the influence of the abrupt onset of colder climate during the glacial period and gave rise to the rather more mesophilic species *P. glauca*, while still however retaining the basic characters of the xerophilic group: a reduced number of culm nodes and more or less abbreviated panicle branchlets. Despite its wide range of variation, *P. glauca* is usually easily recognized by its forming very dense glaucous green tufts without abbreviated vegetative shoots, by the smooth or slightly scabrous culms rising far above the sheaths of the uppermost culm leaves, and by the relatively few (usually 1–3 per branchlet) more or less coloured spikelets.The type specimens of this species, described from Northern Norway, are taller (on average) and possess relatively short ligules (about 1.5 mm long) and relatively laxer panicles with few-flowered (2–3 florets) spikelets. Such specimens are perhaps rather rare, occurring in relatively shaded habitats, but they intergrade very gradually with specimens with denser panicles and larger multiflowered (3–5 florets) spikelets; these are sometimes distinguished under the name "*P. glauca* var. *conferta* (Blytt) Hyl.," but in our opinion scarcely deserve special naming. An interesting variant of *P. glauca* consists of specimens with much reduced ligules (0.2–0.8 mm long), known so far from a few sites in Arctic Siberia (descent into Yenisey Valley at mouth of River Chayka, 16 VIII, No. 2194, Kuznetsov and Reverdatto; north slope of Central Siberian Plateau in Khatanga Basin, Mamontova Hill, 24 VIII 1934, Sambuk). Despite the absence of transitions with respect to this character, such specimens always grow together with typical specimens of *P. glauca* and in considerably lesser numbers, and are apparently of mutational origin. Among the species listed by us as synonyms of this species, *P. Soczawae* Roshev. was described from small but entirely typical specimens of *P. glauca* (Anadyr territory, rocks on River Belaya at foot of Bitcho Range, 1 VIII 1929, Sochava), although in the original description the species was compared by R. Yu. Rozhevits with *P. arctica* and *P. Komarovii* and in the "Flora of the USSR" was placed by him in the series "*Lanatiflorae* Roshev." without any grounds whatsoever. *Poa Ganeschinii*

Roshev. was described from specimens (Khibins Mountains, Kukisvum Valley, among rocks, 24 VIII 1930, Ganeshin) possessing relatively large spikelets arranged in ones to twos on the panicle branchlets (*P. glauca* var. *conferta*), but S. S. Ganeshin on the same day and in the same valley also collected specimens showing all transitions to typical *P. glauca. Poa evenkiensis* Reverd., judging from the isotype present in the herbarium of the Botanical Institute of the USSR Academy of Sciences (Evenkiyskiy National Okrug, Lake Nyakshingda, in marshes, 31 VII 1935, Lomakin), was described from relatively broadleaved specimens collected under moister conditions. Scarcely deserving separation as distinct species are also the next two species (*P. anadyrica* and *P. bryophila*), which really only represent extreme variants with respect to a single character (degree of pubescence of the lemmas).

*Poa glauca* occurs in the more southern subzones of the tundra zone; it plays a prominent role in the vegetational cover only in more continental mountainous districts of the Arctic. It grows on dry stony slopes (more often of southern exposure), on low stony hillocks (in sparse-herb and *Dryas* tundras), also on overgrown talus screes, on rock ledges and in rock crevices, on dry riverine gravelbars and on high sandy shoreline bluffs. At the same time it avoids strongly windswept sites most deficient in snow cover. Near the southern limit of forest the species is common in dry open forest and in the open herbaceous (cryophilic steppe) communities of slopes composed of rocks, stones or stones and fine soil. In its ecology and landform associations *P. glauca* closely approaches the group of cryophilic steppe plants of the Arctic and Subarctic, such as *Erysimum Pallasii, Thymus* spp., *Artemisia lagopus,* etc. It grows in discrete tufts; under favourable conditions it may be found in considerable profusion.

**Soviet Arctic.** Murman (rare); East Bolshezemelskaya Tundra (Vorkuta); Polar Ural; coast of Tazovskaya Bay; Gydanskaya Tundra; lower reaches of Yenisey (north to the River Chayka, 71°53'N); Taymyr (common, north to the lower reaches of the Pyasina and the Lower Taymyra); lower reaches of the Khatanga and Popigay; lower reaches of the Olenek, Lena and Kolyma; district of Bay of Chaun; Ayon Island; Beringian coast of Chukotka Peninsula; Bay of Krest; Anadyr Basin; Penzhina Basin; Bay of Korf. (Map II–50).

**Foreign Arctic.** Arctic Alaska and Canada; Labrador; Canadian Arctic Archipelago (as far as the northern part of Ellesmere Island); all Greenland; Iceland; Spitsbergen; Arctic Scandinavia.

**Outside the Arctic.** Europe (Faroe Islands, Northern Great Britain, and mountains of Scandinavia south to the north shore of Lake Ladoga); Northern Ural (south to 58°50'N); on the Yenisey south to the mouth of the River Kureyka; northern edge of Central Siberian Plateau in the Khatanga Basin; Sayans (rare); Vitimskoye Plateau (rare); Verkhoyansk Range; Stanovoy Range (Lake Toko); coast of Sea of Okhotsk (south to the River Uda and the Shantar Islands); Henteyn Mountains; in North America found in Alaska and a considerable part of Canada (south to British Columbia).

Very closely related species in the mountains of Central Europe (*P. aspera* Gaud. and *P. glaucantha* Gaud.), Southern Siberia (*P. altaica* Trin.) and North America *(P. interior* Rydb.).

**27. *Poa anadyrica*** Roshev. in Bot. mat. Gerb. Bot. inst. AN SSSR XI (1949), 26.

*Poa anadyrica* differs from the preceding species only in lacking a cluster of long tangled hairs on the callus of the lemmas, and has an entirely similar ecology. Scandinavian specimens of *P. glauca* apparently always possess a rather well developed cluster of hairs on the callus, which suggests that *P. anadyrica* is a geographical race native to Arctic Asia and North America. However, it is very probable that in this case we are merely dealing with a certain increase in the variability

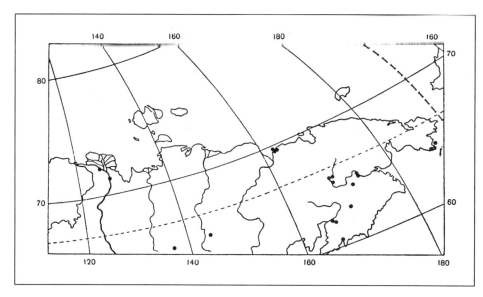

MAP II–51  Distribution of *Poa anadyrica* Roshev.

of *P. glauca* in an eastward direction. Already in the Polar Ural specimens occur with a very weakly developed cluster of hairs on the callus, and it is very probable that in the section *Stenopoa* Dum. this character has by no means such great importance as was attributed to it by R. Yu. Rozhevits.

Like the preceding species growing on rocks, stony and rocky slopes, and dry gravelbars.

**Soviet Arctic.** Lower reaches of Lena (River Atyrkan, Chaytumus settlement); lower reaches of Kolyma (River Medvezhya); SE part of Chukotka Peninsula (Chaplino); Anadyr Basin (common); Penzhina Basin; Bay of Korf. (Map II–51).

**Foreign Arctic.** Range unknown due to confusion with *P. glauca*, but apparently occurring in the arctic part of Alaska and Canada.

**Outside the Arctic.** NE Siberia (Verkhoyansk and Cherskiy Ranges); Alaska and Canada (range not precisely clarified).

28. **Poa bryophila** Trin. in Bull. Sc. Ac. Petersb. I (1836), 69; Grisebach in Ledebour, Fl. ross. IV (1853), 377.

*P. pseudo-glauca* Roshev. et *P. salebrosa* Roshev. in herb., 1948.

Species described from the territory of the USSR (Senyavin Strait, Mertens) but omitted from the "Flora of the USSR," distinguished from *P. glauca* only by the lemmas being short-haired basally between the nerves and, like the preceding species, also having an entirely similar ecology. R. Yu. Rozhevits annotated this species for description from the collections of M. Vellikaynen in Taymyr (north coast of Yamu-Baykura Bay, 17 VII 1947 and 16 VIII 1947) under two different names: "*P. pseudo-glauca* Roshev." and "*P. salebrosa* Roshev.," but these were not described. As in the case of *P. anadyrica*, specimens with the lemmas pubescent between the nerves are apparently completely absent in the western (Scandinavian) part of the range of *P. glauca*, but become rather common in districts east of the Yenisey. From the Polar Ural only one peculiar specimen of *P. bryophila* is so far known (Seida-Labotnanga, 22 VII 1960, M. Mannick), which besides pubescence between the nerves of the lemmas also possesses much reduced leaf ligules (up to 0.8 mm long). Judging from available herbarium material, it may be postulated that at least in some cases pubescence between the nerves of the lemmas in

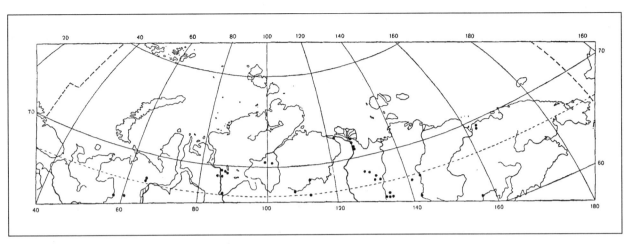

Map II–52 Distribution of *Poa sibirica* Roshev.

*P. bryophila* is partly due to hybridization of *P. glauca* with *P. arctica*.

Growing on rocks, stony or rocky (more rarely sandy) slopes, and dry gravelbars.

**Soviet Arctic.** Polar Ural (Yeletskiy road); lower reaches of Yenisey (Tolstyy Nos); Taymyr (north shore of Lake Taymyr); lower reaches of Lena and Kolyma; Beringian coast of Chukotka Peninsula.

**Foreign Arctic.** Probably occurring in the arctic part of Canada and Alaska.

**Outside the Arctic.** Range unknown due to confusion with *P. glauca*, but apparently occurring in Alaska and the western part of Canada.

**29. Poa trivialis** L., Sp. pl. (1753) 67; Grisebach in Ledebour, Fl. ross. IV (1853), 379; Krylov, Fl. Zap. Sib. II, 296; Perfilev, Fl. Sev. I, 89; Rozhevits in Fl. SSSR II, 386; Hultén, Fl. Al. II, 221; Kuzeneva in Fl. Murm. I, 202; Hylander, Nord. Kärlväxtfl. I, 253.

Ill.: Fl. SSSR II, pl. XXVIII, fig. 17; Fl. Murm. I, pl. LXVII.

Very distinctive, predominantly European species of the forest and forest-steppe zones, with full justification referred recently to the separate section *Coenopoa* Hyland. Scarcely reaching the Arctic.

Growing in moist meadows and on gravelbars, in shrub communities in valleys, and sometimes as an introduction along roads and in settlements.

**Soviet Arctic.** Murman (near the town of Murmansk, Teriberka settlement, Pechenga).

**Foreign Arctic.** Iceland; Arctic Scandinavia.

**Outside the Arctic.** Almost all Europe (in the north ranging east of the Kola Peninsula to Arkhangelsk and Ust-Tsilma on the River Pechora); Caucasus; Asia Minor; Iran; southern part of West Siberia (north to Tobolsk); Southern Krasnoyarsk Kray (north to Yeniseysk, but very sporadically); Central Asia (possibly confused with *P. silvicola* Guss.). As an introduction in Sakhalin, China and Japan, Southern Alaska and the Aleutian Islands, and many parts of Canada and the USA.

**30. Poa sibirica** Roshev. in Izv. Peterb. bot. sada XII (1912), 121; Krylov, Fl. Zap. Sib. II, 277; Perfilev, Fl. Sev. I, 88; Rozhevits in Fl. SSSR II, 380.

*P. pratensis nudiflora* Ledeb., Fl. alt. 1 (1829), 96.

*P. trivialis* var. *altaica* Griseb. in Ledebour, Fl. ross. IV (1853), 380.

Ill.: Fl. SSSR II, pl. XXVIII, fig. 11.

Very distinctive species of the Siberian taiga, in general showing closest affinity to *P. trivialis* L. and appearing to replace that species in Siberia. With respect to its completely bare lemmas it also shows a certain similarity to *P. longifolia* Trin. and to the North American group *Epiles* Hitchc. (*P. epilis* Scribn., *P. Cusickii* Vasey, etc.).

Penetrating the Arctic rather rarely, occurring here in marshy meadows and shrubbery, in NE Siberia often in alder thickets and open larch forest.

**Soviet Arctic.** Polar Ural (rare, Pay-Pudyna, Yeletskiy road); lower reaches of Yenisey (north to Dudinka); lower reaches of Olenek; lower reaches of Lena; northern spurs of Northern Anyuyskiy Range (River Medvezhya). (Map II–52).

**Foreign Arctic.** Not occurring.

**Outside the Arctic.** Urals; West Siberia (very sporadically); Altay; Sayans; East Siberia (of frequent occurrence from east of the Yenisey to the coast of the Sea of Okhotsk); Kamchatka (only in the basin of the Kamchatka River); Tien Shan and mountains of Eastern Kazakhstan; northern part of Mongolia; Northern China.

*31.* **Poa supina** Schrad., Fl. Germ. 1 (1806), 289; Rozhevits in Fl. SSSR II, 379; Hylander, Nord. Kärlväxtfl. I, 252.

*P. annua* var. *supina* (Schrad.) Reichb., Fl. Germ. exc. (1830) 46.

Widely distributed semiweedy species, replacing *P. annua* L. in many mountainous districts of Eurasia and, despite great external similarity, readily distinguished from it by the significantly larger anthers (1.2–2 mm long) and the virtually imperceptible (and bare) intermediate nerves of the lemmas. In the "Flora of the USSR" (II, 379) this species is recorded by R. Yu. Rozhevits only for the Caucasus, but more careful review of available herbarium material has shown that it occurs as an introduced weed in some regions of the European part of the USSR (for example, Vologda and Moscow Oblasts), and that in many districts of Siberia (for example, Krasnoyarsk Kray and Irkutsk Oblast) it occurs considerably more frequently than *P. annua*. Furthermore, according to the data of P. I. Ovchinnikov (Fl. TadzhSSR I, 1957, 179–180), *P. supina* is very common in the mountains of Kazakhstan and apparently in other mountainous districts of Central Asia. Both species belong to the very distinctive section *Ochlopoa* Aschers. et Graebn., which has recently been contrasted with all remaining species of the genus *Poa* as the subgenus *Ochlopoa* (Aschers. et Graebn.) Hyland.

Occurring in the Arctic as an introduced weed along roads and in settlements.

**Soviet Arctic.** Lower reaches of Ob (Salekhard); lower reaches of Yenisey (north to the vicinity of Dudinka).

**Foreign Arctic.** Possibly found as an introduction.

**Outside the Arctic.** Mountainous districts of Western Europe (including Scandinavia), Caucasus, mountains of Central Asia, almost all Siberia with the exception of districts east of the Lena. As an introduction in many other countries of the World.

*32.* **Poa annua** L., Sp. pl. (1753) 68; Grisebach in Ledebour, Fl. ross. IV, 377; Lange, Consp. fl. groenl. 172; Gelert, Fl. arct. 1, 121; Krylov, Fl. Zap. Sib. II, 282; Perfilev, Fl. Sev. I, 87; Rozhevits in Fl. SSSR II, 379; Hultén, Fl. Al. II, 200; Kuzeneva in Fl. Murm. I, 200; Hylander, Nord. Kärlväxtfl. I, 251; Polunin, Circump. arct. fl. 66.

**Ill.:** Fl. SSSR II, pl. XXVIII, fig. 9.

Weedy species distributed almost throughout the World, but possibly completely replaced in many countries by closely related species, such as: the more northern and alpine *P. supina* Schrad. and the more southern, predominantly Mediterranean *P. infirma* H. B. K. Recently and not without good reason *P. annua* has been interpreted as a species of hybrid origin resulting from hybridization of *P. supina* with *P. infirma*, which may partly explain its wide ecological adaptability (Tutin, Origin of *Poa annua* L., Nature 169, 1952, 160).

In the Arctic *P. annua* occurs as an introduced weed along roads and in settlements, sometimes also on shores of waterbodies and on gravelbars.

**Soviet Arctic.** Murman; Bolshezemelskaya Tundra (near settlements); lower reaches of Yenisey (north to Dudinka); Khatanga.

**Foreign Arctic.** Central part of Alaska; Greenland (south of the Arctic Circle); Iceland; Arctic Scandinavia.

**Outside the Arctic.** In almost all countries of the World (cosmopolitan), but far from common everywhere.

### HYBRIDS

Hybrids between closely related species of *Poa* doubtless occur rather frequently, but are far from always easy to recognize due to the insignificance of the morphological differences between the parent species. Such, for instance, are the hybrids *P. pratensis* × *P. alpigena*, *P. pratensis* × *P. subcaerulea*, *P. nemoralis* × *P. palustris*, *P. glauca* × *P. palustris*, *P. arctica* × *P. malacantha* and *P. annua* × *P. supina*. Of rarer occurrence are hybrids between species of different species-groups (series) or sections, which are almost always sterile. The following hybrids of the latter type have been reported in the Soviet Arctic.

*P. arctica* × *P. alpigena*: Lindm. in Lynge, Vasc. pl. N. Z. (1923) 122, tab. 46, fig. 4. Apparently not an uncommon hybrid, usually with viviparous spikelets and then differing from the viviparous variety of *P. alpigena* in having larger, less numerous spikelets and widely diffuse, also less numerous panicle branchlets (however, this hybrid and the viviparous variety of *P. alpigena* intergrade with one another very gradually). Sterile hybrids between these species with non-viviparous spikelets have been noted by us in the Polar Ural; they possess sharper and longer leaf ligules than in *P. alpigena*, and more slender lemmas than in *P. alpigena* which are usually of bright green colour and without pubescence between the nerves.

*P. alpina* × *P. alpigena* [= *P.* × *heryedalica* H. Sm. in Norrl. Handbibl. IX (1920), 159; Hylander, Nord. Kärlväxtfl. I (1953), 255]. This usually viviparous hybrid occurs rather frequently where both parent species grow together. It possesses leaves and panicles more or less intermediate between them and usually forms small clusters of shoots (rarely single shoots) covered by rather numerous sheaths of dead leaves (as in *P. alpina*), but with a few subterranean stolons (as in *P. alpigena*).

*P. alpina* × *P. arctica:* Lindm., l. c. 123, tab. 46, fig. 3, tab. 47, fig. 18–20. A considerably rarer sterile hybrid (according to Lindman also producing a form with viviparous spikelets), known to us only from the Polar Ural (Seida-Labotnangi, 24 VII 1960, Mannick). The stated specimen virtually lacks subterranean stolons and has flat, short-acuminate, rather thickish leaves (as in *P. alpina*), but has spikelets more or less intermediate between the two species and is more reminiscent of *P. alpigena* with respect to the panicle structure (with rather long and slender branchlets directed obliquely upwards).

*P. glauca* × *P. arctica.* Sterile hybrids between these species doubtless occur where they grow together (especially in Taymyr and the lower reaches of the Lena) and have more or less intermediate characters (for example, green colour of the plant and somewhat abbreviated vegetative shoots in specimens resembling *P. glauca* in the structure of the panicles and spikelets, or strongly scabrous panicle branchlets combined with spikelets resembling those of *P. arctica*). Moreover, it is in our opinion highly probable that the species *P. Tolmatschewii* and *P. bryophila* are of hybrid origin, with the characters of *P. arctica* dominating in the first of these species, those of *P. glauca* in the second.

*P. arctica* × *P. paucispicula.* An apparently very rare sterile hybrid known to us only from the lower reaches of the Lena (River Tigiya) and from the district of Senyavin Strait. Externally it is very similar to *P. arctica*, but has smaller (usually abortive) anthers 1–1.5 mm long and the lemmas bare (or almost bare) between the nerves. The very few spikelets are arranged on long slender panicle branchlets, somewhat reminiscent of the Kamchatkan alpine species *P. nivicola* Kom.

## GENUS 23 — Dupontia R. Br. — DUPONTIA

OLIGOTYPIC GENUS (two closely related species) endemic to the Arctic. Very close to the monotypic genus *Arctophila*, which is subarctic but also very widely distributed in the Arctic. Plants (or clones) with characters transitional between *Dupontia* and *Arctophila fulva*, possibly of hybrid origin, occur very rarely. This is an additional indication of the close relationship of the two genera.

*Dupontia* is a purely tundra genus, scarcely penetrating the forest-tundra zone. Such penetration takes place only in East Siberia (along rivers); in Europe the southern boundary of the range runs significantly further north than the southern boundary of the tundra zone. Both species of *Dupontia* are plants of lowland expanses, foreign to specifically montane habitats. They are tundra hygrophytes associated with slightly peaty soils with surplus moisture, also found on sands, silt deposits and silted gravelbars. There are isolated records of *Dupontia* growing in dry mossy tundra; although not a truly coastal plant, *Dupontia* does not avoid wet saline silty shores of maritime lagoons and channels.

The presence of two species of *Dupontia* since the time of Ruprecht's work (Fl. samojed. cisur.) was for long not accepted by all botanists, because both species possess strongly overlapping (but not identical) ranges and display overlapping (so-called transgressive) variation. The situation changed after karyosystematic investigations of the grasses of Spitsbergen (Flovik, Cytological studies of arctic grasses, Hereditas 24, 265, 1938), as a result of which it was revealed that two chromosome numbers are present in the genus *Dupontia*: $2n = 44$ and $2n = 88$. The first form was identified with Ruprecht's *D. psilosantha*, the second with typical *D. Fisheri* (= *D. pelligera* Rupr.). Subsequently Flovik's data were confirmed for Greenland material by other investigators (Jørgensen & al., Fl. pl. Greenl.). In this connection it is very probable that the 44-chromosomed *D. psilosantha* arose from *Arctophila fulva* ($2n = 42$) through an increase of the haploid complement (n) by one chromosome, and *D. Fisheri* from *D. psilosantha* through doubling the number of chromosomes. With respect to lemma pubescence, *D. psilosantha* stands closer to *Arctophila* than to *D. Fisheri*; the similarity of the first species to *Arctophila fulva* is also partly evident in the nature of the panicle (presence of drooping branchlets); however, in the number of florets per spikelet, the width of the lemmas and the width of the leaf blades, *D. Fisheri* is more reminiscent of *A. fulva*. While in North America the range of the octoploid species is situated basically north of that of the tetraploid, for Siberia and Eastern Europe we can rather speak of a wider amplitude in the distribution of the former; in the High Arctic (where occurrence of the tetraploid form is extremely sporadic), the overwhelming predominance of octoploid populations is undoubted at sites where the presence of both forms has been established. The two species do not possess identical ecology; in particular, normally only *D. psilosantha* takes part in the formation of closed low-herb maritime (saltmarsh) communities, and is here represented by its most characteristic form. Possibly the original differentiation of *D. psilosantha* from the *Arctophila* type was associated with the special environment of saltmarsh communities of the Arctic Coast.

With respect to degree of morphological differentiation, *D. Fisheri* and *D. psilosantha* are more reminiscent of two subspecies of a single species, which corresponds to the treatment of these forms by several authors (see below). But the considerable overlap of their ranges, the practical feasibility of dividing the overwhelming majority of specimens of *Dupontia* into two types on the basis of their morphological differences, and finally the difference in multiplication of the basic number of chromosomes (constituting a biological barrier between the tetraploid and octoploid forms)—all this does not allow the term subspecies to be applied to *D. psilosantha* and *D. Fisheri* and compels us to treat them as distinct species.

1. Lemma *bare* (extremely rarely with isolated hairs on basal part of median or lateral nerves). Leaves usually very narrow, involute, *long-acuminate*. Culm below inflorescence and panicle branchlets sometimes bluish gray due to waxy bloom. Panicle frequently interrupted, pyramidal with straight or slightly flexuous strong lateral branchlets, some of which diverge from the rachis almost *at a right angle or even slightly downwards* (more rarely all branchlets directed upwards, appressed to rachis of panicle). Spikelets with 1–2 florets (upper sometimes abortive). Glumes (like lemmas) narrowly lanceolate, *attenuately long-acuminate,* serrate at tip; the upper normally exceeding the lemmas. Bristles at base of lemmas often *few and short.* . . . . . . . . . . . . . . . . . . . . . . . . . . . . .1. **D. PSILOSANTHA** RUPR.
– Lemma more or less *pubescent* (only on basal part of nerves, or also on back inside lateral nerves, or all over basal third to three-quarters; in latter case pubescence of nerves longer). Culm leaves broader, *short–acuminate,* usually with more or less involute margins, more rarely flat and wide; upper culm leaves short. Panicle usually narrow with short or longer flexuous branchlets directed *upwards and appressed to rachis of inflorescence* (more rarely panicle laxish with flexuous branchlets slightly divergent from rachis; very rarely specimens occur with isolated drooping branchlets on lower part of panicle). Spikelets with (1)2–4 florets. Glumes lanceolate, more or less *obtuse* or serrate above, more rarely attenuately acuminate, usually not quite reaching tip of upper floret. Lemmas lanceolate or lanceolate-elliptic. Bristles at base of lemmas *longer and more copious* than in preceding species. . . . . . . . . . . . . . . . . . . . . . . . . .2. **D. FISHERI** R. BR.

1. ***Dupontia psilosantha*** Rupr., Fl. samojed. cisur. (1846) 64; Ledebour, Fl. ross. IV, 386; Lange, Consp. fl. groenl. 165; Holm, Contr. morph. syn. geogr. distr. arct. pl. 9b; Andreyev, Mat. fl. Kanina 158; Böcher & al., Grønl. Fl. 288; Jørgensen & al., Fl. pl. Greenl. 24.

*D. Fisheri* ssp. *psilosantha* Hult., Fl. Al. II (1942), 226; Chernov in Fl. Murm. I, 216, in annotations; A. E. Porsild, Ill. Fl. Arct. Arch. 33.

*D. Fisheri* var. *psilosantha* Trautv., Syll. pl. Sib. bor.-or. (1887) 62; Krylov, Fl. Zap. Sib. II, 304; Perfilev, Fl. Sev. I, 94; Scholander, Vasc. pl. Svalb. 69.

*D. Fisheri* f. *psilosantha* — Nevskiy in Fl. SSSR II (1934), 432.

*D. Fisheri* auct. non R. Br. — Trautvetter, Pl. Sib. bor. 138; Ostenfeld, Fl. arct. 114, pro parte; Tolmachev, Fl. Kolg. 15; id., Obz. fl. N. Z. 149, pro parte; Leskov, Fl. Malozem. tundry 29, pro parte; Kozhevits in Areal 21, pro parte; Karavayev, Konsp. fl. Yak. 56, pro parte; Sokolovskaya & Strelkova in Bot. zhurn. XLV, 3, 376.

*D. micrantha* Holm in Fedde Repert. sp. nov. III, 337; id., Contr. morph. syn. geogr. distr. arct. pl. 9b.

*Graphephorum psilosantha* A. Gray in Transact. Bot. Soc. Canad. (1861) 55.

**Ill.:** Ruprecht, l. c., tab. 6; Scholander, l. c., fig. 42, 1; A. E. Porsild, l. c., fig. 7, c.

The most distinctive and peculiar form of *D. psilosantha* (with respect to its differences from the next species) is found in low-herb grassy communities of silty shores of lagoons of northern seas, usually subject to the action of rising and falling tides but protected from the destructive action of ice driven ashore during summer storms. In the communities just mentioned, *D. psilosantha* may be associated with the obligate arctic halophytes *Calamagrostis deschampsioides, Puccinellia phryganodes, P. tenella, Carex subspathacea* and *C. marina*, which form a complete sward over the ground. In these habitats *D. psilosantha* is a low-growing anthocyanically coloured plant with bluish grey waxy bloom on the panicle branchlets and sparse one-flowered narrow acute spikelets in a pyramidal panicle.

In its Eurasian distribution (contrary to A.E. Porsild's indications for Arctic Canada), *D. psilosantha* gravitates towards sea coasts; a more or less distinct retreat from the coast (by up to 100–200 km) is recorded in the district of the lower course of the Lena.

The fully typical form of *D. psilosantha* also occurs outside maritime saltmarshes in habitats also characteristic for *D. Fisheri*: moist sand and silt deposits in river floodplains; margins of tundra lakes and beds of dried out lakes (turloughs); saucer-shaped depressions and ditchlike fissures in polygonal marshes and various features of tundra marshes with sedge (*Carex stans*) or sedge and cottongrass (*Eriophorum Scheuchzeri*, etc.); in the complex communities of lowland marshy tundras (flat-hummocky, large-hummocky, etc.), *D. psilosantha* is found in microdepressions and sinkholes with surplus moisture; it avoids oligotrophic bogs. On the north coast of Chukotka (Cape Schmidt) reported by B. N. Gorodkov in wet mossy tundra near melting snow (in mid-August); but it usually avoids sites where snow long persists, as well as areas from which the snow is blown off by winter winds. In the lower reaches of the Lena found south of the northern limit of forest in floodplain marshes of tributaries of the Lena (near their mouths). In the habitats just listed less typical specimens, sometimes virtually indistinguishable from *D. Fisheri* in appearance, also occur along with the typical form of *D. psilosantha*.

*Dupontia psilosantha* is not characteristic of high arctic tundras; collections from there are extremely sporadic (New Siberia and Faddeyevskiy Islands, Spitsbergen); specimens from the High Arctic are extremely low-growing, with very depauperate inflorescence; in external appearance they are indistinguishable from high arctic *D. Fisheri*.

There is an interesting gap in the distribution of *D. psilosantha* between the lower reaches of the Yenisey and the Lena, which definitely cannot be considered the result of inadequate collections. In the Anadyr district only *D. psilosantha* is represented (all localities maritime). In many portions of the lower reaches of the Lena and the Kharaulakh, *D. psilosantha* is also the sole form of *Dupontia*.

Investigators of the flora of Arctic America report the presence of a form of *D. psilosantha* in which the lemma possesses a short awn at its tip. The differences in anther size between *D. psilosantha* and *D. Fisheri* sometimes suggested are not confirmed in our material.

**Soviet Arctic.** Eastern part of Murman coast (according to Ye. G. Chernov, Yu. D. Zinserling's records of the discovery of *Dupontia* here apparently refer to the present species); Kanin; Kolguyev (common); Bolshezemelskaya Tundra; Varandey

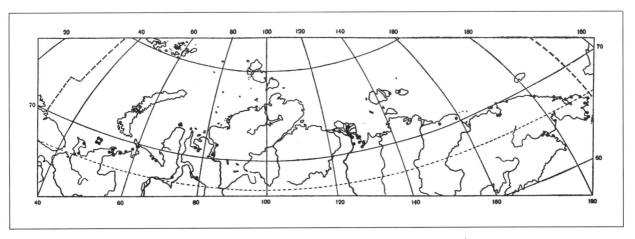

Map II-53  Distribution of *Dupontia psilosantha* Rupr.

Island; Vaygach; South Island of Novaya Zemlya (Belushya Bay); North Island of Novaya Zemlya (southern part); shore of Tazovskaya Bay (Cape Nakhodka); Gydanskaya Tundra (south coast of Gyda-Yam Bay); Yenisey Bay (Golchikha village, Nasonovskiy Island); lower reaches and delta of Lena (south to the River Kuramis); tundras of the coast of Buorkhaya Bay (south to the middle course of the Kharaulakh); New Siberian Islands (New Siberia and Faddeyevskiy Islands); lower reaches of River Alazeya; district of Bay of Chaun (Ayon Island, Ust-Chaun settlement); north coast of Chukotka; Beringian coast of Chukotka Peninsula; Arakamchechen Island; Bay of Krest; mouth of Anadyr and south coast of Anadyr Bay. (Map II–53).

**Foreign Arctic.** Arctic Alaska; arctic coast of Canada; Labrador; Baffin Island (southern half); Greenland (Disko Island and neighbouring part of west coast, northern part of east coast); Spitsbergen; Bear Island.

**Outside the Arctic.** Absent.

2. ***Dupontia Fisheri*** R. Br. in Suppl. to App. Parry's Voyage XI (1824), 290; Ledebour, Fl. ross. IV, 386; Schmidt, Fl. jeniss. 128; Kjellman & Lundström, Phanerogam. N. Z. Waig. 154, pro parte; Kjellman, Phanerogam. sib. Nordk. 117, pro parte; Holm, Nov. Zeml. Veget. 16; Scheutz, Pl. jeniss. 187; Ostenfeld, Fl. arct. 114, pro parte; Simmons, Survey Phytogeogr. 51, pro maxima parte; Holm, Contr. morph. syn. geogr. distr. arct. 9b; Lynge, Vasc. pl. N. Z. 107, pro parte; Krylov, Fl. Zap. Sib. II, 304, pro parte, excl. var. *psilosantha* (under "var. *typica*"); Tolmachev, Fl. Taym. I, 98; Hanssen & Lid, Fl. pl. Franz Josef L. 33; Perfilev, Fl. Sev. I, 94, pro parte, excl. var. *psilosantha*; Scholander, Vasc. pl. Svalb. 68, pro parte, excl. var. *psilosantha*; Nevskiy in Fl. SSSR II, 432, pro parte, excl. f. *psilosantha*; Tolmachev, Obz. fl. N. Z. 149, pro parte; Leskov, Fl. Malozem. tundry 29, pro parte; Hultén, Fl. Al. II, 226; Tikhomirov, Fl. Zap. Taym. 24; Rozhevits in Areal 21, pro parte; Böcher & al., Grønl. Fl. 290; A. E. Porsild, Ill. Fl. Arct. Arch. 33; Karavayev, Konsp. fl. Yak. 56, pro parte; Jørgensen & al., Flow. pl. Greenl. 24; Polunin, Circump. arct. fl. 51, pro parte.

*D. Fisheri* var. *pelligera* Trautv., Syll. pl. Sib. bor.-or. (1887) 61.

*Poa (Dupontia) pelligera* Rupr., Fl. samojed. cisur. (1844) 64.

*Graphephorum Fisheri* A. Gray in Transact. Bot. Soc. Canad. (1861) 55.

*Colpodium humile* Lge. in Holm, Nov. Zeml. Veget. (1885) 16.

*C. Langei* Gand., Nov. Consp. Fl. Eur. (1910) 490.

**Ill.:** Holm, Nov. Zem. Veg. (1885), tab. I, fig. 1, tab. II, fig. 1–8; Vasey, Ill. N. Am. Grass. 2 (1893), tab. 87; Scholander, l. c., fig. 42, 11; Porsild, l. c., fig. 76.

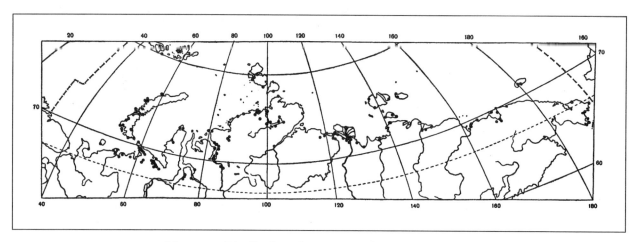

MAP II–54  Distribution of *Dupontia Fisheri* R. Br.

In collections from the Soviet Arctic held in the herbarium of the Botanical Institute, this species is considerably more abundantly represented than the preceding. It is apparently not characteristic of maritime saltmarshes; thus, in the maritime zone on the south coast of Buorkhaya Bay (Arctic Yakutia) *D. Fisheri* was found in a waterlogged trench on a relatively raised area composed of gravel, together with *Carex stans, Ranunculus Pallasii*, etc.; in moist maritime marshes it was everywhere replaced by *D. psilosantha*. Otherwise the ecology of the two species is very similar (see above). In many portions of the tundra zone in Eastern Europe and Siberia, *D. Fisheri* (in contrast to *D. psilosantha*) retreats considerably from the sea coast; in the Taymyr Peninsula it is the sole representative of the genus and is locally very common; in Eastern Europe and West Siberia it is reported in moist shrubby (willow) tundras. While in North America the range of *D. Fisheri* is basically situated further north than the range of *D. psilosantha* (Porsild, Ill. Fl. Arct. Arch.), in Siberia and Eastern Europe we may rather speak of a distinctly greater breadth of the latitudes in which *D. Fisheri* occurs. In particular, *D. Fisheri* is a constant component of high arctic floras, although representing a temperate arctic element in them: on the polar islands of Siberia (Wrangel Island, New Siberian Islands, Severnaya Zemlya), *D. Fisheri* is characteristic of moist streamside alluvia and lowland marshes (or equivalent habitats with surplus moisture in the zone where peat formation is absent).

*Dupontia Fisheri* sometimes produces a form with uniformly leaved culms and wide flat leaves.

The boundary between this species and *D. psilosantha* cannot always be drawn with complete confidence. Thus, individuals occasionally occur in which the lower floret of some spikelets possesses a bare lemma or one with isolated hairs, while the upper floret has a more or less pubescent lemma. For more reliable demarcation of the tetraploid and octoploid races, it is desirable to conduct a study of morphological variation in *Dupontia* with numerous determinations of the chromosome number.

**Soviet Arctic.** East coast of Kanin; Kolguyev (mouth of Krivaya River); Malozemelskaya Tundra (between Pechora Bay and Kolokolkova Bay); Bolshezemelskaya Tundra (northern part); Vaygach; Pay-Khoy; Karsko-Baydaratskaya Tundra; Polar Ural (northern extremity); Novaya Zemlya; Franz Josef Land; Yamal; Belyy Island; Little Yamal; Tazovskaya Bay; Gydanskaya Tundra; lower reaches of Yenisey; coast and islands of Yenisey Bay; Taymyr Peninsula; Severnaya Zemlya (Bolshevik and October Revolution Islands); lower reaches of Olenek; lower reaches and delta of Lena; west and south coasts of Buorkhaya Bay; Yana Delta (Krestyakh); New

Siberian Islands; Cape Svyatoy Nos; Indigirka Delta; coast of East Siberian Sea east of the mouth of the Kolyma; district of Bay of Chaun; Wrangel Island; Beringian coast of Chukotka Peninsula (Uelen, Lawrence Bay); Arakamchechen Island. (Map II–54).

**Foreign Arctic.** Arctic Alaska (northern extremity); entire Canadian Arctic Archipelago; northern mainland coast of Canada from the Mackenzie Delta to the west coast of Hudson Bay (northern part); west and east coasts of Greenland (north of 70°N); Spitsbergen.

**Outside the Arctic.** Absent.

---

GENUS 24   **Arctophila** (Rupr.) Rupr. ex Anderss. — ARCTOPHILA

MONOTYPIC HYDROPHILIC GENUS consisting of a very polymorphic species widespread in arctic and subarctic districts. Despite great similarity in spikelet structure to *Poa, Colpodium* and *Puccinellia*, there are grounds for grouping *Arctophila* not with these genera but with the relatively more primitive, also hydrophilic genus *Scolochloa* Link, which *Arctophila* much resembles in the structure of the vegetative organs. It may be postulated that northern populations of *Scolochloa festucacea* (Willd.) Link, which survived the coldest time of the Quaternary Period in the relatively stable conditions of the aquatic medium, underwent changes in the structure of the panicle and spikelets and gave rise to the genus *Arctophila*. Apart from the structure of the vegetative organs, a relatively primitive character of the latter is the presence of short but conspicuous styles, while in the genera *Poa, Colpodium* and *Puccinellia* the stigmas are almost or completely sessile. Rather close to *Arctophila* is also another small arctic genus *Dupontia*, with which *Arctophila* could even be united as a section or subgenus.

1. ***Arctophila fulva*** (Trin.) Anderss., Gram. Scand. (1852) 49; Gelert, Fl. arct. 1, 118; Lynge, Vasc. pl. N. Z. 102; Nevskiy in Fl. SSSR II, 433; Hultén, Fl. Al. II, 224; Kuzeneva in Fl. Murm. I, 216; Hylander, Nord. Kärlväxtfl. I, 279; A. E. Porsild, Ill. Fl. Arct. Arch. 32; Karavayev, Konsp. fl. Yak. 56; Jørgensen & al., Fl. pl. Greenl. 28; Polunin, Circump. arct. fl. 41.

   *A. pendulina* (Laest.) Anderss. ex Lge., Suppl. fl. dan. III (1874), 4, tab. 126.

   *A. effusa* Lge., l. c. 4 in adnot. and Consp. fl. groenl. 167; Scheutz, Pl. jeniss. 186.

   *A. mucronata* Hack. ex Vasey, Cat. Grass. U. S. (1885) 88; Holm in Greene in Ottawa Nat. XVI (1902), 81.

   *A. brizoides* Holm, l. c. 83.

   *A. chrisantha* Holm, l. c. 84.

   *A. trichopoda* Holm in Fedde, Repert. spec. nov. III (1907), 337.

   *Poa fulva* Trin. in Mém. Ac. Pétersb., sér. 6, 1 (1830), 378.

   *P. trichoclada* Rupr., Fl. samojed. cisur. (1846) 62.

   *P. latiflora* Rupr., l. c. 52.

   *P. Laestadii* Rupr., l. c. 62 (nomen novum for *Glyceria pendulina* Laest.).

   *P. deflexa* Rupr., l. c. 62.

   *P. poecilantha* Rupr., l. c. 63.

   *P. remotiflora* Rupr., l. c. 63.

   *P. similis* Rupr., l. c. 63.

   *P. scleroclada* Rupr., l. c. 63.

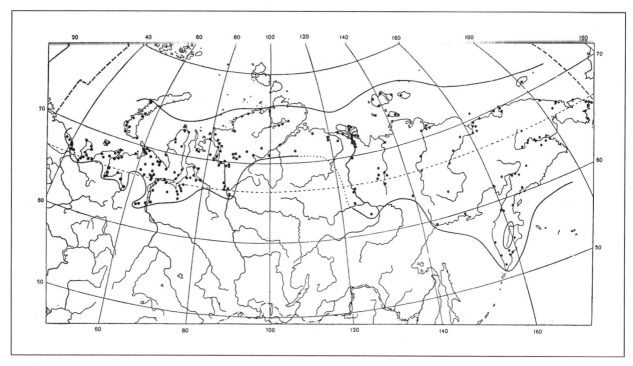

**Map II–55**  Distribution of *Arctophila fulva* (Trin.) Anderss.

*P. pendulina* (Laest.) Vahl, Fl. dan. XIV (1843–1849), tab. 2343.
*Glyceria pendulina* Laest. in Wahlenberg, Fl. Suec., Ed. 2, add. (1833) 1088.
*G. fulva* (Trin.) Fr., Summa Veg. I (1846), 244.
*Colpodium fulvum* (Trin.) Griseb. in Ledebour, Fl. ross. IV (1853), 385; Scribner & Merrill, Grass. Al. 74; Perfilev, Fl. Sev. I, 98.
*C. pendulinum* (Laest.) Griseb., l. c. 386; Krylov, Fl. Zap. Sib. II, 80.
*C. Malmgrenii* Anderss. in Öfvers. Vet. Ac. Stockh. XXIII (1866), 121.
*C. mucronatum* (Hack. ex Vasey) Beal, Grass. N. Amer. II (1896), 558.
*Graphephorum fulvum* (Trin.) A. Gray in Transact. Bot. Soc. Canad. (1861) 55.
*G..pendulinum* (Laest.) A. Gray, l. c. 55.
**Ill.:** Ruprecht, l. c., tab. 4, 5, 6; Vahl, l. c., tab. 2343; Lange, l. c. (1874), tab. 126; Fl. SSSR II, pl. XXXIII, fig. 6; Fl. Murm. I, pl. LXXIII; Porsild, l. c., fig. 12, e; Polunin, l. c. 40.

Very polymorphic circumpolar arctic species, of which particular specimens and populations have been repeatedly described as separate species. The type specimens of *Colpodium fulvum* and *Glyceria pendulina* are of relatively large size, with diffuse panicles and numerous spikelets. High arctic specimens of *Arctophila fulva* are considerably different, especially those from drier habitats which (apart from the small size of the plant) are characterized by a significant reduction in the numbers of culm nodes (to 3–4), of branchlets per node of the panicle (to 2–4), of spikelets on the panicle branchlets, and finally of florets per spikelet (to 2–3). The species *Poa similis*, *P. remotiflora* and *Arctophila effusa* were described from just such specimens, but the typical more southern form of *A. fulva* rather gradually intergrades with the small high arctic form and the characters of the latter listed above are often not correlated with one another; so the latter can at best only be distinguished as a separate subspecies (pronounced ecological-geographical race), *A. fulva* ssp. *similis* (Rupr.) Tzvel. comb. nova (*A. similis* Rupr., l. c.; *Colpodium pendulinum* var. *simile* Griseb., l. c. 386; *C. fulvum* var. *arcticum* Roshev. in Tolmachev, Fl. Taym. I, 98). Sometimes the spikelets of high arctic specimens remain abortive and thus give the impression of being single-flowered. Specimens of *A. fulva* possessing lemmas with a short awn at the tip are known from Alaska and Canada. The species *A. mucronata* and *A trichopoda* were described from such specimens, and it cannot be excluded that they actually represent a distinct ecological-geographical race although Hultén (l. c.), after seeing the type specimen of *A. mucronata*, did not consider it possible to accord this species even the rank of variety. Detailed information on the ecological properties and economic importance of *A. fulva* is given in the monograph by B. A. Tikhomirov (Tr. Dalnevost. fil. AN SSSR, Bot. ser. II, 1937, 672–702).

Growing in the most waterlogged sites on tundras (fissure trenches in polygonal marshes, wet sinkholes, etc.), especially common on the shores of lakes, rivers and streams where it forms littoral beds in water of up to 50–60 cm depth, also on the muddy beds of dried up lakes (turloughs). The strongly anthocyanic colouration of the stems and leaves of *Arctophila* at the beginning and end of the growing season is characteristic.

**Soviet Arctic.** Murman (Ponoy, Lumbovka, only ssp. *fulva*); Kanin; Kolguyev; Malozemelskaya and Bolshezemelskaya Tundras; coast of Kara Sea (Amderma); Vaygach; Novaya Zemlya (south of Krestovaya Bay, both subspecies occurring); Yamal; Obsko-Tazovskiy Peninsula; lower reaches of Yenisey (abundantly); Taymyr; Anabarskaya Tundra; lower reaches of the Olenek, Lena, Yana, Indigirka and Kolyma; New Siberian Islands; Wrangel Island (in the north only ssp. *similis*); Beringian coast of Chukotka Peninsula (ssp. *fulva*); Anadyr Basin; Penzhina Basin; Bay of Korf. (Map II–55).

**Foreign Arctic.** Arctic part of Alaska and Canada; arctic part of Labrador (only the shore of Hudson Bay); Canadian Arctic Archipelago (southern part); SW Greenland (only south of the Arctic Circle); Spitsbergen; Arctic Scandinavia.

**Outside the Arctic.** West Siberia (south on the Ob and its tributaries to 63°N); northern part of Central Siberian Plateau (south to the Lower Tunguska); Yakutia (on the Lena to the mouth of the Vilyuy, also north and northeast of the Aldan); Magadan Oblast; Kamchatka (south to 53°N); Aleutian Islands; Alaska and Canada (south to the upper reaches of the Yukon, Great Slave Lake and the southern part of Hudson Bay).

## GENUS 25 — Colpodium Trin. — COLPODIUM

SMALL GENUS CONTAINING about 20 species, distributed predominantly (with the exception of one species) in the high mountains of Eurasia from the Caucasus and Asia Minor to the Verkhoyansk and Momskiy Ranges of East Siberia. Divisible into five completely distinct species-groups (subgenera), whose relationship to one another is not entirely clear. The substantial morphological similarity between these subgenera, attributable mainly to the lemmas being extensively hyaline and to the general smoothness (absence of acicules) of almost all parts of the plant, could possibly be merely the result of convergent evolution, in which case it would apparently be more correct to consider them all as separate genera. One species of subgenus *Hyalopoa* Tzvel., occupying a position apparently intermediate between the genera *Colpodium* and *Poa*, penetrates the limits of the Soviet Arctic in the lower reaches of the Lena.

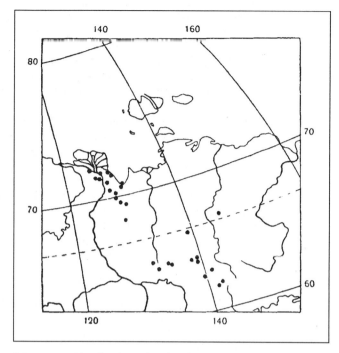

MAP II–56  Distribution of *Colpodium lanatiflorum* (Roshev.) Tzvel.

**1. *Colpodium lanatiflorum* (Roshev.) Tzvel., comb. nova.**
*Poa lanatiflora* Roshev. in Izv. Glavn. bot. sada AN SSSR XXX (1932), 303; id. in Fl. SSSR II, 409; Karavayev, Konsp. fl. Yak. 56.
Ill.: Fl. SSSR II, pl. XXX, fig. 17.

Endemic East Siberian species, referred by R. Yu. Rozhevits to the genus *Poa* but in our opinion considerably more closely related to the small but very distinctive group of *Colpodium* species described by S. A. Nevskiy under the name "*Colpodium* series *Pontica* Nevski," containing besides two predominantly Caucasian species, *C. ponticum* (Bal.) Woron. and *C. lakium* Woron., also the Himalayan species *C. nutans* Griseb. On account of the strong differentiation of this species-group (its very substantial differences from typical species of the genus, which always possess one-flowered spikelets) and its intermediate position between the genera *Colpodium* and *Poa*, we consider it more correct to accord it the rank of a separate subgenus: *Colpodium* subgen. *Hyalopoa* Tzvel., subgen. novum (= *C.* ser. *Pontica* Nevski, Fl. SSSR II, 440) with type *Colpodium ponticum* (Bal.) Woron. (Typus subgeneris). Also noteworthy is the great and not merely superficial resemblance of *C. lanatiflorum*, as well as of other species of this subgenus, to *Arctophila fulva* (Trin.) Anderss., especially to certain specimens of the high arctic race *A. fulva* ssp. *similis* whose sole substantial difference from *C. lanatiflorum* consists of the relatively long styles. The existence of a certain relationship between *C. lanatiflorum* and *Arctophila fulva* is also supported by the fact that specimens of this species are known (Indigirka Basin, Momskiy Range, Ilin-Talynya, 20 VII 1935, Sheludyakova) which possess lemmas with a mucro up to 0.8 mm long at their tip, something never the case in *Poa*. Since different specimens are not so far known from the Momskiy Range while the numerous available specimens from the Verkhoyansk Range and the lower reaches of the Lena do not possess such mucros, the existence of a more eastern ecological-geographical race of *C. lanatiflorum* native to the Momskiy Range may be postulated; this we now describe as the subspecies *C. lanatiflorum* subsp. *momicum* ssp. nova [A subspeciei typica lemmatis apice mucronatis, mucrone ad 0.8 mm lg. differt. Typus: Jacutia, jugum Momicum in systemate fl. Indigirka, Ilinj-Talynja, 20 VII 1935, Scheludjakova. In herb. Inst. Bot. Ac. Sc. URSS (Leningrad) conservatur].

*Colpodium lanatiflorum* is a member of an interesting group of endemic plants of the Verkhoyansk-Kolymsk mountain country, including a whole series of remarkable species with excellent systematic differentiation (*Senecio jacuticus* Schischk., *Corydalis Gorodkovii* Karav., *Androsace Gorodkovii* Ovcz. et Karav.) and one monotypic genus (*Gorodkovia jacutica* Botsch. et Karav.). Like them, *C. lanatiflorum* is very well differentiated from the remaining species of subgenus *Hyalopoa* (which are geographically very remote from it) and also shows an affinity for stony (not carbonate!) soils, especially the eluvia and deluvia of clayey shales and the gravel deposits of mountain rivers. It occurs in the treeless (barren) zone and in the upper (semibarren) portion of the zone of open Dahurian larch forest. Growing both on stony areas without closed turf (riverine gravelbars, temporary runoff channels with fine soil and stones, shale screes) and in closed communities, such as: riverine meadows, nival meadows on areas with fine soil and stones at the foot of slopes, and *Dryas* tundras of dry river terraces, summits and slopes. Occasionally encountered on rocks or on steep sandy shoreline banks of mountain rivers. Occurring both in windswept areas with little winter snow cover and on sites with comparatively late snowmelt (rocky nival meadows, silted gravelbars on the edges of gigantic ice-sheets).

**Soviet Arctic.** SE part of Chekanovskiy Ridge; northern extremity of Kharaulakh Mountains. (Map II–56).

**Foreign Arctic.** Absent.

**Outside the Arctic.** Throughout the Verkhoyansk Range and in the system of the Cherskiy Range.

## GENUS 26    Catabrosa P. B. — BROOK GRASS

SMALL GENUS DISTRIBUTED in the temperate zone of the Northern Hemisphere (one species in Southern South America). One species with restricted distribution in the Atlantic Arctic.

1. ***Catabrosa aquatica*** (L.) Beauv., Agrost. (1812) 97; Grisebach in Ledebour, Fl. ross. IV, 387; Lange, Consp. fl. groenl. 166; Krylov, Fl. Zap. Sib. II, 270; Rozhevits in Fl. SSSR II, 445; Perfilev, Fl. Sev. I, 85; Leskov, Fl. Malozem. tundry 27; Gröntved, Pterid. Spermatoph. Icel. 134; Hultén, Fl. Al. II, 194; id., Atlas, map 201; id., Amphi-atl. pl. 70; Kuzeneva in Fl. Murm. I, 192; Böcher & al., Grønl. Fl. 287; Polunin, Circump. arct. fl. 45.

*Aira aquatica* L., Sp. pl. (1753) 64.
*Poa airoides* Koel. — Ruprecht, Fl. samojed. cisur. 61.
**Ill.:** Beauvois, l. c., tab. 19, fig. 8.

Widely distributed plant of the northern temperate zone, penetrating the Arctic here and there, mainly on river shores.

**Soviet Arctic.** West Murman; lower reaches of Pesha; at mouth of Indiga.
**Foreign Arctic.** West Greenland (north to 70°N); Iceland; Arctic Scandinavia.
**Outside the Arctic.** Widely but unevenly distributed in Europe, South Siberia, the mountains of Central Asia and the Middle East, and North America. Penetrating the far north only in countries neighbouring the Atlantic.

### GENUS 27    Phippsia R. Br. — PHIPPSIA

TYPICALLY ARCTIC GENUS, containing two closely related species which sometimes hybridize with one another. Formally it cannot be considered endemic to the Arctic, since both species occur beyond arctic limits in certain alpine districts of the temperate zone. Both species of *Phippsia* are very typical components of the arctic flora, penetrating arctic districts at high latitudes. One of them reaches the extreme polar limits of land and has a circumpolar distribution, the other is confined exclusively to the arctic part of the USSR.

1. Panicle *dense, almost cylindrical, short* (up to 3 cm long). Lemmas *bare or very slightly pubescent.* Caryopsis *broad, with greatest width above its middle,* not longer than lemma. Anthers 0.4 mm long, *obovoid,* almost as wide as long. Plant always lowly, with culms normally less than 10 cm long (in exceptional cases length of culms reaching 15 cm with panicle length reaching 4.5 cm). . . . . . . . . . . . . . . . . . . . . . . . .1. **PH. ALGIDA** (SOLAND.) R. BR.
— Panicle longer (3–8 cm long), more or less *diffuse*, with its branchlets often *spreading horizontally* or even slightly downcurved. Lemmas *copiously pubescent on nerves with white hairs.* Paleas *densely ciliate on keel.* Caryopsis *narrow, widest below middle,* of same length as or longer than lemma. Anthers *oblong,* 0.5 mm long. Plant often larger, with culms reaching 10–15 cm long (in certain cases in habitats with favourable soil conditions the culms can reach 20–25 cm with panicle length reaching 12 cm — var. *major* Roshev.). . . . . . . . . . . . . . . . .2. **PH. CONCINNA** (TH. FRIES) LINDEB.

    **1. *Phippsia algida*** (Soland.) R. Br. in Suppl. to App. Parry's Voyage XI (1824), 285; Trautvetter, Fl. taim. 18; Gelert, Fl. arct. 1, 101; M. Porsild, Fl. Disko 39; Nevskiy in Fl. SSSR II, 447; Devold & Scholander, Fl. pl. SE Greenl. 143; Seidenfaden & Sørensen, Vasc. pl. NE Greenl. 94; Scholander, Vasc. pl. Svalb. 79; Hanssen & Lid, Fl. pl. Franz Josef L. 34; Hultén, Fl. Al. II, 143; id., Atlas, map 227; Rozhevits in Areal I, 22; Tikhomirov, Fl. Zap. Taym. 25; Chernov in Fl. Murm. I, 218; A. E. Porsild, Ill. Fl. Arct. Arch. 24; Karavayev, Konsp. fl. Yak. 56; Böcher & al., Grønl. Fl. 187; Polunin, Circump. arct. fl. 59.

    *Agrostis algida* Soland., Phipps' Voy. (1774) 200.

    *Catabrosa algida* Th. Fries, Nov. Fl. Suec. Mant. III (1842), 173; Kjellman, Phanerogam. sib. Nordk. 117; Lange, Consp. fl. groenl. 166; Scheutz, Pl. jeniss. 187; Simmons, Survey Phytogeogr. 47; Lynge, Vasc. pl. N. Z. 103; Tolmachev, Fl. Kolg. 14, 40; id., Fl. Taym. I, 94; id., Obz. fl. N. Z. 149; Tolmachev & Pyatkov, Obz. rast. Diksona 154; Perfilev, Fl. Sev. I, 86.

    *Phippsia foliosa* V. Vassil. in Bot. mat. Gerb. Bot. inst. AN SSSR VIII, 5 (1940), 70.

    **Ill.:** Fl. arct. 1, fig. 78; A. E. Porsild, l. c., fig. 3, h; Böcher & al., l. c., fig. 49, L; Polunin, l. c. 58.

Plant very characteristic of the Arctic, especially of high arctic districts. Growing in small tufts or tussocks on poorly vegetated stony, loamy or sandy substrates at sites more or less protected by snow during wintertime. Often abundantly colonizing sites of persistent snowfields, where despite the shortness of the growing season it succeeds in completing all stages of development and produces ripe seeds. Sometimes becoming sterile at sites of the longest persistence of snowfields. Where the growing season has greater duration, it sets seed already by midsummer. Around colonies of nesting birds and near the settlements and temporary

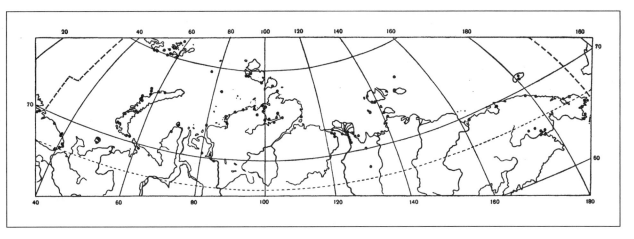

MAP II–57  Distribution of *Phippsia algida* (Soland.) R. Br.

camps of man, it shows itself as nitrophilic and in favourable areas sometimes forms larger tufts with copious bright green foliage and stems reaching 10–15 cm long. It also does not avoid saline areas near sea shores, where it can grow together with *Puccinellia phryganodes, Carex subspathacea* and other halophytic species.

Normally characterized by pale, often rosy colouration of the leaves and stems. Panicles usually pale, but sometimes becoming darker on account of the lilac colour of the lemmas and paleas.

*Phippsia foliosa* V. Vassil., described from the district of the mouth of the Anadyr, represents *Ph. algida* growing in rock fissures where the plant has assumed a rather extenuate form due to developing under somewhat shaded conditions: particularly striking is the extenuation and paleness of the lower parts of the stems. In this sense V. N. Vasilev's statement (Bot. mat. Gerb. Bot. inst. AN SSSR VIII, 72) that "*Phippsia foliosa* mihi differs rather well from both species" of the genus *Phippsia* is correct. The peculiarity of the plants described by him in comparison with individuals of both *Phippsia* species grown under normal conditions is striking. At the same time it is obviously impossible to treat the differences of "*Ph. foliosa*" even as varietal characters, because they were caused by unusual habitat conditions.

Apparently genetically linked with *Ph. algida* are plants described under the name *Puccinellia vaccillans* (Th. Fries) Scholand. (Vasc. pl. Svalb. 95), known mainly from Spitsbergen and reported by Scholander for one point on the NW coast of the North Island of Novaya Zemlya. According to the data presented by Scholander, it seems likely that we are dealing with an intergeneric hybrid *Ph. algida* × *Puccinellia Vahliana*.

**Soviet Arctic.** Murman; Kolguyev; Pay-Khoy; all Novaya Zemlya (more frequent in the north); Franz Josef Land; Northern Yamal; Gydanskaya Tundra; lower reaches of the Yenisey (only in the purely tundra part); Central and Northern Taymyr; Severnaya Zemlya; lower reaches of Olenek and Lena; New Siberian Islands; Bennet Island; coast from the Kolyma to the Bering Strait; Wrangel Island; Beringian coast of Chukotka Peninsula; eastern part of Anadyr Basin. (Map II–57).

**Foreign Arctic.** Arctic Alaska; arctic coast of Canada; Labrador; entire Canadian Arctic Archipelago; all Greenland (from southern extremity to Peary Land); Spitsbergen; Arctic Scandinavia.

**Outside the Arctic.** Mountains of Scandinavia; northern part of Verkhoyansk Range; mountains of Southern Alaska; Rocky Mountains within the state of Colorado.

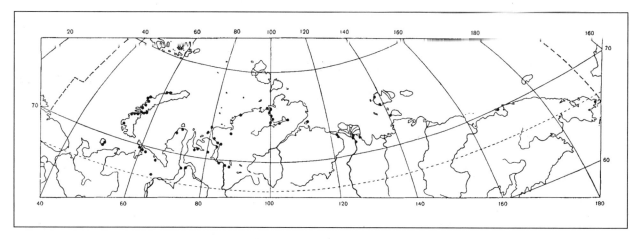

MAP II–58  Distribution of *Phippsia concinna* (Th. Fries) Lindeb.

2. ***Phippsia concinna*** (Th. Fries) Lindeb., Bot. Not. (1898) 155; Scholander, Vasc. pl. Svalb. 80; Hanssen & Lid, Fl. pl. . Franz Josef L. 34; Nevskiy in Fl. SSSR II, 447; Hultén, Atlas, map 228; Tikhomirov, Fl. Zap. Taym. 25; Rozhevits in Areal I, 22; Karavayev, Konsp. fl. Yak. 56; Polunin, Circump. arct. fl. 59.

*Catabrosa concinna* Th. Fries in Öf. Vet.–Ak. Förh. XXVI (1869), 140; Kjellman, Phanerog. sib. Nordk. 117; Scheutz, Pl. jeniss. 187; Gelert, Fl. arct. 1, 116; Lynge, Vasc. pl. N. Z. 105; Tolmachev, Fl. Kolg. 40; id., Fl. Taym. I, 95; id., Obz. fl. N. Z. 149; Perfilev, Fl. Sev. I, 86.

*Phippsia algida* var. *concinna* Richt. — Krylov, Fl. Zap. Sib. II, 198.

*Catabrosa algida* auct. non Th. Fries — Schmidt, Fl. jeniss. 128.

*C. concinna* var. *algidiformis* H. Smith — Lynge, Vasc. pl. N. Z. 105; Tolmachev, Fl. Taym. I, 95.

Ill.: Polunin, l. c. 60.

Common, often abundant plant of slopes without closed turf which are deeply snow covered in wintertime. Growing in small tussocks, with obliquely ascending stems projecting above the foliage and bearing loose divaricate panicles. Sometimes (but more rarely than *Ph. algida*) occurring in slightly saline areas on the sea shore, as well as on alluvial peaty substrate in well-drained tundra hollows. Normally characterized by rather intense reddish colouration and often by dark greyish-lilac panicles due to the intense colouration of the lemmas and paleas.

At sites with looser and more euthermal loamy or sandy-loamy substrates apparently richer in nutritive substances, mainly in the southern part of the range, sometimes forming large tussocks with culms up to 20–26 cm high and loose panicles up to 10–12 cm long. The leaves of this form, f. *major* Roshev., are broader than in typical *Ph. concinna*.

The form described under the name ssp. *algidiformis* H. Smith, which occurs in the same habitats as typical *Ph. concinna* and possesses characters to some extent intermediate between it and *Ph. algida*, is probably of hybrid origin (*Ph. algida* × *Ph. concinna*). It is reported only at sites where the ranges of the two species of *Phippsia* overlap.

**Soviet Arctic.** Kolguyev; northern fringe of Bolshezemelskaya Tundra; Polar Ural; Pay-Khoy; Vaygach; Novaya Zemlya (except the most northern part); Yamal; Obsko-Tazovskiy Peninsula; Gydanskaya Tundra; lower reaches of Yenisey; Taymyr; lower reaches of Olenek and Lena (rarer than *Ph. algida*!); New Siberian Islands; Ayon Island. There are doubtful records for Lawrence Bay on the Beringian coast of the Chukotka Peninsula. (Map II–58).

**Foreign Arctic.** Spitsbergen.
**Outside the Arctic.** Dovrefjell in the mountains of Norway.

## GENUS 27A  Glyceria R. Br. — MANNA GRASS

RELATIVELY SMALL GENUS (about 40 species), distributed mainly in cold and temperate countries of the Northern Hemisphere but also penetrating many tropical districts of Asia, Africa and Australia. So far not found in the Arctic on the continent of Eurasia. The species most closely approaching arctic boundaries in the Soviet Union are *Glyceria lithuanica* (Gorski) Lindm. (as far as the Prepolar Ural inclusively), as well as *G. spiculosa* (Fr. Schmidt) Roshev. and *G. kamtschatica* Kom. (the latter species apparently not differing from *G. orientalis* Kom.) which range north to the Middle Kolyma. It is quite possible that representatives of this genus will be discovered in the basins of the Anadyr and Penzhina. In North America several species of the genus, *G. borealis* (Nash) Batch., *G. maxima* ssp. *grandis* (S. Wats.) Hult., *G. pulchella* (Nash) K. Schum. and *G. striata* ssp. *stricta* (Scribn.) Hult., are reported by Hultén (Fl. Al. II, 228–232) for the Yukon Basin.

## GENUS 28  Puccinellia Parl. — ALKALI GRASS

PREDOMINANTLY HALOPHILIC GENUS containing about 150 species. Very closely approaching the genus *Poa* and, like it, widely distributed in all extratropical countries both of the Northern and Southern Hemispheres. In contrast with *Poa*, the genus *Puccinellia* represents a considerably more integrated and natural group of species which can scarcely be divided into infrageneric subdivisions of the rank of subgenera or sections. Moreover, the viviparous forms characteristic of many arctic and semidesert species of *Poa* are completely lacking in *Puccinellia*. Despite its apparently maritime origin and associated halophily, the genus *Puccinellia* has a rather broad ecological amplitude, being almost completely absent only in forest and marsh communities. In the mountains of Central Asia and South America, species of alkali grass ascend to elevations of 4500–5000 m a. s. l., that is almost to the upper limits of existence of vegetation; the arctic species include several which are high arctic. Much work in clarifying the species composition of the alkali grass flora of the USSR was undertaken by V. I. Krechetovich (Fl. SSSR II, 460–494), who proposed more than 20 new species. Recently the arctic species of the genus have been intensively studied by Sørensen, who described 13 new species from arctic districts alone, of which three were from the territory of the USSR. In his basic work on the genus *Puccinellia*, Sørensen (A Revision of the Greenland Species of Puccinellia Parl., 1953) used characters of the anatomical structure of the leaf epidermis, which in his opinion are good diagnostic characters. We have been unable to verify the constancy of these characters, and their successful use in delimiting closely related species seems to us scarcely possible due to their high dependence on the ecological conditions of the habitat. Also highly variable dependent upon habitat conditions are such characters widely used by Sørensen as the length and thickness of the spikelet pedicels, the length and thickness of segments of the spikelet rachis, the opposite or alternate position of the glumes, as well as differences in the degree of thickness and firmness of the glumes (a character generally usable only if comparative material is available). At the same time, such characters as the size of the anthers and the structure of the margin of the glumes are very substantive and the most stable. Particularly important is accurate measurement of the length of the anthers (this character given in our key for all species), which in alkali grass can almost always be found even in panicles with fruiting spikelets.

Of the 20 species of the genus recorded by us for the Soviet Arctic, the most high-arctic are *P. angustata, P. Vahliana* and *P. colpodioides*. Also included among the species characteristic of the Arctic are *P. phryganodes, P. tenella* and *P. Gorodkovii*. Alkali grasses play an especially prominent role in various types of littoral comunities, where they occur even on banks regularly flooded by tides. Among the arctic species, *P. phryganodes, P. maritima, P. geniculata, P. tenella, P. alascana, P. capillaris, P. coarctata* and *P. vaginata* together with *Calamagrostis deschampsioides* and other species of similar ecology form special vegetational communities of maritime banks and rocks.

1. Spikelets 4–7 mm long, aggregated in more or less condensed panicle 2–6 cm long with completely smooth branchlets; glumes *differing little in length, rather large, almost equalling lemmas or no less than ⅔ as long as them,* often irregularly serrate; segments of spikelet rachis thin, but much thickened at joints; lemmas 3–4.5 mm long, hyaline on distal quarter to third, more or less rosy-violet, with conspicuous, often more or less raised nerves, on basal quarter to third copiously and rather long haired on callus and nerves (usually also between nerves), like the glumes nonciliate on margins and in dry state usually with more or less evident longitudinal folds; paleas *rather copiously hairy on keels, on distal third with a few acicules*; anthers 0.8–1.3 mm long. Small (5–20 cm high) plant of stony tundras of Novaya Zemlya and Taymyr. .................................
................................6. **P. VAHLIANA** (LIEBM.) SCRIBN. ET MERR.
— Glumes *strongly differing in length, the lower of them always less than ⅔ as long as lemma of lowest floret of spikelet;* segments of spikelet rachis usually slightly and more gradually thickened at tips; paleas *scabrous with acicules on distal half and more or less hairy or scabrous basally, or sometimes completely bare and smooth.* ........................................2.

2. Anthers usually linear, *1.2–2.5 mm long;* panicle branchlets smooth or with scattered acicules distally. .......................................3.
— Anthers from oval to oblong-linear, *usually less than 1 mm long, rarely about 1 mm;* panicle branchlets smooth or more or less scabrous with acicules. ................................................................9.

3. Relatively loosely tufted plant of maritime banks more or less flooded by tides, without usual abbreviated vegetative shoots but *possessing strongly*

*elongated creeping vegetative shoots which root at the nodes.* Lemmas dull rosy-violet or greenish *with very narrow hyaline margin, bare or slightly hairy at base of nerves.* . . . . . . . . . . . . . . . . . . . . . . . . . . . . . . . . . . . . . . . . . .4.

— Usually densely tufted plant *without elongated creeping vegetative shoots.* Lemmas more or less rosy-violet with *broad hyaline margin* (equal to about ¼-⅓ of lemma length), *rather copiously hairy near base of nerves.* .6.

4. Panicles 3–10 cm long, with branchlets smooth or slightly scabrous distally; spikelets 7–10 mm long; lemmas 2.8–4.5 mm long, *more or less hairy at base of nerves or on callus;* keels of paleas *scabrous with acicules* which are sometimes modified into hairs on their basal part; anthers 1.5–2.5 mm long. Elongated vegetative shoots almost always with only axillary lateral shoots situated in axils of leaf sheaths. . . . . . . . .1. **P. MARITIMA** (HUDS.) PARL.

— Panicles with completely smooth branchlets; spikelets 4.5–8 mm long; lemmas 2.3–4 mm long, *bare or rarely with isolated short hairs on callus;* keels of paleas *bare and smooth,* rarely with a few (1–3) acicules distally. . . . . . . . . . . . . . . . . . . . . . . . . . . . . . . . . . . . . . . . . . . . . . . . . . . . . . . . . .5.

5. Panicles *usually numerous,* 3–8 cm long; lemmas 2.5–4 mm long; anthers 1.5–2.5 mm long, with *normally developed pollen.* Lateral branches of elongated vegetative shoots *may arise both from axils of leaf sheaths and from significantly above them* (usually from slightly below base of next leaf sheath). . . . . . . . . . . . . . . . . . . . . . . . . . . . . . .3. **P. GENICULATA** (KRECZ.) HULT.

— Panicles *usually few,* 2–5 cm long, occasionally completely absent; lemmas 2.3–3.5 mm long; anthers 1.2–2 mm long, *with abortive pollen* due to which the caryopses do not develop. Lateral branches of the numerous elongated vegetative shoots *almost always arising significantly above axils of leaf sheaths,* usually slightly below base of next leaf sheath. . . . . . . . . . . . . . . . . . . . . . . . . . . . . . . . . . . . . . . .2. **P. PHRYGANODES** (TRIN.) SCRIBN. ET MERR.

6. Panicles 3–8 cm long, diffuse at flowering time and later, with branchlets more or less scabrous distally; spikelets *numerous,* 4–7 mm long, somewhat shining; lemmas *2–2.8 mm long;* paleas on distal part of keels with very short and usually few acicules, bare or with a few hairs basally; anthers *1.2–1.5 mm long.* Plant 10–40 cm high, with leaf blades more or less scabrous above. . . . . . . . . . . . . . . . . . . . . . . . .12. **P. GORODKOVII** TZVEL.

— Spikelets *relatively few* (1–4 per branchlet), 5–8 mm long; lemmas *3–4.5 mm long,* on basal part rather copiously hairy on callus and nerves (often also between nerves); anthers *1.5–2.5 mm long.* . . . . . . . . . . . . . . . . . . . . . .7.

7. Panicles *broadly diffuse at flowering time and later,* 4–10 cm long, with branchlets more or less scabrous distally; lemmas in dry state *without evident longitudinal folds;* paleas with short acicules distally, *bare or with a few hairs basally.* Plant 25–45 cm high; leaf blades *with a few scattered acicules distally.* Lower reaches of Yenisey. . . . . . . . . . . . . . . . . . . . . . . . . . . . . . . . . . . . . . . . . . . . . . . . . . . . .9. **P. JENISSEIENSIS** (ROSHEV.) TZVEL.

– Panicles *condensed or more rarely diffuse;* lemmas in dry state *often with evident longitudinal folds;* paleas with short acicules distally, *rather copiously hairy basally.* Leaf blades *completely smooth.* Far East. . . . . . . . . . . . .8.

8. Plant 5–15 cm high; ligules of upper culm leaves *0.5–1.3 mm long, obtuse.* Panicles *2–5 cm long, usually more or less condensed,* with completely smooth branchlets. . . . . . . . . . . . . . . . . . . . . . . . . . .8. **P. COLPODIOIDES** Tzvel.
– Plant 15–40 cm high; ligules of upper culm leaves *1.8–4 mm long, acuminate.* Panicles *4–10 cm long, usually more or less diffuse,* with branchlets smooth or sparsely scabrous distally. . . . . . . . . . . . . . . . . . . . . . . . . . . . . . . . . . . . . . . . . . . . . . . . . . . . . . . . . . . .7. **P. WRIGHTII** (SCRIBN. ET MERR.) TZVEL.

9. Tips of glumes and lemmas *nonciliate* on margin (without serrations prolonged as cilia), but may be irregularly serrate or apparently erose; panicle branchlets *smooth or only more or less scabrous distally.* Leaf blades smooth distally, or more rarely slightly scabrous with scattered acicules on upper surface and on margin; sheaths of culm leaves usually closed for basal ¼ - ⅓ of their length. . . . . . . . . . . . . . . . . . . . . . . . . . . . . . . . . . . . . . . . .10.
– Tips of glumes and lemmas *minutely ciliate* on margin, with numerous ciliate serrations conspicuous at high magnification (binocular microscope or powerful lens); panicle branchlets *more or less scabrous, or more rarely* (in *P. coarctata* and *P. borealis*) *smooth.* Leaf blades usually more or less scabrous distally, rarely smooth; sheaths of culm leaves closed for basal ⅐ - ¼ of their length. . . . . . . . . . . . . . . . . . . . . . . . . . . . . . . . . . . . . .13.

10. Spikelets 3–6 mm long, rosy-violet or greenish; lemmas *1.6–2.5 mm long, with conspicuous, often more or less raised nerves, bare* at base *or with a few very short hairs on sides of callus;* keels of paleas *bare and smooth, rarely with a few (1–3) acicules distally;* anthers 0.4–0.7 mm long. Small (5–20 cm high) plant of maritime banks and rocks with culms almost not surpassing vegetative shoots. . . . . . . . . . . . . . . . .4. **P. TENELLA** (LGE.) HOLMB.
– Lemmas *2.4–4.5 mm long, usually* (with exception of *P. alascana*) *with inconspicuous nerves, always rather copiously hairy* at base *on nerves and callus;* keels of paleas *always with more or less numerous acicules or hairs.* . . . . . . . . . . . . . . . . . . . . . . . . . . . . . . . . . . . . . . . . . . . . . . . . . . . . . . . . .11.

11. Plant of maritime rocks and banks, 8–30 cm high, occurring only in Far East. Spikelets 4–7 mm long, *pale green, rarely with faint rosy-violet tinge;* lemmas 2.4–3.5 mm long, *with conspicuous, more or less raised nerves,* on margin with very narrow hyaline border; paleas with rather long and numerous acicules on keels; anthers 0.4–0.7 mm long; panicles 2–10 cm long, condensed or more or less diffuse, with completely smooth branchlets. . . . . . . . . . . . . . . . . . . . . . . . . . . . . . . . .5. **P. ALASCANA** SCRIBN. ET MERR.
– Plant usually of habitats more removed from sea shore (not strictly littoral). Spikelets *usually more or less rosy-violet, rarely greenish;* lemmas *with inconspicuous nerves,* hyaline on upper part for almost ¼ - ⅓ of their length. . . . . . . . . . . . . . . . . . . . . . . . . . . . . . . . . . . . . . . . . . . . . . . . . .12.

12. Panicles 3–12 cm long, *diffuse* at flowering time and later, with branchlets *strongly divergent,* smooth or slightly scabrous distally, the longer of them bearing *4–12 spikelets*; lemmas *2.5–3.5 mm long,* rather copiously hairy basally on nerves and callus; paleas distally with scattered acicules on keels, bare or slightly hairy basally; anthers 0.6–0.8 mm long. Plant 10–40 cm high. . . . . . . . . . . . . . . . . . . . . . . . . . . . . . . . . .11. **P. FRAGILIFLORA** SØRENSEN.

– Panicles 2–8 cm long, *more or less condensed* at flowering time and later, *but occasionally with branchlets somewhat divergent;* branchlets smooth or slightly scabrous distally, the longer of them bearing *1–5 spikelets;* lemmas *2.8–4.5 mm long,* copiously hairy basally; paleas distally with more or less long acicules on keels, basally usually with rather long and rather numerous (more rarely only a few) hairs; anthers 0.7–1 mm long. Plant 5–25 cm high. . . . . . . . . . . . . . . . . . . . .10. **P. ANGUSTATA** (R. BR.) RAND ET REDF.

13. Lemmas of lowest floret in spikelet *2.6–4.5 mm long,* usually narrowly rounded or even acuminate at tip. . . . . . . . . . . . . . . . . . . . . . . . . . . . . . . . . . .14.

– Lemmas of lowest floret in spikelet *1.5–2.6 mm long,* usually broadly rounded or truncate at tip. . . . . . . . . . . . . . . . . . . . . . . . . . . . . . . . . . . . . . . . . .17.

14. Plant *15–40 cm high,* with leaf blades *rather strongly scabrous* on upper surface. Panicles 5–15 cm long, at flowering time and later *more or less diffuse;* their branchlets *scabrous for entire length,* the longer of them *with 5–20 spikelets;* lemmas 2.6–4 mm long, like the glumes with conspicuous ciliate serrations at tip; anthers 0.6–0.8 mm long. . . . . . . . . . . . . . . . . . . . .15.

– Plant *5–25 cm high,* with leaf blades *smooth or weakly scabrous* on upper surface. Panicles 2.5–10 cm long, at flowering time and later *usually condensed or with slightly divergent branchlets;* their branchlets *relatively weakly scabrous* (with scattered acicules), *on basal half often smooth,* the longer of them usually *with 1–5 spikelets.* . . . . . . . . . . . . . . . . . . . . . . . . . . .16.

15. Plant of banks and rocks on sea shore; sheaths of culm leaves closed for basal 1/7 - 1/5 *of their length.* Lemmas 2.6–4 mm long, *herbaceous, usually greenish, with relatively narrow, sometimes rosy-violet hyaline border,* relatively weakly hairy at base. Northern European part of USSR. . . . . . . . . . . . . . . . . . . . . . . . . . . . . . . . . . . . . . . . . . . . . . .14. **P. CAPILLARIS** (LILJEBL.) JANSEN.

– Plant of habitats more removed from sea shore; sheaths of culm leaves closed for basal 1/5 - 1/4 *of their length.* Lemmas 2.6–3.5 mm long, *herbaceous-hyaline, usually dull rosy-violet with brighter broad hyaline border, more rarely greenish,* rather copiously hairy at base on nerves and callus. Northern Siberia. . . . . . . . . . . . . . . . . . . . . . . . . . . . . . .17. **P. SIBIRICA** HOLMB.

16. More or less greyish green plant of maritime banks; sheaths of culm leaves closed for basal 1/6 - 1/4 of their length. Lemmas 2.6–3.5 mm long, on basal part *slightly hairy only at base of nerves and on callus, at tip with conspicuous ciliate serrations*; keels of paleas *with scattered acicules but without hairs;* anthers 0.6–0.8 mm long. . . . . .13. **P. VAGINATA** (LGE.) FERN. ET WEATH.

More or less green plants of habitats more removed from sea shore; sheaths of culm leaves closed for basal ¼ - ⅓ of their length. Lemmas 3–4.5 mm long, on basal part *copiously hairy on nerves and callus and usually also between nerves, at tip with inconspicuous ciliate serrations;* keels of paleas *scabrous with acicules distally, with rather numerous hairs basally;* anthers 0.7–1 mm long. . . . . . . . . . . . . . . . . . . . . . . . . . . . . . . . . . . . . . . . . . . . . . . . . . .10. **P. ANGUSTATA** (R. BR.) RAND ET REDF. (see above, couplet 12).

17. Lemmas *(1.3)1.5–1.8(2) mm long,* slightly hairy at base, broadly rounded at tip; paleas with row of rather long acicules on keels; anthers *0.3–0.5 mm long;* panicles 5–20 cm long, more or less diffuse; their branchlets in twos to sevens per node, long and slender. Inland plant of Siberia, 15–50 cm high. . . . . . . . . . . . . . . . . . . . . . . . . . . . . .20. **P. HAUPTIANA** (KRECZ.) KITAGAWA.
– Lemmas *2–2.6 mm long;* anthers *0.6–0.8 mm long.* . . . . . . . . . . . . . . . . . . .18.

18. Plant of maritime banks and rocks in Northern European part of USSR, *8–30 cm high.* Panicles 3–10 cm long, *usually more or less condensed, more rarely more or less diffuse;* their branchlets arranged usually in ones to threes per node, often with spikelets almost to base, sometimes distinctly abbreviated; lemmas 2.2–2.6 mm long, *slightly short-haired* at base, relatively narrowly rounded at tip; keels of paleas *with relatively few short acicules.* . . . . . . . . . . . . . . . . . . . . . . . . . . . . . . . . . . . . . . . . . . . . . . . . . . . . . . . . . . .19.
– Plant of habitats more removed from sea shore (not strictly littoral), but may also be found on sea shore as introduced weed, *15–50 cm high.* Panicles 5–25 cm long, *at flowering time and later broadly diffuse;* their branchlets in twos to sevens per node, longer, often with numerous spikelets; lemmas 2–2.5 mm long, *rather copiously long-haired* on nerves at base, usually broadly rounded at tip; keels of paleas *with rather numerous acicules usually modified into cilia or hairs basally.* . . . . . . . . . . . . . . .20.

19. Panicle branchlets *scabrous* with acicules *for almost their whole length;* glumes and lemmas on average thinner with broader hyaline border. Leaf blades *with rather numerous acicules* on upper surface. . . . . . . . . . . . . . . . . . . . . . . . . . . . . . . . . . . . . . . . . . . . . . . . . . . . . . . . . . .15. **P. PULVINATA** (KRECZ.) TZVEL.
— Panicle branchlets *smooth or with scattered acicules only on distal part.* Leaf blades on upper surface *with relatively few scattered acicules,* sometimes completely smooth. . . . . . . . . . . . . .16. **P. COARCTATA** FERN. ET WEATH.

20. Panicle branchlets *usually smooth basally* (and sometimes for their whole length); lemmas usually more or less rosy-violet, rarely greenish; paleas *usually hairy basally.* NE Siberia. . . . . . . . . . . . . . . .19. **P. BOREALIS** SWALLEN.
— Panicle branchlets *scabrous for their whole length;* lemmas greenish or rosy-violet; paleas *usually without hairs basally, but sometimes with acicules prolonged as cilia.* Northern European part of USSR. . . . . . . . . . . . . . . . . . . . . . . . . . . . . . . . . . . . . . . . . . . . . . . . . . . . . . . . . . . . .18. **P. DISTANS** (L.) PARL.

1. ***Puccinellia maritima*** (Huds.) Parl., Fl. Ital. (1848) 370, pro parte; Hultén, Atlas, map 224; Hylander, Nord. Kärlväxtfl. I, 268; Sørensen, Revis. Greenl. sp. Puccinellia 61; Polunin, Circump. arct. fl. 73.

   *Poa maritima* Huds., Fl. Angl. (1762) 35.

   *P. arenaria* var. γ Trin. in Mém. Ac. Pétersb., sér. 6, 1 (1830), 390.

   *Glyceria maritima* (Huds.) Wahlb., Fl. Gothob. (1820) 17; Lange, Consp. fl. groenl. 168; Gelert, Fl. arct. 1, 126, pro parte.

   *Atropis maritima* (Huds.) Griseb. in Ledebour, Fl. ross. IV, 389, pro parte; Perfilev, Fl. Sev. I, 95; Krechetovich in Fl. SSSR II, 470; Kuzeneva in Fl. Murm I, 220.

   **Ill.:** Fl. SSSR II, pl. XXXV, fig. 1; Hylander, l. c., fig. 34, a, b; Sørensen, l. c., fig. 52, 75, 76.

   European littoral species, belonging to the relatively primitive and rather isolated species-group called the series "*Littorales* Krecz." by V. I. Krechetovich (Fl. SSSR II, 470) and the "*Maritima* Group" by Sørensen (l. c. 89). All species of this group are strictly littoral halophytes and grow under conditions of regular flooding by sea tides. Among them *P. maritima* is one of the most characteristic plants of Western European maritime saltmarshes and scarcely penetrates the Arctic. The most northern specimens of this species are distinguished by an especially great development of stolonlike vegetative shoots and few comparatively small panicles, in these and other characters (smooth or almost smooth panicle branchlets, almost bare lemmas, etc.) significantly approaching the next species which is a very characteristic arctic plant. Such specimens, sometimes distinguished as the separate form or variety *P. maritima* f. *arenaria* (Fr.) Holmb. (Bot. Notis., 1916, 253), are connected with the usual more southern specimens by a series of gradual transitions. It is very probable that the most northern localities for *P. maritima* (for example, the isolated locality in the north of the Kola Peninsula), which usually appear as intrusions in the range of *P. phryganodes,* are in fact secondary, the result of transport of seeds or vegetative parts of this species with ballast or other cargoes. Thus Sørensen (l. c. 61–62) considered all localities for *P. maritima* in Southern Greenland to be the result of introductions, noting that Greenlandic specimens of this species often occur only in a vegetative state or, if they flower, do so very late and do not succeed in ripening seed. Vegetative specimens of *P. maritima* in the majority of cases are readily distinguished from *P. phryganodes* by the manner of branching of the stolonlike vegetative shoots: their lateral branches arise directly from the axils of the leaf sheaths, not significantly above them (near the base of the next leaf sheath) as in *P. phryganodes.* However, specimens intermediate between the two species with respect to this character occasionally occur (with both axillary and extraaxillary lateral shoots).

   Occurring on maritime banks and in saltmarshes more or less flooded by tides.

   **Soviet Arctic.** Murman [only one locality, apparently of introduced origin (near mouth of River Lumbovka, 25 VIII 1928, No. 905, Zinserling), only vegetative specimens].

   **Foreign Arctic.** Southern Greenland (south of 62°N; probably as an introduction); Iceland.

   **Outside the Arctic.** Atlantic coast of Europe from Northern Norway (south of 70°N) to Portugal, including Great Britain, Ireland and the Faroe Islands; coasts of Baltic Sea to Estonia and Finland; coast of White Sea (Solovetskiye Islands, mouth of Severnaya Dvina, Karelia; in the north recorded by Hultén for the district of Kandalaksha). East coast of North America south from the Gulf of St. Lawrence (only as an introduction); introduced to other countries.

2. ***Puccinellia phryganodes*** (Trin.) Scribn. et Merr., Grass. Al. (1910) 78; Lynge, Vasc. pl. N. Z. 126; Hanssen & Holmboe, Vasc. Pl. Bear Isl. 218; Tolmachev & Pyatkov, Obz. rast. Diksona 155; Scholander, Vasc. pl. Svalb. 95; Hultén, Fl. Al. II, 236; Tikhomirov, Fl. Zap. Taym. 26; Hultén, Atlas, map 225; Hylander, Nord. Kärlväxtfl.

I, 269; Sørensen, Revis. Greenl. sp. Puccinellia 51, 83; A. E. Porsild, Ill. Fl. Arct. Arch. 38; Polunin, Circump. arct. fl. 73.

*P. vilfoidea* (Anders.) A. et D. Löve in Bot. Notis. CXIV, 1 (1961), 35.

*P. vilfoidea* ssp. *asiatica* Hadač et A. Löve in A. et D. Löve, l. c. 36.

*Poa phryganodes* Trin. in Mém. Ac. Pétersb., sér. 6, 1 (1830), 389.

*Atropis angustata* auct. non R. Br. — Grisebach in Ledebour, Fl. ross. IV, 390. pro parte.

*A. vilfoidea* (Anders.) Richt., Pl. Eur. 1 (1890), 92; Perfilev, Fl. Sev. I, 95.

*A. maritima* var. *vilfoidea* (Anders.) Fedtsch. et Fler., Fl. Yevr. Ross (1910) 127.

*A. phryganodes* (Trin.) Steffen in Beih. Bot. Centralbl. XLIV, 2 (1928), 330; Krechetovich in Fl. SSSR II, 470; Kuzeneva in Fl. Murm. I, 220.

*Catabrosa vilfoidea* Anders. in Öfv. Vet. Akad. Förh. XIX (1862) 254.

*Glyceria vilfoidea* (Anders.) Fries in Öfv. Vet. Akad. Förh. XXVI (1869), 139; Lange, Consp. fl. groenl. 170; Holm, Contr. morph. syn. geogr. distr. arct. pl. 10, b.

*G. reptans* Krok in Bot. Notis. (1899) 140.

*G. maritima* auct. non Wahlb. — Gelert, Fl. arct. 1, 126, pro parte.

**Ill.:** Fries, l. c., tab. 4; Holm, l. c., fig. B; Hylander, l. c., fig. 31, c; Sørensen, l. c., fig. 44–51, 89–96; Porsild, l. c., fig. 10, c; Polunin, l. c. 68.

Circumpolar arctic littoral species, formed from ancestral forms of the *P. maritima* type as a result of the onset of much colder climate during the Quaternary Period. Characteristic of this species is the complete transition to vegetative reproduction with the aid of the copiously developed stolonlike vegetative shoots, whose lateral branches are distinctly thickened and can easily break off and give rise to new individuals. Panicles are formed relatively rarely and, according to the investigations of a whole series of authors, are always sterile due to constant abortion of the pollen grains. In the vegetative state *P. phryganodes* is easily overlooked by collectors, and this is probably to a considerable degree the reason for the small number of localities at present known for this species in the territory of the USSR. Thus, in the sole portion of the coast of the Barents Sea visited by us in the summer of 1953 near the settlement of Dalniye Zelentsy, this species was found in abundance on banks in Yarnyshnaya Bay, where it formed dense carpetlike lawns with its interlaced stolonlike shoots always below the high tide mark and extending seawards below all other littoral species. Only on more elevated portions of the banks was it associated with isolated specimens of such species as *Carex subspathacea, Puccinellia capillaris, Triglochin maritima* and *Plantago Schrenkii*. With much difficulty we succeeded in finding here some small panicles of *P. phryganodes* with abortive anthers. According to Sørensen's (l. c. 52–60) carefully conducted investigations, *P. phryganodes* does not remain constant throughout its extensive range, but is divisible into at least four weakly marked geographical races. Specimens from Spitsbergen and Novaya Zemlya (Sørensen's "Spitsbergen type") have the most reduced spikelets with relatively small glumes and lemmas. Thence a vicariant sequence of three other races runs in an eastward direction: the "Fennoscandian type" (north coast of Eurasia from Norway to East Siberia), the "Beringian type" (Chukotka Peninsula and Alaska), and the "Greenland type" (Greenland, Canadian Arctic Archipelago and north coast of America east of Alaska). According to Sørensen, there are rather substantial anatomical differences between these four races. It is interesting that the Spitsbergen and Greenland types are the most different from one another, from which Sørensen drew the conclusion that there was no direct connection between them in the past. Very recently Sørensen's Spitsbergen type has been named as the separate species *Puccinellia vilfoidea* (Anders.) A. et D. Löve, with which the Fennoscandian type has been united as the distinct subspecies "ssp. *asiatica* Hadač et A. Löve." However, the material of *P. phryganodes* available to us, of which admittedly only a few specimens possess developed panicles, does not demonstrate such differences between

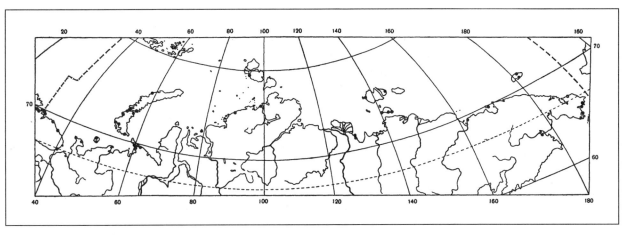

MAP II–59  Distribution of *Puccinellia phryganodes* (Trin.) Scribn. et Merr.

Asian and American specimens that would permit us to refer them to two separate species. Apparently it is more correct to accept Sørensen's types as subspecies of equal status, subordinate to the single species *P. phryganodes*. In this case the subspecies of the Soviet Arctic can, according to Sørensen, be distinguished as follows.

1. Spikelets with 2–4 florets; glumes relatively small, the upper 1.5–2 mm long, with obscure lateral nerves evident only near their base; lemmas 2.5–3 mm long; keels of paleas without papillae. Epidermis of upper surface of culm leaves smooth or with scarcely noticeable papillae, that of leaves of vegetative shoots also smooth without tubercles. . . . . . . . . . . . . . . . . . . . . . . . . . . . . . . . . . . .
   **P. PHRYGANODES** SSP. **VILFOIDEA** (ANDERS.) TZVEL., comb. nova. = *Catabrosa vilfoidea* Anders., l. c. = Spitsbergen type.
– Spikelets with 2–6 florets; glumes larger, the upper 2–3 mm long, with 3 distinct nerves; lemmas usually 3–4 mm long; keels of paleas with minute papillae. Epidermis of upper surface of leaves with minute papillae or tubercles. . . .
   . . . . . . . . . . . . . . . . . . . . . . . . . . . . . . . . . . . . . . . . . . . . . . . . . . . . . . . . . . . . . . . . . . . . .2.

2. Epidermal tissue somewhat irregularly meshed; its cells tuberculately inflated on outer surface but without or with scarcely noticeable papillae. Spikelets with 3–6 florets; lemmas usually obtuse or obtusish, with relatively broad whitish hyaline border. . .**P. PHRYGANODES** SSP. **PHRYGANODES** = Beringian type.
– Epidermal tissue regularly meshed; its cells not inflated on outer surface but with well expressed papillae. Spikelets with 2–4 florets; lemmas usually more or less acuminate, with very narrow hyaline border. . . . . . . . . . . . . . . . . . . . . . . . .
   **P. PHRYGANODES** SSP. **ASIATICA** (HADAČ ET A. LÖVE) TZVEL., comb. nov. = *P. vilfoidea* ssp. *asiatica* Hadač et A. Löve, l. c. (this subspecies described from Vaygach Island: Sinus Warnek, 14 VIII 1907, Ekstam) = Fennoscandian type.

Like the preceding species, *P. phryganodes* occurs on sandy, clayey or stony banks periodically flooded by tides, sometimes in the mouths of larger rivers.

**Soviet Arctic.** Murman (near settlement of Dalniye Zelentsy, 30 VII 1953, No. 188, Tsvelev; recorded by Hultén for the Rybachiy Peninsula, the vicinity of Murmansk and near the settlement of Varzino); Kolguyev; northern part of Kanin (partly specimens transitional to *P. maritima*); Yugorskiy Peninsula (at NW entrance to Belkovskiy Shar, 13 VIII 1934, Tolmachev); Karsk-Baydaratsk coast (at Amderma settlement); Vaygach Island; Novaya Zemlya [South Island and district of Matochkin Shar, only vegetative specimens; recorded by Sørensen also for the

North Island (Arkhangelskaya Bay)]; islands in Bay of Yenisey; Taymyr (Cape Sterlegov and mouth of Lower Taymyra, only vegetative specimens); Severnaya Zemlya; coast of Buorkhaya Bay (Tiksi, etc.); New Siberian Islands; coast of East Siberian Sea near the setlement of Medvezhiy; Ayon Island; Wrangel Island; Lawrence Bay; mouth of Anadyr. (Map II–59).

**Foreign Arctic.** Arctic Alaska; arctic coast of Canada; Labrador (south to 52°N); Canadian Arctic Archipelago; entire coast of Greenland; Spitsbergen; Arctic Scandinavia (north of 69°N).

**Outside the Arctic.** Only a few localities in Northern Europe known: coast of Gulf of Bothnia in Finland (about 65°N), vicinity of town of Kandalaksha (according to Hultén), and Sosnovets Island on the SE coast of the Kola Peninsula; Kuzova Islands and vicinity of Sumskiy Posad settlement in Karelia (only vegetative specimens transitional to *P. maritima*); in North America occurring along the entire coast of Hudson Bay.

3. ***Puccinellia geniculata*** (Krecz.) Hult., Fl. Al. X, Suppl. (1950) 1917; Sørensen, Revis. Grœnl. sp. Puccinellia 58; Krechetovich in Fl. SSSR II, 471, in synonymy.
*Atropis angustata* auct. — Grisebach in Ledebour, Fl. ross. IV, 390, pro parte.
*A. geniculata* Krecz. in Fl. SSSR II (1934), 471, 758.
*Poa geniculata* Turcz. ex Krecz. in Fl. SSSR II, 471, 758, in synonymy.

East Siberian littoral species, extremely close to *P. phryganodes* and completely replacing that species on the coast of the Sea of Okhotsk. In distinction from *P. phryganodes*, *P. geniculata* possesses according to Sørensen (l. c.) anthers with normally developed pollen, and there are in fact copiously fruiting specimens among the material of this species available to us. Furthermore, *P. geniculata* possesses relatively less numerous stolonlike shoots, whose lateral branches often arise from the axils of the leaf sheaths as in *P. maritima*. Among the subspecies of *P. phryganodes*, the closest approach to *P. geniculata* is shown by "ssp. *phryganodes*" (Sørensen's Beringian type), specimens of which usually possess rather numerous panicles and better developed anthers than in the other subspecies. Sørensen drew from this the conclusion that the whole complex *P. phryganodes* s. l. is of Beringian origin, considering that the contemporary subspecies of that species were formed in relatively recent times (during or after migrations of an ancestral form of the *P. geniculata* type eastwards and westwards from the Bering Strait on the north coasts of Eurasia and North America). This hypothesis well explains the existence of rather profound morphological differences between the Spitsbergen and Greenlandic subspecies of *P. phryganodes* (Sørensen's Spitsbergen type and Greenland type), but one should not forget the very close relationship of *P. phryganodes* in Northern Europe with *P. maritima*, a still more primitive species than *P. geniculata* and one which, in our opinion, is no less qualified to be recognized as the ancestor of *P. phryganodes*. Moreover, Sørensen's hypothesis is also weakened by the complete absence of species close to *P. geniculata* south of the Sea of Okhotsk and on Kamchatka, as well as on the west coast of North America, whereas *P. maritima* in Southern Europe is very closely approached by such Mediterranean maritime species as *P. iberica* (Wolley-Dod) Tzvel. comb. nova (= *Atropis iberica* Wolley-Dod in Journ. Bot. London, VII, 1914, 14) and *P. festuciformis* (Host) Parl. It is even quite probable that the penetration of the districts of the Bering and Okhotsk Seas by *P. phryganodes* s. l. took place in relatively recent times, and that *P. geniculata* is a species which has secondarily reverted to generative reproduction.

Like the preceding species, *P. geniculata* grows on maritime banks more or less flooded by tides, often in the mouths of larger rivers.

**Soviet Arctic.** Lower reaches of Penzhina (Ust-Penzhino settlement, 3 VIII 1960, Kildyushevskiy); Koryakia (Cape Olyutorskiy, 6 VIII 1960, Katenin and Shamurin).

**Foreign Arctic.** Absent.

**Outside the Arctic.** Coast of Sea of Okhotsk (Bay of Nikolay, Shantar Islands, vicinities of Ayan, Okhotsk and Magadan); recorded by Sørensen for Alaska (Qiqertariaq, No. 1069, A. E. Porsild).

4. *Puccinellia tenella* (Lge.) Holmb. in M. Porsild, Meddl. om Grønl. LXIII (1926), 45; Karavayev, Konsp. fl. Yak. 57.

*P. paupercula* (Holm) Fern. et Weath. in Rhodora XVIII (1916), 18, pro parte.

*P. laeviuscula* Krecz. in Fl. SSSR II (1934), 483, 761, in synonymy.

*P. pumila* auct. non Hitchc. — Hultén, Fl. Al. II, 237, pro parte.

*P. Langeana* (Berlin) Sørensen in Hultén, Fl. Al. X, Suppl. 1710; id., Revis. Greenl. sp. Puccinellia 20; Tsvelev in Bot. mat. Gerb. Bot. inst. AN SSSR XVI, 51; A. E. Porsild, Ill. Fl. Arct. Arch. 37; Polunin, Circump. arct. fl. 73, pro parte.

*Glyceria tenella* Lge. in Kjellman & Lundström, Vega–Exp. Vetensk. Jaktt. 1 (1882), 313; Gelert, Fl. arct. 1, 129.

*G. Langeana* Berlin in Öfv. Vet.-Akad. Förh. 7 (1884), 79.

*G. paupercula* Holm in Fedde, Repert. Sp. Nov. III (1907), 337.

*Atropis tenella* (Lge.) Simmons, Survey Phytogeogr. (1913) 52; Perfilev, Fl. Sev. I, 96.

*A. laeviuscula* Krecz., l. c. 483, 761.

*A. paupercula* (Holm) Steffen in Beih. Bot. Centralbl. LVII, Abt. B (1937), 386.

**Ill.:** Lange, l. c., tab. 6; Gelert, l. c., fig. 95; Fl. SSSR II, pl. XXXVIII, fig. 17; Sørensen, l. c. (1953), pl. 1, fig. 81, 55–56; A. E. Porsild, l. c., fig. 10, f.

Arctic littoral species, with almost circumpolar distribution but absent from East Greenland and Spitsbergen. Apparently an arctic derivative of the widespread North Pacific species *P. pumila* (Vasey) Hitchc., as is especially confirmed by its predominantly Eastern Arctic distribution. Probably sometimes overlooked on account of its small size and specific habitat requirements, this species has only recently been found at a whole series of points on the arctic coast of Siberia. It was described by Lange (l. c.) from two herbarium sheets of collections by the Swedish polar expedition on the ship "Vega"; one sheet was collected on Novaya Zemlya near the Bay of Rogachev, the other on Cape Greben on Vaygach (In sinu Rogatschew insularum Nova-Semlja et promontorium Grebeni insula Waigatsch, VII 1875, Kjellman et Lundström). Since no type of the species was designated and the first of the herbarium sheets (from Novaya Zemlya) is now lost, the specimens on the second herbarium sheet from Vaygach (Waigatsch, cape Grebeni, 30–31 VII 1875, Kjellman et Lundström) preserved in the Uppsala herbarium in Sweden must be considered the lectotype of *P. tenella*. In our opinion this is the only correct selection of the type of *P. tenella*, and it will remain valid in the event of rediscovery of the lost specimen from Novaya Zemlya. However, Sørensen (l. c., 1953, 20–26, 81–82) on the basis of rather complicated indirect considerations reached the conclusion that the lost specimen from Novaya Zemlya belonged to another species (described by V. I. Krechetovich under the name "*P. pulvinata*"), which therefore should be called "*P. tenella*." On this assumption Sørensen selected a neotype of "*P. tenella*" from Vaygach (Waigatsch, sinus Ljamtschina, 19 VIII 1907, Ekstam), fragments of which were kindly sent to us from Stockholm by Hultén. V. I. Krechetovich did not know specimens of *P. tenella* from further west than the Chukotka Peninsula, so it is not surprising that typical *P. tenella* from the Chukotka Peninsula was described by him as a new species *P. laeviuscula* Krecz., while under the name "*Atropis tenella*" he described another species (*P. coarctata*) in the "Flora of the USSR." We also did not know of specimens of the true *P. tenella* from further west than Dikson, and consequently (l. c., 1954) accepted Krechetovich's interpretation but changed his *P. laeviuscula* to *P. Langeana*. But now that it is known with full certainty that the specimens from Vaygach which Lange had before him belong not to *P. coarctata* and not to "*P. pulvinata*" but to the species known

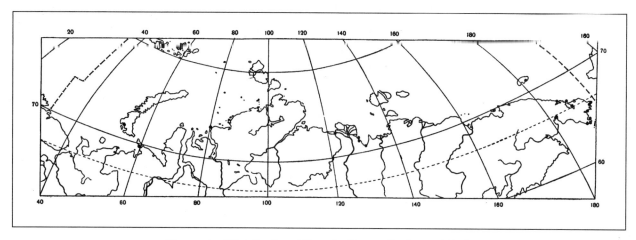

MAP II–60  Distribution of *Puccinellia tenella* (Lge.) Holmb.

under the names "*P. Langeana*," "*P. paupercula*" and "*P. laeviuscula*," Sørensen's arguments in favour of the new interpretation of "*P. tenella*" adopted by him seem to us completely unconvincing. We particularly emphasize the total similarity (in every detail) of the plant illustrated by Lange with *P. tenella* in our sense, but not with Sørensen's neotype. Such characters in Lange's original diagnosis as "ramis laevibus," "spiculis glabris" and "palea inf. anguste albomarginata, leviter erosa" also do not agree with Sørensen's neotype but with *P. tenella* in our sense. It is very significant, for example, that Sørensen's neotype (as generally in *P. capillaris*) has the panicle branchlets strongly scabrous for almost their entire length, while in Lange's diagnosis reference is made to completely smooth panicle branchlets.

Sørensen divides his species *P. Langeana* into three subspecies: *P. Langeana* ssp. *typica* Sørensen (l. c., 1950, 1710), *P. Langeana* ssp. *asiatica* Sørensen (l. c.) and *P. Langeana* ssp. *alascana* (Scrib. et Merr.) Sørensen (l. c.). The last of them in our opinion represents a separate species, *P. alascana*. The remaining two subspecies, whose names should be changed to *P. tenella* ssp. *tenella* (= *P. Langeana* ssp. *asiatica*) and *P. tenella* ssp. *Langeana* (Berlin) Tzvel. comb. nova (= *P. Langeana* ssp. *typica*), are ecological-geographical races with very weak morphological differentiation (ssp. *tenella* being Asian, ssp. *Langeana* American). The differences between them according to Sørensen amount to the following: In ssp. *tenella* the spikelets have a faint rosy-violet tinge or are greenish, and all borne on pedicels; the glumes are usually obtuse; the lemmas are also rather obtuse, usually with a few very short hairs at base, more rarely bare. In ssp. *Langeana* the spikelets are usually rosy-violet, and some on the sides of the branchlets almost sessile; glumes usually acuminate; lemmas rather acute, always bare at base.

The complete ranges of both subspecies are given by Sørensen (l. c., 1953, fig. 114). The first of them penetrates North America only in the north of Alaska, where it occurs together with "ssp. *Langeana*"; the latter subspecies apparently does not cross into Asia at all, since Sørensen's single record of it for Chukotka was based on the type specimens of *P. laeviuscula* Krecz. which belong rather to "ssp. *tenella*" than to "ssp. *Langeana*."

*Puccinellia tenella* is a strictly littoral halophyte, growing on maritime banks and rocks usually more or less flooded by tides.

**Soviet Arctic.** Novaya Zemlya (recorded for the Bay of Rogachev); Vaygach (Cape Greben); Taymyr (Dikson); coast of Olenek Bay (near Stannakh-Khocho settlement) and Buorkhaya Bay (Tiksi); Ayon Island; Wrangel Island; Chukotka Peninsula (Cape Vankarem, Kolyuchin Bay, Kolyuchin Island, Lawrence Bay,

Senyavin Strait, Arakamchechen Island, Provideniye Bay, Bay of Krest). Only "ssp. *tenella.*" (Map II–60).

**Foreign Arctic.** Arctic Alaska (only north of the Bering Strait); shores of Hudson Bay (south to 53°N); Labrador (south to 50°N); Canadian Arctic Archipelago (only Baffin and Devon Islands); West Greenland (only between 65° and 72°N). With the exception of Alaska, where both subspecies occur, only "ssp. *Langeana.*"

**Outside the Arctic.** Occurring only on the southern part of Hudson Bay, in Labrador and on the island of Newfoundland. Only "ssp. *Langeana.*"

5. ***Puccinellia alascana*** Scribn. et Merr. in Contrib. U. S. Nat. Herb. XIII, 3 (1910), 78; Swallen in Journ. Wash. Ac. Sc. XXXIV, 21; Tsvelev in Bot. mat. Gerb. Bot. inst. AN SSSR XVI (1954), 52.

*P. pumila* auct. — Hitchcock in Amer. Journ. Bot. XXI, 129, pro parte; Hultén, Fl. Al. II, 237, pro parte.

*P. Langeana* ssp. *alascana* (Scribn. et Merr.) Sørensen in Hultén, Fl. Al. X, Suppl. 1710; id., Revis. Greenl. sp. Puccinellia 25.

*Atropis alascana* (Scribn. et Merr.) Krecz. in Fl. SSSR II (1934), 480, quoad nomen.

**Ill.:** Sørensen, l. c., 1953, fig. 11.

Littoral Chukotkan-Alaskan species undoubtedly related to the preceding, but readily distinguished from it by the lemmas being rather copiously hairy on their basal part, by the paleas being strongly scabrous on their keels, and by the broader leaf blades. Sørensen (l. c.) united this species with *P. tenella* as a third subspecies, but the differences between these species are incomparably more profound and more constant than those between the subspecies *P. tenella* ssp. *tenella* and *P. tenella* ssp. *Langeana*. The distinctness of *P. alascana* is confirmed by the fact that both *P. alascana* and *P. tenella* were collected simultaneously by B. N. Gorodkov on rocks of Kolyuchin Island without any transitional specimens whatsoever between them. The record of *P. alascana* for Kamchatka (Fl. SSSR II, 480) is erroneous and refers to *P. pumila* (Vasey) Hitchc.

Growing mainly on maritime rocks, more rarely on gravelly or sandy banks on the sea coast.

**Soviet Arctic.** Kolyuchin Island (rocks near bird colony, 25 VII 1938, Gorodkov).

**Foreign Arctic.** West coast of Alaska (near Nome).

**Outside the Arctic.** Southern Alaska (Alaska Peninsula, Nunivak Island); St. Matthew Island, Pribilof Islands; Aleutian Islands.

6. ***Puccinellia Vahliana*** (Liebm.) Scribn. et Merr. in Contrib. U. S. Nat. Herb. XIII, 3 (1910), 78; Lynge, Vasc. pl. N. Z. 126; Scholander, Vasc. pl. Svalb. 103.

*Poa Vahliana* Liebm. in Fl. Dan. XLI (1845), 4.

*Glyceria Vahliana* (Liebm.) Fries in Öfv. Vet.-Akad. Förhandl. XXXVI (1869), 140; Lange, Consp. fl. groenl. 171; Holm, Nov. Zem. Veg. 15; Gelert, Fl. arct. 1, 126.

*G. ? Kjellmanii* Lge. in Kjellman et Lundström, Vega-Exp. Vetensk. Jaktt. 1 (1882), 314.

*Atropis Vahliana* (Liebm.) Richt., Pl. Eur. (1890) 92; Simmons, Survey Phytogeogr. 53; Perfilev, Fl. Sev. I, 96.

*Colpodium Vahlianum* (Liebm.) Nevski in Fl. SSSR II (1934), 436; Sørensen, Revis. Greenl. sp. Puccinellia 18, 72; A. E. Porsild, Ill. Fl. Arct. Arch. 32; Polunin, Circump. arct. fl. 46.

**Ill.:** Liebmann, l. c., tab. 2401; Kjellman & Lundström, Vega-Exp. Vetensk. Jaktt., tab. 7; Sørensen, l. c., fig. 3–5, 53–54; Porsild, l. c., fig. 9, a, b, c; Polunin, l. c. 47.

Amphiatlantic high arctic species, occupying a relatively isolated position in the genus. S. A. Nevskiy (l. c.) transferred it to the genus *Colpodium* Trin., but the critical review we have conducted of that small but very polymorphic genus has shown the complete absence of any clear relationship whatsoever between its species and *P. Vahliana*. At the same time the latter species shows a considerably clearer relationship with such species of alkali grasses as *P. alascana, P. Wrightii, P. angustata* and *P. colpodioides*, and through them with other species of the genus *Puccinellia*. The placement of *P. Vahliana* in the genus *Puccinellia* is also supported by the structure of the lemmas, which in this species have five almost uniformly expressed nerves without a prominent keel (while in *Colpodium* species the lemmas are more or less keeled like in *Poa*, with 3–5 nerves but in the latter case the intermediate nerves are scarcely evident). Furthermore, the keels of the paleas in *P. Vahliana* always have acicules on their distal part, while the genus *Colpodium* is characterized by the complete absence of aciculate trichomes. According to Sørensen's investigations (l. c. 19), this species (despite a certain peculiarity) shows closer similarity to *P. tenella* and *P. alascana* also in the anatomical structure of the leaves than to species of the genus *Colpodium*. It should also be noted that such characters of *P. Vahliana* as the relatively large glumes of almost the same length and the strong thickening of the segments of the spikelet rachis at the joints cannot at all be considered specific characters of *Colpodium*, because that genus also includes species with very small and very unequal glumes and with weakly expressed joints of the spikelet rachis. The other significant peculiarity of *P. Vahliana*, the elongate hilum of the caryopsis (usually equal to about one-third of its length), also cannot be considered a sufficient basis for transferring this species to *Colpodium*, because there are also species in that genus with a small, broadly oval hilum and there are also other species in *Puccinellia* with an elongate, faintly coloured hilum like in *P. Vahliana* (for example, the Central Asian alpine species *P. subspicata* Krecz.). Differences in the degree of membranosity of the lemmas between *P. Vahliana* and other species of alkali grasses (highly membranous lemmas being characteristic of *Colpodium*) are scarcely detectable, and in general this character cannot be considered a good generic character because highly membranous lemmas can appear convergently in representatives of the most diverse genera and species-groups of grasses.

Until recently *P. Vahliana* was not known from Arctic Siberia. The sole record of this species in the literature for Dikson Island (Tolmachev & Pyatkov, Obz. rast. Diksona 155) is very doubtful, because *P. Vahliana* is recorded there as a littoral plant. However, some sheets of *P. Vahliana* have been found by us among the collections of B. A. Tikhomirov and M. I. Vellikaynen from the Lower Taymyra, which considerably extends the range of this species in an eastward direction. It cannot even be excluded that this species, like *P. angustata*, will eventually prove to be circumpolar.

In contrast with many other species of alkali grasses, *P. Vahliana* always occurs at some distance from the sea shore in more or less rocky habitats, on rocky or stony slopes and in rock debris, often near icefields or snowfields. There are indications that the species is a calciphile.

**Soviet Arctic.** Novaya Zemlya (North Island and district of Matochkin Shar, north of 73°N); Taymyr (Lower Taymyra, vicinity of River Bunge, mossy tundra with *Dryas* patches, 11 VIII 1948, Tikhomirov and Vellikaynen).

**Foreign Arctic.** Arctic Canada (south to entrance of Hudson Bay and Southampton Island); Labrador (only NW extremity, Port Burwell); Canadian Arctic Archipelago; NW and East Greenland (north of 69°N); Spitsbergen.

**Outside the Arctic.** Occurring only in Canada in the Mackenzie Mountains (about 63°N).

7. ***Puccinellia Wrightii*** (Scribn. et Merr.) Tzvel. comb. nova.
*Colpodium Wrightii* Scribn. et Merr. in Contrib. U. S. Nat. Herb. XIII, 3 (1910), 74; Hultén, Fl. Al. II, 223; Polunin, Circump. arct. fl. 47.
*Poa Wrightii* (Scribn. et Merr.) Hitchc. in Amer. Journ. Bot. II (1915), 309.

**Ill.:** Polunin, l. c. 47.

Still inadequately studied Beringian species, known so far only from a few sites. Like the preceding species, it was formerly referred to the genus *Colpodium*; but it undoubtedly much more closely approaches species of alkali grasses of Sørensen's "*Langeana* Group" and "*Pumila* Group," and in anther size it is close to such arctic North American species as *P. arctica* (Hook.) Fern. et Weath. and *P. poacea* Sørensen. In floret structure *P. Wrightii* also significantly approaches *P. Vahliana*, but is well distinguished from that species both by the greater size of the plant as a whole and by the significantly larger anthers. In the herbarium of the Botanical Institute of the USSR Academy of Sciences there is only an isotype of this species from Arakamchechen Island (Arakamtchetchene Isl., 1853–1856, Wright), identified as "*Glyceria arctica* Hook." and containing shoots of *P. Wrightii* which are partly somewhat deviant in spikelet structure towards the more southern North Pacific species *P. pumila* (Vasey) Hitchc. [= *P. kurilensis* (Takeda) Honda]. Identical with the isotype are rather numerous specimens of *P. Wrightii* collected by V. A. Gavrilyuk also on the Chukotka Peninsula in the vicinity of Chaplino Hot Springs. At the same site as well as on sand and gravel deposits near the mouth of a stream, some further specimens were collected (3 IX 1956 and 7 VIII 1957, Gavrilyuk) which are externally similar to *P. Wrightii,* but differ from this species in having smaller anthers (1.2–1.5 mm long), paleas almost smooth on their keels, and lemmas without the obvious longitudinal folds characteristic both of *P. Vahliana* and of *P. Wrightii*. These specimens occupy an apparently intermediate position between *P. Wrightii* and the just mentioned *P. pumila* and might readily be assumed to be hybrids between these species, were it not that typical *P. pumila* has so far not been found on the Chukotka Peninsula. The anthers of these specimens are obviously defective and apparently contain a high percentage of abortive pollen, something which supports the possibility of hybrid origin. But for final resolution of this question more abundant material is required.

Growing on stony slopes of riverine and maritime terraces, also on sand and gravel deposits in river mouths and on the sea coast.

**Soviet Arctic.** Chukotka Peninsula (Arakamchechen Island and near Chaplino settlement).

**Foreign Arctic.** West coast of Alaska (only in district of Bering Strait, Nome, Port Clarence, etc.).

**Outside the Arctic.** So far not recorded.

8. ***Puccinellia colpodioides*** Tzvel., sp. nova.—Planta perennis dense caespitosa 5–15 cm alta; culmi glabri et laeves, nodis 1–2; folia parum canescente-viridia; laminae 0.6–1.5 mm lt., vulgo plus minusve convolutae, rarius planae, glabrae et laeves; ligulae 0.5–1.3 mm lg. obtusae glabrae et laeves. Paniculae 2–5 cm lg. plus minusve contractae vel paulo diffusae, ramis vulgo binis laevibus spiculis 1–3 instructis; spiculae 5–7.5 mm lg., 2–4-florae, roseoviolaceae; glumae tenues obtusae non ciliolatae, inferior circa 1.8 mm lg. uninervia, superior circa 2.5 mm lg. trinervia; lemmata 3–4 mm lg., obtusa vel acutiuscula, late hyalino-marginata, apice non ciliolata, obscure 5-nervia, in parte superiore parum carinata, in parte inferiore nervis et callo (saepe et inter nervos) sat copiose pilosa; paleae lemmatis subaequales, carinae superiore tertia parte parce et breviter spinulosae infra plus minusve pilosae interdum subglabrae; antherae 1.5–2.5 mm lg. Grana ignota.

Typus speciei: Sibiria, ins. Wrangelii, prope sinum Rogers, 9 VIII 1938, Gorodkov. In Herb. Inst. Bot. Ac. Sc. URSS (Leningrad) conservatur.

High arctic species, with respect to anther size and a series of other characters showing undoubted affinity with *P. Wrightii* and the two American arctic species *P. arctica* (Hook.) Fern. et Weath. and *P. poacea* Sørensen (of which we have not seen specimens). All these species have significantly greater overall plant size (15–40 cm high) and larger (5–10 cm long) usually diffuse panicles with more or less scabrous (rarely smooth in *P. Wrightii*) branchlets, whereas the panicles of our new species are small and always more or less condensed with completely smooth branchlets. Moreover, the panicle branchlets in *P. arctica* and *P. poacea* (judging from the diagnioses of these species) are arranged in threes to sevens on the lower nodes, while in *P. colpodioides* they are arranged in ones to twos on all nodes. The ligules of the upper culm leaves of *P. colpodioides* are very short (0.5–1.3 mm long) and obtuse, while in *P. Wrightii* they are 1.8–4 mm long and rather acute. Also noteworthy is the great and far from superficial resemblance of *P. colpodioides* to the large-anthered Central Asian alpine species *P. subspicata* Krecz., which differs from *P. colpodioides* only in having completely smooth and bare lemmas and paleas.

*Puccinellia colpodioides* is so far known only from rather numerous specimens from Wrangel Island, where it was first collected in 1938 by B. N. Gorodkov who identified this species as "*Colpodium Wrightii.*" One of the specimens collected by him (Wrangel Island, Polar Station on Rodgers Bay, lichen-moss polar desert on coastal terrace, 9 VIII 1938, Gorodkov) is selected by us as the type of this species.

Judging from B. N. Gorodkov's data, this species grows on relatively dry stony or rocky areas of mossy or moss-lichen tundras, and is apparently a calciphile.

**Soviet Arctic.** Wrangel Island.

Not found in the **Foreign Arctic** or **outside the Arctic.**

9. ***Puccinellia jenisseiensis*** (Roshev.) Tzvel., comb. nova; Krechetovich in Fl. SSSR II (1934), 471, in synonymy.

*Atropis jenisseiensis* Roshev. in Izv. Bot. sada AN SSSR XXX (1932), 300, pro parte; Krechetovich, l. c. 471.

Still inadequately studied species, so far known only from the numerous type specimens (15 of them in the herbarium of the Botanical Institute of the USSR Academy of Sciences) from the Zherevskiy Sands on the Yenisey (Shoreline slopes and bluffs at Zherevskiy Sands on Yenisey, 24–25 VIII 1914, Kuznetsov and Reverdatto). The sole specimen from another locality (Tundra at Cape Selyakinskiy below mouth of River Kheta, 11 VIII 1914, Kuznetsov and Reverdatto) cited by R. Yu. Rozhevits in the original description of this species has proved to belong to *P. sibirica*, a species quite different from *P. jenisseiensis*. Thanks to its large anthers (1.4–2.2 mm long) and peculiar habitus, *P. jenisseiensis* occupies a rather isolated position among the Siberian species of alkali grasses, showing some affinity only with *P. Wrightii* although with respect to a series of other characters the species approaches *P. angustata* to a high degree. The imperfect development of the anthers (all plants being apparently sterile) suggests a hybrid origin of this species; but in the lower reaches of the Yenisey there is no other species of alkali grass possessing anthers more than 1 mm long, so in this case one of the parent species would have to be assumed to be *P. macranthera* Krecz. distributed in saltflats of the southern part of Krasnoyarsk Kray (including the upper reaches of the Yenisey) and Prebaikalia. The possibility of intergeneric hybridization also cannot be excluded in this case.

According to the indication on the label of the type specimens, growing on poorly vegetated slopes on the bank of the Yenisey and in ravines descending into its valley.

**Soviet Arctic.** Lower reaches of Yenisey (at Zherevskiy Sands).

Not found in the **Foreign Arctic** or **outside the Arctic.**

10. ***Puccinellia angustata*** (R. Br.) Rand et Redf., Fl. of Mount Desert Isl. (1894) 181; Lynge, Vasc. pl. N. Z. 125; Hanssen & Lid, Fl. pl. Franz Josef L. 37; Tolmachev, Fl. Taym. I, 98; Scholander, Vasc. pl. Svalb. 94; Tikhomirov, Fl. Zap. Taym. 25;

Sørensen, Revis. Greenl. sp. Puccinellia 28, 77; A. E. Porsild, Ill. Fl. Arct. Arch. 35; Karavayev, Konsp. fl. Yak. 57; Polunin, Circump. arct. fl. 72.

*P. taimyrensis* Roshev. in Bot. mat. Gerb. Bot. inst. AN SSSR XI (1949), 27; Tikhomirov, l. c. 26.

*P. Palibinii* Sørensen, l. c. 74.

*P. contracta* (Lge.) Sørensen, l. c. 77.

*Poa angustata* R. Br. in Suppl. to App. Parry's Voyage XI (1824), 291.

*Glyceria angustata* (R. Br.) Fries, Mant. III (1842), 176; Lange, Consp. fl. groenl. 171; Gelert, Fl. arct. 1, 128.

*G. vaginata* var. *contracta* Lge. in Kjellman, Vega Exp. Vet. Jaktt. 1 (1882), 273.

*G. gracilis* Palib. in Bull. Jard. Bot. Petersb. III (1903), 46, nomen nudum.

*Atropis angustata* (R. Br.) Griseb. in Ledebour, Fl. ross. IV, 390, quoad nomen; Perfilev, Fl. Sev. I, 96; Krechetovich in Fl. SSSR II, 472.

Ill.: Fl. SSSR II, pl. XXXV, fig. 4; Sørensen, l. c., fig. 12, 15–17, 23–25, 61–62; Porsild, l. c., fig. 10, a; Polunin, l. c. 71.

Circumpolar high arctic species, absent however from Wrangel Island, the Chukotka Peninsula and Arctic Alaska. Although material of this species from the Soviet Arctic displays considerable polymorphism, we have not been able to divide it satisfactorily into any units of lesser content and therefore here accept as synonyms of *P. angustata* the species *P. taimyrensis*, *P. Palibinii* and *P. contracta* described by R. Yu. Rozhevits (l. c.) and Sørensen (l. c.). A fragment of an isotype of *P. angustata* present in the herbarium of the Botanical Institute of the USSR Academy of Sciences possesses large relatively dull coloured spikelets on rather long and slender pedicels, lemmas copiously hairy on their basal part, and panicle branchlets scabrous for almost their entire length. Among specimens collected in the Soviet Arctic, many from Taymyr and Severnaya Zemlya are identical with the isotype, while the most different from it are higher arctic specimens from Novaya Zemlya and Franz Josef Land, which possess more brightly coloured spikelets, less copiously hairy lemmas and almost completely smooth panicle branchlets. The latter specimens, which also occur in many other districts both of the Eurasian and American Arctic, were apparently assumed by R. Yu. Rozhevits to be the true *P. angustata*, since specimens described by him as *P. taimyrensis* (Taymyr, at mouth of Lower Taymyra, bare clayey soil of slump, 31 VIII 1946, Tikhomirov) are identical with the fragment of the isotype of *P. angustata*, and the differences of *P. taimyrensis* from *P. angustata* given by him (scabrous panicle branchlets, copiously hairy lemmas) are, to the contrary, very characteristic of the latter species. With respect to Sørensen's species, of which basic material (including the type specimen of *P. contracta*) was kindly sent to us from the herbaria in Stockholm, Lund and Uppsala, we were also unable to find substantial differences between them and *P. angustata*. The specimens of *P. contracta* (Cape Greben on Vaygach, Oleniy Island on Novaya Zemlya, vicinity of Khabarova settlement, Preobrazheniye Island, vicinity of Irkaypiy and Pitlekay settlements; type of species: Sibiria arctica, Pitlekaj, 28 IX 1878, Kjellman, in Uppsala herbarium) differ obviously from specimens of *P. angustata* from the European Arctic and show a very close resemblance both to *P. taimyrensis* and to the fragment of the isotype of *P. angustata*. More distinct from typical specimens of *P. angustata* are specimens of *P. Palibinii* (Novaya Zemlya, Mezhdusharskiy Island, Karmakuly, Bezymyannaya Bay, valleys of Matochkin Shar River, Pomorskoye, Serebryanka Bay, Krestovaya Bay; type of species: Novaya Zemlya, Pomorskaya, 27 VIII 1921, Lynge, in herbarium of City of Oslo, isotype in Leningrad), which possess almost smooth panicle branchlets and glumes virtually nonciliate on margin (in typical *P. angustata* they are slightly ciliate on margin). In many respects these are similar to the specimens from the European High Arctic mentioned above, differing from them only by their looser tufts, softer leaves and laxer panicle, that is only in characters which could be produced by growing under moister conditions (probably the result of transport of seeds of this species from higher and drier areas of tundra). Sørensen considers this species endemic to Novaya Zemlya, but in our opinion, if it is recognized as a separate species despite its being connected with *P. angustata* by numerous transitions, at least the majority of specimens of *P. angustata* from Greenland, Spitsbergen, Novaya Zemlya and Franz Josef Land should be referred to it. Its range in that case would be predominantly amphiatlantic, while the range of *P. angustata* s. str. would extend basically from Vaygach to the Canadian Arctic Archipelago. But for completely accurate delimitation of these two species, considerably more material is required than we presently have available. We think it useful to present here a fragment of the key to the identification of species of *Puccinellia* from Novaya Zemlya and Vaygach from the work of Sørensen cited above.

1. Spikelets all pedicelled; pedicels slender. Glumes thin throughout, translucent. Rachilla slender.

2. Glumes, especially the 1st, tapering toward the base, lanceolate or lanceolate-oblong, entire. Panicle branches glabrous, the pedicels sometimes with a few spinules. Anthers 1.0–1.4 mm long. Culms geniculate. Plant strikingly slender and soft-leaved; loosely caespitose. .................**P. PALIBINII** SØRENSEN.

2a. Glumes dilated at base, ovate-orbicular, strongly erose-ciliolate. Panicle branches scabrous, at least in their distal part. Anthers 0.8–1.0 mm long. Culms erect or prostrate. Moderately robust plant; densely caespitose. ............................................**P. CONTRACTA** (LGE.) SØRENSEN.

1a. Distal spikelets sessile or subsessile; pedicels vigorous. Glumes translucent only toward the margin. Rachilla stout.

3. Pedicels not or slightly thickened. Glumes approximately opposite-inserted, purple, scarious, corroded or ± erose-ciliolate at margin. Panicle contracted. .. ....................................**P. ANGUSTATA** (R. BR.) RAND ET REDF.

3a. Pedicels evidently thickened, claviform. Glumes ± alternate-inserted, herbaceous and greenish or yellowish at base, hyaline-bordered, entire or coarsely dentate not erose-ciliolate. Panicle spreading. .....**P. FRAGILIFLORA** SØRENSEN.

As we have already indicated above (see the general remarks on the genus *Puccinellia*), many of the characters given in this key appear to us insufficiently constant (changing in response to ecological conditions of the habitat) and not at all convenient for practical use. The size of the anthers in *P. Palibinii* is apparently somewhat exaggerated, since in all specimens seen by us they are 0.8–1 mm long (not 1–1.4 mm as stated in the key). With reference to the systematic position of *P. contracta* and *P. Palibinii*, Sørensen considers the first of these species closely related to *P. angustata* although showing certain points of resemblance to *P. Vahliana* (lemmas with some indication of longitudinal folds in the dry state and segments of spikelet rachis much thickened at joints), while *P. Palibinii* occupies an apparently intermediate position between *P. angustata* and *P. fragiliflora*.

*Puccinellia angustata* usually grows in more or less moist clayey or stony areas of tundra away from the sea shore, but occasionally near it. A.E. Porsild (l. c.) also indicates that this species is not a littoral plant, but usually grows in moist clayey sites on tundra.

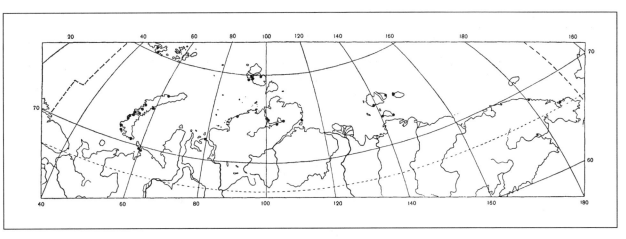

MAP II–61 Distribution of *Puccinellia angustata* (R. Br.) Rand et Redf.

**Soviet Arctic.** Yugorskiy Peninsula (Khabarovo settlement); Vaygach (Cape Greben and Varnek Bay); Novaya Zemlya (Oleniy and Mezhdusharskiy Islands, Karmakuly settlement, Bezymyannaya Bay, district of Matochkin Shar, Serebryanka Bay, Krestovaya Bay, Belushya Bay, Mashigin Bay, East Krestovyy Island, Pakhtusov and Lichutin Islands, Russkaya Gavan); Franz Josef Land; Gydanskaya Tundra (near Cape Leskin, 13 VIII 1926, Tolmachev); Taymyr (basin of Lake Taymyr and north of it); Severnaya Zemlya; Preobrazheniye Island; New Siberian Islands; coast of East Siberian Sea between the Bays of Chaun and Kolyuchin (Irkaypiy and Pitlekay). (Map II–61).

**Foreign Arctic.** Arctic coast of Canada between Alaska and Hudson Bay; Canadian Arctic Archipelago (west to Banks Island, south to the Arctic Circle); Greenland (except southern part, on west coast south to the Arctic Circle, on east coast to 70°N); Spitsbergen. Recorded for Arctic Alaska (Hooker & Arnett, Beechy's voyage, 1841, 132), but apparently erroneously.

**Outside the Arctic.** Not occurring.

11. ***Puccinellia fragiliflora*** Sørensen, Revis. Greenl. sp. Puccinellia (1953) 73.
   *P. sibirica* var. *lenensis* Holmb. in Bot. Notis. (1927) 207.
   *P. Andersonii* auct. non Swall. — Polunin, Circump. arct. fl. (1959) 72, pro parte.
   **Ill.:** Sørensen, l. c., pl. 10, fig. 13–14.

Predominantly Siberian subarctic species, first reported by Holmberg (l. c.) as a variety of *P. sibirica* but well distinguished from that species by the smooth (or almost smooth) panicle branchlets, as well as by the absence of marginal cilia from the tips of the glumes and lemmas. Considerably later it was described by Sørensen (l. c.) as a separate species, *P. fragiliflora*, showing in his opinion closest relationship with *P. Andersonii* Swall., a widespread species in the American Arctic, and appearing to replace it in Eurasia. According to Sørensen, these two species (united by him in the "*Andersonii* Group") share a series of anatomical peculiarities which distinguish them from *P. angustata* (Sørensen's "*Angustata* Group"). Specimens of *P. Andersonii* are completely lacking from the herbarium of the Botanical Institute of the USSR Academy of Sciences, but (judging from data in the literature) there are rather substantial morphological differences between that species and *P. fragiliflora*. At the same time *P. fragiliflora* shows undoubted affinity also with *P. angustata*, differing from that species by the (on average) larger size of the plant as a whole, smaller lemmas, and broadly diffuse panicle branchlets with more numerous spikelets. The two specimens cited by Sørensen in the original description of this species were sent at our request from the herbaria of

Uppsala and Lund: one from NE Siberia (Irkajpij, 12–15 IX 1878, Kjellman) is identical to typical specimens of *P. sibirica* var. *lenensis* Holmb. (the herbarium of the Botanical Institute contains one of the authentic specimens of this variety: ad fl. Lena prope pag. Krestjach, 13 VIII 1898, H. Nilsson-Ehle); the other from Novaya Zemlya (Matotchkin Shar, 13 VIII 1905, Ekstam, type of the species) differs in the dwarfish size of the plant as a whole (about 10 cm high) and was possibly introduced to Novaya Zemlya from another more southern part of the species' range.

Growing on sands and gravelbars in the lower reaches of the larger rivers and near the sea coast (but not on maritime banks), usually in the vicinity of beach ridges and on the slopes of sandy mounds (baydzharakhs).

**Soviet Arctic.** Novaya Zemlya (Matochkin Shar, possibly as an introduction); Taymyr (mouth of River Bunge on the Lower Taymyra, lower reaches of River Yamu-Nera); coast of Olenek Bay and lower reaches of Olenek (north of 72°N); lower reaches of Lena (near mouth of Lena, Tit-Ary Island, near Krestyakh settlement, according to Holmberg also Kumakhsurt and Balaganakh); coast of East Siberian Sea (Irkaypiy).

Not occurring in the **Foreign Arctic** or **outside the Arctic.**

12. *Puccinellia Gorodkovii* Tzvel., sp. nova. — Planta perennis dense caespitosa 10–40 cm alta; culmi glabri et laeves saepe decumbentes nodis 2–3; folia canescente-viridia; laminae 0.7–2.5 mm lt., planae vel laxe convolutae, supra et marginibus scabriusculae, subtus laeves; ligulae 1.5–2.5 mm lg., glabrae et laeves, saepe acutiusculae. Paniculae 3–8 cm lg., late diffusae, ramis vulgo binis gracilibus subflexuosis inferne laevibus superne plus minusve scabriusculis spiculis 2–8 instructis, pedicellis gracilibus et sat longis; spiculae 4–6.5 mm lg., 3–4-florae, plus minusve roseoviolaceae, paulo nitidae; glumae tenues, apice non ciliolatae, inferior circa 1 mm lg. uninervia, superior circa 1.5 mm lg. vulgo trinervia; lemmata 2–2.8 mm lg., obtusa vel acutiuscula, late hyalino-marginata, apice non ciliolata, obscure 5-nervia, nervis prope basin (et callo) pilosis; paleae lemmatis vulgo sublongiores, carinae superiore tertia parte spinulosae infra glabrae et laeves rarius parce pilosae; antherae 1–1.5 mm lg. Grana 1.4–2 mm lg. hilo oblongo-ovali.

Typus speciei: Sibiria, Taimyr, prope ostium fl. Taimyra inferioris, 2 IX 1946, No. 20, Gorodkov. In Herb. Inst. Bot. Ac. Sc. URSS (Leningrad) conservatur.

The species just described is so far known only from the Lower Taymyra, from a single herbarium sheet collected by B. N. Gorodkov and numerous collections (13 sheets) of B. A. Tikhomirov and M. I. Vellikaynen. However, since the samples of the latter collectors only contain late collected specimens (often with the florets completely shed), we have selected as type of this species the single herbarium sheet collected by B. N. Gorodkov which contains well developed spikelets. This species possesses broadly diffuse panicles like the preceding species, but differs from it and from the majority of other arctic species of *Puccinellia* in having significantly larger anthers (1.2–1.5 mm long) while the lemmas are of relatively small size (2–2.8 mm long). In this respect *P. Gorodkovii* seems to approach a species widespread in more southern districts of Yakutia (reaching as far north as Verkhoyansk), which is usually identified as "*P. tenuiflora* (Griseb.) Scribn. et Merr." but is considerably closer and perhaps even identical with *P. chinampoensis* Ohwi (Acta Phytotax. et Geob. IV, 1935, 31) described from Korea and China.

Growing on relatively dry sandy or stony slopes with more or less closed turf on mounds in river mouths and near the sea coast.

**Soviet Arctic.** Taymyr (mouth of Lower Taymyra, left bank, polar desert on summit of mound on shore of bay, 2 IX 1946, No. 20, Gorodkov, type of the species; Lower Taymyra, Granite Point, south slope at foot of terrace, 2 IX 1948, Tikhomirov and Vellikaynen).

Not occurring in the **Foreign Arctic** or **outside the Arctic.**

13. **Puccinellia vaginata** (Lge.) Fern. et Weath. in Rhodora XVIII (1916), 14; Sørensen, Revis. Greenl. sp. Puccinellia 46; A. E. Porsild, Ill. Fl. Arct. Arch. 38; Polunin, Circump. arct. fl. 73.
*Glyceria vaginata* Lge., Fl. Dan., Fasc. 44 (1858), tab. 2583; id., Consp. fl. groenl. 168.
Ill.: Lange, l. c. (1858), tab. 2583; Sørensen, l. c., fig. 41, 87–88.

This arctic littoral species of Greenland and North America was until recently not recorded at all in Eurasia, but we have found a specimen from East Siberia (Yakutia, coast of Olenek Bay near Stannakh-Khocho settlement, gravelbar in salt-marsh, 25 VII 1956, Tolmachev, Polozova and Yurtsev), which is identical with three specimens of *P. vaginata* from Greenland present in the herbarium of the Botanical Institute of the USSR Academy of Sciences (Godhavn, 1905, G. Kleist; Nordfiord, Nordre Kagsimavit, 31 VII 1902, M. Porsild; Tasiussaq, 4 VIII, Wulff). Duplicates of the latter specimens held in Scandinavian herbaria were cited by Sørensen (l. c. 123–124) as belonging to *P. vaginata,* so the correctness of their identification cannot be doubted. The new locality for *P. vaginata* can scarcely be attributed to introduction, and it cannot be excluded that the species may be more widespread on maritime banks of Arctic Siberia and perhaps here replaces the European littoral species *P. capillaris* and *P. pulvinata*. In external appearance *P. vaginata* significantly approaches *P. angustata,* but is readily distinguished from that species by the tips of the glumes and lemmas having a strongly ciliate margin, the considerably weaker pubescence of the lemmas, and several other characters which show that it is no less closely related to *P. capillaris.* Sørensen (l. c. 89) refers this species to a separate monotypic group ("*Vaginata* Group").

Growing on sandy and gravelly banks on the sea coast, but usually above high tide mark.

**Soviet Arctic.** Coast of Olenek Bay.

**Foreign Arctic.** Arctic Canada between Alaska and Hudson Bay (especially abundantly in the district of the Mackenzie Delta); Labrador (only NE coast); Canadian Arctic Archipelago (known so far only from Baffin, Bylot and Southampton Islands, absent from a considerable part of the archipelago); West Greenland (from 68° to 79°N); East Greenland (from 70° to 75°N).

**Outside the Arctic.** Coast of Hudson Bay (isolated locality near mouth of Churchill River, about 59°N).

14. **Puccinellia capillaris** (Liljebl.) Jansen in Fl. Neerl. 1, 2 (1951), 69, in adnot.; Tsvelev in Sp. rast. Gerb. Fl. SSSR XIV (1957), 79.
*P. suecica* (Holmb.) Holmb. in Bot. Notis. (1916) 254.
*P. retroflexa* auct. — Holmberg in Lindman, Svensk Fanerogamfl. (1918) 97; Hultén, Atlas, map 226; Hylander, Nord. Kärlväxtfl. I, 266, pro parte; non *Poa retroflexa* Curt., 1777.
*Festuca capillaris* Liljebl., Utk. Sv. Fl., Ed. 2 (1798), 48.
*Glyceria intermedia* Klinggr., Fl. v. Preuss. (1848) 491.
*Atropis suecica* Holmb. in Bot. Notis. (1908) 245; Krechetovich in Fl. SSSR II, 478; Kuzeneva in Fl. Murm. I, 221.
*A. distans* var. *capillaris* (Liljebl.) Hack. in Kneucker, Gram. Exs. Lief. XXXI, 916.
*A. convoluta* auct. non Griseb. — Perfilev, Fl. Sev. I, 96.
Ill.: Fl. SSSR II, pl. XXXVI, fig. 10.

Northern European littoral species, perhaps representing a northern littoral race of the widespread *P. distans* but, on the other hand, still more closely approaching such arctic littoral species as *P. vaginata* and *P. coarctata.* In Western European (mainly Scandinavian) literature this species was until recently known under the name "*Puccinellia retroflexa* (Curt.) Holmb.," based on "*Poa retroflexa* Curt." (Fl. Londin., 1777, tab. 45). But "*Poa retroflexa* Curt." was not the name of a newly described species but merely an illegitimate new name for *Poa distans* L., and is consequently a synonym of the latter species without any qualifications. Holmberg (l. c., 1918, 97) considered this name a prior name for *P. capillaris* only on the basis of a weak resemblance of the illustration of *Poa retroflexa* to that species (in the figure the lemmas appear longer and more acute than in *P. distans*). This reasoning cannot be considered at all well founded; details of the spikelet structure can scarcely have been represented accurately by the artist in the general illustration of the plant when the enlarged spikelet inserted in the same plate (assumed by Holmberg to belong to a different plant) as well as the description provided refer to *P. distans* rather than to *P. capillaris*. *Puccinellia capillaris* is basically distributed outside the Arctic, and the localities for it situated within arctic limits are to a considerable degree the result of introductions. In North America this species is apparently replaced by closely related species (*P. coarctata, P. vaginata*, etc.), but occasionally occurs as an introduction.

Growing on sandy or gravelly maritime banks, but usually above high tide mark. Readily dispersed with ballast.

**Soviet Arctic.** Murman (Murmansk, Iokanga, Dalniye Zelentsy, Lumbovka); Kanin (southern part); Timanskaya Tundra (near mouth of Indiga); Karskaya Tundra (River Kara near mouth of River Bolshaya Vanuyta, 22 VIII 1909, No. 662, Sukachev; probably introduced specimens, somewhat resembling *P. sibirica*).

**Foreign Arctic.** Arctic Scandinavia (rare).

**Outside the Arctic.** Coasts of the White and Baltic Seas; east coasts of Norwegian and North Seas. In Great Britain and Holland apparently only as an introduction; occasionally introduced in many other countries of Europe and North America.

15. ***Puccinellia pulvinata*** (Krecz.) Tzvel., comb. nova; Krechetovich in Fl. SSSR II (1934), 478, in synonymy.

*P. retroflexa* var. *pulvinata* (Fr.) Holmb. in Lindman, Svensk Fanerogamfl. (1918) 97, pro parte.

*P. retroflexa* auct. — Hultén, Atlas, map 226, pro parte; Hylander, Nord. Kärlväxtfl. I, 267, pro parte.

*P. tenella* auct. non Holmb. — Sørensen, Revis. Greenl. sp. Puccinellia 80; Polunin, Circump. arct. fl. 73, pro parte.

*Glyceria distans* var. *pulvinata* Fries, Mant. II (1839), 11, pro parte.

*G. tenella* f. *pumila* Lge. in Holm, Nov.-Zem. Veg. . (1885) 16.

*Atropis distans* var. *arctica* auct. — Fedchenko & Flerov, Fl. Yevr. Ross. 126.

*A. distans* var. *capillaris* f. *contracta* Hack. in Kneucker, Gram. Exs. Lief. XXXI (1915), 916.

*A. pulvinata* Krecz. in Fl. SSSR II (1934), 478, 761; Kuzeneva in Fl. Murm. I, 221.

**Ill.:** Fl. SSSR II, pl. XXXVI, fig. 11.

Northern European littoral species, accepted by V. I. Krechetovich (l. c.) and Sørensen (l. c.) as a distinct species although so closely approaching the preceding species that the morphological boundaries between them are scarcely detectable. All the same, it likely constitutes a weakly differentiated more northern race of *P. capillaris* characterized mainly by the lemmas being shorter and more obtuse than in *P. capillaris* and by the panicles being (on average) smaller and more condensed, although it cannot be excluded that at least in some cases introduced specimens of *P. distans* (which readily hybridizes both with *P. capillaris* and *P. coarctata*) were also involved in the origin of specimens of this species. We have already presented reasons above (see the remarks on *P. tenella*) in support of the interpretations of the species *P. tenella* and *P. pulvinata* we have accepted. Here it need only be noted that the fact that Lange (l. c.) referred very small underdeveloped specimens of "*Glyceria tenella* f. *pumila* Lge." to *P. tenella* although they actually belong to *P. pulvinata* cannot be considered an argument of any weight in favour of a different interpretation of the latter species. Probably Lange simply failed to notice the pubescence on the lemmas of these specimens, and his statement that this form differs from *P. tenella* by the more obscure nerves of the lemmas ("magis obsolete nervata") speaks more in favour of our interpretation of this species. In his monograph of the Greenlandic species of alkali grasses, Sørensen (l. c. 22) says that *Atropis pulvinata* Krecz. is not identical with *Glyceria distans* var. *pulvinata* Fries, indicating that under the latter name Fries (l. c.) described typical specimens of *P. capillaris*, not *Atropis pulvinata* Krecz. whose type cannot be Fries' specimens. Among the numerous exsiccates of Fries' variety, specimens of *P. capillaris* do in fact predominate, but there are also specimens of two other closely related species (*P. pulvinata* and *P. coarctata*) apparently of introduced origin. Krechetovich selected one of these very specimens as the type of *P. pulvinata* (with label: *Heleochloa distans pulvinata* Fr. = *Poa maritima* Wahlb. Lapp. Hall. bor. Varberg. = Jul. Leg. Fries), and it does in fact belong to this species. Because only a small part of the herbarium material of Fries' "var. *pulvinata*" belongs to *Atropis pulvinata* Krecz., we consider V. I. Krechetovich to be the author of this species and that he described it independently of Fries.

Like the preceding species, growing on maritime banks, more rarely on maritime rocks, usually above high tide mark. Readily dispersed with ballast.

**Soviet Arctic.** Murman (common on entire coast); Kanin (only on northern part); Bolshezemelskaya Tundra (Belkovskiy Shar); Karsk-Baydaratsk coast (at Amderma setlement); Vaygach; Novaya Zemlya (Petukhovskiy Shar and Bay of Karmakuly according to Sørensen).

**Foreign Arctic.** Arctic Scandinavia.

**Outside the Arctic.** Almost entire coast of Scandinavian Peninsula and Finland, but apparently only as an introduction in the south; coast of White Sea (recorded by Sørensen, mainly specimens transitional to *P. capillaris*). Introduced with ballast to other countries.

16. ***Puccinellia coarctata*** Fern. et Weath. in Rhodora XVIII (1916), 13; Sørensen, Revis. Greenl. sp. Puccinellia 42, 83; Hultén, Amphi-atl. pl. 280.

   *P. retroflexa* ssp. *borealis* Holmb. in Bot. Notis. (1926) 182.

   *P. capillaris* var. *vaginata* auct. — Tsvelev in Sp. rast. Gerb. Fl. SSSR XIV, 79; non *Glyceria vaginata* Lge.

   *P. tenella* auct. non Holmb. — Polunin, Circump. arct. fl. 73, pro parte.

   *Glyceria Borreri* auct. non Bab. — Lange, Consp. fl. groenl. 168.

   *Atropis tenella* auct. non Richt. — Krechetovich in Fl. SSSR II, 483; Kuzeneva in Fl. Murm. I, 221.

   Ill.: Fernald & Weatherby, l. c., pl. 115, fig. 28–32; Fl. SSSR II, pl. XXXVII, fig. 16; Sørensen, l. c., fig. 28–32, 71–74.

   Amphiatlantic arctic littoral species, in our opinion extremely close to the two preceding species but differing from them in having smooth or slightly scabrous panicle branchlets and less membranous, relatively small lemmas. According to Sørensen (l. c. 88) it belongs to the separate species-group "*Coarctata* Group" and shows closer relationship to *P. fasciculata* (Torr.) Bickn., a more southern and likewise amphiatlantic species of this group, than to the species *P. capillaris* and *P. pulvinata*, of which the first was only doubtfully included in the "*Coarctata* Group" as a deviant type by Sørensen and the second was (also doubtfully) included in the "*Pumila* Group." However, the differences between *P. coarctata* and *P. fasciculata*, despite the sometimes considerable external resemblance of these species, seem to us rather substantial and constant, while in Northern Scandinavia and the Kola Peninsula *P. coarctata* blends with *P. capillaris* and *P. pulvinata* (producing numerous specimens transitional to those species) to such an extent that delimiting them there becomes very difficult. Despite the considerable probability that this blending of the three stated species there is secondary and the result of too fre-

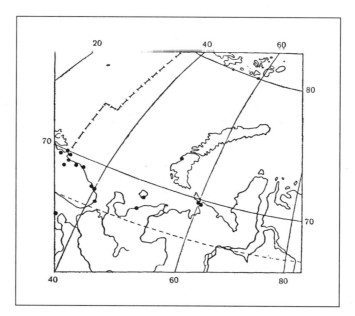

MAP II–62  Distribution of *Puccinellia coarctata* Fern. et Weath.

quent introductions of *P. capillaris* and *P. pulvinata* into the distributional range of *P. coarctata* (with consequent interspecific hybridization), it is now scarcely possible to deny the assumption that a very close relationship exists between all these species. We note that Holmberg (l. c.), who occupied himself for a long time with the study of this difficult genus, subordinated this species to *P. capillaris* as a more northern variety ("*P. retroflexa* var. *borealis* Holmb."). V. I. Krechetovich (l. c.) described *P. coarctata* under the erroneous name "*Atropis tenella.*"

Growing on maritime banks and rocks, but only occasionally descending below high tide mark. According to our observations in the district of Dalniye Zelentsy settlement on the Barents Sea, this species like the two preceding occurs usually in the form of separate tufts on stony (often with large stones) or sandy banks together with such more or less halophilic species as *Carex subspathacea*, *Festuca arenaria*, *Stellaria humifusa*, *Cochlearia groenlandica*, *Plantago Schrenkii* and *Ligusticum scoticum*. On maritime rocks *P. coarctata* usually accompanies only a few species, such as *Tripleurospermum phaeocephalum* and *Rhodiola arctica*, which however never descend as close to the sea as does *P. coarctata*.

**Soviet Arctic.** Murman (common on the entire coast from the Rybachiy Peninsula to the mouth of the Ponoy); Malozemelskaya Tundra (Cape Svyatoy Nos); Kolguyev; Yugorskiy Shar and Vaygach (Cape Greben according to Sørensen); Novaya Zemlya (Pukhovaya Bay). (Map II–62).

**Foreign Arctic.** Labrador (NE part); Greenland (south of 78°N); Jan Mayen Island; Iceland; Arctic Scandinavia.

**Outside the Arctic.** Solovetskiye Islands; Labrador (south to the Gulf of St. Lawrence); Newfoundland; Faroe Islands; Scandinavia (southern boundary unclarified, occurring as an introduction apparently on almost the entire coast).

17. ***Puccinellia sibirica*** Holmb. in Bot. Notis. (1927) 206, excl. var.; Polunin, Circump. arct. fl. 73, pro parte.

*Glyceria distans* auct. non Wahlb. — Scheutz, Pl. jeniss. 187, pro parte.

*Atropis sibirica* (Holmb.) Krecz. in Fl. SSSR II (1934), 479. pl. XXXVI, fig. 12.

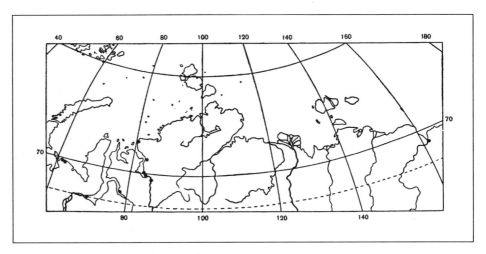

MAP II–63  Distribution of *Puccinellia sibirica* Holmb.

Ill.: Fl. SSSR II, pl. XXXVI, fig. 12; Polunin, l. c. 71.

Siberian arctic species, occupying a more eastern range in comparison with the three preceding species. Showing undoubted and rather close affinity, on the one hand, with *P. capillaris* (but unlike it not a littoral species), on the other hand with *P. borealis* Swall. which approaches *P. distans*. The most typical specimens of *P. sibirica*, for example the specimen cited by Holmberg with label "Sibiria, Jenisei, Tolstonos, 8 IX 1876, Brenner (Lectotypus speciei)," are readily distinguished from *P. capillaris* by the rosy-violet spikelets (in *P. capillaris* the lemmas may be tinged with rosy-violet only near their tips), while specimens with greenish or scarcely tinted spikelets often have a very great external similarity to *P. capillaris* and differ from the latter only in having distinctly thinner lemmas with more broadly hyaline margins.

*Puccinellia sibirica* usually grows in habitats more or less removed from the sea shore: on poorly vegetated slopes of mounds and riverine terraces, on sands and gravelbars in river valleys, etc.

**Soviet Arctic.** Karsk-Baydaratsk coast (mouth of River Nynze-Yaga and near Amderma settlement); lower reaches of Ob and Taz (Nangi Island in the Ob Delta, Cape Nakhodka in Tazovskaya Bay); lower reaches of Yenisey (Tolstyy Nos, Selyakino, Dudinka; according to Holmberg also Mezenkin settlement and Yefremov Rock); lower reaches of Kolyma (right bank of Kolyma near Panteleikha settlement, 25 VII 1950, Nepli, not entirely typical specimen). (Map II–63).

**Foreign Arctic.** Absent.

**Outside the Arctic.** Not found so far.

18. ***Puccinellia distans*** (L.) Parl., Fl. Ital. (1848) 367; Hultén, Atlas, map 223; Hylander, Nord. Kärlväxtfl. I, 266; Polunin, Circump. arct. fl. 73, pro parte.
    *Poa distans* L., Mant. Pl. I (1767), 32.
    *P. retroflexa* Curt., Fl. Londin. 1 (1777).
    *Glyceria distans* (L.) Wahlb., Fl. Ups. (1820) 36.
    *Atropis distans* (L.) Griseb. in Ledebour, Fl. ross. IV (1853), 388; Perfilev, Fl. Sev. I, 96; Krechetovich in Fl. SSSR II, 484; Kuzeneva in Fl. Murm. I, 222.
    *A. distans* var. *typica* Trautv. in Acta Hort. Petrop. 1 (1871), 282; Krylov, Fl. Zap. Sib. II, 310, pro parte.
    Ill.: Curtis, l. c., tab. 45; Polunin, l. c. 71.

Widely distributed but predominantly European semiweedy species, occurring in the Arctic only as a rare introduction along roads and in settlements. Apparently occasionally also introduced on the sea coast, where it readily hybridizes with the closely related littoral species *P. capillaris*, *P. pulvinata* and *P. coarctata*.

**Soviet Arctic.** So far known only as an introduced weed from the vicinity of Murmansk and Vorkuta.

**Foreign Arctic.** Apparently not reported.

**Outside the Arctic.** Almost all Europe south of 63°N (further north only a few introduced localities in Scandinavia and the districts of Arkhangelsk and Kandalaksha); West Siberia (north to Salekhard and the Poluy); East Siberia and Central Asia (only as a rare introduction); North America (from Southern Canada to Mexico). Introduced in many other countries of both hemispheres.

**19. *Puccinellia borealis*** Swall. in Journ. Wash. Ac. Sc. XXXIV (1944), 19; Hultén, Fl. Al. X, Suppl. (1950) 1712.

Subarctic species very close to the preceding and possibly even derived from introduced specimens of *P. distans* which have become highly modified under the influence of the harsh arctic climate. The most typical specimens of *P. borealis* from Alaska (six specimens were kindly sent at our request from the Stockholm herbarium) possess panicle branchlets strongly scabrous for their entire length and relatively large anthers (0.6–0.8 mm long). Specimens from the most eastern parts of the Soviet Arctic (district of Bay of Chaun) are identical to these, but specimens from the district of Tiksi Bay already have the panicle branchlets smooth basally and relatively smaller anthers (0.4–0.6 mm long). Almost completely smooth panicle branchlets with anthers about 0.5 mm long are shown by the sole specimen from Taymyr available to us. Thus, the majority of Siberian specimens of *P. borealis* are not identical with typical Alaskan material of this species, and may be distinguished as a separate subspecies: *P. borealis* ssp. *neglecta* Tzvel., ssp. nova [A subspeciei typica paniculae ramulis in parte inferiore glabris et antheris minoribus 0.4–0.6 mm lg. differt. Typus subspeciei: Taimyr, in ripa boreali lac. Taimyr prope jug. Byrranga, cap. Sabler, in declivitate collis arenariae, 17 VII 1948, Tichomirov et Vellikainen. In Herb. Inst. Bot. Ac. Sc. URSS (Leningrad) conservatur].

Growing on poorly vegetated slopes of riverine terraces, on slopes of mounds (baydzharakhs), and as a semiweedy plant along roads and near settlements.

**Soviet Arctic.** Taymyr [north coast of Lake Taymyr, southern foothills of Byrranga Range, Cape Sabler, slope of sandy mound (baydzharakh), 17 VII 1948, Tikhomirov and Vellikaynen, type of ssp. *neglecta*]; district of Tiksi (Muostakh Island); lower reaches of the Indigirka (near Ozhogino settlement) and the Kolyma (on the Bolshaya Kuropatochya River); Ayon Island and district of Bay of Chaun.

**Foreign Arctic.** Arctic Alaska (district of Bering Strait).

**Outside the Arctic.** Alaska (Yukon Basin and south of it).

**20. *Puccinellia Hauptiana*** (Krecz.) Kitagawa in Rep. Inst. Sc. Res. Manchukuo 1 (1937), 255; Krechetovich in Fl. SSSR II, 485, 763, in synonymy; Hultén, Fl. Al. X, Suppl. 1712; Karavayev, Konsp. fl. Yak. 57.

*P. filiformis* V. Vassil. in Bot. mat. Gerb. Bot. inst. AN SSSR XI (1940), 50; non Keng, 1938.

*P. tenuiflora* auct. — Hultén, Fl. Al. II, 238, pro parte.

*Atropis Hauptiana* Krecz. in Fl. SSSR II, 385, 763.

*Poa Hauptiana* Trin. ex Krecz., l. c., in synonymy.

**Ill.:** Fl. SSSR II, pl. XXXV, fig. 21.

Widely distributed often semiweedy species, replacing *P. distans* in Siberia and only penetrating arctic limits in a few places. Differing from *P. distans* mainly in having smaller lemmas and smaller anthers (0.3–0.5 mm long).

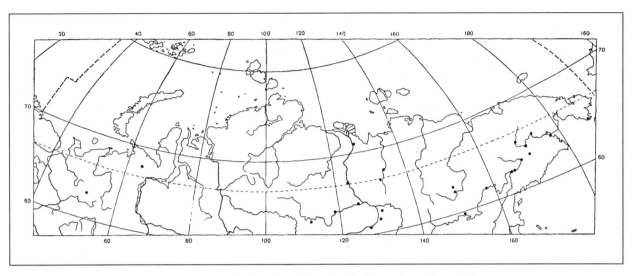

MAP II–64 Distribution of *Puccinellia Hauptiana* (Krecz.) Kitagawa.

In NE Siberia (Kheta Basin on shore of River Medvezhya, on the Indigirka near Ozhogino settlement, and in the Srednekolymsk district), there sometimes occur, together with typical specimens of *P. Hauptiana*, specimens with somewhat elongate lemmas with acuminate tips (2–2.5 mm long). These possibly belong to *P. interior* Sørensen (Hultén, l. c. 1713) described from Alaska, a species possibly of hybrid origin which appears to occupy an intermediate position between *P. Hauptiana* and *P. borealis*.

Occurring usually in slightly solonetzic or saline meadows, on gravelbars and shores of waterbodies, sometimes also as a weed along roads and in settlements.

**Soviet Arctic.** Bolshezemelskaya Tundra (as introduced weed in Vorkuta district); lower reaches of Lena (near Kyusyur settlement, 28 VII 1957, Yurtsev); Provideniye Bay; Anadyr Basin (rather frequent, there are specimens from the vicinity of the settlements of Anadyr, Markovo, Yeropol and Ust-Belaya); lower reaches of Penzhina (settlements of Kamenskoye and Ust-Penzhino). (Map II–64).

**Foreign Arctic.** Apparently absent.

**Outside the Arctic.** European part of USSR (only in Leningrad and more eastern regions as a rare introduction); considerable part of Asia from the northern provinces of China and Mongolia in the south to arctic districts of Siberia in the north (northern boundary running through Berezovo on the Ob, Yeniseysk on the Yenisey, Kyusyur on the Lena, and Ust-Belaya on the Anadyr); more southern districts of Alaska; possibly also occurring in Canada and northwestern states of the USA.

## GENUS 29 — Festuca L. — FESCUE

FOR IDENTIFYING FESCUES, especially species of section *Ovinae* to which six of the species described below (Nos. 3–8) belong, it is extremely important to study details of a transverse section of the leaf blade under the microscope. For better comparison we recommend taking the middle leaf of the living leaves of a vegetative shoot and cutting the blade at about the middle of its length. In describing the section certain specific terms should be used, as explained in the legend to Fig. 1. Leaf sections of *F. ovina* and *F. polesica* are so characteristic that they enable these species to be identified with only a cursory glance at the preparation. *Festuca auriculata, F. kolymensis* and *F. brachyphylla* are not so easily distinguished by leaf section. They are also extremely similar externally. Unfortunately, we have not succeeded in distinguishing these species by using some characters which are extremely stable in certain other fescues, such as the degree of closure of the sheath and the nature of the trichomes on the upper leaf surface and on the keels of the palea. Such characters as the height of the plant, the panicle size, the spikelet colour, and the pubescence of the culm, spikelets and outer leaf surface have a generally facultative character in all fescues of section *Ovinae* and therefore possess very little taxonomic value.

The chromosome number also does not possess great importance for distinguishing arctic species of fescues. In species which have been repeatedly studied cytologically (like *F. rubra, F. polesica, F. ovina* and *F. brachyphylla*), 2–3 (or even more) different chromosome numbers have been obtained for each (from $2n = 14$ to $2n = 56$). Especially great variation occurs (quite understandably, of course) in viviparous forms, for which the numbers 21, 35 and 49 are also known. The value of the cytological data published in the literature is much reduced by the circumstance that incredible confusion reigns over identifications of material and species concepts in section *Ovinae*.

This confusion also much impedes citation of literature, especially foreign literature for which we have almost no herbarium documentation.

In contemporary foreign literature *Festuca vivipara* (L.) Smith is usually recognized as a distinct species. But it is well known that vivipary occurs in many genera of grasses (*Poa, Calamagrostis, Deschampsia*, etc.). It is also known in *F. rubra*, as mentioned particularly by Scholander. It is obvious that the taxonomic significance of such a character, which has evolved in parallel in many taxa, is extremely problematic and should not in any circumstance be evaluated a priori without correlation with other characters. Unfortunately even Scholander and the Löves, who wrote about *F. vivipara* in some detail, completely neglected to study those characters to which the greatest importance was attached by all systematists who made special investigations of the *Ovinae* group of fescues (starting with Hackel, Monographia Festucarum europaearum, 1882), namely characters of the anatomical structure of the leaf blade. One needs only to look at leaf sections for it to become immediately obvious that *F. vivipara* is a purely artificial conglomerate consisting (at least in the European Arctic) of viviparous biotypes of *F. brachyphylla* and *F. ovina*.

As soon as the name "*F. vivipara*" is removed, a significant geographical parallelism is revealed in the distribution of the viviparous forms of *F. brachyphylla* and *F. ovina* (maps II–68, II–72): vivipary is frequent in the western half of the ranges of both species and almost completely absent in the eastern half (in the case of *F. ovina*, with reference mainly to the northern fringe of its range). Only on the Pacific Ocean itself do viviparous forms begin to become frequent again. It is fitting here to recall the work of Wycherley (The distribution of the viviparous grasses in Great Britain, Journ. Ecology 2, 1953), who showed that in Great Britain "*F. vivipara*" is distributed in districts with especially high precipitation, while *F. ovina* with normal reproduction by seed is almost absent in such districts.

In the district of the Lower Yenisey, viviparous forms of *F. auriculata* apparently also occur. However, this cannot yet be confirmed with complete certainty, because the material is rather little and because not every specimen of *F. auriculata* can be distinguished from *F. brachyphylla* by the vegetative organs without possibility of error.

1. Leaves both of generative and vegetative shoots *flat, 2–5 mm wide, with 12–24 nerves* (viewed against the light or in section), convoluted in bud stage. Ligule of leaf truncate (transversely cut off), nonciliate. *Rachis of spikelet smooth.* Florets without awns, 6–8 mm long. . . . . . . . . . . . . . . . . . . . . . . . . . . . . . . . . . . . . . . . . . . . . . . . . . . . . . . . . . . . . . . . . . . .1. **F. PRATENSIS** HUDS.
   – Leaves of vegetative shoots *doubly folded lengthwise* and in this form *0.4–1.3 mm wide, rarely almost flat* (but then still conspicuously keeled) *and up to 2 mm wide;* 3–13 nerves; leaves folded lengthwise in bud stage. Ligule truncate or more often bilobed (with rounded or more or less extended lobes on sides), normally short-ciliate on margin. *Rachis of spikelet scabrous with acicules or hairy.* . . . . . . . . . . . . . . . . . . . . . . . . . . . . .2.

2. *Tufts loose.* At least some (often all) *vegetative shoots extravaginal:* they perforate the protective leaf sheath, grow at first more or less horizontally and arcuately (plagiotropically, diageotropically) and bear a series of scales which gradually transform into normal leaves. In transverse leaf section, 3–7 ribs with deep (two-thirds of leaf thickness) incisions between them; if 3 ribs these all triangular, if more than 3 some of them sometimes trapezoidal. 3–5(7) weak sclerenchymatous bundles beneath outside epidermis (see Fig. 1, *2–3*). . . . . . . . . . . . . . . . . . . . . . . . . . . . . . . . . . . . . . .3. **F. RUBRA** L.
   – *Tufts very dense* (somewhat loosened only if sprouting in moss). *All shoots entirely intravaginal,* directed upwards right from their origin. Leaves with developed blade preceded by only a single bicarinate membranous prophyll. Leaf appearing otherwise in section. . . . . . . . . . . . . . . . . . . . . . . . . . . .3.

3. *Plant tall (40–80 cm).* Leaf section very characteristic (see Fig. 1, *1*). Branches of inflorescence *obtusely angled or rounded,* smooth or slightly scabrous. Spikelets more or less very dark brownish-violet. Lemma 6–8 mm long, *more or less keeled for its whole length.* Anthers 3–4 mm long. . . . . . . . . . . . . . . . . . . . . . . . . . . . . . . . . . . . . . . . . . . . . . . . . . . . . . . . . . . . . . . .2. **F. ALTAICA** TRIN.
   – *Plant shorter.* Leaf of different structure in section. Branches of inflorescence *with sharp edges,* more or less sharply scabrous. Lemma (excluding

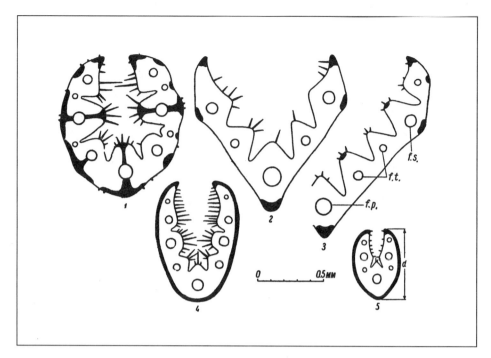

**FIGURE 1** Schematic transverse sections of leaf blade of vegetative shoots of *Festuca*. (Epidermis not shown; sclerenchyma shown *in black*).
f. p. — vascular bundle of first size (median); f. s. — vascular bundle of second size; f. t. — vascular bundle of third size; d — diameter of blade. *1 — F. altaica* Trin.; *2,3 — F. rubra* L.; *4 — F. polesica* Zapal.; *5 — F. ovina* L.

awn) 3–6 mm long, rounded on back and *more or less keeled only at its very tip*. Anthers less than 3 mm long. . . . . . . . . . . . . . . . . . . . . . . . . . . . . . . . . . . .4.

4. Leaf blade convex on sides (especially obviously in living state), in section oval in general shape. *Mechanical tissue* (sclerenchyma) *forming continuous sheath of uniform thickness beneath outside epidermis* (occasionally breaks observed in it). . . . . . . . . . . . . . . . . . . . . . . . . . . . . . . . . . . . . . . . . . .5.
– Leaf blade more or less flat on sides, in dry state often sunken. *Mechanical tissue* (sclerenchyma) *not forming continuous sheath, but grouped into several (3 or more) cords* (on midline and margins of blade, often also beneath certain vascular bundles). . . . . . . . . . . . . . . . . . . . . . . . . . . . . . . . . .6.

5. Leaves *erect, firm, almost wiry*, 0.6–1.2 mm in diameter, on inside (upper) surface with (3)5(7) ribs and dense trichomes 0.05–0.1 mm long (see Fig. 1, *4*). . . . . . . . . . . . . . . . . . . . . . . . . . . . . . . . . . . . . . . . . . . . . . .4. **F. POLESICA** ZAPAL.
– Leaves *soft, often arched or drooping*, 0.3–0.7 mm in diameter, on inside surface with only one rib and trichomes up to 0.03 mm long (see Fig. 1, *5*). . . . . . . . . . . . . . . . . . . . . . . . . . . . . . . . . . . . . . . . . . . . . . . . . . . . . . . . .5. **F. OVINA** L.

6. In leaf section mechanical tissue *with 5–7 rather weak cords*. Anthers *0.4–1.0 mm long*. . . . . . . . . . . . . . . . . . . . . . . . . . . . . . .8. **F. BRACHYPHYLLA** SCHULT.

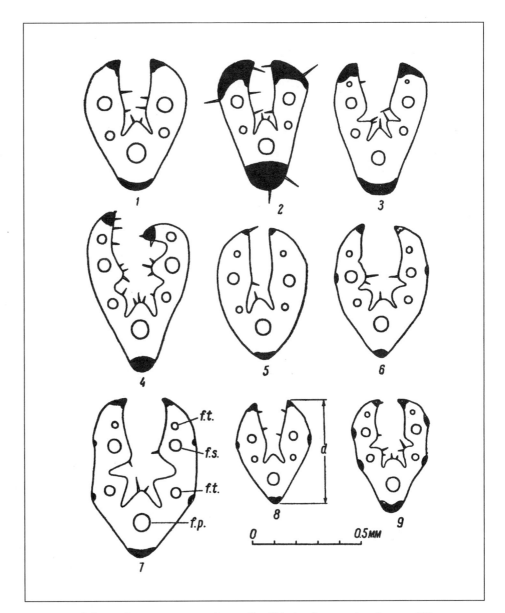

**FIGURE 2** Schematic transverse sections of leaf blade of vegetative shoots of *Festuca*. (Lettering the same as in Fig. 1).
*1–3* — *F. kolymensis* Drob.; *4–6* — *F. auriculata* Drob.; *7–9* — *F. brachyphylla* Schult.

– In leaf section mechanical tissue *with more or less 3 cords* (occasionally with additional weaker cords). *Anthers 1.8–2.6 mm long.* ............... 7.

7. Leaves *more or less firm, mostly glaucous.* Cords of mechanical tissue mostly rather strong. Summit of midrib situated at about half total height of section. Vascular bundle of second size significantly closer to leaf margin than to median bundle. Upper bundles of third size absent or distinctly more weakly developed than lower (see Fig. 2, *1–3*). ...................
................................................. 6. **F. KOLYMENSIS** DROB.

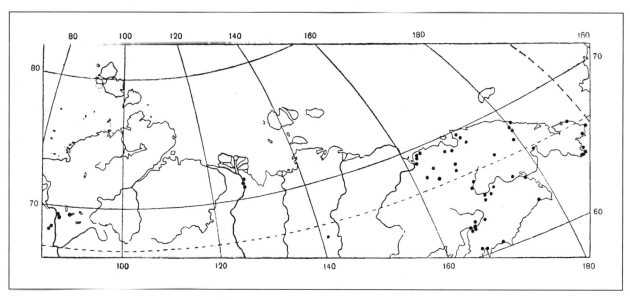

Map II–65  Distribution of *Festuca altaica* Trin.

– Leaves *soft, mostly green*. Cords of mechanical tissue more often weak. Summit of midrib distinctly below middle of total height of section. Vascular bundle of second size situated almost midway between leaf margin and median bundle. Upper and lower bundles of third size uniformly developed. ................................... 7. **F. AURICULATA** DROB.

1. ***Festuca pratensis*** Huds., Fl. angl. (1762) 37; Krylov, Fl. Zap. Sib. II, 328; Perfilev, Fl. Sev. I, 97; Krechetovich & Bobrov in Fl. SSSR II, 530; Gröntved, Pterid. Spermatoph. Icel. 137; Hultén, Atlas, map 241; Chernov in Fl. Murm. I, 230; Löve & Löve, Consp. Icel. Fl. 161; Polunin, Circump. arct. fl. 54.

    *F. elatior* L., Sp. pl. (1753) 75, partim ? — nomen ambiguum.

    **Ill.:** Reichenbach, Ic. fl. Germ. 1, tab. 70, fig. 165; Fl. SSSR II, pl. XXXIX, fig. 9; Fl. Murm. I, pl. LXXVI.

    Meadow plant of the forest zone. In water meadows and secondary meadows near habitations occasionally penetrating the Arctic Region.

    **Soviet Arctic.** So far found only in West Murman (River Kolosioki in the Pechenga district).

    **Foreign Arctic.** Greenland (occasionally as an introduction); Norway (to 70°N); Iceland (introduced?).

    **Outside the Arctic.** Western Europe; Asia Minor; Caucasus; entire European part of USSR; forest and forest-steppe zones of West Siberia; mountains of Central Asia; South-Central Siberia. Introduced to the southern Soviet Far East and to boreal North America.

2. ***Festuca altaica*** Trin. in Ledebour, Fl. alt. 1 (1829), 109; Grisebach in Ledebour, Fl. ross. IV, 354; Schmidt, Fl. jeniss. 126; Trautvetter, Syll. pl. Sib. bor.-or. 135; Scheutz, Pl. jeniss. 184; Krylov, Fl. Zap. Sib. II, 330; Komarov, Fl. Kamch. I, 186; Krechetovich & Bobrov in Fl. SSSR II, 528; Hultén, Fl. Al. II, 239; Raup, Bot. SW Mackenzie 110; Karavayev, Konsp. fl. Yak. 58; Polunin, Circump. arct. fl. 53.

    **Ill.:** Ledebour, Ic. fl. ross., tab. 288.

    Preferring well drained habitats, both open (slopes, dry meadows on ridges, dry herbacous or low-shrub tundras, dwarf birch carr, etc.) and moderately shaded (floodplain poplar groves, willow carr with mixed herbs), far more rarely in moister sites (collected near Markovo on the Anadyr in alder carr with sedges and cottongrass).

**Soviet Arctic.** Lower reaches of Yenisey (near boundary of forest zone, crossing it on the left bank); lower reaches of Lena (subzone of open larch forest); lower reaches of Kolyma and north slope of Northern Anyuyskiy Range; district of Bay of Chaun; Chukotka Range and north coast of Chukotka (Cape Schmidt); Chukotka Peninsula; Anadyr and Penzhina Basins; coast of Koryakia. (Map II–65).

**Foreign Arctic.** Alaska (district of Mackenzie River).

**Outside the Arctic.** Dzhungarskiy Alatau; Altay; Sayans; Central Siberian Plateau; Prebaikalia and Transbaikalia; Mongolia; mountainous districts of Yakutia; Magadan Oblast; Kamchatka; all Alaska; Yukon; British Columbia.

3. ***Festuca rubra*** L., Sp. pl. (1753) 74; Grisebach in Ledebour, Fl. ross. IV, 352; Schmidt, Fl. jeniss. 126; Trautvetter, Pl. Sib. bor. 134; id., Syll. pl. Sib. bor.-or. 60; Scheutz, Pl. jeniss. 183; Drobov, Predst. sekts. *Ovinae* v Yakut. 165; Lynge, Vasc. pl. N. Z. 109; Tolmatchev, Contr. fl. Vaig. 126; Tolmachev & Pyatkov, Obz. rast. Diksona 155; Krylov, Fl. Zap. Sib. II, 323; Tolmachev, Fl. Taym. I, 100; Leskov, Fl. Malozem. tundry 31; Scholander, Vasc. pl. Svalb. 72; Perfilev, Pl. Sev. I, 98; Krechetovich & Bobrov in Fl. SSSR II, 517; Gröntved, Pterid. Spermatoph. Icel. 138; Hultén, Fl. Al. II, 178; id., Atlas, map 242; Chernov in Fl. Murm. I, 225; A. E. Porsild, Ill. Fl. Arct. Arch. 40; Karavayev, Konsp. fl. Yak. 58; Polunin, Circump. arct. fl. 54.

*F. arenaria* Osbeck in Retzius, Suppl. prodr. fl. Scand. 1 (1805), 4; Andreyev, Mat. fl. Kanina 158; Krechetovich & Bobrov, l. c. 520; Leskov, l. c. 30; Chernov, l. c. 228.

*F. Richardsonii* Hook., Fl. Bor. Amer. II (1840), 250; Löve & Löve, Consp. Icel. Fl. 102.

*F. cryophila* Krecz. et Bobr. in Fl. SSSR II (1934), 519, 769; Tolmachev, Obz. fl. N. Z. 149; Chernov, l. c. 226; Tikhomirov, Fl. Zap. Taym. 28; Karavayev, l. c. 58.

*F. eriantha* auct. — Krechetovich & Bobrov, l. c. 523; ? Honda & Tatewaki in Tokyo Bot. Mag. 42 (1928), 185.

*F. egena* Krecz. et Bobr. in Fl. SSSR II (1934), 523, 768; Karavayev, l. c. 58.

**Ill.:** Fl. SSSR II, pl. XL, fig. 14, 15, 16, 18, 19; Fl. Murm. I, pl. LXXV; Porsild, l. c., fig. 12, a, b.

*Festuca rubra* is highly variable in a series of readily detected external characters and is one of the most polymorphic species among boreal grasses. Consequently there have been repeated attempts to divide it into several species or to separate from it various selected populations as distinct species (listed above are only those names of interest for our region). But all these attempts have been insufficiently founded and have achieved little success. No one has succeeded in drawing morphological, ecological, cytological or geographical dividing lines within *F. rubra* clearly and convincingly. Therefore in the present state of knowledge it is not possible to treat *F. rubra* as other than a single species. This naturally does not in any way deny that there are a great number of biotypes and populations differing from one another in various characters within *F. rubra*, nor of course does it deny the possibility that the species may be convincingly subdivided in the future.

It must however be emphasized that the impression of unusual polymorphism in *F. rubra* is based mainly on observation of readily detected characters, such as the appearance of the tuft, the length and colour of the leaves, the length of the stolons, culm height, panicle size, spikelet pubescence, etc. But if attention is paid to certain other characters which are not so readily detectable but no less genotypically constant, for example details of the structure of the leaf blade, then other species of fescues, such as *F. auriculata*, *F. kolymensis* or the steppe *F. sulcata*, prove to be not in the slightest less polymorphic than *F. rubra*.

Among the forms of *F. rubra* with pretensions of specific distinctness, the greatest recognition has been accorded to the American *F. Richardsonii* Hook. and

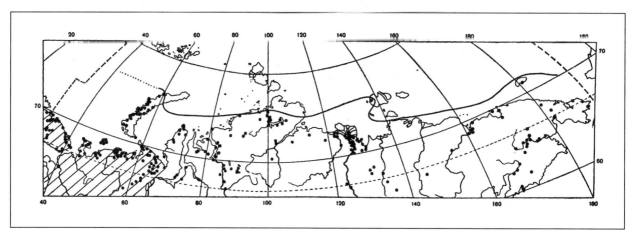

MAP II–66  Distribution of *Festuca rubra* L.

apparently almost as much to *F. cryophila* Krecz. et Bobr., because both of them possess pubescent lemmas. In fact, among (for example) all the material so far accumulated in the herbaria of the Botanical Institute of the USSR Academy of Sciences and Moscow University, there is not a single specimen of *F. rubra* from either Spitsbergen or Novaya Zemlya with completely bare lemmas. But already on Vaygach and Kolguyev forms with bare lemmas have also been collected. Further south forms with bare lemmas become more abundant and, for instance in Moscow Oblast, very sharply predominate. But plants with lemma pubescence identical to that in Novaya Zemlya also occur, though infrequently, in Moscow, Ryazan and other central Oblasts of the USSR. Evidently we are dealing with intraspecific geographical variation of a clinal type.

There is analogous variation also in other characters: as one moves northwards, one encounters with increasing frequency plants with short leaves, a shorter and more compact panicle, more tightly arranged florets and hence shorter and broader spikelets, and violet colouration of the spikelets and culms; conversely the pubescence of the sheaths, which is almost always conspicuous in southern plants, weakens to the north. Due to this clinal nature of the variation, it is difficult to consider the arctic forms *F. cryophila* or *F. Richardsonii* as subspecies: where to draw the boundaries between these subspecies is unknown.

With respect to *F. egena* or *F. aucta*, these are even more difficult to delimit: the characters attributed to them (especially differences in anther size) are completely unstable and do not even possess any geographical pattern to speak of. It is evident that, should one wish to retain and name the various taxa described within *F. rubra*, they should be recognized as varieties in the sense of Ascherson & Graebner.

The number of these varieties could be still more increased because *F. rubra* has a strong tendency to form local ecotypes. The author is aware through personal observations of a whole series of very interesting local forms differing in a series of special characters from the limestones of Tulskaya Oblast, from oak woods on the Oka, from rocks on the shore of the White Sea, and from mountainous districts of Central Asia; all these are genotypically determined and retain their characters in cultivation, and all could be described as separate species with just as much justification as *F. Richardsonii, F. egena*, etc. The diversity of forms of *F. rubra* in the forest region and the forest-steppe is undoubtedly very much greater than in the Arctic. It is precisely the paucity of opportunities for morphological diversification which produces the impression that arctic plants are something distinct. They are simply less diverse.

*Festuca rubra* is a plant of a predominantly pioneering nature and therefore occurs most often on fresh alluvia, cliffs, screes and rocks, in various open communities, usually under conditions of rather good drainage. It also grows in meadows, but usually only if there is no continuous closed turf and the herbage is not too dense; more rarely on tundras and dry barrens.

**Soviet Arctic.** Murman; Kanin; Kolguyev; Malozemelskaya Tundra; lower reaches of Pechora; Bolshezemelskaya Tundra; Polar Ural, shore of Baydaratskaya Bay; Pay-Khoy; Vaygach; Novaya Zemlya (except the northern half of the North Island); Yamal (including Belyy Island); shores of Ob Sound and Tazovskaya Bay; Gydanskaya Tundra; lower reaches of Yenisey; Taymyr (except a great part of the north coast); Khatanga Basin and Khatanga Bay; Anabarskaya Tundra; lower reaches of Olenek and Lena; Chekanovskiy Ridge; Kharaulakh Mountains; New Siberian Islands; lower reaches of Kolyma; district of Bay of Chaun (including Ayon Island); Wrangel Island; Chukotka Range; Chukotka Peninsula; Bay of Krest; Anadyr and Penzhina Basins. (Map II–66).

**Foreign Arctic.** Alaska; Mainland Canada; Labrador; Canadian Arctic Archipelago (Victoria and Banks Islands); West Greenland (southern part and Disko Island); East Greenland; Iceland; Spitsbergen; Arctic Scandinavia.

**Outside the Arctic.** All Europe; North Africa; all Asia except deserts and tropical districts; North America to Mexico.

4. ***Festuca polesica*** Zapal., Consp. fl. Galic. (1905) 303; Krechetovich & Bobrov in Fl. SSSR II, 508; Hultén, Atlas, map 240; Chernov in Fl. Murm. I, 225.

*F. sabulosa* Lindb. fil. in Sched. ad pl. Finland. exc. (1906) 23; Rozhevits & Shishkin in Fl. Len. obl. I, 153.

*F. Beckeri* (Hack.) Smirn. in Mayevskiy, Fl. Sr. Ross. (1933) 143; Perfilev, Fl. Sev. I, 99.

*F. ovina* subsp. *Beckeri* Hack., Monogr. Festuc. eur. (1882) 100.

**Ill.:** Smirnov, Mater. k sist. i geogr. srednerusskikh ovsyanits 102, fig. 3; Fl. Murm. I, pl. LXXIV, fig. 3.

Open wind-eroded unstable sands.

**Soviet Arctic.** Known only from the Malozemelskaya Tundra (10 km east of the lower reaches of the Neruta).

**Foreign Arctic.** Not occurring.

**Outside the Arctic.** Distributed on open, poorly stabilized sands from Northern Serbia and Hungary through the steppe and forest-steppe zones of the European part of the USSR to Northern Kazhakhstan; frequent on the shores of the Baltic Sea from Denmark to Turku and Leningrad; in the forest zone on the larger sand dunes (with wide gaps) from Poland to Ust-Shchugor, Velikiy Ustyug and the south of the Kola Peninsula (River Varzuga).

5. ***Festuca ovina*** L., Sp. pl. (1753) 73; Smirnov in Mayevskiy, Fl. Sr. Ross. (1933) 142; id., Mater. k sist. i geogr. srednerusskikh ovsyanits 93; Perfilev, Fl. Sev. I, 95; Krechetovich & Bobrov in Fl. SSSR II, 503; Hultén, Atlas, map 239; Chernov, Fl. Murm. I, 223.

*F. supina* auct. [an Schur, Enumer. pl. Transsilv. (1866) 184 ?]. — Drobov, Predst. sekts. *Ovinae* v Yakut. 153; Krylov, Fl. Zap. Sib. II, 317; Andreyev, Mat. fl. Kanina 158; Krechetovich & Bobrov in Fl. SSSR II, 504; Leskov, Fl. Malozem. tundry 31; Chernov in Fl. Murm. I, 224; Karavayev, Konsp. fl. Yak. 58.

*F. vivipara* Smith, Fl. Brit. I (1800), 114 (pro parte ?); Scholander, Vasc. pl. Svalb. 73, pro parte (respecting semiviviparous plants from continental Norway); Löve & Löve, Consp. Icel. Fl. 96, pro parte.

**Ill.:** Syreyshchikov, Ill. fl. Mosk. I, 147; Smirnov, Mater k sist. i geogr. srednerusskikh ovsyanits 102, fig. 1; Fl. Murm. I, pl. LXXIV, fig. 1, 2.

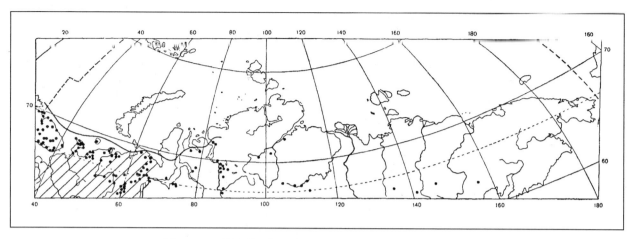

MAP II–67  Distribution of *Festuca ovina* L.

There is a rather widespread tendency in the literature to separate arctic and alpine plants under the name "*F. supina* Schur" from plants in larch or pine forest and dry valleys of the forest zone. But there is absolutely no serious foundation for this. Naturally in *F. ovina,* as in any other nonapomictic species, a whole series of differences between individuals and local populations can be observed at any given locality, especially in mountainous districts. If little material is available, these differences can appear to be constant and to possess a certain geographical or ecological pertinence. However, study of a large quantity of material shows that the range of variation is roughly one and the same in different districts and that the species as a whole appears extremely monolithic and homogenous over the whole extent of its huge range.

In general *F. ovina* undoubtedly gravitates towards the forest zone and does not occur in the High Arctic. It grows mainly in sunny sites under conditions of low or moderate moisture and adequate drainage. It readily colonizes poor sandy, stony or rocky substrates, even on sites trampled by livestock, but fares less well within a strongly developed moss cover. Except at the extreme limits of its range, it is a common and abundant plant.

**Soviet Arctic.** Murman Coast (with adjacent islands); Kanin; Kolguyev; Malozemelskaya and Bolshezemelskaya Tundras; Polar Ural; Pay-Khoy; district of Ob Sound (southern part) and Tazovskaya Bay; Gydanskaya Tundra; lower reaches of Yenisey; Khatanga Basin. Not found further east in the Soviet Arctic. *F. ovina* f. *vivipara:* South Island of Novaya Zemlya (Belushya Bay); both shores of Yugorskiy Shar; lower reaches of Yenisey; Taymyr (watershed of the Dudypta and the Yangoda in the Pyasina Basin). Since the species is rather widespread in Kamchatka, its discovery in the southern part of Koryakia cannot be excluded. (Maps II-67, II-68).

**Foreign Arctic.** Labrador (?); Iceland; Arctic Scandinavia.

**Outside the Arctic.** England; France; all Scandinavia; Central Europe and mountains of Balkan Peninsula (except probably Greece); European part of USSR and West Siberia (roughly within the range of spruce); Altay; Sayans; Khangai Mountains in Mongolia; mountainous districts of East Siberia; Kamchatka; Far East. *F. ovina* f. *vivipara*: Great Britain; Subarctic Scandinavia; Khibins.

6. ***Festuca kolymensis*** Drob., Predst. sekts. *Ovinae* v Yakut. (1915) 155; Krechetovich & Bobrov in Fl. SSSR II, 511; Karavayev, Konsp. fl. Yakut. 58.

*F. pseudosulcata* Drob., Predst. sekts. *Ovinae* v Yakut. (1915) 156; Krechetovich & Bobrov in Fl. SSSR II, 504; Karavayev, l. c. 57.

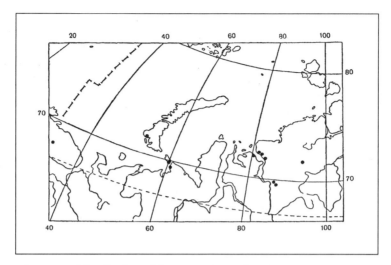

MAP II–68  Distribution of *Festuca ovina* L. f. *vivipara*.

*F. lenensis* Drob., Predst. sekts. *Ovinae* v Yakut. (1915) 158; Krechetovich & Bobrov in Fl. SSSR II, 513; Karavayev, l. c. 58.

Ill.: Drobov, Predst. sekts. *Ovinae* v Yakut., pl. V, fig. 5, 6; Fl. SSSR II, pl. XL, fig. 9.

Although the type of *F. kolymensis* is not preserved, the meaning of the name can be reconstructed with complete certainty. *Festuca kolymensis* was described from the Lower Kolyma; V. P. Drobov associated with this name forms with a high degree of development of mechanical tissue in the leaves (see fig. 6 of his work and our Fig. 2, *2*). Such plants can absolutely not be accepted as *F. auriculata* or *F. brachyphylla,* and there are no other species with similar leaf structure in the North-East. *Festuca kolymensis* has been repeatedly collected again on the Lower Kolyma.

*Festuca kolymensis* is widespread on the steppes of East Siberia and Transbaikalia. It is entirely probable that some of the fescues described by V. V. Reverdatto (Materials for the recognition of Siberian species of the genus *Festuca,* I–III, 1927, 1928, 1936) for the Altay and the southern part of Krasnoyarsk Kray also represent this species. *Festuca kolymensis* is absent from the steppes of West Siberia. We have also not succeeded in finding it in material from the Urals. The extremely interesting collections of V. S. Govorukhin from the steppe-covered limestone rocks of Ylych and Shchugor have proved to represent *F. sulcata* (Hack.) Nym. The latter species, characteristic of the steppes of Europe and West Siberia, is undoubtedly closely related to *F. kolymensis.* Occasionally occurring specimens of *F. kolymensis* with 3-ribbed leaf blade (see Fig. 2, *3*) are especially similar to *F. sulcata.* It is noteworthy that there is also, apparently, found in boreal North America under rather xeric conditions a species possessing leaf structure almost identical to *F. kolymensis* but differing in the shorter (1.0–1.5 mm) anthers. It is impossible to name this species due to the completely obscure systematics of the section *Ovinae* in America. To this species belongs the specimen published by Hitchcock (Am. Grass., No. 476) under the name *F. saximontana,* as well as specimens from the Winnipeg Valley (Bourgeau, 1859) and from Burt Lake in Michigan (Ehlers, No. 2009), etc.

Occasionally occurring in the Arctic on southfacing stony slopes, and sometimes on well drained ridges in river valleys. Known only from the more continental sector of Arctic Siberia, from the lower reaches of the Lena to the Chukotka Range, also in the basins of the Anadyr and Penzhina.

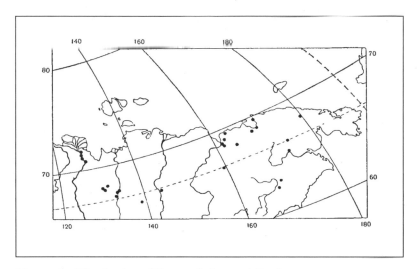

MAP II–69  Distribution of *Festuca kolymensis* Drob.

**Soviet Arctic.** Lower reaches of Lena (north to 72°N); lower reaches of Kolyma and the adjacent part of the Northern Anyuyskiy Range; district of Bay of Chaun (Ust-Chaun, Pevek); Chukotka Range (Amguema Basin); Anadyr Basin; Penzhina Basin. (Map II–69).

**Foreign Arctic.** Apparently absent.

**Outside the Arctic.** Yakutia; Magadan Oblast; Prebaikalia and Transbaikalia; Mongolia; Altay (?).

7. ***Festuca auriculata*** Drob., Predst. sekts. *Ovinae* v Yakut. (1915) 159; Krechetovich & Bobrov in Fl. SSSR II, 511; Karavayev, Konsp. fl. Yakut. 58.

**Ill.:** Drobov, Predst. sekts. *Ovinae* v Yakut., pl. V, fig. 9.

Difficulties may arise in distinguishing *F. auriculata* and *F. brachyphylla* in material lacking anthers, because both species are virtually identical in external morphology and, with respect to the anatomical structure of the leaf, their ranges of variation come into contact or even somewhat overlap. *Festuca brachyphylla* usually differs from *F. auriculata* in the position of the vascular bundle of second size and in the degree of development of the upper bundles of third size (compare Figs. 2, *4–6* and *7–9)*, and in these characters is somewhat reminiscent of *F. kolymensis*. But in the disposition of the mechanical tissue it is, to the contrary, *F. auriculata* which is often reminiscent of *F. kolymensis*. However, specimens of *F. auriculata* with leaf structure identical to *F. brachyphylla* occasionally also occur.

*Festuca auriculata* is, unlike *F. ovina* and *F. kolymensis,* a true barrens and arctic species (to a greater degree a barrens rather than an arctic species, while *F. brachyphylla* is the reverse).

In the tundra zone distributed only in continental districts with mountainous relief adjacent to high subarctic mountains. Growing in dry areas with little snow cover, most often on stony substrate or on sand and gravel deposits of rivers; more common on sites devoid of vegetational turf, but also occurring in closed cryophilic-steppe communities, dry mountain meadows, *Dryas-* and *Alectoria-* tundras, and dry open larch forests; only exceptionally encountered in association with mesophilic mosses.

Especially in the lower reaches of the Lena very common in sparse herb and *Dryas* -carpet tundras of stony summits, ridges and windswept southfacing slopes, on overgrown talus screes and on rock ledges, normally replacing or quantitatively greatly predominant over *F. brachyphylla* in these habitats; occurring also on dry

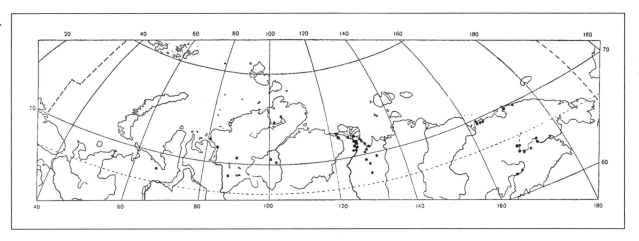

MAP II–70  Distribution of *Festuca auriculata* Drob.

riverine gravelbars and shoreline cliffs; very often together with meadowy steppe or cryophilic steppe plants (sometimes together with *F. kolymensis*). Entering the northern part of the forest zone (and the zone of open forest below the barrens) on rocks, in dry mountain meadows, dry larch forest on southfacing slopes, and dry gravelbars; sometimes growing in glades within thickets of cedar-pine stlanik (on stony summits and slopes). Climbing high in the mountains: to over 700 m a. s. l. on Mount Sokuydakh in the lower reaches of the Lena (70°30'N), to over 2000 m in the Suntar-Khayata Range (about 63°N).

Races very close to *F. auriculata* exist in Central Asia, the Iranian mountains, the Caucasus and the Alps. But it is not yet possible to present a well-founded treatment of this group as a whole and to delineate the boundary between *F. auriculata* and these southern races. Also extremely close to *F. auriculata* is the Altaic *F. Kryloviana* Reverd.; perhaps there also exist forms transitional between the two. But we are forced to leave open for the present the question of the relations between these species, because a complete review of the fescues of the Altay and Sayans is really indispensible for resolving the matter. Apparently the Uralian specimens identified by P. L. Gorchakovskiy, K. N. Igoshina and the present author as *F. Kryloviana* should be referred to *F. auriculata*.

**Soviet Arctic.** Polar Ural (upper reaches of River Khuuta); coast of Yenisey Bay; Taymyr (mountainous districts remote from coast); lower reaches of Lena and Olenek (Chekanovskiy Ridge, Kharaulakh Mountains); lower reaches of Kolyma (mountainous right bank, Northern Anyuyskiy Range); district of Bay of Chaun (Pevek); Anadyr Basin; lower reaches of Penzhina. (Map II–70).

**Foreign Arctic.** Apparently absent.

**Outside the Arctic.** Urals (occasionally on barrens and rocks); barrens of the Verkhoyansk-Kolymsk mountain country, Prebaikalia, the Okhotsk Coast, the Sikhote-Alin Range and Kamchatka. Closely related forms in the Altay and Sayans (barrens).

8. ***Festuca brachyphylla*** Schult., Mant. III, Add. 1 (1827), 646; Scholander, Vasc. pl. Svalb. 69; Hultén, Fl. Al. II, 241; A. E. Porsild, Ill. Fl. Arct. Arch. 40; Polunin, Circump. arct. fl. 52.

*F. brevifolia* R. Br. in Suppl. to Parry's Voyage XI (1824), 289; Drobov, Predst. sekts. *Ovinae* v Yakut. 161; Krylov, Fl. Zap. Sib. II, 313; Tolmachev, Fl. Taym. I, 100; Perfilev, Fl. Sev. I, 98; Krechetovich & Bobrov in Fl. SSSR II, 514; Chernov in Fl. Murm. I, 225; Karavayev, Konsp. fl. Yak. 58; non *F. brevifolia* Muhl., Descr. (1817) 158.

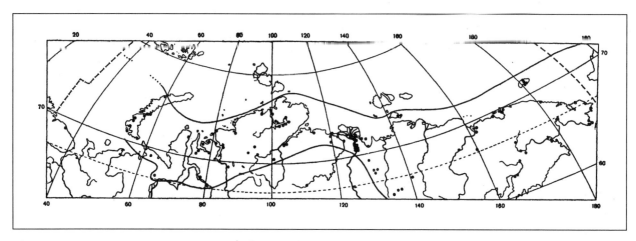

**MAP II–71** Distribution of *Festuca brachyphylla* Schult.

*F. ovina* ssp. *brevifolia* Hack., Monogr. Festuc. (1882) 117.
*F. ovina* var. *brevifolia* Lynge, Vasc. pl. N. Z. 108.
*F. baffinensis* Polunin, Bot. Can. E Arctic 1 (1940), 91; id., Circump. arct. fl. 52; A. E. Porsild, Vasc. pl. W Can. Arch. 84; id., Ill. Fl. Arct. Arch. 40.
*F. hyperborea* Holmen, Cytol. fl. Peary Land (1952) 26.
**Ill.**: Drobov, l. c., pl. VI, fig. 11–13; Fl. SSSR II, pl. LX, fig. 10–11; Scholander, l. c., fig. 39, a; Fl. Murm. I, pl. LXXXIV, fig. 4; Polunin, Circump. arct. fl. 53.

At several sites in the Eurasian Arctic (Spitsbergen; Novaya Zemlya, Gubin Bay; Wrangel Island) specimens have been found with rather dense culm pubescence, a character described for *F. baffinensis* Polunin. But to recognize these specimens as a separate species different from the rest of our plants would be too hasty. They do not differ from *F. brachyphylla* either in general habitus or in leaf structure. N. Polunin gives anther measurements for his species as 0.3–0.5 mm (l. c., 1940) or 0.3–0.6 mm (l. c., 1959). According to our material the range of measurements in *F. brachyphylla* is 0.4–0.8 (rarely 1.0) mm. The differences are of extremely doubtful significance, especially since the number of specimens of *F. baffinensis* studied by N. Polunin was not very great. Out of two specimens from Wrangel Island possessing pubescent stems, in one the anthers proved to be 0.4–0.6 mm long, in the other about 0.7 mm. In specimens of *F. baffinensis* from Canada (Southampton and Cornwall Islands collected by Tikhomirov in 1959) the anthers proved to be 0.5–0.6 mm long, while the leaf structure was once again typical of *F. brachyphylla*. Thus, the lack of foundation for elevating occasionally occurring specimens of *P. brachyphylla* with pubescent stems to the rank of a distinct species is evident enough.

Arctic-alpine species, the most "tundra-loving" of all fescues. Growing both on dry and on highly moist but adequately aerated substrates (stony, rocky, sandy or clayey, with varying degree of winter snow cover); normally not occurring in oligotrophic communities. Most often growing in areas with little competition from other plants, although not avoiding closed communities (streamside tundra meadows, certain kinds of mossy tundras with shallow depth of moss cover); in mossy tundras most often associated with convex patches of bare loam.

Thus, in the lower reaches of the Lena it occurs much more rarely than *F. auriculata* in stony tundras with sparse herbs or *Dryas* carpets on snowless summits and windswept southwestern slopes, and on rocks and overgrown screes; but it is more common on convex bare patches in dry stony polygonal tundras, on loamy patches in moss-herb and moist moss-sedge (with *Carex arctosibirica*) tundras, and on the tops of mounds in *Cassiope* tundras. Often growing on riverine sands and gravelbars, and in low willow-moss and herbaceous communities in valleys.

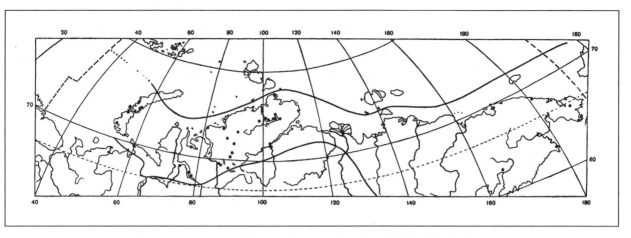

MAP II–72 Distribution of *Festuca brachyphylla* Schult. f. *vivipara*. Continuous lines: boundaries of distribution of *F. brachyphylla* s.l.

Entering the northern part of the forest zone only in East Siberia on gravelbars, moist shoreline slopes and rocks; sometimes also growing in mossy larch woods and dwarf birch carr in valleys (more often on bare patches), in ashes of burntover forest, and as a weed along roads and near the camps of reindeer-herders.

**Soviet Arctic.** Murman Coast (Gavrilovo); Polar Ural; Pay-Khoy (upper reach of River Talata); Novaya Zemlya; Yamal (northern part); district of Tazovskaya Bay; Gydanskaya Tundra; lower reaches of Yenisey, Sibiryakov Island; Dikson Island; Taymyr; recorded by Ye. S. Korotkevich for Severnaya Zemlya (October Revolution Island); Khatanga; Anabarskaya Tundra; Chekanovskiy Ridge, lower reaches of Olenek and Lena; Kharaulakh Mountains; New Siberian Islands (Bolshoy Lyakhovskiy Island); district of lower reaches of Kolyma; district of Bay of Chaun; Ayon Island; Wrangel Island; Chukotka; Anadyr Basin; Penzhina Basin. *F. brachyphylla* f. *vivipara*: west coast of Baydaratskaya Bay; Vaygach; Novaya Zemlya; district of Tazovskaya Bay; Gydanskaya Tundra; lower reaches of Yenisey; Taymyr; district of Bay of Chaun (Pevek); north coast of Chukotka (Cape Schmidt); SE extremity of Chukotka Peninsula; Penzhina Basin; Koryak Highlands. Absent between Khatanga and the Bay of Chaun. (Maps II–71, II–72).

**Foreign Arctic.** Alaska; Arctic Canada; Labrador; Canadian Arctic Archipelago; Greenland (ubiquitous); Iceland; Spitsbergen. *F. brachyphylla* f. *vivipara*: together with the normal form but apparently rare in the western part of North America.

**Outside the Arctic.** High Ural Mountains (south to Denezhkin Rock); Altay; Sayans; Prebaikalia; Central Siberian Plateau (northern edge); Verkhoyansk-Kolymsk mountain country; Kamchatka; Pamir (one locality); North America (in the high mountains of the Rocky Mountain complex south at least to Utah).

### ❧ GENUS 30    *Zerna* Panz. — PERENNIAL BROME GRASS

RELATIVELY SMALL (about 50 species) genus distributed predominantly in cold and temperate countries of the Northern Hemisphere. Formerly usually united with the genus *Bromus* as a section or subgenus, although morphologically quite distinct and showing considerably greater differences from *Bromus* s. str. than those between many other generally accepted genera of grasses. The majority of species of *Bromus* are steppe or desert ephemerals and most diverse in the Mediterranean, while the genus *Zerna* contains predominantly forest and meadow species and is especially richly represented in North America (28 species) where species of *Bromus* only occur as introductions. Only a few species of *Zerna* penetrate the Arctic, and none among them reach the High Arctic. The four species in the Soviet Arctic are very closely related to one another and are sometimes accepted as subspecies or varieties of a single species, *Z. inermis* (Leyss.) Lindm. s. l. All of them are characterized by the presence of more or less elongate subterranean stolons (rhizomes).

1. Plant 30–80 cm high; culms with 3–5 *bare nodes*; sheaths bare; leaf blades 3–10 mm wide, usually hairless on both surfaces, more rarely sparsely hairy on upper surface. Glumes bare; lemmas *hairless or more or less covered with short hairs only on lower third*, awnless or with short awns (up to 4 mm long) at tip; paleas usually with only very short acicules on keels. . . . . . . . . . . . . . . . . . . . . . . . . . . . . . . . . . . . . . . . . . . . . . . . . . . 1. Z. INERMIS (LEYSS.) LINDM.
   – Culms almost always *more or less hairy on or near nodes*, very rarely bare; sheaths more or less hairy or bare; leaf blades more or less hairy on upper or both surfaces, more rarely bare. Lemmas *more or less hairy on sides* (along submarginal nerves) *from base to half or more of their length*, usually also more or less hairy on back (along median nerve), usually with awns 1–6 mm long (rarely awnless) at tip; paleas usually with both short and longer (often almost ciliate) acicules on keels. . . . . . . . . . . . . . . . . . . . . . .2.

2. Glumes *with scattered hairs over their whole surface*; lemmas *hairy over their whole surface or almost so* (except for small portions of their distal quarter). Plant 20–40 cm high; stems with 2–3 approximated nodes on their lower third, bare below panicle; sheaths bare or sparsely hairy. . . . . . . . . . . . . . . . . . . . . . . . . . . . . . . . . . . . . . . . . . . . . . .4. Z. ARCTICA (SHEAR) TZVEL.
   – Glumes *bare, or very rarely short-ciliate only on keel and lateral nerves*; lemmas *bare between dorsal and submarginal strips of hairs from tip to half or more of their length*. . . . . . . . . . . . . . . . . . . . . . . . . . . . . . . . . . . . . . .3.

3. Culms below panicle *bare*; leaf sheaths *bare or sparsely hairy*; leaf blades *entirely bare or more or less hairy on upper surface*, very rarely sparsely hairy on both surfaces. . . . . . . . . . . . . . . . .2. Z. PUMPELLIANA (SCRIBN.) TZVEL.
   – Culms below panicle *more or less hairy*; sheaths and leaf blades *very densely hairy* (virtually velvety) *on both surfaces*. . . . . . . . . . . . . . . . . . . . . . . . . . . . . . . . . . . . . . . . . . . . . . . . . . . . . . . .3. Z. IRCUTENSIS (KOM.) NEVSKI.

1. **Zerna inermis** (Leyss.) Lindm., Svensk Fanerogam. fl. (1918) 101; Nevskiy in Tr. Sredneaz. gos. univ., ser. 8B, XVII (1934), 18.

    *Bromus inermis* Leyss., Fl. Hal. (1761) 16; Grisebach in Ledebour, Fl. ross. IV, 357; Krylov, Fl. Zap. Sib. II, 335, pro parte; Perfilev, Fl. Sev. I, 100; Nevskiy & Sochava in Fl. SSSR II, 558; Leskov, Fl. Malozem. tundry 32; Hultén, Fl. Al. II, 250; id., Atlas, map 255; Hylander, Nord. Kärlväxtfl. I, 351; Kuzeneva in Fl. Murm. I, 235; Karavayev, Konsp. fl. Yak. 59.

    *B. inermis* var. *grandiflora* Rupr., Fl. samojed. cisur. 61.

    *B. inermis* var. *glabra* Trautv., Pl. Sib. bor. 135.

    *B. inermis* ssp. *inermis* — Wagnon in Brittonia VII (1952), 464.

    Ill.: Fl. Murm. I, pl. LXXVII.

    Predominantly European boreal species, distributed in many other countries of both hemispheres as a result of importation (as a fodder plant) or introduction. Only penetrating the Arctic in a few districts.

    Growing on sands and gravelbars in river valleys, on relatively poorly vegetated clayey or sandy slopes, and sometimes as an introduction along roads and near settlements.

    **Soviet Arctic.** Murman (recorded by Hultén for the Rybachiy Peninsula and the vicinity of Ponoy settlement); Kanin (southern part); Malozemelskaya and Bolshezemelskaya Tundras; Polar Ural (rare); lower reaches of Ob (near Salekhard and on the Poluy); Gydanskaya Tundra; lower reaches of Yenisey (north to Dudinka).

    **Foreign Arctic.** Beringian coast of Alaska (near Nome, introduced).

    **Outside the Arctic.** Almost all Europe; Middle East and Central Asia; Mongolia; Northern China; Siberia north to Salekhard, Dudinka, the northern part of the Central Siberian Plateau and east to the Lena (further east than the Lena only as an introduction); considerable part of North America from Southern Alaska and Newfoundland to Mexico (only as an imported or introduced plant).

2. **Zerna Pumpelliana** (Scribn.) Tzvel., comb. nova.

    *Z. vogulica* (Socz.) Nevski in Tr. Sredneaz. gos. univ., ser. 8B, XVII (1934), 17.

    *Z. Richardsonii* (Link) Nevski in Tr. Sredneaz. gos. univ., ser. 8B, XVII (1934), 17, quoad pl.

    *Z. occidentalis* Nevski in Tr. Sredneaz. gos. univ., ser. 8B, XVII (1934), 18.

    *Bromus ciliatus* auct. non L. — Grisebach in Ledebour, Fl. ross. IV (1853), 358; Scheutz, Pl. jeniss. 184.

    *B. inermis* var. *ciliata* (L.) Trautv., Pl. Sib. bor. (1877) 135; id., Syll. pl. Sib. bor.-or. (1888) 61, quoad pl.

    *B. Pumpellianus* Scribn. in Bull. Torr. Bot. Cl. XV (1889), 9; Hultén, Fl. Al. II, 251; Polunin, Circump. arct. fl. 41.

    *B. inermis* var. *villosus* Kryl., Fl. Alt. (1914) 1683.

    *B. sibiricus* Drob. in Tr. Bot. muz. Peterb. Ak. nauk XII (1914), 229; Perfilev, Fl. Sev. I, 100; Nevskiy & Sochava in Fl. SSSR II, 561; Govorukhin, Fl. Urala 129; Karavayev, Konsp. fl. Yak. 59.

    *B. inermis* var. *sibiricus* (Drob.) Kryl., Fl. Zap. Sib. II, 335.

    *B. vogulicus* Socz. in Dokl. AN SSSR, ser. A, 7 (1929), 167; Perfilev, l. c. 100; Nevskiy & Sochava, l. c. 561; Govorukhin, l. c. 130.

    *B. sibiricus* var. *taimyrensis* Roshev. in Tolmachev, O fl. nakh. v tsentr. chasti Taym. (1930) 108 and in Fl. Taym. I, 100.

    *B. Richardsonii* auct. non Link — Nevskiy & Sochava, l. c. 562; Karavayev, l. c. 59.

    *B. uralensis* Goworuch., Fl. Urala (1937) 129, 531.

    *B. Julii* Goworuch., l. c. 129, 532.

    *B. inermis* ssp. *Pumpellianus* (Scribn.) Wagnon in Rhodora LII (1950), 211, pro parte.

    *B. arcticus* var. *vogulica* (Socz.) Karav., Konsp. fl. Yak. (1958) 59.

**Ill.**: Fl. SSSR II, pl. XLIII, fig. 3; Tolmachev, Fl. Taym. I, fig. 21; Govorukhin, l. c., fig. 37.

Subarctic species widely distributed in more northern and mountainous districts of Siberia and North America, where it almost completely replaces *Z. inermis*. Displaying much polymorphism, sometimes significantly approaching the preceding species although specimens transitional between them are not too frequent. North American material of *Z. Pumpelliana* available to us is very similar to Siberian and likewise very variable with respect to the size of the plant as a whole and the degree of pubescence of the lemmas, leaf blades and sheaths. Consequently, there are in our opinion insufficient grounds for referring Siberian specimens to a separate species (*B. sibiricus* Drob.), as was done in Volume II of the "Flora of the USSR" and many subsequent works. The characters given by the author of that species, V. P. Drobov, as distinguishing it from *B. Pumpellianus*, such as smaller spikelet size, narrower leaf blades and more weakly pubescent lemmas, do not withstand any critique on the basis of the very extensive herbarium material now available, and geographical differences unsupported by sufficiently constant morphological differences provide no convincing argument in favour of recognizing *B. sibiricus* as a separate species. In connection with the high variability of *Z. Pumpelliana* and the possibility of dividing this species into infraspecific units, great importance attaches to selection of a lectotype of *B. sibiricus*, whose type was not designated by the author in his description. We have selected the lectotype of this species (Yakutsk Oblast, stony shore of River Amga near ferry on Ust-Maya road, 2 VIII 1912, No. 572, Drobov) on the basis of the following considerations. V. P. Drobov indicated that he began his investigation of the perennial species of the genus *Bromus* L. s. l. in connection with study of his own collections of this species from Yakutia. These collections, like G. I. Dolenko's collections from Yakutia in the same year, bear determinations by V. P. Drobov (as "*Bromus sibiricus* m.") which are earlier than such determinations of the bulk of the remaining material reviewed by him. Among the specimens collected by V. P. Drobov, only two specimens with the label cited above belong to his variety "var. *glaber* Drob." (Tr. Bot. muz. Peterb. Ak. nauk XII, 232), the most widespread variety of the species and undoubtedly accepted by him as the typical variety. The type specimens of *B. sibiricus* selected by us thus belong to the variety *Z. Pumpelliana* var. *glabra* (Drob.) Tzvel. comb. nova; they possess bare leaf blades and sheaths, more or less condensed panicles and relatively weakly hairy lemmas, like the majority of more southern and more western specimens of the species. We have not seen the type specimens of *B. Pumpellianus*, described from the Rocky Mountains (Montana State), but they apparently possess more copiously hairy lemmas like the majority of specimens of this species from the East Siberian Arctic, which can thus be accepted as the typical variety of *Z. Pumpelliana*. Specimens with hairy sheaths and leaf blades, *Z. Pumpelliana* var. *pollita* (Drob.) Tzvel. comb. nova (= *B. sibiricus* var. *pollitus* Drob.), occur occasionally throughout the range of the species but are perhaps more common in the mountains of Yakutia, where they sometimes significantly approach the closely related species *Z. ircutensis*. Specimens of another of V. P. Drobov's varieties, *Z. Pumpelliana* var. *flexuosa* (Drob.) Tzvel. comb. nova (= *B. sibiricus* var. *flexuosus* Drob.), differing from *Z. Pumpelliana* var. *glabra* only in the more widely diffuse panicles often with more or less flexuous branchlets, occur mainly in more southern districts; they undergo such a gradual transition to specimens with more abbreviated panicle branchlets that they scarcely deserve separation as a distinct taxonomic unit, although they were erroneously identified by S. A. Nevskiy and V. B. Sochava with the North American species *Z. Richardsonii* (Link) Nevski. The latter species, like the closely related also North American *Z. ciliata* (L.) Henr., differs from *Z. Pumpelliana* not so much in panicle structure as in its ability to grow in tufts without subterranean stolons. Although a considerable majority of the Siberian specimens formerly determined as "*B.*

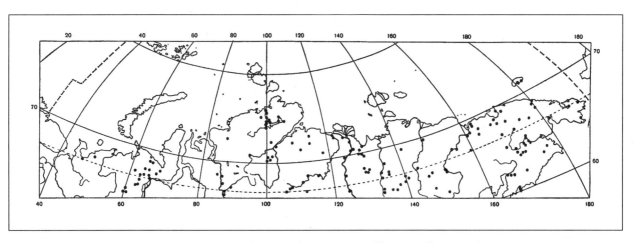

MAP II–73  Distribution of *Zerna Pumpelliana* (Scribn.) Tzvel.

*Richardsonii*" undoubtedly belong to *Z. Pumpelliana*, it cannot be excluded that there is an inadequately studied tufted species near *Z. Richardsonii* (but hardly identical with that species) in districts of East Siberia along the Pacific Ocean. Arctic and high alpine specimens of *Z. Pumpelliana* are of relatively small size and have culms with the nodes more or less approximated on their lower part. Such specimens from the Urals were described as *Bromus vogulicus* Socz., *B. uralensis* Goworuch. and *B. Julii* Goworuch. The differences existing between the type specimens of these species have proved to be far less substantial than they seemed to be when only very limited Uralian material was available, on which account we prefer to include these names in the list of synonyms of *Z. Pumpelliana*. The type specimen of *B. uralensis* (from Mount Akhtas-Syupa-Pol at the sources of the River Unya), like the majority of other Uralian specimens, has especially strongly abbreviated spikelet pedicels. Still earlier S. A. Nevskiy drew attention to this same character, and proposed the species *Z. occidentalis* Nevski also basically from Uralian material. The type specimens of *B. vogulicus* (from the upper reaches of the River Kozhima) combine rather short panicle branchlets with awnless lemmas and sparsely hairy leaf sheaths, but there are specimens very similar to them from the same district with rather long awns and bare sheaths, upon which V. S. Govorukhin based his second species *B. Julii* (from the sources of the River Grube-Yu, a tributary of the River Khulga).

*Zerna Pumpelliana* grows in meadows and on sands and gravelbars in river valleys, on stony or sandy grassy slopes, sometimes in valley shrub communities and as an introduction along roads.

**Soviet Arctic.** Malozemelskaya Tundra (on the Rivers Pesha and Indiga); Bolshezemelskaya Tundra (mainly the eastern part); Polar Ural (frequent); Karsko-Baydaratskaya Tundra (on the Kara); lower reaches of Yenisey; Taymyr (basins of Lake Taymyr and the Pyasina); lower reaches of Olenek and Lena (from southern part of delta); Kharaulakh Mountains and coast of Buorkhaya Bay; lower reaches of Kolyma (to mouth); east of the mouth of the Kolyma throughout Chukotka to the district of the Bering Strait, including Wrangel Island; Anadyr and Penzhina Basins; Koryakia (in the districts of Olyutorskiy and Korf Bays). (Map II–73).

**Foreign Arctic.** Arctic Alaska (Yukon Basin); NW part of Canada (Mackenzie River Basin to its mouth).

**Outside the Arctic.** In the Urals south to 60°N; Altay; Sayans; Central Siberian Plateau; further east throughout the mountainous districts of East Siberia south, apparently, to Korea and Northern Japan; southern part of Alaska; Rocky Mountains (south to 36°N) and disjunctly in the American Great Lakes district.

3. ***Zerna ircutensis*** (Kom.) Nevski in Tr. Sredneaz. gos. univ., ser. 8B, XVII (1934), 17.
   *Bromus ircutensis* Kom. in Bot. mat. Gerb. Peterb. bot. sada II (1921), 130; Nevskiy & Sochava in Fl. SSSR II, 560.
   *B. arcticus* ssp. *ornans* (Kom.) Karav., Konsp. fl. Yak. (1958) 59, quoad pl.
   **Ill.:** Fl. SSSR II, pl. XLIII, fig. 5.

   East Siberian species extremely close to the preceding and possibly merely constituting an extreme variant of it. But the few known specimens of *Z. ircutensis* have a definite association with mountainous districts of Prebaikalia, Yakutia (east of the Lena) and Chukotka, where there also occur the most pubescent specimens of the preceding species which have possibly originated as the hybrids *Z. Pumpelliana* × *Z. ircutensis*. Moreover, a very similar species, *Z. ornans* (Kom.) Nevski, differing from *Z. ircutensis* only in having hairy glumes and lemmas hairy over almost their whole surface, is known from Kamchatka and the Okhotsk Coast.

   Growing on sands and gravelbars in river valleys, more rarely on the sea shore.

   **Soviet Arctic.** Known only from the district of the Bay of Chaun (pebble beach of Bay of Chaun at Pevek settlement, 18 VIII 1937, Yakovlev).
   **Foreign Arctic.** Apparently not occurring.
   **Outside the Arctic.** Mountains of Prebaikalia; Cherskiy Range.

4. ***Zerna arctica*** (Shear) Tzvel., comb. nova.
   *Bromus purgans* auct. non L. — Grisebach in Ledebour, Fl. ross. IV (1853), 361.
   *B. arcticus* Shear in Scribn. et Merr., Grass. Al. (1910) 83.
   *B. Pumpellianus* var. *arcticus* (Shear) Porsild in Rhodora XLI (1939), 182; Hultén, Fl. Al. II, 251; A. E. Porsild, Ill. Fl. Arct. Arch. 40.
   *B. inermis* ssp. *Pumpellianus* var. *arcticus* (Shear) Wagnon in Rhodora LII (1950), 211; id. in Brittonia VII (1952), 466.

   Subarctic Beringian species also very close to *Z. Pumpelliana* and sometimes united with that species as a variety. However, its range in North America is distinctly more northern than the range of *Z. Pumpelliana* although not completely separated from the range of the latter, as can be judged from the monograph of the American species of *Zerna* cited above (Wagnon, l. c., 1952). In the Soviet Arctic *Z. arctica* apparently only occurs in the Far East, although specimens of *Z. Pumpelliana* approaching this species (with glumes ciliate on the nerves) are also known from Northern Yakutia.

   Growing on sands and gravelbars in river valleys and near the sea shore.

   **Soviet Arctic.** Chukotka Peninsula (Lawrence Bay), Anadyr Basin (River Yanraveyveyem, near the sea coast).
   **Foreign Arctic.** Arctic Alaska (Beringian coast and Yukon Basin); NW part of Canada (rather frequent).
   **Outside the Arctic.** Only the southern part of Alaska and the upper reaches of the Yukon and Mackenzie.

◆ GENUS 31     **Bromus** L. — BROME GRASS

RELATIVELY SMALL GENUS (about 50 species), distributed mainly in the temperate and warmer countries of Eurasia and Africa, especially in the Mediterranean. Many species have been introduced to America, Australia and New Zealand, where they have now spread widely. The majority of species of the genus are steppe or desert ephemerals and only a few occur in the forest zone. In the Soviet Arctic only two species of widespread agricultural weeds are reported as introductions.

1. Panicles widely diffuse with long branchlets which often bear 2–4 spikelets; *spikelets with more or less distinct rosy-violet tinge; lemmas always with rather long awns (3–8 mm long); anthers 3–4 mm long.* Sheaths of lower leaves usually more or less hairy. . . . . . . . .1. **B. ARVENSIS** L.
- Panicles more or less diffuse or condensed; their branchlets usually with one (more rarely two) spikelets; *spikelets pale green; lemmas awnless or with awns 1–6 mm long; anthers 1.5–2.5 mm long.* Sheaths of lower leaves bare or slightly hairy. . . . . . . . . . . . . . . . . . . . . . . . . . . . . .2. **B. SECALINUS** L.

    *1.* ***Bromus arvensis*** L., Sp. pl. (1753) 77; Grisebach in Ledebour, Fl. ross. IV, 362; Krylov, Fl. Zap. Sib. II, 337; Perfilev, Fl. Sev. I, 101; Krechetovich & Vvedenskiy in Fl. SSSR II, 576; Hultén, Atlas, map 250; Kuzeneva in Fl. Murm. I, 236; Hylander, Nord. Kärlväxtfl. I, 354.

    Ill.: Fl. Murm. I, pl. LXXVIII, fig. 3.

    Occurring in the Arctic only as a rare introduction along roads and near settlements; outside arctic limits in various agricultural crops.

    **Soviet Arctic.** Murman (vicinities of Pechenga and Teriberka).
    **Foreign Arctic.** Arctic Scandinavia (as introduction, to 70°N).
    **Outside the Arctic.** Almost all Europe; South Siberia; Caucasus; Middle East. As an introduction in many countries of both hemispheres.

    *2.* ***Bromus secalinus*** L., Sp. pl. (1753) 76; Grisebach in Ledebour, Fl. ross. IV, 364; Krylov, Fl. Zap. Sib. II, 336; Perfilev, Fl. Sev. I, 100; Krechetovich & Vvedenskiy in Fl. SSSR II, 576; Hultén, Fl. Al. II, 253; id., Atlas, map 260; Hylander, Nord. Kärlväxtfl. I, 355; Kuzeneva in Fl. Murm. I, 235.

    Ill.: Fl. SSSR II, pl. XLII, fig. 8; Fl. Murm. I, pl. LXXVIII, fig. 1–2.

    Like the preceding species, occurring in the Arctic only as an introduction along roads and near settlements; outside arctic limits a weed of various crops, especially rye.

    **Soviet Arctic.** Murman (vicinity of Teriberka).
    **Foreign Arctic.** Arctic Alaska (near Nome and in the Yukon Basin); Arctic Scandinavia (to 69°30'N).
    **Outside the Arctic.** Almost all Europe; South Siberia; Caucasus and Middle East; North Africa. As an introduction in many countries of both hemispheres.

## GENUS 32    Nardus L. — MATGRASS

MONOTYPIC GENUS DISTRIBUTED mainly in Europe.

> 1. ***Nardus stricta*** L., Sp. pl. (1753) 53; Grisebach in Ledebour, Fl. ross. IV, 324; Lange, Consp. fl. groenl. 154; Gelert, Fl. arct. I, 132; Krylov, Fl. Zap. Sib. II, 342; Nevskiy in Fl. SSSR II, 587; Perfilev, Fl. Sev. I, 101; Devold & Scholander, Fl. pl. SE Greenl. 142; Seidenfaden, Vasc. pl. SE Greenl. 90; Gröntved, Pterid. Spermatoph. Icel. 142; Kuzeneva in Fl. Murm I, 240; Hultén, Atlas, map 249; id., Amphi-atl. pl. 120; Böcher & al., Grønl. Fl. 273; Polunin, Circump. arct. fl. 59.
>
> Ill.: Fl. Murm. I, pl. LXXIX.
>
> Characteristic plant of meadows in dry valleys of the temperate north and at subalpine elevations. In this type of habitat penetrating the Arctic in its Atlantic districts, mainly on mountain slopes.
>
> **Soviet Arctic.** Murman (more or less ubiquitous). East of the White Sea not reaching the northern limits of forest.
>
> **Foreign Arctic.** South Greenland (to 65°30'N); Iceland; Arctic Scandinavia.
>
> **Outside the Arctic.** Greater part of Europe outside the USSR; Middle East; Caucasus; European part of USSR, except the southeastern steppe region; in the east not reaching beyond the eastern foothills of the Urals. Occasionally reported as an introduction in Siberia.

## GENUS 33    Roegneria C. Koch — RHIZOMELESS WHEAT GRASS

GENUS WIDESPREAD IN cold and temperate countries of both hemispheres, especially diversely represented in mountainous districts of Asia and North America. The number of species apparently reaches 100–120. Right until most recent times this genus, like the genus *Elytrigia* Desv., was frequently united with the genus *Agropyron* Gaertn., although S. A. Nevskiy demonstrated the existence of quite fundamental differences between these genera. For despite the great external resemblance between species of *Roegneria* and *Elytrigia* (many species of *Roegneria* were sometimes even united with *E. repens* as varieties or forms!), there is now no doubt that *Roegneria* has considerably greater affinity with the genus *Elymus* L. (= *Clinelymus* Nevski) than with the genus *Elytrigia*, which in its turn shows a rather close relationship with the genus *Leymus* Hochst. Therefore, if *Roegneria* is not recognized as a separate genus, then it should be united first with the genus *Elymus*, not with *Elytrigia* and *Agropyron*. The relative ease of hybridization of species of *Roegneria* with species of *Hordeum* even suggests the possibility of a hybrid origin of such genera as *Elymus* and *Asperella*, whose species are in many respects very reminiscent of hybrids between *Roegneria* and *Hordeum* spp. which arise at the present time and which can sometimes be fertile and rather stable with respect to morphological characters. To the contrary, hybrids between *Roegneria* and species of *Agropyron* or *Elytrigia* are extremely rare and always sterile. Interspecific hybrids in the genus *Roegneria* occur rather rarely, something probably explained by the fact that all (or almost all) species of this genus are facultative selfpollinators. This is the explanation for the tendency of *Roegneria* to form morphologically stable, often narrowly endemic races which usually differ in only one or a few characters and are even sometimes distributed in only a single district. Similar races are often united as subspecies or varieties of larger polytypic species, whose boundaries, however, remain rather blurred. Much work in clarifying the species of *Roegneria* in the flora of the USSR was undertaken by S. A. Nevskiy, who in general introduced many substantial changes in the systematics of the tribe *Hordeae*. Recently the genus *Roegneria* has been intensively studied by Melderis (The short-awned species of the genus *Roegneria* of Scotland, Iceland and Greenland, Svensk Bot. Tidskrift, Bd. 44, H. 1, 1950).

Of the 14 species included by us in the present "Flora," only a few (for example, *R. villosa* and *R. borealis)* are relatively widespread in the Arctic, being arctic or subarctic species. The majority consists of boreal species only just penetrating the arctic zone, where they grow in the most protected sites, most often on sands and gravelbars in river valleys and in valley shrub communities.

1. Segments of spikelet rachis *covered only with minute fine acicules, always without hairs* (but callus of lemmas with rather long hairs!); glumes with 4–5 nerves, usually slightly shorter than (not less than ⅔ as long as) lemmas; lemmas (excluding callus) bare and smooth, with very short acicules only on nerves on distal part, usually with short awn (1–4 mm long) at tip; anthers 1–2 mm long; spikes 4–10 cm long, erect or more rarely slightly nodding, rather compact, sometimes with rosy-violet tinge. Plant 20–60 cm high; culm nodes usually short-haired, rarely bare; leaf blades 1.5–5 mm wide. (*R. borealis* s. l.). . . . . . . . . . . . . . . . . . . . . . . . . . . . . . . . . . . . . . . .2.
– Segments of spikelet rachis *with longer acicules transitional to hairs along whole length of segments, or more rarely only on their distal part*. . . . . . . . .4.

2. Leaf blades *bare but more or less scabrous.* . . . . . . . .14. **R. SCANDICA** NEVSKI.
– Leaf blades *more or less hairy.* . . . . . . . . . . . . . . . . . . . . . . . . . . . . . . . . . . . . .3.

3. Leaf blades *short-haired on both surfaces.* . . . . . . . . . . . . . . . . . . . . . . . . . . . .
 . . . . . . . . . . . . . . . . . . . . . . . . . . . . . . . . . . .13. **R. KRONOKENSIS** (KOM.) TZVEL.
– Leaf blades *hairy only on upper surface.* . . .12. **R. BOREALIS** (TURCZ.) NEVSKI.

4. Lemmas *more or less scabrous* with acicules, *with long curved-divergent awn (2–3 cm long)* at tip; glumes with 3 nerves, usually ⅖ - ½ as long as lemmas, with mucro or awn at tip up to 3–4 mm long; anthers 1–2 mm long; spikes 8–20 cm long, *nodding, with more or less flexuous rachis and widely spaced spikelets.* Plant 30–80 cm high; leaf blades 2.5–9 mm wide, usually with sparse hairs on upper surface, more rarely bare. . . . . . . . . . . . . .
 . . . . . . . . . . . . . . . . . . . . . . . . . . . . . . . . .1. **R. CONFUSA** (ROSHEV.) NEVSKI.
– Awns of lemmas *straight or slightly flexuous, usually considerably shorter (up to 0.8 cm long),* more rarely (in *R. canina*) up to 1.8 cm but then spikes *relatively compact, erect or slightly nodding,* and lemmas *bare and smooth* over almost their whole surface. . . . . . . . . . . . . . . . . . . . . . . . . . . . . . . . . . . .5.

5. Lemmas *with rather long slightly flexuous awns (1–1.8 cm long)* at tip, *bare and smooth,* with very short acicules only on nerves on distal part; glumes

usually not less than ⅔ as long as lemmas, with 4–5 nerves, with mucro or awn at tip up to 5 mm long; anthers 2–2.5 mm long. Plant 30–80 cm high; leaf blades 4–10 mm wide, bare or sparsely hairy on upper surface. ........
................................................2. **R. CANINA** (L.) NEVSKI.

– Lemmas *awnless or with short straight awns up to 0.5 cm long, more rarely* (in *R. jacutensis*) *up to 0.8 cm* but then lemmas *more or less hairy or scabrous on basal half.* ..............................................6.

6. Anthers *1.8–2.5 mm long;* glumes *only slightly* shorter than lemmas, with 4–5 nerves, rather gradually narrowing at tip into mucro or short awn up to 3 mm long; lemmas with sparse acicules (occasionally transitional to short hairs) over their whole or almost their whole surface, with awn 1–5 mm long at tip; spikes 7–18 cm long, *compact,* erect, *often more or less secund.* Plant 25–80 cm high; leaf blades *4–10 mm wide,* covered with rather long hairs on upper surface. ...........3. **R. MUTABILIS** (DROB.) HYL.

— Anthers *1–1.8 mm long;* glumes ½ - ⅔ as long as lemmas, with 3–5 nerves, often more abruptly acuminate at tip; spikes *usually looser, two-sided.* Leaf blades *bare or more rarely hairy* (in *R. turuchanensis, R. subfibrosa* and *R. hyperarctica*) but then usually narrower. .......................7.

7. Lemmas *hairy on callus and sometimes also basally on sides,* usually with very short acicules on nerves near tip, otherwise bare and smooth, usually awnless, more rarely with awns up to 2 mm long; spikes 6–20 cm long, usually more or less nodding, more rarely erect. Plant 25–80 cm high. ...8.

— Lemmas *covered on basal half or over almost their whole surface with sparse or densely arranged hairs or elongated acicules;* glumes bare on inside, rarely with a few very short hairs. .............................9.

8. Glumes almost always with 3 nerves, usually little more than half as long as lemmas, *more or less hairy on inside* (usually with numerous hairs); lemmas (except callus) bare; anthers 1.2–1.8 mm long; spikes usually rather compact. Leaf blades bare but more or less scabrous, 2.5–8 mm wide......................................4. **R. FIBROSA** (SCHRENK) NEVSKI.

— Glumes with 3–5 nerves, ½ - ⅔ as long as lemmas, *bare on inside;* lemmas *usually more or less hairy basally on sides;* anthers 1–1.5 mm long; spikes sometimes with spikelets widely spaced on lower part. Leaf blades usually bare but more or less scabrous, more rarely with sparse hairs on upper surface, 2–7 mm wide. .......................5. **R. SUBFIBROSA** TZVEL.

9. Plant *10–35 cm high,* forming dense tufts; leaf blades rather narrow (1–4 mm wide), often with involute margins; nodes of culm *usually short-pubescent,* rarely almost or completely bare. Spikes 3–10 cm long, erect and relatively compact, *usually with rosy-violet tinge,* more rarely greenish; lemmas usually more or less hairy, more rarely scabrous, usually with awn 1–5 mm long at tip; glumes ½ - ⅔ as long as lemmas, with 3–5 nerves, *on distal part relatively broad then abruptly acuminate,* sometimes with 1–2 lateral teeth, sometimes more or less hairy; anthers 1–1.5 mm long. (*R. violacea* s. l.). ................................................10.

–   Plant *25–80 cm high,* forming dense or loose tufts, sometimes with prostrate shoot bases which root at the nodes; leaf blades 2–6 mm wide; culm nodes *usually bare, rarely with a few short hairs.* Spikes 4–20 cm long, erect or nodding, sometimes with spikelets widely spaced on lower part, *usually greenish,* more rarely with faint rosy-violet tinge; glumes usually little more than half as long as lemmas, with 3–4 nerves, *on distal part relatively gradually attenuate and acuminate,* sometimes more or less hairy. (*R. macroura* s. l.). . . . . . . . . . . . . . . . . . . . . . . . . . . . . . . . . . . . . . . . . . . . . . . .11.

10. Leaf blades with short hairs on both surfaces. . . . . . . . . . . . . . . . . . . . . . . . . . . . . . . . . . . . . . . . . . . . . . . . . . . . . . . . . . . .11. **R. HYPERARCTICA** (POLUNIN) TZVEL.
–   Leaf blades bare but more or less scabrous with acicules. . . . . . . . . . . . . . . . . . . . . . . . . . . . . . . . . . . . . . . . . . . . . . . . . . . . . . . . . . . . .10. **R. VILLOSA** V. VASSIL.

11. Leaf blades *hairy* on upper surface. Spikes *relatively compact, 4–9 cm long;* glumes with relatively broad hyaline border; lemmas awnless or more rarely with awns up to 2–3 mm long. . . . . . . . . . . . . . . . . . . . . . . . . . . . . . . . . . . . . . . . . . . . . . . . . . . . . . . . . . . . . .9. **R. TURUCHANENSIS** (REVERD.) NEVSKI.
–   Leaf blades *bare but more or less scabrous.* Spikes *looser, 8–20 cm long,* with spikelets often widely spaced on lower part. . . . . . . . . . . . . . . . . . . . . .12.

12. Lemmas at tip *with awns 3–8 mm long;* anthers 1.2–1.8 mm long. . . . . . . . . . . . . . . . . . . . . . . . . . . . . . . . . . . . . . . . . . . . . .8. **R. JACUTENSIS** (DROB.) NEVSKI.
–   Lemmas *awnless or more rarely with awns up to 2–3 mm long;* anthers 1–1.6 mm long. . . . . . . . . . . . . . . . . . . . . . . . . . . . . . . . . . . . . . . . . . . . . . . . .13.

13. Stems *scabrous* below spikes. . . . . . . . . . . . . . . . . . . . . .7. **R. NEPLIANA** V. VASSIL.
–   Stems *smooth* below spikes. . . . . . . . . . . . .6. **R. MACROURA** (TURCZ.) NEVSKI.

   *1. Roegneria confusa* (Roshev.) Nevski in Fl. SSSR II (1934), 605; Karavayev, Konsp. fl. Yak. 59.
      *Agropyron confusum* Roshev. in Bot. mat. Glavn. bot. sada RSFSR V (1924), 150.
      East Siberian boreal species, like other species of section *Roegneria* (= *Clinelymopsis* Nevski, l. c. 603) showing considerably closer affinity to the genus *Elymus* than do the remaining species occurring in the Arctic which belong to section *Cynopoa* Nevski. In particular, this species is in many respects very similar to the widespread Siberian *Elymus sibiricus* L. and could quite well also be referred to the genus *Elymus*. Perhaps it would be generally more correct to refer the whole section *Roegneria* to the genus *Elymus*, while retaining as the separate genus *Goulardia* Husnot only the species of section *Cynopoa*, although the existence of a close affinity between them and the species of section *Roegneria* is also not subject to doubt.
      *Roegneria confusa* penetrates the Arctic only in extreme NE Siberia.
      Growing mainly in shrub communities of river valleys (willow carr, alder thickets, etc.), sometimes also occurring on sands and gravelbars, also reported in the district of the Bay of Korf on stony slopes and rocks.
      **Soviet Arctic.** Between the lower reaches of the Kolyma and the Bay of Chaun; Anadyr Basin (rather frequent); Penzhina Basin; vicinity of Bay of Korf. (Map II–74).
      **Foreign Arctic.** Absent.

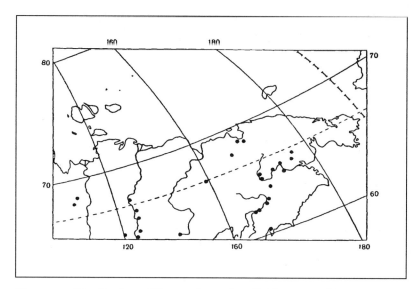

MAP II–74 Distribution of *Roegneria confusa* (Roshev.) Nevski.

**Outside the Arctic.** Mountainous districts of East Siberia from the Sayans, Northern Mongolia and Preamuria in the south to the Kureyka River, the vicinity of Verkhoyansk, the Momskiy Range and the lower reaches of the Kolyma in the north; Kamchatka.

2. ***Roegneria canina*** (L.) Nevski in Fl. SSSR II (1934), 617; Chernov in Fl. Murm. I, 242; Hylander, Nord. Kärlväxtfl. I, 372; Karavayev, Konsp. fl. Yak. 60.
*Triticum caninum* L., Sp. pl. (1753) 86; Grisebach in Ledebour, Fl. ross. IV, 340; Scheutz, Pl. jeniss. 183.
*Agropyron caninum* (L.) Beauv., Agrost. (1812) 146; Krylov, Fl. Zap. Sib. II, 348; Perfilev, Fl. Sev. I, 102; Hultén, Atlas, map 265.
Ill.: Fl. Murm. I, pl. LXXX, fig. 1, pl. LXXXI, fig. 1.

Predominantly European boreal forest species, however spreading eastwards to the upper reaches of the Lena and the Pamir-Alay Mountains. Penetrating arctic limits only in a few districts.

Growing mainly in valley shrub communities and birch woods, sometimes in meadows and on gravelbars in river valleys and near the sea coast.

**Soviet Arctic.** Murman (near Murmansk and between the settlements of Gremikha and Lumbovka, also recorded by Hultén for the Rybachiy Peninsula, Pechenga and the Ponoy); lower reaches of Yenisey (near the Yermilovskiy Station, 25 VII 1914, Reverdatto).

**Foreign Arctic.** Arctic Scandinavia.

**Outside the Arctic.** Almost all Europe (north to Arctic Scandinavia, the mouth of the Severnaya Dvina and Ust-Tsilma on the Pechora); Caucasus; Middle East; mountains of Central Asia; West Siberia (north to the basin of the Severnaya Sosva); further east very sporadically to the upper reaches of the Lena (Olekminsk district) and the Sayan Mountains.

3. ***Roegneria mutabilis*** (Drob.) Hyl. in Ups. Univ. Arskr. 7 (1945), 36; id., Nord. Kärlväxtfl. I (1953), 376.
*R. angustiglumis* (Nevski) Nevski in Fl. SSSR II (1934), 618; Chernov in Fl. Murm. I, 244; Karavayev, Konsp. fl. Yak. 60.
*Triticum caninum* β *altaicum* Griseb. in Ledebour, Fl. ross. IV (1853), 340.
*Agropyron mutabile* Drob. in Tr. Bot. muz. Peterb. Ak. nauk XVI (1916), 88, pro parte

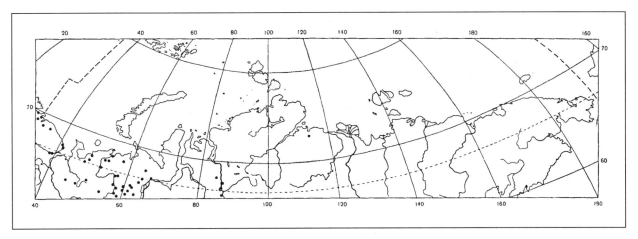

MAP II–75 Distribution of *Roegneria mutabilis* (Drob.) Hyl.

(var. *scabrum* Drob.); emend. Vestergren in Holmb., Skand. Fl. II (1926), 271; Krylov, Fl. Zap. Sib. II, 350; Hultén, Atlas, map 271; Polunin, Circump. arct. fl. 34.

*A. angustiglume* Nevski in Izv. Bot. sada AN SSSR XXX (1932), 615; Perfilev, Fl. Sev. I, 104; Leskov, Fl. Malozem. tundry 32; Hultén, Fl. Al. II, 257.

**Ill.:** Fl. Murm. I, pl. LXXX, fig. 3, pl. LXXXI, fig. 3.

Predominantly Siberian boreal (or even subarctic) species, known in recent Soviet botanical literature under the name "*R. angustiglumis* (Nevski) Nevski." But this name is superfluous, because before S. A. Nevskiy's work Vestergren (l. c.) affixed the name "*Agropyron mutabile* Drob." to one of the varieties of that originally polytypic species, "*A. mutabile* var. *scabrum* Drob." which agrees fully in content with S. A. Nevskiy's species. In the more western part of its range *R. mutabilis* retains, like most other species of the genus, considerable constancy of characters, usually varying only in the overall size of the plant and the density of the disposition of acicules on the lemmas. In East Siberia the polymorphism of the species distinctly increases, partly in connection with the appearance here of a whole series of very similar ecological-geographical races differing from *R. mutabilis* usually in only one or two characters. These include *R. transbaicalensis* (Nevski) Nevski with bare leaf blades and usually longer hairlike acicules on the lemmas, which is widespread in Prebaikalia and reaches north on the Yenisey to Khantayka; and *R. kamczadalorum* Nevski with bare and almost smooth (except at tip) lemmas, which is distributed in Kamchatka and the Okhotsk region. Both these species may also be found within arctic limits. Specimens of *R. mutabilis* from the vicinity of the Bay of Korf possess lemmas covered with much elongated hairlike acicules, as in *R. transbaicalensis,* but bare leaf blades; possibly they belong to a distinct ecological-geographical race. On the Kola Peninsula and in the basin of the Severnaya Dvina, where *R. mutabilis* sometimes occurs together with *R. canina,* hybrids between these two species are known but not yet reported from the Arctic.

Growing in floodplain meadows and shrub communities, frequently also on gravelbars and sands in river valleys, sometimes on relatively dry grassy slopes of maritime and riverine terraces.

**Soviet Arctic.** Murman (on the Tuloma, Iokanga and Ponoy; recorded by Hultén for Pechenga); Malozemelskaya Tundra (on the northern part of Timanskiy Ridge on the Pesha, Indiga and Sula); Bolshezemelskaya Tundra; Polar Ural (only southern part); lower reaches of Anabar (limestone cliffs on Anabar near mouth of River Sodamega, 5 IX 1932, No. 411, Sochava); vicinity of Bay of Korf (not entirely typical specimens). (Map II–75).

**Foreign Arctic.** Arctic Alaska (near Nome); Arctic Scandinavia (rather frequent).

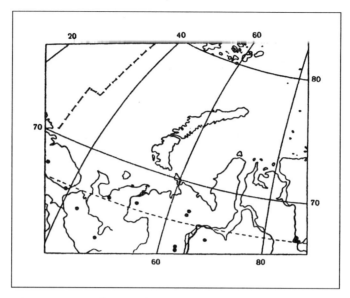

MAP II–76  Distribution of *Roegneria fibrosa* (Schrenk) Nevski.

**Outside the Arctic.** Northern Fennoscandia north of the Arctic Circle; Kola Peninsula; basins of Pinega, Mezen and Pechora; Urals (to the Southern Ural inclusively); West Siberia (north to Berezovo on the Ob); East Siberia (north to Igarka on the Yenisey, the lower reaches of the Anabar and the southern part of the Verkhoyansk Range); Kamchatka; northern part of Mongolia; Alaska.

4. ***Roegneria fibrosa*** (Schrenk) Nevski in Fl. SSSR II (1934), 625; Hylander, Nord. Kärlväxtfl. I, 379.

*Triticum fibrosum* Schrenk in Bull. Phys.-Math. Ac. Petersb. III, 209; Grisebach in Ledebour, Fl. ross. IV, 338.

*Agropyron fibrosum* (Schrenk) Nevski in Tr. Glavn. bot. sada SSSR XXIX (1930), 538; Perfilev, Fl. Sev. I, 103; Leskov, Fl. Malozem. tundry 32; Hultén, Atlas, map 267.

**Ill.:** Hylander, l. c., fig. 53, G-J.

Boreal species, widely but rather sporadically distributed in the European part of the USSR and West Siberia. It is the ecological-geographical race with the most western range among a large group of closely related Siberian and North American species of *Roegneria* which also includes (apart from the next five species) the well-known fodder plant "rhizomeless wheatgrass" [*R. pauciflora* (Schwein.) Hyl. = *R. trachycaulum* (Link) Nevski]. The latter species is represented in North America by a large number of still inadequately investigated forms which perhaps deserve species rank; some of these are particularly close to *R. fibrosa,* but readily distinguished from it by having glumes bare on the inside. *Roegneria fibrosa* enters the Arctic only in the north of the European part of the USSR and on the Ob.

Growing in floodplain meadows, on relatively dry grassy slopes of riverine terraces, in willow carr, and sometimes on sands and gravelbars in river valleys.

**Soviet Arctic.** Malozemelskaya Tundra (on the Pesha; recorded by A. I. Leskov also for meadows on the Indiga and Belaya); Bolshezemelskaya Tundra; lower reaches of Ob (on the Poluy). (Map II–76).

**Foreign Arctic.** Not occurring.

**Outside the Arctic.** Finland; Kola Peninsula (Lake Umbozero and near the settlement of Tetrino); Karelia (near the north shore of Lake Onega); basins of the Severnaya Dvina, Mezen and Pechora (rather frequent); central district of European part of USSR (very sporadically, west to Kalinin, Tula, Orel and Voronezh Oblasts); West

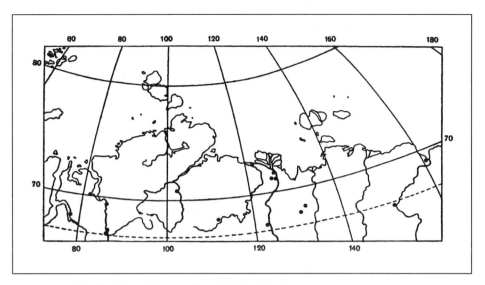

**Map II-77**  Distribution of *Roegneria subfibrosa* Tzvel.

Siberia (rather frequent, north to the Poluy and Berezovo on the Ob); North and East Kazakhstan; middle part of Yenisey Basin (very rare, north to the mouth of the Kureyka).

5. ***Roegneria subfibrosa*** Tzvel., sp. nova.

Planta perennis, 25–80 cm alta, viridis, caespitosa; culmi 2–4 nodis, glabri et laeves, rarius sub panicula scabriusculi; vaginae glabrae et laeves; ligulae 0.4–1.4 mm lg.; laminae 2–7 mm lt., planae, plus minusve scabrae, supra interdum sparse pilosae. Spicae 6–20 cm lg., bilaterales, vulgo leviter curvatae, in parte inferiore spiculis saepe valde distantibus (internodis inferioribus 1–2 cm lg.); spiculae plerumque virides, 2–4-florae; glumae subaequales, 4–8 mm lg., 3–5-nerves, extus scabriusculae, intus glabrae et laeves, apice sat subito acutatae vel aristulatae (aristula ad 1 mm lg.); lemmata 8–12 mm lg., glumis duplo vel sesqui longiora, apice acutata, rarius breviter aristata (arista ad 2 mm lg.), glabra et laevia, solum callo et basi ab utroque latere breviter pilosa et prope apicem secus nervis scabriusculae; palea lemmatis paulo brevior carinis scabris; rhachilla pubescens; antherae 1–1.5 mm lg.

Typus speciei: Sibiria, obl. Jacutsk, Balaganach, in arenosis ad fluminem, 15 VII 1898, Nilsson in Herb. Inst. Bot. Ac. Sc. URSS (Leningrad) conservatur.

A specie proxima — *R. macroura* praesertim lemmatis solum prope basin paulo pilosis differt.

East Siberian boreal species, occupying an apparently intermediate position between *R. fibrosa* and *R. macroura* and doubtless closely related to these species. From the former *R. subfibrosa* differs mainly in having the glumes bare on the inside, smaller anthers and the lemmas more or less hairy at the base; from the latter in having the lemmas hairy only at their very base. *Roegneria subfibrosa* is also rather close to the very polymorphic species widespread in North America *R. pauciflora* (Schwein.) Hyl., from which it differs in having the lemmas more or less hairy at the base and, apparently, also in having the spikelet rachis hairy. *Roegneria subfibrosa* is especially common in the lower reaches of the Lena, where typical *R. macroura* is completely absent, and like the latter species also produces a special ecological form on shifting river deposits with elongate prostrate bases of the shoots and widely spaced spikelets. Specimens with leaf blades sparsely hairy on the upperside occur almost throughout the range of the species. Their origin and taxonomic rank is still not entirely clear to us, and it cannot be excluded that they are hybrids between *R. subfibrosa* and *R. borealis,* a species widespread in the same districts. Provisionally they may be described as the variety *R. subfibrosa* var. *pilosa* Tzvel. var. nova = *Agropyron angustissimum* Drob. in herb. (A var. typica foliis supra sparse pilosis differt. Typus: Jakutia, ad fl. Aldan prope pag. Usty-Amginskoje, 3 VIII 1925, No. 1046, Drobov et Tarabukin).

Growing on sands and gravelbars of river valleys, more rarely in floodplain meadows and on grassy slopes, especially abundant on sandbanks along the bed of the Lena.

**Soviet Arctic.** Cape Nakhodka on Tazovskaya Bay; lower reaches of Yenisey in vicinity of Dudinka; lower reaches of Olenek; lower reaches of Lena; Tiksi; lower reaches of Kolyma. (Map II–77).

**Foreign Arctic.** Not occurring.

**Outside the Arctic.** Northern part of Central Siberian Plateau; Verkhoyansk Range; Lena and Kolyma Basins.

6. ***Roegneria macroura*** (Turcz.) Nevski in Fl. SSSR II (1934), 627; Karavayev, Konsp. fl. Yak. 60.

*Triticum macrourum* Turcz. in Steudel, Syn. pl. glum. I (1855), 343 et Fl. baic.-dahur. II, 1 (1856), 346.

*Agropyron sericeum* Hitchc. in Amer. Journ. Bot. II (1915), 309; Hultén, Fl. Al. II, 260.

*A. macrourum* (Turcz.) Drob. in Tr. Bot. muz. Peterb. Ak. nauk XVI (1916), 86; Polunin, Circump. arct. fl. 33.

*A. Tugarinovii* Reverd. in Sist. zam. po mat. gerb Tomsk. univ. 4 (1932), 1.

*A. Nomokonovii* M. Pop., Fl. Sredn. Sib. I (1957), 114, Russian diagnosis.

**Ill.:** Drobov, l. c., pl. 9, fig. 20; Fl. SSSR II, pl. XLV, fig. 2; Polunin, l. c. 35.

East Siberian boreal species, entering many districts of the Siberian Arctic. Usually described (for example by S. A. Nevskiy in the "Flora of the USSR") as having subterranean stolons, quite inappropriately for the genus *Roegneria*; but the copious herbarium material now available shows that the formation of strongly elongated prostrate shoot bases which root at the nodes (resembling true rhizomes in appearance) occurs far from universally and is associated exclusively with development of the plant on shifting river deposits or windblown sands which constantly drift over the tufts. In such cases elongation of the shoot bases usually occurs simultaneously with correlated elongation of the lower internodes of the spikes, so that the lower spikelets appear widely spaced from one another. In habitats with more vegetated or more stable soil cover *R. macroura,* like other species of *Roegneria,* forms dense tufts without any "stolons" and its spikes become more compact. As S. A. Nevskiy already reported, the more high arctic specimens of *R. macroura* rather sharply differ in habitus from typical more southern specimens of this species. They possess denser tufts and shorter, more compact, often violet-coloured spikes; in many cases they closely approach *R. villosa* V. Vassil., differing from that species only in normally lacking awns on the lemmas and in having broader leaf blades and bare (or rarely almost bare) culm nodes. Such specimens transitional between these two species include the duplicates of the type specimens of *Agropyron Tugarinovii* Reverd. (Yenisey, vicinity of Dudinka, meadow on shore of Dudinka, 16.VIII 1914, Reverdatto) present in the herbarium of the Botanical Institute of the USSR Academy of Sciences; these were referred by S. A. Nevskiy to *R. jacutensis* but in our opinion are significantly closer to *R. macroura.* The duplicates of another specimen cited in the original description of that species (Yenisey, mouth of River Leninskaya, steep clayey slopes of bluffs, 19 VII, Reverdatto) possess slightly hairy lemmas with short awns and much reduced anthers without normally developed pollen and are apparently of hybrid origin. Judging from the original diagnoses, *Agropyron sericeum* Hitchc. from Alaska and

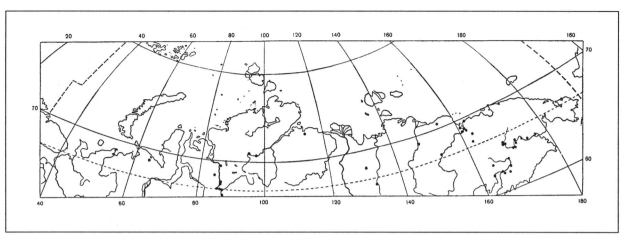

MAP II–78 Distribution of *Roegneria macroura* (Turcz.) Nevski.

*A. Nomokonovii* M. Pop. from the Upper Lena can be included among the synonyms of *R. macroura*, but we have not seen specimens of them. *Roegneria macroura* occasionally hybridizes with *Hordeum jubatum* L., forming intergeneric hybrids which are rather constant with respect to morphological characters: *Roegneria macroura* × *Hordeum jubatum* = × *Horderoegneria chatangensis* (Roshev.) Tzvel. comb. nova = *Elymus chatangensis* Roshev. [Izv. Bot. sada AN SSSR XXX (1932), 778], known from the Khatanga Basin and the lower reaches of the Kolyma. Similar hybrids with *Hordeum jubatum* can apparently be formed by other species of *Roegneria* close to *R. macroura*.

Growing mainly on sands and gravelbars in river valleys, more rarely in floodplain meadows and valley shrub communities (willow carr, alder thickets, etc.).

**Soviet Arctic.** Karskaya Tundra (sandy shore of Kara in its middle course, 20 VII 1909, No. 649, Sukachev); lower reaches of Yenisey (north to the vicinity of Dudinka); River Kheta; lower reaches of Anabar; lower reaches of Alazeya and Kolyma (rather frequent); Anadyr and Penzhina Basins (rather frequent). (Map II–78).

**Foreign Arctic.** Arctic Alaska (district of Bering Strait).

**Outside the Arctic.** Central Siberian Plateau; Prebaikalia; greater part of Yakutia (but absent from a considerable part of the Lena Basin); north and west coasts of Sea of Okhotsk; Alaska.

7. ***Roegneria Nepliana*** V. Vassil. in Bot. mat. Gerb. Bot. inst. AN SSSR XVI (1954), 56; Karavayev, Konsp. fl. Yakutia 60.

This species is so far known only from the type specimen (Northern Anyuyskiy Range, Umkaveyem River Basin, 4 IX 1950, Nepli), which is now apparently lost. Although the author compares it with the species *R. mutabilis* and *R. transbaicalensis*, the characters indicated in the original diagnosis agree fully with *R. macroura* except that the stem is scabrous below the panicle. The latter character occurs occasionally in isolated specimens of certain other species of *Roegneria*, so that its systematic importance is apparently not great. Consequently the distinctness of *R. Nepliana* needs confirmation from more abundant material.

Growing apparently on sands and gravelbars in river valleys.

**Soviet Arctic.** Coast east of the mouth of the Kolyma.

Unknown in the **Foreign Arctic** or **outside the Arctic.**

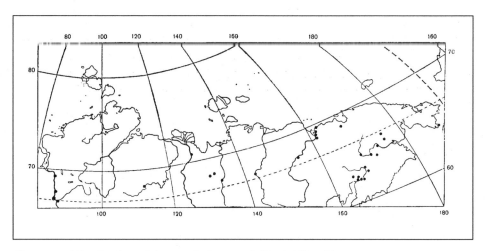

MAP II–79  Distribution of *Roegneria jacutensis* (Drob.) Nevski.

**8. *Roegneria jacutensis*** (Drob.) Nevski in Fl. SSSR II (1934), 625; Karavayev, Konsp. fl. Yak. 60.

*Agropyron jacutense* Drob. in Tr. Bot. muz. Peterb. Ak. nauk XVI (1916), 94.

**Ill.:** Drobov in Tr. Bot. muz. Peterb. Ak. nauk XVI, pl. 9, fig. 5.

East Siberian boreal species, very close to *R. macroura* and identical to it in ecology. Its range coincides to a considerable extent with that of *R. macroura*, so that *R. jacutensis* could be interpreted as an awned variety of that species. However, the almost complete lack of specimens intermediate between these species with respect to awn length, as well as the somewhat larger anthers of *R. jacutensis*, can serve as confirmation of the distinctness of this species.

Growing on sands and gravelbars in river valleys, floodplain meadows, and sometimes in valley shrub communities.

**Soviet Arctic.** Lower reaches of Yenisey (vicinity of Dudinka); lower reaches of Lena (district of Bulun settlement); lower reaches of Kolyma (to its mouth) and further east to the district of the Bay of Chaun; Anadyr and Penzhina basins (rather frequent). (Map II–79).

**Foreign Arctic.** Not occurring.

**Outside the Arctic.** Central Siberian Plateau (basins of Lower Tunguska and Vilyuy, southern part of Olenek Basin); basins of Yana (district of Verkhoyansk and Batagay-Alyta settlement) and Kolyma; Aldan Basin.

**9. *Roegneria turuchanensis*** (Reverd.) Nevski in Fl. SSSR II (1934), 626; Karavayev, Konsp. fl. Yak. 59.

*Agropyron turuchanense* Reverd. in Sist. zam. po mat. gerb. Tomsk. univ. 4 (1932), 2.

Apparently a subarctic species extremely close to *R. macroura* (especially to smaller arctic specimens of that species), but differing in having leaf blades hairy on the upperside and glumes with more broadly hyaline margins.

Growing on sands and gravelbars of river valleys, as well as in floodplain meadows.

**Soviet Arctic.** Bolshezemelskaya Tundra (eastern part, on the Adzva, Usa and Korotaikha and in the vicinity of Vorkuta); Polar Ural; Karsk-Baydaratsk coast (lower reaches of Kara); lower reaches of Yenisey (vicinity of Dudinka); lower reaches of Yana (recorded by M. N. Karavayev but possibly in error for *R. villosa*).

**Foreign Arctic.** Absent.

**Outside the Arctic.** Prepolar Ural (on the Kozhym).

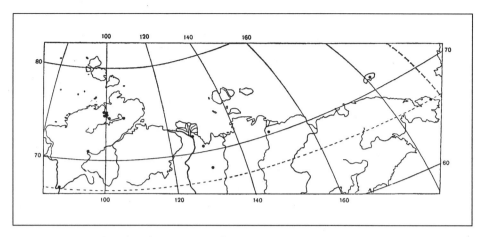

MAP II–80  Distribution of *Roegneria villosa* V. Vassil.

**10. *Roegneria villosa*** V. Vassil. in Bot. mat. Gerb. Bot. inst. AN SSSR XVI (1954), 57.
*Agropyron violaceum* auct. non Vasey — Tolmachev, Fl. Taym. I, 101.

East Siberian arctic species belonging to the systematically very complex group of short-awned species of *Roegneria* which most deeply penetrates the Arctic. With respect to the structure of the spikes, glumes and lemmas, this group approaches the group of closely related species united by S. A. Nevskiy as the series *Borealis* Nevski (*R. borealis* s. l.). At the same time there is no doubt that they are no less closely related to species of the group *R. macroura* s. l. Among the species of this most arctic group in the genus, *R. villosa* in our opinion rather closely approaches the Greenlandic species *R. violacea* (Hornem.) Meld. and the Rocky Mountain species *R. latiglumis* (Scribn. et Sm.) Nevski, although it cannot be identified with them. Melderis in his special work on the short-awned species of *Roegneria* (The short-awned species of the genus *Roegneria* of Scotland, Iceland and Greenland, Svensk Bot. Tidskrift, Bd. 44, H. 1, 1950, 132–166), united *R. violacea* with *R. latiglumis,* and also proposed the variety *R. borealis* var. *hyperarctica* (Polunin) Meld. from Arctic America and Greenland; in many respects the latter is closer to *R. villosa,* but differs from it in having the leaf blades hairy on both sides. In our opinion there is just as much basis for subordinating this variety to *R. violacea,* as did its author N. Polunin, because, while the differences between "var. *hyperarctica*" and *R. borealis* and closely related species are completely definite and constant, the differences between this variety and *R. violacea* (hairy culm nodes, form of glumes, etc.) are far from so constant. Moreover, as shown by the map of the distribution of *R. violacea* and *R. borealis* var. *hyperarctica* in Greenland presented by Melderis, "var. *hyperarctica*" has a more northern range than has *R. violacea* and appears to be a higher arctic race of the latter species. Typical *R. borealis* is completely absent from Greenland and apparently from the whole of the American Arctic. Like "var. *hyperarctica*" (which we choose to treat as a distinct species), *R. villosa* displays much polymorphism, and is possibly divisible into several local races with more restricted range since material from any given small district almost always displays much constancy of characters. The type specimens of this species (Chukotka Peninsula, River Chegitun, 8 VIII 1938, Trushkovskiy), which vary somewhat in the degree of pubescence of the glumes and lemmas, are of relatively large size and possess bare or very slightly hairy culm nodes, approaching *R. macroura* in these respects. In contrast to these specimens, many collected by B. A. Tikhomirov and M. I. Vellikaynen from the basin of Lake Taymyr and often identical to the type specimens of *R. villosa* with respect to other characters, possess densely hairy culm nodes. The single dwarf (6–15 cm high) specimen of *R. villosa* from Wrangel

Island (upper reaches of Red Flag River, stony-clayey polar desert on limestones, 20 VIII 1938, Gorodkov) possesses bare culm nodes but is rather different in appearance from typical specimens of this species, although identical with certain Taymyr specimens. Finally, on the northern part of the Verkhoyansk Range collections have been made of plants both with bare and with densely hairy culm nodes. The glumes of *R. villosa*, as also in many species of the group *R. macroura* s. l., may be either scabrous or more or less hairy, but they are especially copiously hairy in the specimen from Wrangel Island.

Growing on open stony slopes, gravelbars and sands in river valleys, sometimes on rocks (in the lower reaches of the Lena).

**Soviet Arctic.** Basin of Lake Taymyr (north almost to the mouth of the Lower Taymyra); lower reaches of Lena (near Tit-Ary Island) and northern part of Verkhoyansk Range (Kharaulakh Mountains); lower reaches of Indigirka (Allaykhovsk district on the River Arangas); Chukotka (on River Chegitun); Wrangel Island. (Map II–80).

**Foreign Arctic.** Unknown.

**Outside the Arctic.** Yenisey, northern part of Central Siberian Plateau in the Kheta Basin; Verkhoyansk Range.

11. ***Roegneria hyperarctica*** (Polunin) Tzvel., comb. nova.

*R. borealis* var. *hyperarctica* (Polunin) Meld. in Svensk Bot. Tidskr. XLIV, 1 (1950), 161; Böcher & al., Grønl. Fl. 302.

*Agropyron violaceum* auct. non Vasey — Gelert, Fl. arct. 1, 133, pro parte.

*A. violaceum* var. *hyperarcticum* Polunin in Bull. Nat. Mus. Canad. 92, Bot. ser. XXIV (1940), 95.

*A. latiglume* auct. non Rydb. — Hultén, Fl. Al. II, 258, pro parte ?; A. E. Porsild, Ill. Fl. Arct. Arch. 40, pro parte.

*A. boreale* auct. non Drob. — Polunin, Circump. arct. fl. 33, pro parte.

**Ill.:** Melderis, l. c., fig. 1; Polunin, Circump. arct. fl. 33.

Arctic species so far known in the Soviet Arctic only from the Pyasina Basin in Taymyr. Apparently widely distributed in the American Arctic (described from Baffin Island), although its range there is still inadequately clarified due to confusion with closely related species. Thus, Hultén and Porsild unite it with *R. latiglumis* (Scribn. et Sm.) Nevski, while Melderis (l. c.) subordinates it to *R. borealis* as a variety but considers *R. latiglumis* a synonym of *R. violacea* (Hornem.) Meld. Recently N. Polunin (l. c., 1959), probably following Melderis, also includes it in *R. borealis*, with which we cannot agree. *Roegneria hyperarctica* differs from *R. villosa* only in having the leaf blades hairy on both surfaces, but this character is apparently completely constant since in all the (rather numerous) specimens of *R. villosa* available to us the leaf blades are more or less scabrous but without hairs. Comparison of the specimens from the Pyasina with two reliably determined (cited by Melderis) specimens of *R. hyperarctica* present in the herbarium of the Botanical Institute of the USSR Academy of Sciences shows their considerable similarity despite the wide gap in the range, although the dwarfish dimensions of the Pyasina specimens possibly suggest the existence here of a distinct local race.

Like the preceding species, apparently growing on stony slopes and on sands and gravelbars in river valleys.

**Soviet Arctic.** Taymyr (right bank of Pyasina near mouth of Tareya, 15 VIII 1934, Vinogradova).

**Foreign Arctic.** Alaska (district of Bering Strait); NW part of Canada west of Hudson Bay; Labrador (about 60°N); Canadian Arctic Archipelago (mainly in western and northern part, on Baffin Island only in extreme north-west, on Ellesmere Island reaching north of 80°N); Greenland (on the west coast between the Arctic Circle and 72°N, on the east coast between 70 and 75°N).

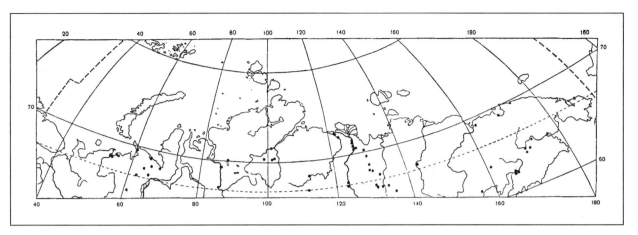

MAP II–81  Distribution of *Roegneria borealis* (Turcz.) Nevski.

**Outside the Arctic.** Southern Alaska; Mackenzie Mountains and east of the Mackenzie River Basin to Hudson Bay (possibly many records, especially for the cordillera, refer to *R. latiglumis*).

**12. *Roegneria borealis*** (Turcz.) Nevski in Fl. SSSR II (1934), 624; Melderis in Svensk Bot. Tidskr. XLIV, 1, 161, excl. var.; Hylander, Nord. Kärlväxtfl. I, 379, pro parte; Karavayev, Konsp. fl. Yak. 60.
*Triticum borealis* Turcz., Fl. baic.-dahur. II, 1 (1856), 345.
*T. violaceum* auct. non Hornem. — Scheutz, Pl. jeniss. 183.
*Agropyron violaceum* auct. non Vasey — Gelert, Fl. arct. 1, 133, pro parte.
*A. boreale* (Turcz.) Drob. in Tr. Bot. muz. Ak. nauk XVI (1916), 84; Polunin, Circump. arct. fl. 33, pro parte.
*A. mutabile* var. *glabrum* Drob., l. c. 87, pro parte.
*A. repens* var. *Hornemannii* Kryl., Fl. Zap. Sib. II, 353, pro parte.
*A. latiglume* ssp. *eurasiaticum* Hult., Fl. Al. II (1942), 259; id., Atlas, map 269, pro parte.
Ill.: Drobov in Tr. Bot. muz. XVI, pl. 9, fig. 1.

Subarctic species widely distributed in the north of the Eurasian continent but apparently absent from Arctic America. Readily distinguished from the two preceding species by the bare and smooth lemmas (with very short acicules only on the nerves near the tip) and by the peculiarity of the spikelet rachis (without hairs, covered only with minute acicules). Displaying considerable constancy of morphological characters, varying only with respect to the overall size of the plant and the width of the leaf blades.

Growing on relatively dry stony slopes and rocks, on sands and gravelbars in river valleys, and sometimes in floodplain meadows and valley shrub communities.

**Soviet Arctic.** Bolshezemelskaya Tundra (River Pay-Yaga, near mouth of Pechora, Vangurey Range, River Adzva, vicinity of Vorkuta); Polar Ural; Pay-Khoy; Karsko-Baydaratskaya Tundra (lower reaches of Kara and Pyderata); lower reaches of Yenisey (vicinity of Dudinka); Taymyr (upper reaches of Pyasina); lower reaches of Olenek (north to 72°40'N); lower reaches of Lena and Kharaulakh Mountains (rather frequent); between lower reaches of Kolyma and Bay of Chaun (rather frequent); Anadyr Basin (rather frequent); Penzhina Basin (rather frequent). (Map II–81).

**Foreign Arctic.** Arctic Scandinavia.

**Outside the Arctic.** Prepolar Ural; northern part of Central Siberian Plateau (south to the basins of the Lower Tunguska and the Vilyuy); Yenisey Ridge; Altay (eastern part) and Sayans; Prebaikalia; mountainous districts of Yakutia and mountains along Okhotsk Coast; Kamchatka (only the basin of Bolshoye Nachikinskoye Lake); also occurring together with *R. scandica* in the mountains of Fennoscandia.

13. **Roegneria kronokensis** (Kom.) Tzvel., comb. nova.
    *R. borealis* (Turcz.) Nevski in Fl. SSSR II (1934), 624, pro parte.
    *Agropyron kronokense* Kom. in Fedde, Repert. sp. nov. XIII (1915), 87.

    This species differs from the preceding only in having leaf blades with short but copious pubescence on both surfaces; but, since it shows a well defined geographical association, we think it possible to retain it as a distinct species under V. L. Komarov's name.

    Growing on stony slopes and rocks.

    **Soviet Arctic.** Basin of Penzhina Bay (on rocks on River Palmatkina in its middle course, 8 VIII 1930, Sochava).
    **Foreign Arctic.** Not occurring.
    **Outside the Arctic.** Kamchatka (basins of Kronotskoye Lake and the River Kamchatka).

14. **Roegneria scandica** Nevski in Fl. SSSR II (1934), 624 and in Tr. Bot. inst. AN SSSR, ser. I, 2 (1936), 54; Chernov in Fl. Murm. I, 245.
    *Triticum violaceum* f. *subalpinum* Neum., Sv. Fl. (1901) 726.
    *Agropyron latiglume* ssp. *subalpinum* (Neum.) Vestergren in Holmb., Scand. Fl. II (1926), 272.
    *A. latiglume* ssp. *eurasiaticum* Hult., Fl. Al. II (1942), 259; id., Atlas, map 269, pro parte.
    Ill.: Fl. SSSR II, pl. XLV, fig. 1; Fl. Murm. I, pl. LXXX, fig. 2, pl. LXXXI, fig. 2.

    Subarctic Eurasian species whose sole constant difference from *R. borealis* is that the leaf blades are bare but more or less scabrous on both surfaces. The ranges of these species coincide to a considerable extent, for which reason many contemporary authors (for example, Hultén and Melderis) consider it possible to unite them as a single species. It is not entirely clear to us whether there are any differences in their distribution in the mountains of Scandinavia, where both species occur, but in the Kola Peninsula only *R. scandica* is so far known.

    Growing on stony slopes and rocks, or more rarely on riverine gravelbars.

    **Soviet Arctic.** Murman (near Cape Orlov north of the mouth of the Ponoy); Bolshezemelskaya Tundra (right bank of River Kara, River Gnet-Yu, 29 VII 1959, Rebristaya); lower reaches of Olenek; basin of Penzhina Bay (Kamennyy Range in the Oklan Valley, 7 VII 1930, Sochava).
    **Foreign Arctic.** Arctic Scandinavia.
    **Outside the Arctic.** Mountains of Fennoscandia; Karelia (limestone rocks on shore of Lake Segozero); Central Siberian Plateau (Kheta and Olenek Basins); Yana and Indigirka Basins (north to vicinities of Verkhoyansk and Oymyakon); Kolyma Basin (north to lower reaches of Omolon). Possibly occurring in the North American cordillera [in the herbarium of the Botanical Institute of the USSR Academy of Sciences there is a specimen from there very similar to *R. scandica* (Wyoming, Teton Pass, 13 VII 1901, No. 246, Merrill and Wilcox) identifed as "*Agropyron violaceum*"].

## GENUS 34    Elytrigia Desv. — WHEAT GRASS

GENUS WIDELY DISTRIBUTED in temperate countries of the Northern and parts of the Southern Hemisphere, containing apparently over 100 species (their number cannot be counted satisfactorily due to a lack of precise information regarding the disposition of species among the closely related genera originally united as the genus *Triticum* L., later *Agropyron* Gaertn., as well as due to very different interpretations of the content of species).

Species of *Elytrigia* are most readily distinguished from the externally rather similar species of *Roegneria* by the size of the anthers, which in *Roegneria* are usually 1–2.5 mm long but in *Elytrigia* 3.5–5 mm long. Only two species of the genus, *E. repens* and *E. jacutorum,* just penetrate the limits of the Soviet Arctic. Recently B. A. Yurtsev has found in Northern Yakutia (Sakkyryr district, southern part of Kular Range on left bank of River Ulakhan-Sakkyryr, 24 VIII 1959, Yurtsev) a very interesting further species of the genus, *E. villosa* (Drob.) Tzvel. comb. nova = *Brachypodium villosum* Drob. [Tr. Bot. muz. Ak. nauk XII (1914), 175, pl. 13] = *Agropyron Karawaewii* Smirn. = *Roegneria Karawaewii* (Smirn.) Karav., which was previously known only from a few sites in more southern Yakutia and is most closely related to the widespread North American *E. dasystachya* (Hook.) A. et D. Löve = *Agropyron dasystachyum* (Hook.) Scribn.

1. Plant 30–80 cm high, forming rather *dense tufts without* elongated subterranean *stolons;* leaf blades *1.5–3 mm wide,* loosely folded lengthwise or flat, rather copiously hairy on upper surface. Lemmas at tip *with long curved-divergent awns 1.2–2 cm long.* . . . .2. **E. JACUTORUM** (NEVSKI) NEVSKI.
- Plant 20–80 cm high, *with long* subterranean *stolons* (rhizomes), usually *not forming tufts*; leaf blades *3–8 mm wide,* usually flat, bare or more rarely sparsely hairy on upper surface. Lemmas at tip *with short straight awns up to 0.6 cm long* or completely awnless. . . . . .1. **E. REPENS** (L.) NEVSKI.

    ***1. Elytrigia repens*** (L.) Nevski in Tr. Sredneaz. gos. univ., ser. 8B, XVII (1934), 61 and in Tr. Bot. inst. AN SSSR, ser. I, 2 (1936), 85; Hylander, Nord. Kärlväxtfl. I, 366.
    *Triticum repens* L., Sp. pl. (1753) 86; Grisebach in Ledebour, Fl. ross. IV, 340, pro parte; Trautvetter, Pl. Sib. bor. 133; id., Syll. pl. Sib. bor.-or. 60; Scheutz, Pl. jeniss. 183.
    *Agropyron repens* (L.) Beauv., Agrost. (1812) 102; Gelert, Fl. arct. 1, 132; Krylov, Fl. Zap. Sib. II, 351, pro parte; Perfilev, Fl. Sev. I, 103; Nevskiy in Fl. SSSR II, 652; Leskov, Fl. Malozem. tundry 32; Hultén, Fl. Al. II, 260; id., Atlas, map 272; Chernov in Fl. Murm. I, 246; Karavayev, Konsp. fl. Yak. 60.
    **Ill.:** Fl. SSSR II, pl. XLVII, fig. 6; Fl. Murm. I, pl. LXXXII; Polunin, l. c. 35.

    Species widespread almost throughout the Eurasian continent, entering the Arctic only in a few districts. Occasionally forming hybrids with *Leymus arenarius* (L.) Hochst. Displaying very great polymorphism, but relatively constant within the Arctic.

    Growing on relatively dry grassy slopes, on gravelbars and sands in river valleys, often also as an introduction along roads and near settlements.

    **Soviet Arctic.** Murman (sporadically along the entire coast); Kanin (southern part); Malozemelskaya Tundra (recorded for the Belaya River).
    **Foreign Arctic.** Arctic Scandinavia.

**Outside the Arctic.** Almost all Eurasia, with the northern boundary in the USSR running through Murman, Southern Kanin, Ust-Tsilma on the Pechora, the basin of the Severnaya Sosva and the River Synya (tributary of the Ob), the basin of the Lower Tunguska, and the vicinities of Verkhoyansk and Srednekolymsk; North Africa. As an introduction in the majority of other extratropical countries of both hemispheres.

2. ***Elytrigia jacutorum*** (Nevski) Nevski in Tr. Bot. inst. AN SSSR, ser. I, 2 (1936), 78.
   *Agropyron aegilopoides* Drob. in Tr. Bot. muz. Peterb. Ak. nauk XII (1914), 46, respecting Yakutian plants.
   *A. jacutorum* Nevski in Izv. Bot. sada AN SSSR XXX (1932), 502; id. in Fl. SSSR II (1934), 636; Karavayev, Konsp. fl. Yak. 60.

   East Siberian boreal species, only just penetrating the Arctic.
   Growing on relatively dry stony slopes, more rarely on gravelbars or in more or less sparse shrub communities.

   **Soviet Arctic.** Lower reaches of Kolyma (between Pokhodsk settlement and the Panteleyevskaya farmstead, 28 VII 1905, Shulga; right bank of Panteleikha near Panteleikha settlement, 6 & 26 VII 1950, Nepli).
   **Foreign Arctic.** Not occurring.
   **Outside the Arctic.** Almost all Yakutia; replaced in Prebaikalia by the very closely related species *E. Gmelinii* (Schrad.) Nevski.

## GENUS 35 — Leymus Hochst. — WILD RYE

SMALL GENUS (apparently up to 50 species) widely distributed in temperate countries of the Northern Hemisphere. The species belonging here were for long referred to the genus *Elymus* L., and indeed are often united with it at the present time as the separate section *Psammelymus* Griseb. However, typical species of the genus *Elymus* (*E. sibiricus* L., *E. nutans* Griseb., etc.) and species of the genus *Leymus* show very substantial morphological differences both in the structure of the generative organs and in their growth form (species of *Elymus* lacking subterranean stolons). These genera also differ rather profoundly with respect to caryology. In particular, species of *Elymus* (like the genus *Roegneria* C. Koch) hybridize very readily with perennial barleys (genus *Hordeum* L.) and can even produce fertile hybrids, while hybrids between the genus *Leymus* and *Hordeum* are extremely rare and always sterile. There would be considerably more justification for uniting the genera *Elymus* and *Roegneria*, which are universally treated as separate genera more on the strength of tradition than on the basis of any substantial morphological or other differences. In distinction from the genus *Elymus*, the genus *Leymus* is apparently of littoral origin. Thus, the most "primitive" species of this genus, *L. arenarius* and *L. mollis*, are littoral species, while the numerous inland species are secondary; these were separated by S. A. Nevskiy as the distinct genus *Aneurolepidium* Nevski, but in our opinion have an unbroken connection with typical species of *Leymus* [through *L. angustus* (Trin.) Pilger and species close to it]; their evolution took place on the wide, more or less saline expanses exposed by the retreat of the ancient Mediterranean Sea.

Of the four closely related species of *Leymus* occurring in the Soviet Arctic, three are littoral while the fourth (*L. interior*) is widespread on riverine sands and gravelbars away from the sea coast. No high arctic species are included among them, although the species *L. arenarius* and *L. interior* reach many islands of the Arctic Ocean. Three East Siberian species (*L. mollis, L. villosissimus* and *L. interior*) are connected to one another by rather numerous transitional specimens apparently of hybrid origin. In the monograph of the genus by Bowden (Cytotaxonomy of section Psammelymus of the genus Elymus, Canad. Journ. of Bot. XXXV, 1957, 964–973), they are treated as subspecies of a single species, *E. mollis* s. l. Both the species *L. arenarius* and *L. mollis* are also so closely connected to one another that, despite the differences in chromosome number (2n = 56 in *L. arenarius,* 2n = 28 in the remaining species), we think it more correct either to accept all four species as distinct (as we do in the present work) or to consider all four as subspecies of a single polytypic species, *L. arenarius* s.l.

1. Plant 40–100 cm high, with well developed thickened rhizomes; culms below spikes *bare and smooth*; leaf blades 5–12 mm wide. Spikes 10–25 cm long, *with rachis usually covered with acicules or hairs only on ribs;* spikelets 15–25 mm long; glumes 12–25 mm long, of almost same length as spikelets and longer than lemmas, *bare or slightly hairy only on distal part and on margins;* lemmas with short hairs, those on distal part often transitional to acicules. . . . . . . . . . . . . . . . . . . . . . . . . . . .1. **L. ARENARIUS** (L.) HOCHST.
– Culms below spikes *to greater or lesser extent* (sometimes only at very base of spike) *covered with hairs or shorter setae*. Spikes *with shortly but copiously hairy rachis*; glumes *usually hairy over almost entire outer surface*. . . . 
. . . . . . . . . . . . . . . . . . . . . . . . . . . . . . . . . . . . . . . . . . . . . . . . . . . . . . . . . . .2.

2. Spikes 5–10 cm long; spikelets *10–18 mm long,* often with brownish-violet tinge, copiously clothed with rather long hairs; glumes *5–12 mm long, usually shorter than lemmas, more rarely of almost equal length*. Plant of gravelbars and sands away from sea shore, 20–70 cm high, with relatively thin rhizomes. . . . . . . . . . . . . . . . . . . . . . . . . . . . . .4. **L. INTERIOR** (HULT.) TZVEL.
– Spikelets *14–25 mm long;* glumes *12–22 mm long, often longer than lemmas*. Plant of sands and gravelbars near sea shore. . . . . . . . . . . . . . . . . . . . .3.

3. Plant *25–60 cm high*, with relatively thin rhizomes. Spikes *5–11 cm long, copiously clothed with rather long hairs*, often with brownish-violet tinge. . 
. . . . . . . . . . . . . . . . . . . . . . . . . . . . . . . . . .3. **L. VILLOSISSIMUS** (SCRIBN.) TZVEL.
– Plant *50–100 cm high*, with more thickened rhizomes. Spikes *11–25 cm long, copiously clothed with shorter hairs*, pale green. . . . . . . . . . . . . . . . . . . . .
. . . . . . . . . . . . . . . . . . . . . . . . . . . . . . . . . . . . . . . .2. **L. MOLLIS** (TRIN.) PILGER.

*1.* **Leymus arenarius** (L.) Hochst. in Flora XXXI (1848), 118; Tsvelev in Bot. mat. Gerb. inst. AN SSSR XX, 429.

*Elymus arenarius* L., Sp. pl. (1753) 83; Grisebach in Ledebour, Fl. ross. IV, 695; Gelert, Fl. arct. 1, 134; Nevskiy in Fl. SSSR II, 695; Hultén, Atlas, map 273; Kuzeneva in Fl.

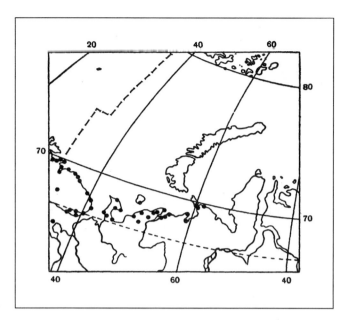

MAP II–82 Distribution of *Leymus arenarius* (L.) Hochst.

Murm. I, 252; Hylander, Nord. Kärlväxtfl. I, 370; Bowden in Canad. Journ. Bot. XXXV, 960; Polunin, Circump. arct. fl. 51, pro parte.

Ill.: Fl. SSSR II, pl. XLIX, fig. 8; Fl. Murm. I, pl. LXXXIII; Bowden, l. c., fig. 12–13; Polunin, l. c. 50.

Northern European littoral species, now distributed both in maritime and inland districts of many other countries of the Northern Hemisphere through introduction and planting as a sand stabilizer. It displays relatively constant morphological characters, somewhat varying only in the degree of pubescence of the glumes which in this species are on average firmer than in the next three species. Outside the Arctic but in its immediate proximity (SE part of Kola Peninsula, Solovetskiye Islands, delta of Severnaya Dvina) sterile intergeneric hybrids are known: *Leymus arenarius* × *Elytrigia repens* = × *Leymotrigia Bergrothii* (Lindb. fil.) Tzvel. comb. nova = × *Tritordeum Bergrothii* Lindb. fil. in Meddl. Soc. Faun. et Fl. Fenn. XXXII (1906), 21 = × *Elymotrigia Bergrothii* (Lindb. fil.) Hyl. in Bot. Notis. (1953) 358. These hybrids are very similar to *L. arenarius* in general form, but significantly approach *Elytrigia repens* in spikelet structure; they can persist for a long time, multiplying by rhizomes and forming large clones.

Growing usually on maritime sands and gravelbars, but always above the zone of tidal flooding. Occurring also on sandy shores of the larger lakes of Northern Europe outside the Arctic.

**Soviet Arctic.** Murman (on almost the whole coast, but rather sporadically); Kanin; Kolguyev; northern part of Malozemelskaya Tundra; Bolshezemelskaya Tundra (lower reaches of Pechora and shore of Pechora Bay); Yugorskiy Peninsula (Khabarovo, Amderma); Vaygach. (Map II–82).

**Foreign Arctic.** Iceland; Arctic Scandinavia; SW Greenland (as introduction, north to Disko Island).

**Outside the Arctic.** Sea coasts of Northern Europe west to Northern France and Ireland; shores of Lakes Ladoga and Onega. In North America only as an introduced or imported plant at a few sites on the east coast of Canada and the USA; also introduced in many other countries.

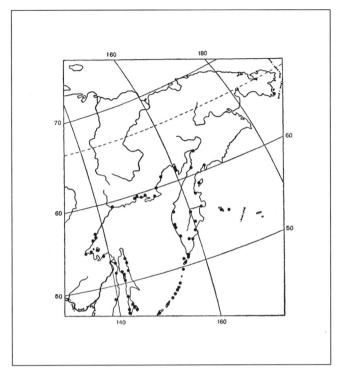

MAP II–83 Distribution of *Leymus mollis* (Trin.) Pilger.

2. **Leymus mollis** (Trin.) Pilger in Bot. Jahrb. LXX, V (1945), 6; Tsvelev in Bot. mat. Gerb. Bot. inst. AN SSSR XX, 429.

*Elymus mollis* Trin. in Sprengel, Neue Entdeck. II (1821), 72; Grisebach in Ledebour, Fl. ross. IV, 332; Gelert, Fl. arct. 1, 133, pro parte; Nevskiy in Fl. SSSR II, 695.

*E. arenarius* β *villosus* E. Mey., Pl. Labrad. (1830) 20; Lange, Consp. fl. groenl. 154.

*E. arenarius* ssp. *mollis* (Trin.) Hult., Fl. Kamtch. 1 (1927), 153; id., Fl. Al. II, 268, map 195, pro parte; A. E. Porsild, Ill. Fl. Arct. Arch. 41.

*E. mollis* ssp. *mollis* — Bowden in Canad. Journ. Bot. XXXV (1957), 964.

*E. arenarius* auct. non L. — Polunin, Circump. arct. fl. 51, pro parte.

Ill.: Fl. SSSR II, pl. XLIX, fig. 9; Porsild, l. c., fig. 13, a; Bowden, l. c., fig. 7.

North American and East Asian littoral species, like the preceding species also occurring on the shores of very large inland lakes. Scarcely reaching the Soviet Arctic. On Kamchatka and the Okhotsk Coast connected by specimens with transitional characters both with *L. villosissimus* and *L. interior*, differing from both these species by the greater size of the whole plant and the spikes as well as by the shorter pubescence of the spikelets.

Growing on sands and gravelbars on the sea coast above the zone of tidal flooding, sometimes also in the mouths of the larger rivers.

**Soviet Arctic.** Only the mouth of the Penzhina (Ust-Penzhino settlement, 6 VII 1960, No. 75/7, Kildyushevskiy) and the Bay of Korf (vicinity of Kultushnoye settlement, 30 VII 1960, Katenin). (Map II–83).

**Foreign Arctic.** Arctic Alaska (only in the district of the Bering Strait); Labrador (on entire coast); SW and South Greenland (north on west coast to 72°N, on east coast to 64°N); Iceland.

**Outside the Arctic.** East coast of Asia from Kamchatka to the Korean Peninsula; Japan; Sakhalin; Aleutian Islands; west coast of North America (from Bering Strait to

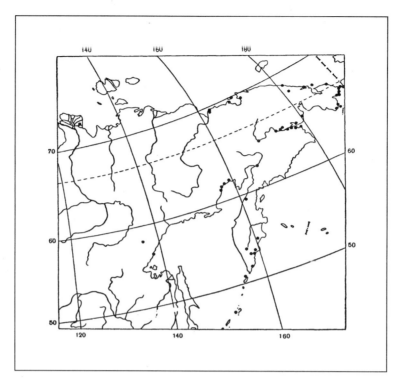

MAP II–84  Distribution of *Leymus villosissimus* (Scribn.) Tzvel.

40°N); shores of Lake Superior and lakes in the upper reaches of the Mackenzie River; shores of Hudson Bay; east coast of North America (from Labrador south to 42°N).

**3. Leymus villosissimus** (Scribn.) Tzvel. in Bot. mat. Gerb. Bot. inst. AN SSSR XX (1960), 429.

*Elymus villosissimus* Scribn. in U. S. Dept. Agricult. Div. Agrost. Bul. XVII (1899), 326; Nevskiy in Fl. SSSR II, 696, pro parte; Karavayev, Konsp. fl. Yak. 61, pro parte.

*E. mollis* auct. non Trin. — Gelert in Ostenfeld, Fl. arct. 1, 133, pro parte.

*E. arenarius* ssp. *mollis* var. *villosissimus* (Scribn.) Hult., Fl. Al. II (1942), 270.

*E. mollis* ssp. *villosissimus* (Scribn.) A. Löve in Bot. Notis. (1950) 33; Bowden in Canad. Journ. Bot. XXXV, 971.

*E. arenarius* ssp. *mollis* auct. — A. E. Porsild, Ill. Fl. Arct. Arch. 41, pro parte.

*E. arenarius* auct. non L. — Polunin, Circump. arct. fl. 51, pro parte.

Ill.: Scribner, l. c., fig. 622; Fl. SSSR II, pl. XLIX, fig. 10; Bowden, l. c., fig. 10.

Littoral species replacing *L. mollis* in more northern districts of East Asia and North America. Sometimes entering the lower reaches of the larger rivers, where it occurs on sands in river valleys, sometimes together with *L. interior* and with specimens of hybrid origin intermediate between the two species. On the north coast of East Siberia only a few isolated localities are known for *L. villosissimus*, the most westerly of them being situated in the Lena Delta.

Like the preceding species, growing on sands and gravelbars on the sea coast and in the lower reaches of the larger rivers.

**Soviet Arctic.** Lena Delta; coasts of East Siberian and Chukchi Seas east of the mouth of the Kolyma (mouth of River Medvezhya, Bay of Chaun, Ayon Island, district of Valkaray settlement, Cape Vankarem, Kolyuchin Island, becoming very common further east and in the district of the Bering Strait); Chukotka Peninsula; lower

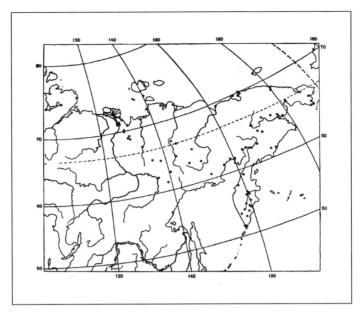

MAP II–85  Distribution of *Leymus interior* (Hult.) Tzvel.

reaches of Anadyr; near mouth of Penzhina (mainly specimens transitional to other species). (Map II–84).

**Foreign Arctic.** Arctic Alaska; arctic coast of Canada west of Hudson Bay (south to 62°N); arctic part of Labrador (only north of 62°N); Canadian Arctic Archipelago (north to 72°N).

**Outside the Arctic.** Coast of Sea of Okhotsk north from Ayan, mainly in the district of Gizhiga Bay; Kamchatka (rare and mainly specimens transitional to other species).

**4. *Leymus interior* (Hult.) Tzvel., comb. nova.**

*Elymus villosissimus* auct. non Scribn. — Tolmachev, Fl. Taym. I, 101; Karavayev, Konsp. fl. Yak. 61, pro parte.

*E. interior* Hultén, Fl. Al. II (1942), 270.

*E. mollis* ssp. *interior* (Hult.) Bowden in Canad. Journ. Bot. XXXV (1957), 951.

Ill.: Bowden, l. c., fig. 11.

Inland subarctic species, rather widely distributed in mountainous districts of NE Siberia. Hultén (l. c.), who described this species from A. I. Tolmachev's Taymyr collections (East Taymyr, SE shore of Lake Taymyr, 24 VII 1928, No. 434, Tolmachev), quite rightly emphasized that, unlike the three preceding species, this species is not coastal. The sole maritime locality for this species in the Tiksi district is apparently secondary and the result of transport of seeds from the Kharaulakh Mountains. Specimens of *L. interior* from the Tiksi coast are completely typical and, despite the maritime habitat, do not show a gradual transition to *L. villosissimus* which is absent from that portion of the coast. This and the fact that *L. interior* is completely absent from North America (where *L. villosissimus* is widespread) testifies to the undoubted distinctness of these two species, despite the existence of a very close relationship between them. The most typical specimens of *L. interior* (including all specimens from Taymyr) possess glumes 5–8 mm long, distinctly shorter than the lemmas. But specimens sometimes occur with longer glumes, which however are usually shorter and narrower on their distal part than in *L. villosissimus* and *L. mollis*.

Characteristic plant of mountain river valleys in NE Siberia; entering the Arctic in association with mountainous districts which, as a rule, represent continuations of more southern mountain chains. In high subarctic mountains the plant is basically native to the zone of open larch forest below the barrens and the lower portion of the zone of tundra barrens; it climbs extremely high in river valleys (to 1900 m in the Suntar-Khayata Range); mountain rivers carry isolated plants (pieces of rhizome capable of rooting) to low elevations, where however the species does not play a conspicuous role.

*Leymus interior* colonizes the upper (rarely flooded) portions of sand and gravel deposits of mountain rivers (sometimes among isolated bushes of *Salix kolymensis* and *S. alaxensis*), and as it spreads forms extensive rhizomatous clones; it grows in pure stands or together with *Zerna Pumpelliana* and *Helictotrichon dahuricum*; as the river bed deepens and the soils are dealkalinized, these subalpine floodplain meadows are replaced by lichenaceous and mossy shrub communities (*Betula exilis, Salix Krylovii, Rhododendron parvifolium*). In the Arctic the species also grows along mountain rivers on sand or sand-and-gravel deposits; more rarely it occurs on the talus of valley outcrops of clayey shales. East of the Lena Delta the species is locally common on dry sandy maritime terraces of the Bykovskiy Peninsula.

**Soviet Arctic.** Taymyr (basin of Lake Taymyr); lower reaches of Lena (north of Kyusyur settlement) and district of Buorkhaya Bay; further east in the lower reaches of the Kolyma; district of Bay of Chaun; Wrangel Island; Chukotka Peninsula (rather frequent); Anadyr and Penzhina Basins. (Map II–85).

**Foreign Arctic.** Not reported.

**Outside the Arctic.** Verkhoyansk Range and NE Siberia to the east of it (including Kamchatka) south to the Dzhugdzhur and Pribrezhnyy Ranges.

## GENUS 36    Hordeum L. — BARLEY

SMALL GENUS (about 20 species) distributed in temperate and subtropical latitudes, mainly in regions with dry climate. One species, removed by some botanists to the separate (monotypic) genus *Critesion* Rafin., penetrates arctic limits in East Siberia and NW North America.

1. ***Hordeum jubatum*** L., Sp. pl. (1753) 85; Grisebach in Ledebour, Fl. ross. IV, 239; Trautvetter, Pl. Sib. bor. 132; id., Fl. rip. Kolym. 74; Macoun & Holm, Vasc. pl. 8A; Holm, Contr. morph. syn. geogr. distr. arct. pl. 68, b; Hultén, Fl. Al. II, 266; Karavayev, Konsp. fl. Yak. 61; Polunin, Circump. arct. fl. 57.

    *Critesion jubatum* (L.) Nevski in Fl. SSSR II, 721.

    Occurring within the forest-tundra and the extreme south of the true tundra zone. Growing on unstabilized deposits on river shores, on dry grassy slopes, sometimes among willow thickets in valleys, occasionally reported near buildings.

    **Soviet Arctic.** Lower reaches of Olenek and Lena (in forest-tundra); lower reaches of Kolyma and neighbouring forest-tundra (common); coast of Bay of Chaun; Anadyr Basin (common in districts remote from the sea).

    **Foreign Arctic.** West coast of Alaska; western part of Canadian arctic coast.

    **Outside the Arctic.** Widespread in NE Siberia (basically in Yakutia), reaching the Khatanga Basin in the west and the coast of the Sea of Okhotsk in the east. Occurring as an introduction in West Siberia, the European part of the USSR and the south of the Soviet Far East. Widespread in North America from the Yukon Basin to Labrador and Newfoundland, south to California and the middle part of the Mississippi Basin.

# APPENDIX I

# Summary of Data on the Geographical Distribution of Vascular Plants of the Soviet Arctic

SUPPLEMENTARY TO THE information provided in the text of this work, we present below a tabular summary of data on the geographical distribution of the species of plants treated in the present volume of this Flora (see Tables 1–2). This summary is given on the basis of the preliminary subdivision of the Soviet Arctic into districts published briefly in our article in the "Botanicheskiy Zhurnal" (1956) (see Map I–1). Along with data on the distribution of plants in these districts, we also include brief information on their distribution in foreign arctic districts. For the convenience of readers we repeat here the characterization of the districts adopted, supplementing it with corresponding characterization of districts of the Foreign Arctic.

1. *Murman District:* from the Norwegian frontier to the neck of the White Sea.

2. *Kanin-Pechora District:* Kanin Peninsula, Malozemelskaya and Timanskaya Tundras, lower reaches of River Pechora, Bolshezemelskaya Tundra (except Ural foothills and Pay-Khoy), Kolguyev Island.

3. *Polar Ural District:* Polar Ural, upper reaches of River Usa, district south of Baydaratskaya Bay.

4. *Yugorskiy District:* Pay-Khoy, Vaygach Island.

5. *Novaya Zemlya.*

6. *Franz Josef Land.*

7. *Ob-Tazovskiy District:* Yamal Peninsula, tundra from the mouth of the Ob to the Taz, Gydanskaya Tundra (except left bank of Yenisey); islands north of Yamal and the Gydanskaya Tundra.

8. *Yenisey District:* lower reaches of Yenisey with the strip watered by tributaries of the Yenisey; shores of the very narrow part of Yenisey Bay; Sibiriakov Island.

9. *Taymyr District:* Taymyr Peninsula from the Yenisey-Pyasina watershed to the Khatanga River; Severnaya Zemlya and other islands north of Taymyr.

10. *Anabar-Olenek District:* from the Khatanga River to the lower reaches of the Olenek inclusively; Begichev and Preobrazheniye Islands.

11. *Lena District:* delta and lower course of Lena River (west to the watershed with the Olenek, east to Buorkhaya Bay).

12. *Yana-Kolyma District:* from Buorkhaya Bay to the eastern boundary of the Yakut ASSR; New Siberian Islands, De Long Islands.

13. *North Chukotka District:* Chukotka north of the Anadyr watershed, east to Kolyuchin Bay; Ayon Island, Wrangel Island.

14. *Beringian Chukotka District:* Chukotka east of Kolyuchin Bay and Krest Bay, with adjacent small islands.

15. *Anadyr District:* Anadyr River Basin and neighbouring areas.

16. *Koryak District:* Koryak Range and adjacent coast; district of Penzhina River, Bay of Penzhina and Bay of Gizhiga.

17. *Arctic part of Alaska.*

18. *Arctic mainland of Canada west of Hudson Bay.*

19. *Arctic part of Labrador Peninsula* (including Arctic Quebec).

20. *Canadian Arctic Archipelago.*

21. *North-West Greenland:* districts of Greenland north of 80° on the east coast and 74° on the west coast.

22. *South-West Greenland:* west coast of Greenland from 74°N to its southern extremity.

23. *East Greenland:* east coast of Greenland from 80°N to its southern extremity.

24. *Iceland* (with Jan Mayen).

25. *Svalbard* (Spitsbergen, Bear Island).

26. *Arctic Scandinavia.*

## TABLE 1

# The Distribution of Vascular Plants of the Soviet Arctic

*Polypodiaceae–Butomaceae*

---

**KEY**

+ indicates the existence of reliable records of the presence of the species in the relevant district;

• indicates the absence of such records.

# FLORA OF THE RUSSIAN ARCTIC

| PLANT NAMES | \| SOVIET ARCTIC \| | | | | | | | | | | | | | | | | \| FOREIGN ARCTIC \| | | | | | | | | | |
|---|---|---|---|---|---|---|---|---|---|---|---|---|---|---|---|---|---|---|---|---|---|---|---|---|---|---|
| | Murman | Kanin-Pechora | Polar Ural | Yugorskiy | Novaya Zemlya | Franz Josef Land | Ob-Tazovskiy | Yenisey | Taymyr | Anabar-Olenek | Lena | Yana-Kolyma | North Chukotka | Beringian Chukotka | Anadyr | Koryak | Arctic Alaska | NW Canada | Labrador | Canadian Archipelago | NW Greenland | SW Greenland | East Greenland | Iceland | Svalbard | Arctic Scandinavia |
| **I. POLYPODIACEAE** | | | | | | | | | | | | | | | | | | | | | | | | | | |
| *Woodsia glabella* | + | + | + | • | • | • | • | + | + | + | + | • | • | + | + | + | + | + | + | + | + | + | + | + | + | + |
| *Woodsia alpina* | + | + | • | • | • | • | • | • | • | • | + | • | • | + | • | + | + | • | + | + | • | + | + | + | • | + |
| *Woodsia ilvensis* | + | • | • | • | • | • | + | • | + | + | + | • | + | + | + | + | + | + | + | + | + | • | + | + | • | + |
| *Cystopteris filix-fragilis* | + | • | + | + | + | • | • | + | + | + | + | + | + | + | + | + | + | + | + | + | + | • | + | + | ? | + |
| *Cystopteris Dickieana* | • | + | + | + | + | • | • | + | + | + | • | + | • | + | + | • | • | • | • | + | + | ? | ? | • | + | ? |
| *Cystopteris montana* | + | + | • | • | • | • | • | • | • | • | • | • | • | • | + | • | + | + | • | + | + | + | • | + | • | + |
| *Dryopteris filix-mas* | + | • | • | • | • | • | • | • | • | • | • | • | • | • | • | • | • | • | • | • | • | + | + | • | • | + |
| *Dryopteris fragrans* | • | • | + | + | • | • | • | + | + | + | + | + | + | + | + | + | + | + | + | + | + | + | + | + | • | • |
| *Dryopteris austriaca* | + | + | • | • | • | • | • | • | • | • | • | • | • | • | + | + | • | • | • | • | + | + | + | • | • | + |
| *Dryopteris spinulosa* | + | + | • | • | • | • | • | • | • | • | • | • | • | • | • | • | • | • | • | • | • | • | • | • | • | + |
| *Thelypteris phegopteris* | + | • | • | • | • | • | • | • | • | • | • | • | • | • | + | • | + | • | • | • | + | + | + | • | • | + |
| *Gymnocarpium dryopteris* | + | + | • | • | • | • | + | + | • | • | • | • | + | • | • | • | + | + | + | • | • | • | + | + | • | + |
| *Gymnocarpium continentale* | • | • | • | • | • | • | • | • | • | • | + | • | • | + | • | • | • | • | • | • | • | • | • | • | • | • |
| *Polystichum lonchitis* | + | • | • | • | • | • | • | • | • | • | • | • | • | • | • | • | • | • | • | • | + | + | + | + | • | + |
| *Athyrium filix-femina* | + | + | • | • | • | • | • | • | • | • | • | • | • | • | + | • | • | • | • | • | • | + | + | + | • | + |
| *Athyrium alpestre* | + | + | • | • | • | • | • | • | • | • | • | • | • | • | • | • | • | • | • | • | + | + | + | + | • | + |
| *Asplenium viride* | + | • | + | • | • | • | + | • | • | • | • | • | • | • | • | • | + | • | • | • | + | + | + | • | • | + |
| *Cryptogramma crispa* | + | • | • | • | • | • | • | • | • | • | • | • | • | • | • | • | • | • | • | • | • | • | + | + | • | + |
| *Cryptogramma Stelleri* | • | • | • | • | • | + | • | • | • | • | • | + | • | • | + | • | • | • | • | • | • | • | • | • | • | • |
| *Polypodium vulgare* | + | • | • | • | • | • | • | • | • | • | • | • | • | • | • | • | • | • | • | • | • | • | + | + | • | + |
| **II. OPHIOGLOSSACEAE** | | | | | | | | | | | | | | | | | | | | | | | | | | |
| *Botrychium lunaria* | + | + | + | • | • | + | • | • | • | • | + | • | • | • | + | + | • | • | • | • | • | + | + | + | + | + |
| *Botrychium boreale* | + | + | • | • | • | • | + | • | • | • | • | + | • | • | + | • | • | • | • | • | + | + | • | • | • | • |
| **III. EQUISETACEAE** | | | | | | | | | | | | | | | | | | | | | | | | | | |
| *Equisetum hiemale* | + | • | • | • | • | • | • | • | • | • | • | • | • | • | • | • | • | • | • | • | + | • | + | • | • | + |
| *Equisetum variegatum* | + | + | + | • | + | • | + | + | + | + | + | + | + | + | + | + | + | + | + | + | + | + | + | + | + | + |
| *Equisetum scirpoides* | + | + | + | + | + | • | + | • | • | + | • | + | • | + | + | + | + | + | + | + | + | • | • | + | + | + |
| *Equisetum limosum* | + | + | • | + | • | • | + | + | • | • | + | • | • | + | • | + | • | • | • | • | • | • | + | • | • | + |
| *Equisetum palustre* | + | + | + | + | • | • | + | • | • | • | + | • | • | • | + | • | • | • | • | • | • | + | • | • | • | + |
| *Equisetum silvaticum* | + | + | + | • | • | • | + | • | • | • | • | • | • | + | • | + | • | • | + | • | • | + | • | + | • | + |
| *Equisetum pratense* | + | + | + | • | • | • | + | • | • | ? | • | • | • | + | • | + | • | • | + | • | + | + | + | • | • | + |
| *Equisetum arvense* | + | + | + | + | + | • | + | + | + | + | + | + | + | + | + | + | + | + | + | + | + | + | + | + | + | + |
| **IV. LYCOPODIACEAE** | | | | | | | | | | | | | | | | | | | | | | | | | | |
| *Lycopodium selago* | + | + | + | + | • | + | + | • | • | • | • | + | + | + | + | + | + | • | + | + | • | + | + | + | ? | + |
| *Lycopodium selago arcticum* | • | + | + | • | • | • | + | + | + | + | + | + | + | + | • | • | • | + | + | • | ? | • | + | • | + | • |

TABLE 1 · 311

| PLANT NAMES | Murman | Kanin-Pechora | Polar Ural | Yugorskiy | Novaya Zemlya | Franz Josef Land | Ob-Tazovskiy | Yenisey | Taymyr | Anabar-Olenek | Lena | Yana-Kolyma | North Chukotka | Beringian Chukotka | Anadyr | Koryak | Arctic Alaska | NW Canada | Labrador | Canadian Archipelago | NW Greenland | SW Greenland | East Greenland | Iceland | Svalbard | Arctic Scandinavia |
|---|---|---|---|---|---|---|---|---|---|---|---|---|---|---|---|---|---|---|---|---|---|---|---|---|---|---|
| | \multicolumn{16}{c}{SOVIET ARCTIC} | | | | | | | | | | | | | | | | \multicolumn{10}{c}{FOREIGN ARCTIC} | | | | | | | | | | |
| *Lycopodium annotinum* | + | + | • | • | • | • | + | + | • | • | • | • | • | • | • | • | + | • | • | • | • | + | + | + | • | + |
| *Lycopodium pungens* | + | + | + | • | • | • | + | + | • | • | + | + | • | + | + | + | + | • | • | • | • | + | + | + | • | + |
| *Lyc. clavatum monostachyon* | + | + | + | • | • | • | + | + | • | • | • | • | • | + | + | + | • | • | • | • | • | + | • | + | • | + |
| *Lycopodium complanatum* | + | + | • | • | • | • | + | + | • | • | • | • | • | • | • | • | + | • | • | • | • | • | • | • | • | + |
| *Lycopodium tristachyum* | + | • | • | • | • | • | • | • | • | • | • | • | • | • | • | • | • | • | • | • | • | + | + | • | • | • |
| *Lycopodium alpinum* | + | + | + | + | • | • | + | + | • | • | • | + | + | + | + | + | + | • | • | • | + | + | + | + | • | + |

**V. SELAGINELLACEAE**

| Plant | | | | | | | | | | | | | | | | | | | | | | | | | | |
|---|---|---|---|---|---|---|---|---|---|---|---|---|---|---|---|---|---|---|---|---|---|---|---|---|---|---|
| *Selaginella selaginoides* | + | + | + | • | • | • | • | + | • | • | • | • | • | • | • | • | + | • | • | • | • | + | + | + | • | + |
| *Selaginella sibrica* | • | • | • | • | • | • | • | • | + | + | + | + | + | + | + | • | + | • | • | • | • | • | • | • | • | • |

**VI. PINACEAE**

| Plant | | | | | | | | | | | | | | | | | | | | | | | | | | |
|---|---|---|---|---|---|---|---|---|---|---|---|---|---|---|---|---|---|---|---|---|---|---|---|---|---|---|
| *Picea excelsa* | + | • | • | • | • | • | • | • | • | • | • | • | • | • | • | • | • | • | • | • | • | • | • | • | • | + |
| *Picea obovata* | + | + | + | • | • | • | + | + | • | • | • | • | • | • | • | • | • | • | • | • | • | • | • | • | • | • |
| *Larix sibirica* | + | + | + | • | • | • | + | + | • | • | • | • | • | • | • | • | • | • | • | • | • | • | • | • | • | • |
| *Larix dahurica* | • | • | • | • | • | • | • | • | + | + | + | • | • | + | + | + | • | • | • | • | • | • | • | • | • | • |
| *Pinus silvestris lapponica* | + | • | • | • | • | • | • | • | • | • | • | • | • | • | • | • | • | • | • | • | • | • | • | • | • | + |
| *Pinus pumila* | • | • | • | • | • | • | • | • | • | • | + | + | • | + | + | • | • | • | • | • | • | • | • | • | • | • |

**VII. CUPRESSACEAE**

| Plant | | | | | | | | | | | | | | | | | | | | | | | | | | |
|---|---|---|---|---|---|---|---|---|---|---|---|---|---|---|---|---|---|---|---|---|---|---|---|---|---|---|
| *Juniperus sibirica* | + | + | + | + | • | • | + | + | • | • | • | • | • | + | + | + | • | • | • | • | • | +? | +? | +? | • | + |

**VIII. SPARGANIACEAE**

| Plant | | | | | | | | | | | | | | | | | | | | | | | | | | |
|---|---|---|---|---|---|---|---|---|---|---|---|---|---|---|---|---|---|---|---|---|---|---|---|---|---|---|
| *Sparaganium simplex* | + | + | • | • | • | • | + | • | • | • | • | • | • | • | • | • | • | • | • | • | • | • | • | • | • | + |
| *Sparaganium affine* | + | • | • | • | • | • | • | + | • | • | + | • | • | • | • | • | • | • | • | • | • | • | + | + | • | + |
| *Sparaganium minimum* | • | + | • | • | • | • | + | • | • | • | • | • | + | + | • | • | • | • | • | • | • | • | + | • | + | + |
| *Sparaganium hyperboreum* | + | + | • | + | • | • | + | + | • | • | + | + | + | • | + | + | + | • | + | • | • | + | • | + | • | + |

**IX. POTAMOGETONACEAE**

| Plant | | | | | | | | | | | | | | | | | | | | | | | | | | |
|---|---|---|---|---|---|---|---|---|---|---|---|---|---|---|---|---|---|---|---|---|---|---|---|---|---|---|
| *Potamogeton filiformis* | + | • | • | • | • | • | • | • | + | • | • | • | • | • | • | • | + | + | + | + | • | + | + | + | • | + |
| *Potamogeton subretusus* | • | • | • | • | • | • | + | • | • | • | • | • | • | • | • | • | • | • | • | • | • | • | • | • | • | • |
| *Potamogeton vaginatus* | • | • | • | • | • | • | + | • | • | • | • | • | • | • | • | • | + | + | • | • | • | • | • | • | • | • |
| *Potamogeton pectinatus* | • | + | • | + | • | • | + | • | • | • | • | • | • | • | • | • | • | • | • | • | • | • | • | • | • | + |
| *Potamogeton subsibiricus* | • | • | • | • | • | • | • | + | • | • | • | • | • | • | • | • | • | • | • | • | • | • | • | • | • | • |
| *Potamogeton pusillus* | + | + | • | • | • | • | • | • | • | • | • | • | • | • | • | • | • | • | • | • | • | • | • | + | • | + |
| *Potamogeton alpinus* | + | + | • | • | • | • | + | + | • | • | • | • | • | • | • | • | • | • | • | • | • | • | • | + | • | + |
| *Potamogeton alpinus tenuifolius* | • | • | • | • | • | • | • | • | • | • | + | • | • | • | • | + | + | • | + | • | • | + | • | • | • | • |
| *Potamogeton natans* | • | • | • | • | • | • | + | • | • | • | • | • | • | • | • | • | • | • | • | • | • | • | • | + | • | + |
| *Potamogeton gramineus* | + | + | • | • | • | • | + | • | • | • | • | • | • | • | • | • | • | • | • | • | • | • | • | + | • | + |

| PLANT NAMES | Murman | Kanin-Pechora | Polar Ural | Yugorskiy | Novaya Zemlya | Franz Josef Land | Ob-Tazovskiy | Yenisey | Taymyr | Anabar-Olenek | Lena | Yana-Kolyma | North Chukotka | Beringian Chukotka | Anadyr | Koryak | Arctic Alaska | NW Canada | Labrador | Canadian Archipelago | NW Greenland | SW Greenland | East Greenland | Iceland | Svalbard | Arctic Scandinavia |
|---|---|---|---|---|---|---|---|---|---|---|---|---|---|---|---|---|---|---|---|---|---|---|---|---|---|---|
| *Potamogeton praelongus* | + | + | • | • | • | • | + | + | • | • | • | • | • | • | • | • | • | + | + | • | • | • | • | + | • | + |
| *Potamogeton perfoliatus* | + | + | • | • | • | • | • | + | • | • | • | + | • | • | • | • | + | • | • | • | • | • | • | + | • | + |
| *Zostera marina* | + | • | • | • | • | • | • | • | • | • | • | • | • | • | • | + | + | • | • | • | • | • | + | • | • | + |
| **X. JUNCAGINACEAE** | | | | | | | | | | | | | | | | | | | | | | | | | | |
| *Triglochin maritimum* | + | + | • | • | • | • | • | • | • | • | + | • | • | + | • | + | • | • | + | • | • | • | • | + | • | + |
| *Triglochin palustre* | + | + | + | • | • | • | • | • | • | • | • | • | + | • | + | + | • | • | + | • | • | • | + | + | • | + |
| *Scheuchzeria palustris* | + | • | • | • | • | • | • | • | • | • | • | • | • | • | • | • | • | • | • | • | • | • | • | • | • | + |
| **XI. ALISMATACEAE** | | | | | | | | | | | | | | | | | | | | | | | | | | |
| *Alisma plantago-aquatica* | + | • | • | • | • | • | • | • | • | • | • | • | • | • | • | • | • | • | • | • | • | • | • | • | • | • |
| **XII. BUTOMACEAE** | | | | | | | | | | | | | | | | | | | | | | | | | | |
| *Butomus umbellatus* | • | + | • | • | • | • | • | • | • | • | • | • | • | • | • | • | • | • | • | • | • | • | • | • | • | • |

## TABLE 2

# The Distribution of Vascular Plants of the Soviet Arctic

*Gramineae*

---

**KEY**

+    indicates the existence of reliable records of the presence of the species in the relevant district;

•    indicates the absence of such records.

# XIII. GRAMINEAE

| PLANT NAMES | Murman | Kanin-Pechora | Polar Ural | Yugorskiy | Novaya Zemlya | Franz Josef Land | Ob-Tazovskiy | Yenisey | Taymyr | Anabar-Olenek | Lena | Yana-Kolyma | North Chukotka | Beringian Chukotka | Anadyr | Koryak | Arctic Alaska | NW Canada | Labrador | Canadian Archipelago | NW Greenland | SW Greenland | East Greenland | Iceland | Svalbard | Arctic Scandinavia |
|---|---|---|---|---|---|---|---|---|---|---|---|---|---|---|---|---|---|---|---|---|---|---|---|---|---|---|
| *Typhoides arundinacea* | + | + | • | • | • | • | • | + | • | • | • | • | • | • | + | • | • | • | • | • | • | • | • | • | • | + |
| *Anthoxanthum alpinum* | + | + | + | • | • | • | • | + | • | • | • | • | • | • | • | • | • | • | • | • | • | • | + | + | • | + |
| *Hierochloë alpina* | + | + | + | + | + | • | + | + | + | + | + | + | + | + | + | + | + | + | + | + | + | + | + | + | • | + |
| *Hierochloë pauciflora* | • | • | + | + | + | • | + | + | + | + | + | + | + | + | + | + | + | + | + | + | + | + | + | • | • | • |
| *Hierochloë odorata* | + | + | + | • | • | • | + | • | • | • | • | + | • | + | + | + | + | + | + | + | + | + | + | + | • | + |
| *Milium effusum* | + | + | + | • | • | • | • | • | • | • | • | • | • | • | • | • | • | • | • | • | • | • | • | + | • | + |
| *Phleum commutatum* | + | + | + | + | • | • | • | • | • | • | • | • | • | • | • | • | • | + | • | • | • | + | + | + | • | + |
| *Phleum pratense* | + | + | • | • | • | • | • | • | • | • | • | • | • | • | • | • | • | • | • | • | • | + | + | + | • | + |
| *Alopecurus pratensis* | + | + | + | + | + | • | + | + | • | • | • | • | • | + | • | • | • | • | • | • | • | • | • | + | • | + |
| *Alopecurus arundinaceus* | + | + | • | • | • | • | • | • | • | • | • | • | • | • | • | • | • | • | • | • | • | • | • | • | • | + |
| *Alopecurus glaucus* | • | • | • | • | • | • | • | • | • | + | + | • | • | • | + | • | • | + | • | • | • | • | • | • | • | • |
| *Alopecurus alpinus* | • | • | + | + | + | + | + | + | + | + | + | + | + | + | • | + | + | + | + | + | + | + | + | + | + | • |
| *Alopecurus Stejnegeri* | • | • | • | • | • | • | • | • | • | • | • | • | • | ? | + | + | + | • | • | • | • | • | • | • | • | • |
| *Alopecurus geniculatus* | + | • | • | • | • | • | + | • | • | • | • | • | • | • | • | • | • | • | • | • | • | • | • | + | • | + |
| *Alopecurus aequalis* | + | + | + | + | • | • | + | • | • | • | • | • | • | • | + | + | + | ? | + | + | • | + | + | + | • | + |
| *Alopecurus amurensis* | • | • | • | • | • | • | • | • | • | • | • | • | • | + | + | • | • | • | • | • | • | • | • | • | • | • |
| *Arctagrostis latifolia* | + | + | + | + | + | • | + | + | + | + | + | + | + | + | + | + | + | + | + | + | + | + | + | + | + | + |
| *Arctagrostis arundinacea* | • | • | • | • | • | • | + | + | + | + | + | + | + | + | + | + | + | • | • | • | • | • | • | • | • | • |
| *Agrostis gigantea* | + | + | + | • | • | • | • | • | • | • | • | • | • | • | • | • | • | • | • | • | • | • | • | + | • | + |
| *Agrostis stolonifera* | + | + | + | • | • | • | • | • | • | • | • | • | • | • | • | • | + | • | • | • | • | + | + | + | • | + |
| *Agrostis straminea* | + | + | • | + | • | • | • | • | • | • | • | • | • | • | • | • | • | • | • | • | • | • | • | ? | • | + |
| *Agrostis tenuis* | + | • | • | • | • | • | • | • | • | • | • | • | • | • | • | • | • | • | • | • | • | + | + | + | • | + |
| *Agrostis canina* | + | • | • | • | • | • | • | • | • | • | • | • | • | • | • | • | • | • | • | • | • | + | + | + | • | + |
| *Agrostis Trinii* | • | • | • | • | • | • | • | + | • | • | • | • | + | • | + | + | • | • | • | • | • | • | • | • | • | • |
| *Agrostis anadyrensis* | • | • | • | • | • | • | • | • | • | • | • | + | + | + | + | + | + | • | • | • | • | • | • | • | • | • |
| *Agrostis borealis* | + | + | + | + | • | • | • | ? | • | • | • | • | • | • | • | • | • | + | + | + | + | • | + | + | • | + |
| *Agrostis clavata* | • | • | • | • | • | + | • | • | • | • | • | • | + | • | • | + | • | • | • | • | • | • | • | • | • | • |
| *Agrostis scabra* | • | • | • | • | • | • | • | • | • | • | • | • | • | • | + | + | + | • | • | • | • | • | • | • | • | • |
| *Calamagrostis epigeios* | + | + | • | • | • | • | + | • | • | • | • | • | • | • | • | • | • | • | • | • | • | • | • | • | • | + |
| *Calamagrostis canescens* | + | • | • | • | • | • | • | • | • | • | • | • | • | • | • | • | • | • | • | • | • | • | • | • | • | + |
| *Calamagrostis Langsdorffii* | + | + | + | + | • | • | + | + | • | • | + | + | + | + | + | + | + | + | + | + | • | + | + | • | • | • |
| *Calamagrostis angustifolia* | • | • | • | • | • | • | • | • | • | • | • | • | • | + | + | + | + | + | • | • | • | • | • | • | • | • |
| *Calamagrostis lapponica* | + | + | + | + | • | • | + | • | • | + | + | + | + | + | + | + | + | + | + | + | • | + | + | • | • | • |
| *Calamagrostis neglecta* | + | + | + | + | + | • | + | + | + | • | + | + | + | + | + | + | + | + | + | • | + | + | + | + | + | + |
| *Calamagrostis Holmii* | ? | + | + | + | • | • | + | + | + | + | + | + | + | + | + | + | + | + | + | + | + | • | • | • | • | • |
| *Calamagrostis deschampsioides* | + | + | + | • | • | • | • | • | • | + | • | + | + | + | + | + | + | + | • | • | • | • | • | • | • | • |
| *Calamagrostis Korotkyi* | • | • | • | • | • | • | • | • | • | • | • | • | • | • | • | • | • | • | • | • | • | • | • | • | • | • |
| *Calamagrostis purpurascens* | • | • | • | • | • | • | • | • | + | + | + | + | • | + | + | + | + | + | • | + | + | • | + | + | • | • |

TABLE 2

| PLANT NAMES | SOVIET ARCTIC | | | | | | | | | | | | | | | | FOREIGN ARCTIC | | | | | | | | | |
|---|---|---|---|---|---|---|---|---|---|---|---|---|---|---|---|---|---|---|---|---|---|---|---|---|---|---|
| | Murman | Kanin-Pechora | Polar Ural | Yugorskiy | Novaya Zemlya | Franz Josef Land | Ob-Tazovskiy | Yenisey | Taymyr | Anabar-Olenek | Lena | Yana-Kolyma | North Chukotka | Beringian Chukotka | Anadyr | Koryak | Arctic Alaska | NW Canada | Labrador | Canadian Archipelago | NW Greenland | SW Greenland | East Greenland | Iceland | Svalbard | Arctic Scandinavia |
| *Calamagrostis arctica* | • | • | • | • | • | • | • | • | • | • | • | • | + | + | • | • | • | • | • | • | • | • | • | • | • | • |
| *Calamagrostis sesquiflora* | • | • | • | • | • | • | • | • | • | • | • | • | • | + | • | + | • | • | • | • | • | • | • | • | • | • |
| *Apera spica-venti* | + | • | • | • | • | • | • | • | • | • | • | • | • | • | • | • | • | • | • | • | • | • | • | • | • | + |
| *Vahlodea atropurpurea* | + | + | + | • | • | • | • | • | • | • | • | • | • | • | • | • | + | + | ? | • | + | + | • | • | • | + |
| *Deschampsia flexuosa* | + | + | + | • | • | • | + | • | • | • | • | • | • | • | • | • | • | • | + | • | • | + | + | + | • | + |
| *Deschampsia caespitosa* | + | + | + | + | • | • | • | • | • | • | • | • | • | • | • | • | +? | +? | + | + | • | • | • | • | + | • |
| *Deschampsia glauca* | + | + | + | + | • | • | + | + | + | • | + | • | • | • | • | • | • | • | • | • | • | • | • | ? | • | + |
| *Deschampsia anadyrensis* | • | • | • | • | • | • | • | • | • | • | • | • | • | + | • | • | • | • | • | • | • | • | • | • | • | • |
| *Deschampsia Sukatschewii* | • | • | • | • | • | • | • | + | + | • | + | + | • | + | + | + | + | • | • | • | • | • | • | • | • | • |
| *Deschampsia borealis* | • | + | + | + | + | • | + | + | + | + | + | + | + | + | • | + | • | • | • | ? | • | ? | ? | • | ? | • |
| *Deschampsia brevifolia* | • | + | + | + | + | • | • | + | + | + | • | • | • | • | • | • | + | + | • | + | • | • | • | • | +? | • |
| *Deschampsia alpina* | + | • | • | • | + | + | • | • | • | • | • | • | • | • | • | • | • | • | + | + | • | + | + | + | + | + |
| *Deschampsia obensis* | +? | + | + | + | • | • | + | + | + | + | + | + | + | • | • | • | • | • | • | • | • | • | • | • | • | • |
| *Trisetum sibiricum* | • | + | + | + | • | • | • | + | • | • | • | • | • | + | • | • | • | • | • | • | • | • | • | • | • | • |
| *Trisetum sibiricum ssp. litoralis* | • | + | • | • | • | • | + | + | + | + | + | + | • | • | • | • | • | • | • | • | • | • | • | • | • | • |
| *Trisetum subalpestre* | • | • | • | • | • | • | • | + | + | • | + | • | • | • | • | • | • | • | • | • | • | • | • | • | • | + |
| *Trisetum spicatum* | + | + | + | + | + | • | + | + | + | + | + | + | + | + | + | + | + | + | + | + | + | + | + | • | + | + |
| *Trisetum molle* | • | • | • | • | • | • | • | • | • | + | + | + | • | + | + | + | + | • | + | • | • | • | • | • | • | • |
| *Trisetum molle ssp. alascanum* | • | • | • | • | • | • | • | • | • | • | • | • | • | + | • | • | + | • | • | • | • | • | • | • | • | • |
| *Helictotrichon dahuricum* | • | • | • | • | • | • | • | • | • | + | • | + | • | + | + | • | • | • | • | • | • | • | • | • | • | • |
| *Helictotrichon Krylovii* | • | • | • | • | • | • | • | • | • | • | • | • | + | • | • | • | • | • | • | • | • | • | • | • | • | • |
| *Beckmannia eruciformis* | + | • | • | • | • | • | • | • | • | • | • | • | • | • | • | • | • | • | • | • | • | • | • | • | • | • |
| *Beckmannia syzigachne* | • | • | • | • | • | • | + | • | • | • | + | • | • | + | + | + | + | • | • | • | • | • | + | • | • | • |
| *Phragmites communis* | + | + | • | • | • | • | • | • | • | • | • | • | • | • | • | • | • | • | • | • | • | • | • | • | • | + |
| *Molinia coerulea* | + | • | • | • | • | • | • | • | • | • | • | • | • | • | • | • | • | • | • | • | • | • | • | + | • | + |
| *Koeleria seminuda* | • | • | • | • | • | • | • | • | • | + | • | • | • | • | • | • | • | • | • | • | • | • | • | • | • | • |
| *Koeleria asiatica* | • | + | + | + | • | • | + | + | + | + | + | + | + | • | + | • | • | • | • | • | • | • | • | • | • | • |
| *Koeleria Pohleana* | • | + | • | • | • | • | • | • | • | • | • | • | • | • | • | • | • | • | • | • | • | • | • | • | • | • |
| *Melica nutans* | + | • | • | • | • | • | • | • | • | • | • | • | • | • | • | • | • | • | • | • | • | • | • | • | • | + |
| *Pleuropogon Sabinii* | • | • | + | + | + | + | • | + | + | • | + | + | + | • | • | • | • | + | + | + | + | • | + | • | + | • |
| *Dactylis glomerata* | + | • | • | • | • | • | • | • | • | • | • | • | • | • | • | • | • | • | • | • | • | • | • | • | • | + |
| *Poa eminens* | • | • | • | • | • | • | • | • | • | • | • | • | + | + | + | + | + | • | + | • | • | • | • | • | • | • |
| *Poa Trautvetteri* | • | • | • | • | • | • | • | • | • | • | + | • | • | • | • | • | • | • | • | • | • | • | • | • | • | • |
| *Poa platyantha* | • | • | • | • | • | • | • | • | • | • | • | • | • | • | • | • | • | + | • | • | • | • | • | • | • | • |
| *Poa malacantha* | • | • | • | • | • | • | • | • | • | • | • | • | + | + | + | + | + | • | • | • | • | • | • | • | • | • |
| *Poa lanata* | • | • | • | • | • | • | • | • | • | • | + | + | • | + | + | + | • | • | • | • | • | • | • | • | • | • |
| *Poa arctica* | + | + | + | + | + | + | + | + | + | + | + | + | + | + | + | + | + | + | + | + | + | + | + | • | + | + |
| *Poa Tolmatchewii* | • | • | • | +? | • | • | • | + | • | • | + | • | • | • | • | • | • | • | + | + | + | + | • | • | + | • |
| *Poa pratensis* | + | + | + | + | • | • | + | + | + | + | + | + | + | + | + | + | ? | ? | • | • | ? | ? | ? | • | ? | ? |

| PLANT NAMES | \multicolumn{16}{c}{SOVIET ARCTIC} | \multicolumn{10}{c}{FOREIGN ARCTIC} |

| PLANT NAMES | Murman | Kanin-Pechora | Polar Ural | Yugorskiy | Novaya Zemlya | Franz Josef Land | Ob-Tazovskiy | Yenisey | Taymyr | Anabar-Olenek | Lena | Yana-Kolyma | North Chukotka | Beringian Chukotka | Anadyr | Koryak | Arctic Alaska | NW Canada | Labrador | Canadian Archipelago | NW Greenland | SW Greenland | East Greenland | Iceland | Svalbard | Arctic Scandinavia |
|---|---|---|---|---|---|---|---|---|---|---|---|---|---|---|---|---|---|---|---|---|---|---|---|---|---|---|
| *Poa sublanata* | • | • | • | • | • | • | + | + | + | + | • | • | • | • | + | + | • | • | • | • | • | • | • | • | • | • |
| *Poa alpigena* | + | + | + | + | + | + | + | + | + | + | + | + | + | + | + | + | + | + | + | + | + | + | + | + | • | + |
| *Poa subcaerulea* | + | • | • | • | • | • | • | • | • | • | • | • | • | + | • | • | • | • | • | • | • | • | • | • | • | + |
| *Poa angustifolia* | ? | • | • | • | • | • | • | • | • | • | • | • | • | • | • | • | • | • | • | • | • | • | • | • | • | • |
| *Poa alpina* | + | + | + | + | + | • | • | + | + | • | • | + | • | + | • | • | + | + | + | + | • | + | + | + | + | + |
| *Poa abbreviata* | • | • | + | + | + | • | • | • | + | • | • | • | • | • | • | • | + | + | + | + | + | + | + | • | + | • |
| *Poa leptocoma* | • | • | • | • | • | • | • | • | • | • | • | • | • | + | • | • | • | • | • | • | • | • | • | • | • | • |
| *Poa paucispicula* | • | • | • | • | • | • | • | + | • | + | • | + | + | + | + | + | + | • | • | • | • | • | • | • | • | • |
| *Poa pseudoabbreviata* | • | • | • | • | • | • | • | + | • | + | • | + | • | + | • | • | + | • | • | • | • | • | • | • | • | • |
| *Poa compressa* | + | • | • | • | • | • | • | • | • | • | • | • | • | • | • | • | • | • | • | • | • | • | • | • | • | • |
| *Poa nemoralis* | + | + | + | • | • | • | + | • | • | • | • | • | • | + | • | • | • | • | • | • | • | • | + | + | + | +? |
| *Poa Tanfiljewii* | • | + | • | • | • | • | • | • | • | • | • | • | • | • | • | • | • | • | • | • | • | • | • | • | • | • |
| *Poa lapponica* | + | + | • | • | • | • | • | • | • | • | • | • | • | • | • | • | • | • | • | • | • | • | • | • | • | +? |
| *Poa palustris* | + | + | • | • | • | • | + | + | • | • | + | + | • | + | + | • | • | • | • | • | • | • | • | • | • | +? |
| *Poa stepposa* | • | • | • | • | • | • | • | + | • | • | + | • | + | + | • | • | • | • | • | • | • | • | • | • | • | • |
| *Poa filiculmis* | • | • | • | • | • | • | • | • | • | • | + | • | • | + | + | • | • | • | • | • | • | • | • | • | • | • |
| *Poa botryoides* | • | • | • | • | • | • | • | • | • | • | + | • | • | + | • | • | • | • | • | • | • | • | • | • | • | • |
| *Poa glauca* | + | + | + | • | • | • | + | + | + | + | + | + | + | + | + | + | + | + | + | + | + | + | + | + | + | + |
| *Poa anadyrica* | • | • | • | • | • | • | • | • | • | • | • | + | • | + | + | + | ? | ? | • | • | • | • | • | • | • | • |
| *Poa bryophila* | • | • | + | • | • | • | • | + | + | • | + | • | + | • | • | • | ? | ? | • | • | • | • | • | • | • | • |
| *Poa trivialis* | + | • | • | • | • | • | • | • | • | • | • | • | • | • | • | • | • | • | • | • | • | • | • | • | + | + |
| *Poa sibirica* | • | • | + | • | • | • | • | + | • | + | + | • | • | • | • | • | • | • | • | • | • | • | • | • | • | • |
| *Poa supina* | • | • | • | • | • | • | + | + | • | • | • | • | • | • | • | • | • | • | • | • | • | • | • | • | • | • |
| *Poa annua* | + | + | • | • | • | • | • | + | + | • | • | • | • | • | • | • | • | • | • | • | • | • | + | + | + | + |
| *Dupontia psilosantha* | ? | + | • | + | + | • | + | + | • | + | + | + | + | + | + | • | + | + | • | + | + | • | + | + | + | + |
| *Dupontia Fisheri* | • | + | + | + | + | + | + | + | + | + | + | + | + | • | + | + | + | + | + | + | • | + | + | + | • | + |
| *Arctophila fulva* | + | + | + | + | • | + | + | + | + | + | + | + | + | + | + | + | + | + | • | + | • | + | • | • | • | • |
| *Colpodium lanatiflorum* | • | • | • | • | • | • | • | • | • | • | • | + | • | • | • | • | • | • | • | • | • | • | • | • | • | • |
| *Catabrosa aquatica* | + | + | • | • | • | • | • | • | • | • | • | • | • | • | • | • | • | • | • | • | • | + | • | + | • | + |
| *Phippsia algida* | + | + | • | + | + | + | + | + | + | + | + | + | + | + | • | + | + | + | + | + | + | + | • | + | + | + |
| *Phippsia concinna* | • | + | + | + | + | • | + | + | + | + | + | + | + | ? | • | • | • | • | • | • | • | • | • | • | + | • |
| *Puccinellia maritima* | + | • | • | • | • | • | • | • | • | • | • | • | • | • | • | • | • | • | • | • | • | • | + | + | + | + |
| *Puccinellia phryganodes* | + | + | • | + | + | • | + | + | + | + | + | + | + | + | + | + | + | + | + | + | • | + | + | + | + | + |
| *Puccinellia geniculata* | • | • | • | • | • | • | • | • | • | • | • | • | • | • | • | + | • | • | • | • | • | • | • | • | • | • |
| *Puccinellia tenella* | • | • | + | + | • | • | • | + | + | • | + | + | • | + | • | • | + | • | + | + | • | + | • | • | • | • |
| *Puccinellia alascana* | • | • | • | • | • | • | • | • | • | • | • | • | • | + | • | + | • | • | • | • | • | • | • | • | • | • |
| *Puccinellia Vahliana* | • | • | • | + | + | • | • | • | + | • | • | • | • | • | • | • | • | + | + | + | + | • | • | • | + | • |
| *Puccinellia Wrightii* | • | • | • | • | • | • | • | • | • | • | • | • | + | • | + | • | • | • | • | • | • | • | • | • | • | • |
| *Puccinellia colpodioides* | • | • | • | • | • | • | • | • | • | • | • | + | • | • | • | • | • | • | • | • | • | • | • | • | • | • |

TABLE 2    317

| PLANT NAMES | SOVIET ARCTIC | | | | | | | | | | | | | | | | FOREIGN ARCTIC | | | | | | | | |
|---|---|---|---|---|---|---|---|---|---|---|---|---|---|---|---|---|---|---|---|---|---|---|---|---|---|
| | Murman | Kanin-Pechora | Polar Ural | Yugorskiy | Novaya Zemlya | Franz Josef Land | Ob-Tazovskiy | Yenisey | Taymyr | Anabar-Olenek | Lena | Yana-Kolyma | North Chukotka | Beringian Chukotka | Anadyr | Koryak | Arctic Alaska | NW Canada | Labrador | Canadian Archipelago | NW Greenland | SW Greenland | East Greenland | Iceland | Svalbard | Arctic Scandinavia |
| *Puccinellia jenisseiensis* | · | · | · | · | · | · | · | + | · | · | · | · | · | · | · | · | · | · | · | · | · | · | · | · | · | · |
| *Puccinellia angustata* | · | · | · | + | + | + | + | · | + | · | · | + | + | · | · | · | ? | + | · | + | + | · | + | · | + | · |
| *Puccinellia fragiliflora* | · | · | · | + | · | · | · | + | + | + | · | + | · | · | · | · | · | · | · | · | · | · | · | · | · | · |
| *Puccinellia Gorodkovii* | · | · | · | · | · | · | · | · | + | · | · | · | · | · | · | · | · | · | · | · | · | · | · | · | · | · |
| *Puccinellia vaginata* | · | · | · | · | · | · | · | · | · | + | · | · | · | · | · | · | + | + | · | · | + | + | · | · | · | · |
| *Puccinellia capillaris* | + | + | · | + | · | · | · | · | · | · | · | · | · | · | · | · | · | · | · | · | · | · | · | · | · | + |
| *Puccinellia pulvinata* | + | + | · | + | + | · | · | · | · | · | · | · | · | · | · | · | · | · | · | · | · | · | · | · | · | + |
| *Puccinellia coarctata* | + | + | · | + | + | · | · | · | · | · | · | · | · | · | · | · | · | + | · | · | + | + | + | · | · | + |
| *Puccinellia sibirica* | · | · | · | + | · | · | + | + | + | · | · | + | · | · | · | · | · | · | · | · | · | · | · | · | · | · |
| *Puccinellia distans* | + | + | · | · | · | · | · | · | · | · | · | · | · | · | · | · | · | · | · | · | · | · | · | · | · | · |
| *Puccinellia borealis* | · | · | · | · | · | · | · | · | + | · | + | + | + | · | · | · | + | · | · | · | · | · | · | · | · | · |
| *Puccinellia Hauptiana* | · | · | · | · | · | · | · | · | · | + | · | · | · | + | + | · | · | · | · | · | · | · | · | · | · | · |
| *Festuca pratensis* | + | · | · | · | · | · | · | · | · | · | · | · | · | · | · | · | · | · | · | · | · | + | + | + | · | + |
| *Festuca altaica* | · | · | · | · | · | · | · | + | · | · | + | + | + | + | + | + | + | · | · | · | · | · | · | · | · | · |
| *Festuca rubra* | + | + | + | + | + | · | + | + | + | + | + | + | + | + | + | + | + | + | + | + | + | + | + | + | + | + |
| *Festuca polesica* | · | + | · | · | · | · | · | · | · | · | · | · | · | · | · | · | · | · | · | · | · | · | · | · | · | · |
| *Festuca ovina* | + | + | + | + | + | · | + | + | + | · | · | · | · | · | · | · | ? | · | · | · | · | · | + | + | · | + |
| *Festuca kolymensis* | · | · | · | · | · | · | · | · | · | + | + | + | · | + | + | · | · | · | · | · | · | · | · | · | · | · |
| *Festuca auriculata* | · | · | + | · | · | · | · | + | + | + | + | + | · | + | + | · | · | · | · | · | · | · | · | · | · | · |
| *Festuca brachyphylla* | + | · | + | + | + | · | + | + | + | + | + | + | + | + | + | + | + | + | + | + | + | + | + | + | + | · |
| *Zerna inermis* | + | + | + | · | · | · | + | · | · | · | · | · | · | · | · | · | + | · | · | · | · | · | · | · | · | · |
| *Zerna Pumpelliana* | · | + | + | + | · | · | · | + | + | + | + | + | + | + | + | + | · | · | · | · | · | · | · | · | · | · |
| *Zerna ircutensis* | · | · | · | · | · | · | · | · | · | · | · | · | + | · | · | · | · | · | · | · | · | · | · | · | · | · |
| *Zerna arctica* | · | · | · | · | · | · | · | · | · | · | + | + | · | + | + | · | · | · | · | · | · | · | · | · | · | · |
| *Bromus arvensis* | + | · | · | · | · | · | · | · | · | · | · | · | · | · | · | · | · | · | · | · | · | · | · | · | · | + |
| *Bromus secalinus* | + | · | · | · | · | · | · | · | · | · | · | · | · | · | · | · | · | + | · | · | · | · | · | · | · | + |
| *Nardus stricta* | + | · | · | · | · | · | · | · | · | · | · | · | · | · | · | · | · | · | · | · | · | + | + | + | · | + |
| *Roegneria confusa* | · | · | · | · | · | · | · | · | · | · | · | + | · | + | + | · | · | · | · | · | · | · | · | · | · | · |
| *Roegneria canina* | + | · | · | · | · | · | + | · | · | · | · | · | · | · | · | · | · | · | · | · | · | · | · | · | · | + |
| *Roegneria mutabilis* | + | + | + | · | · | · | + | · | + | · | · | · | · | + | · | · | + | · | · | · | · | · | · | · | · | + |
| *Roegneria fibrosa* | · | + | + | · | · | + | · | · | · | · | · | · | · | · | · | · | · | · | · | · | · | · | · | · | · | · |
| *Roegneria subfibrosa* | · | · | · | · | · | + | + | · | + | + | + | · | · | · | · | · | · | · | · | · | · | · | · | · | · | · |
| *Roegneria macroura* | · | · | · | + | · | · | · | + | · | + | · | + | · | + | + | + | · | · | · | · | · | · | · | · | · | · |
| *Roegneria Nepliana* | · | · | · | · | · | · | · | · | · | · | · | + | · | · | · | · | · | · | · | · | · | · | · | · | · | · |
| *Roegneria jacutensis* | · | · | · | · | · | · | · | + | · | · | + | + | + | + | · | · | · | · | · | · | · | · | · | · | · | · |
| *Roegneria turuchanensis* | · | + | + | + | · | · | + | · | · | · | ? | · | · | · | · | · | · | · | · | · | · | · | · | · | · | · |
| *Roegneria villosa* | · | · | · | · | · | · | · | + | · | + | + | + | · | · | · | · | · | · | · | · | · | · | · | · | · | · |
| *Roegneria hyperarctica* | · | · | · | · | · | · | · | + | · | · | · | · | · | · | · | · | + | + | + | + | · | + | + | · | · | · |
| *Roegneria borealis* | · | + | + | + | · | · | + | · | · | + | + | + | · | + | · | · | · | · | · | · | · | · | · | · | · | + |

# FLORA OF THE RUSSIAN ARCTIC

| PLANT NAMES | \multicolumn{16}{c}{SOVIET ARCTIC} | \multicolumn{10}{c}{FOREIGN ARCTIC} |
|---|---|---|---|---|---|---|---|---|---|---|---|---|---|---|---|---|---|---|---|---|---|---|---|---|---|---|
|  | Murman | Kanin-Pechora | Polar Ural | Yugorskiy | Novaya Zemlya | Franz Josef Land | Ob-Tazovskiy | Yenisey | Taymyr | Anabar-Olenek | Lena | Yana-Kolyma | North Chukotka | Beringian Chukotka | Anadyr | Koryak | Arctic Alaska | NW Canada | Labrador | Canadian Archipelago | NW Greenland | SW Greenland | East Greenland | Iceland | Svalbard | Arctic Scandinavia |
| *Roegneria kronokensis* | • | • | • | • | • | • | • | • | • | • | • | • | • | • | • | + | • | • | • | • | • | • | • | • | • | • |
| *Roegneria scandica* | + | + | • | • | • | • | • | • | • | + | • | • | • | • | • | + | • | • | • | • | • | • | • | • | • | + |
| *Elytrigia repens* | + | + | • | • | • | • | • | • | • | • | • | • | • | • | • | • | • | • | • | • | • | • | • | • | • | + |
| *Elytrigia jacutorum* | • | • | • | • | • | • | • | • | • | • | • | + | • | • | • | • | • | • | • | • | • | • | • | • | • | • |
| *Leymus arenarius* | + | + | • | + | • | • | • | • | • | • | • | • | • | • | • | • | • | • | • | • | • | + | • | + | • | + |
| *Leymus mollis* | • | • | • | • | • | • | • | • | • | • | • | • | + | • | + | • | + | • | + | • | • | + | + | + | • | • |
| *Leymus villosissimus* | • | • | • | • | • | • | • | • | • | + | + | + | + | + | + | + | + | + | • | • | • | • | • | • | • | • |
| *Leymus interior* | • | • | • | • | • | • | • | + | • | + | + | + | + | + | • | • | • | • | • | • | • | • | • | • | • | • |
| *Hordeum jubatum* | • | • | • | • | • | • | • | • | + | + | + | + | • | + | • | + | + | • | • | • | • | • | • | • | • | • |

# Index of Plant Names

*Abies* Mill. 57
   *Ledebourii* Rupr. 60
   *obovata* Rupr. 59
   *orientalis* Schrenk 59
   *sibirica* Ldb. 57
*Acrostichum alpinum* Bolton 6
   *ilvense* L. 6
*Agropyron aegilopoides* Drob. 300
   *angustiglume* Nevski 289
   *angustissimum* Drob. 292
   *boreale* (Turcz.) Drob. 297
   *boreale* auct. 296
   *caninum* (L.) Beauv. 288
   *confusum* Roshev. 287
   *dasystachyum* (Hook.) Scribn. 299
   *fibrosum* (Schrenk) Nevski 290
   *jacutense* Drob. 294
   *jacutorum* Nevski 300
   *Karawaewii* Smirn. 299
   *kronokense* Kom. 298
   *latiglume* auct. 296
      ssp. *eurasiaticum* Hult. 297–98
      ssp. *subalpinum* (Neum.) Vestergren 298
   *macrourum* (Turcz.) Drob. 292
   *mutabile* Drob. 288
      var. *glabrum* Drob. 297
      var. *scabrum* Drob. 289
   *Nomokonovii* M. Pop. 292
   *repens* (L.) Beauv. 299
      var. *Hornemannii* Kryl. 297
   *sericeum* Hitchc. 292
   *Tugarinovii* Reverd. 292
   *turuchanense* Reverd. 294
   *violaceum* auct. 295–97
      var. *hyperarcticum* Polunin 296
*Agrostis* L. 93, **119**
   *abakanensis* Less. ex Trin. 126
   *alba* auct. 122
      var. *gigantea* (Roth.) Griseb. 121
      var. *maritima* auct. 123
      var. *salina* Pohle ex Tolm. 123
      var. *stolonifera* (L.) Sm. 122
   *albida* Trin. 122
   *algida* Soland. 234
   *alpina* auct. 125
   *anadyrensis* Socz. 121, 125
   *Bakeri* Rydb. 126
   *borealis* Hartm. 121, 125
   *bottnica* Murb. 126
   *canescens* Web. 132
   *canina* L. 121, 124
   *canina* var. *rubra* Trautv. 124
   *capillaris* auct. 123
   *clavata* Trin. 121, 126
   *delicatula* Pourr. 124
   *diffusa* Host. ex Bess. 121
   *geminata* Trin. 125
   *gigantea* Roth 120, 121
   *hiemalis* (Walt.) Britt., Sterns. et Pogg. 126–27
   *hiemalis* auct. 127
   × *lapponica* Mont. 124
   *laxiflora* (Michx.) Richards. 127
   *laxiflora* auct. 126
   *maritima* Lam. 122
   *maritima* auct. 123
   *Michauxii* Trin. 127
   × *Murbeckii* Fouill. 124
   *prorepens* (Koch) Golub. 122
   *rubra* auct. 125
   *rupestris* auct. 125
   *scabra* Willd. 121, 126–27
   *spica-venti* L. 148
   *stolonifera* L. 120, 122
   *stolonifera* var. *prorepens* Koch 122–23
   *stolonizans* Bess. 122
   *straminea* Hartm. 120, 123
   *Syreitschikovii* Smirn. 124
   *tenuis* Sibth. 120, 123
   *Trinii* Turcz. 121, 124
   *viridissima* Kom. 125
   *vulgaris* With. 124
*Aira alpina* L. 160
   *alpina* Liljebl. 100
   *aquatica* L. 233
   *arctica* Spreng. 159
   *atropurpurea* Wahlb. 149
   *brevifolia* M. B. 155
   *brevifolia* (R. Br.) Lge. 159
   *caespitosa* L. 153
      var. *borealis* Trautv. 158

**Numbers in bold indicate pages with maps.**

var. *brevifolia* (R. Br.) Trautv. 159
var. *submutica* Trautv. 156
*coerulea* L. 174
*flexuosa* L. 152
*laevigata* Sm. 161
*magellanica* Hook. f. 148
*montana* L. 152
*Sukatschewii* Popl. 156
Alisma L. 83
*latifolium* Gilib. 83
*plantago* L. 83
*plantago–aquatica* L. 83
Allosorus *crispus* Bernh. 21
*gracilis* Presl 22
*Stelleri* Rupr. 22
Alopecurus L. 92, 106
*aequalis* Sobol. 107, 113
ssp. *aristulatus* (Michx.) Tzvel. 114
*alpinus* Sm. 107, 110, **112**
var. *borealis* (Trin.) Griseb. 111
var. *elatus* Roshev. 109
var. *glaucus* (Less.) Kryl. 109
var. *scoticus* Griseb. 111
var. *Stejnegeri* (Vasey) Hult. 113
*altaicus* (Griseb.) Petr. 111
*amurensis* Kom. 107, 114
*antarcticus* Vahl 113
*aristulatus* Michx. 113
*arundinaceus* Poir. 108
var. *exserens* (Griseb.) Marss. 109
*behringianus* Gand. 111
*borealis* Trin. 111
*brachystachyus* M. B. 110
*fulvus* Sm. 113
var. *sibiricus* Kryl. 114
*glaucus* Less. 107, 109, **110**
var. *altaicus* Griseb. 111
*geniculatus* L. 107, 113
*pratensis* L. 108, **109**
var. *alpestris* Wahlenb. 108
*pseudobrachystachyus* Ovcz. 109
*Roshevitzianus* Ovcz. 109
*ruthenicus* Weinm. 109
*ruthenicus* var. *exserens* Griseb. 109
*Stejnegeri* Vasey 107, 113
*tenuis* Kom. 109
*ventricosus* Pers. 108
Anthoxanthum L. 91, 98
*alpinum* A. et D. Löve 98, **99**
*odoratum* L. 98
ssp. *alpinum* Löve 98
var. *glabrescens* Celak. 98
Apera Adans. 93, 148

*spica-venti* (L.) Beauv. 148
Arctagrostis Griseb. 93, 115
*anadyrensis* V. Vassil. 115
*angustifolia* Nash 117
*aristulata* Petr. 115
*arundinacea* (Trin.) Beal 115, 117, **118**
*arundinacea* auct. 115
*caespitans* V. Vassil. 118
*calamagrostidiformis* V. Vassil. 118
*glauca* Petr. 115, 117
*festucacea* Petr. 117
*latifolia* (R. Br.) Griseb. 115, **117**
var. *angustifolia* (Nash) Hultén 118
f. *aristulata* (Petr.) Roshev. 116
var. *arundinacea* (Trin.) Griseb. 117
var. *arundinacea* f. *parviflora* Reverd.
ex Kryl. 117
ssp. *gigantea* (Turcz. ex Griseb.)
Tzvel. 116
*macrophylla* Nash 117
*parviflora* (Reverd. ex Kryl.) Petr. 118
*poaeoides* Nash 119
*stricta* Petr. 115
*tenuis* V. Vassil. 118
*Tilesii* (Griseb.) Petr. 115, 118
*viridula* V. Vassil. 118
*ursorum* (Kom.) Kom. ex Roshev. 118
Arctophila (Rupr.) Anderss. 97, 229
*brizoides* Holm 229
*chrisantha* Holm 229
*effusa* Lge. 229
*fulva* (Trin.) Anderss. 229, **230**
ssp. *similis* (Rupr.) Tzvel. 230, 232
*mucronata* Hack. ex Vasey 229
*pendulina* (Laest.) Anderss. ex Lge. 229
*trichopoda* Holm 229
Arundo *calamagrostis* L. 132
*epigeios* L. 131
*groenlandica* Schrank 138
*Langsdorffii* Link 133
*lapponica* Wahlb. 136
*neglecta* Ehrh. 138
*phragmites* L. 174
*stricta* Timm 138
Atropis *alascana* (Scribn. et Merr.) Krecz.
249
*angustata* (R. Br. ) Griseb. 253
*angustata* auct. 244, 246
*convoluta* auct. 257
*distans* (L.) Griseb. 261
var. *capillaris* (Liljebl.) Hack. 257
var. *capillaris* f. *contracta* Hack. 258
var. *typica* Trautv. 261

*geniculata* Krecz. 246
*Hauptiana* Krecz. 262
*iberica* Wolley-Dod 246
*jenisseiensis* Roshev. 252
*laeviuscula* Krecz. 247
*maritima* (Huds.) Griseb. 243
    var. *vilfoidea* (Anders.) Fedtsch. et Fler. 244
*paupercula* (Holm) Steffen 247
*phryganodes* (Trin.) Steffen 244
*pulvinata* Krecz. 258
*sibirica* (Holmb.) Krecz. 260
*suecica* Holmb. 257
*tenella* (Lge.) Simmons 247
*tenella* auct. 259
*Vahliana* (Liebm.) Richt. 249
*vilfoidea* (Anders.) Richt. 244
*Aspidium alpestre* Hoppe 20
  *dilatatum* Sw. 15
  *dryopteris* (L.) Baumg. 17–18
  *filix-mas* (L.) Sw. 13
  *fragrans* Sw. 14
  *lonchitis* (L.) Sw. 19
  *spinulosum* Sw. 16
    ssp. *dilatatum* Roep. 15
*Asplenium* L. 2, 21
  *filix-femina* Bernh. 20
  *viride* Huds. 21
*Athyrium* Roth 3, 19
  *alpestre* (Hoppe) Rylands 19–20
    var. *americanum* Butt. 20
    var. *gaspense* 20
  *filix-femina* (L.) Roth 19
    ssp. *cyclosorum* (Rupr.) Chr. 19
*Avena agrostidea* Fries 167
  *flavescens* L. 165
  *flavescens* auct. 166–67
    var. *agrostidea* Trautv. 166
  *Krylovii* N. Pavl. 172
  *mollis* Michx. 169
  *planiculmis* Turcz. 171
    ssp. *dahurica* Kom. 171
  *sesquiflora* (Trin.) Griseb. 147
  *subspicata* Clairv. 168
*Avenastrum dahuricum* Roshev. 171
  *Krylovii* Roshev. 172
*Avenella flexuosa* (L.) Parl. 152

*Beckmannia* Host 90, 173
  *eruciformis* (L.) Host 173
    var. *baicalensis* Kusn. 173
  *syzigachne* (Steud.) Fern. 173

*Botrychium* Sw. 23
  *boreale* (Fr.) Milde 23, 24
  *lunaria* (L.) Sw. 23
    var. *boreale* Fr. 24
*Bromus* L. 95, 283
  *arcticus* Shear 282
    ssp. *ornans* (Kom.) Karav. 282
    var. *vogulica* (Socz.) Karav. 279
  *arvensis* L. 283
  *ciliatus* auct. 279
  *inermis* Leyss. 279
    var. *ciliata* (L.) Trautv. 279
    var. *glabra* Trautv. 279
    var. *grandiflora* Rupr. 279
    ssp. *inermis* 279
    ssp. *Pumpellianus* (Scribn.) Wagnon 279
    ssp. *Pumpellianus* var. *arcticus* (Shear) Wagnon 282
    var. *sibiricus* (Drob.) Kryl. 279
    var. *villosus* Kryl. 279
  *ircutensis* Kom. 282
  *Julii* Goworuch. 279
  *Pumpellianus* Scribn. 279
    var. *arcticus* (Shear) Porsild 282
  *purgans* auct. 282
  *Richardsonii* auct. 279
  *secalinus* L. 283
  *sibiricus* Drob. 279
    var. *glaber* Drob. 280
    var. *flexuosus* Drob. 280
    var. *pollitus* Drob. 280
    var. *taimyrensis* Roshev. 279
  *uralensis* Goworuch. 279
  *vogulicus* Socz. 279
*Butomus* L. 85
  *umbellatus* L. 85

*Calamagrostis* Adans. 93, 127
  *alascana* Kearney 136
  *angustifolia* Kom. 131, 135
  *arctica* Vasey 129, 146
  *arundinacea* auct. 144
  *borealis* Laest. 138
  *bracteolata* V. Vassil. 142
  *Bungeana* Petr. 140
  *caespitosa* V. Vassil 144
  *canadensis* (Michx.) Beauv. 135
    ssp. *Langsdorffii* (Link) Hult. 133
    var. *Langsdorffii* (Link) Inman 133
  *canescens* (Web.) Roth 131–32, 135
  *cinnoides* (Muhl.) Bart. 137

*confinis* auct. 136
*confusa* V. Vassil. 133
*Czekanowskiana* Litw. 144
*czukczorum* Socz. 135
*deschampsioides* Trin. 128, 142, **143**
    var. *Churchilliana* Polunin 144
    ssp. *macrantha* Piper 143
*elata* Blytt 133
*epigeios* (L.) Roth 128, 131
*evenkiensis* Reverd. 140
*flexuosa* Rupr. 133
*foliosa* Kearney 147
*fusca* Kom. 133
*grandis* Petr. 133
*groenlandica* (Schrank) Kunth 138
*Henriettae* Petr. 136
*hirsuta* V. Vassil. 135
*Holmii* Lge. 130, 140, **142**
*inexpansa* A. Gray 138
*inopia* Litw. 142
*jakutensis* Petr. 138
*kalarica* Tzvel. 146
*kolgujewensis* Gand. 138
*kolymaensis* Kom. 140
*Korotkyi* Litw. 129, 144
*lanceolata* Roth 132
*Langsdorffii* (Link.) Trin. 131–32, **134**
    var. *angustifolia* (Kom.) Jarosch. 135
*lapponica* (Wahlb.) Hartm. 130, 136, **137**
    var. *groenlandica* Lge. 137
    var. *optima* Hartm. 136
*magadanica* V. Vassil. 135
*micrantha* Kearney 138
*monticola* Petr. ex Kom. 144
*neglecta* (Ehrh.) Gaertn., Mey. et Scherb. 131, 138, **140**
    f. *arctica* Roshev. 138
    var. *borealis* (Laest.) Trautv. 138
*neglecta* ssp. *groenlandica* (Schrenk) Matuszk. 139
*neglecta* var. *micrantha* (Kearney) Stebbins 138
*neglecta* spp. *neglecta* 139
*ochotensis* V. Vassil. 138
*phragmitoides* Hartm. 133
*purpurascens* R. Br. 129, 144
    ssp. *arctica* (Vasey) Hult. 146
    var. *arctica* (Vasey) Kearney 146
*purpurea* auct. 133
*Reverdattoi* Golub. 138
*scabra* C. Presl. 132
*sesquiflora* (Trin.) Tzvel. 129, 147
*sibirica* Petr. 136
*Steinbergii* Roshev. 140
*stricta* (Timm) Koel. 138
    var. *borealis* (Laest.) Hartm. 138
*Sugawarae* Ohwi 144
*sylvatica* auct. 144
*tenuis* V. Vassil. 135
*Turczaninowii* Litw. 144
*Tweedyi* Scribn. 144–45
*unilateralis* Petr. 133
*uralensis* Litw. 155
*urelytra* Hack. 147
*Vaseyi* Beal. 146
*vilnensis* Bess. 132
*wiluica* Litw. 144
*yukonensis* Nash 144
*Catabrosa* Beauv. 91, 233
*algida* Th. Fries 234
*algida* auct. 236
*aquatica* (L.) Beauv. 233
*concinna* Th. Fries 236
    var. *algidiformis* H. Smith 236
*vilfoidea* Anders. 244–45
*Colpodium* Trin. 95, 231
*fulvum* (Trin.) Griseb. 230
    var. *arcticum* Roshev. 230
*humile* Lge. 227
*lakium* Woron. 232
*lanatiflorum* (Roshev.) Tzvel. 210, **232**
    ssp. *momicum* Tzvel. 232
*Langei* Gand. 227
*latifolium* R. Br. 115
*Malmgrenii* Anderss. 230
*mucronatum* (Hack. ex Vasey) Beal 230
*nutans* Griseb. 232
*pendulinum* (Laest.) Griseb. 230
    var. *simile* Griseb. 230
*ponticum* (Bal.) Woron. 232
*Tilesii* Griseb. 118
*Vahlianum* (Liebm.) Nevski 249
*Wrightii* Scribn. et Merr. 250
*Critesion jubatum* (L.) Nevski 306
*Cryptogramma* R. Br. 2, 21
*crispa* (L.) R. Br. 21
*Stelleri* (Gmel.) Prantl 21, 22
*Cystopteris* Bernh. 3, 7
*Dickieana* Sims. 8–9, **11**
*filix-fragilis* (L.) Borbas 8, **9**
*fragilis* (L.) Bernh. 8
    ssp. *Dickieana* Hiyt. 10
    var. *Dickieana* Moore 10
*fragilis* Simmons 10

    *montana* (Lam.) Desv. 8, 12
    *montana* Link. 12
    *regia* (L.) Presl. var. *Dickieana* Milde 10

*Dactylis* L. 94, 181
    *glomerata* L. 181
*Deschampsia* Beauv. 96, 150
    *alpina* (L.) Roem. et Schult. 151, 160, **161**
    *anadyrensis* V. Vassil. 152, 156
    *antarctica* Desv. 150
    *arctica* (Spreng.) Ostenf. 159
        var. *borealis* (Trautv.) Kryl. 158
    *atropurpurea* (Wahlb.) Scheele 149
    *beringensis* Hult. 157, 162
    *Biebersteiniana* Roem. et Schult. 155
    *borealis* (Trautv.) Roshev. 152, 158, **159**
        var. *glacialis* Roshev. 158
    *bottnica* (Wahlb.) Trin. 162
    *brevifolia* R. Br. 152, 159, **160**
    *brevifolia* auct. 158
        γ *pumila* Griseb. 155
    *caespitosa* (L.) Beauv. 151, 153, **154**
        var. *grandiflora* Trautv. 162
        f. *minor* Kom. 157
        var. *minor* auct. 158
        ssp. *orientalis* Hult. 157
        var. *glauca* (Hartm.) Sam. 154
    *flexuosa* (L.) Trin. 150, 152, **153**
        var. *montana* (L.) Gremli 152
        ssp. *montana* (L.) A. Löve 152
    *glauca* Hartm. 152, 154, **155**
    *Komarovii* V. Vassil. 156
    *laevigata* (Sm.) Roem. et Schult. 160
    *Mackenzieana* Raup 162
    *mezensis* Senjan.-Korcz. et Korcz. 162
    ×*Neumaniana* (Dörfl.) Hyl. 162
    *obensis* Roshev. 151, 162, **163**
    *paramushirensis* Honda 157
    *pumila* Ostenf. 155
    *pumila* (Stev.) Fom. et Woron. 155
    *pumila* auct. 154
    *Sukatschewii* (Popl.) Roshev. 152, 156, **157**
        ssp. *minor* (Kom.) Tzvel. 157
        ssp. *orientalis* (Hult.) Tzvel. 156–57
        ssp. *submutica* (Trautv.) Tzvel. 156
        ssp. *Sukatschewii* 156
    *Wibeliana* (Sond.) Parl. 162
*Digraphis arundinacea* Trin. 98
*Dryopteris* Adans. 3, 12
    *austriaca* (Jacq.) Woynar 13, 15
    *continentalis* Petrov 18
    *crassirhizoma* Nakai 14
    *disjuncta* (Ldb.) Mort. 18
    *dilatata* A. Gray 15
    *euspinulosa* (Diels) Fomin 16
    *filix-mas* (L.) Schott 13
    *fragrans* (L.) Schott 13, **14**
    *Linnaeana* C. Chr. 18
    *phegopteris* (L.) C. Chr. 17
    *pulchella* (Salisb.) Hayek 18
        var. *continentalis* Petrov 18
    *Robertiana* (Hoffm.) C. Chr. 18
    *spinulosa* (Muell.) O. Ktze. 13, **16**
        ssp. *dilatata* Aschers. 15
        ssp. *euspinulosa* Aschers. 16
*Dupontia* R. Br. 97, 224
    *Fisheri* R. Br. 224–25, 227, **228**
        var. *pelligera* Trautv. 227
        f. *psilosantha* Nevski 225
        ssp. *psilosantha* Hult. 225
        var. *psilosantha* Trautv. 225
    *micrantha* Holm 226
    *pelligera* Rupr. 227
    *psilosantha* Rupr. 224–25, **227**

*Elymus arenarius* L. 301
    ssp. *mollis* var. *villosissimus* (Scribn.) Hult. 304
    β *villosus* E. Mey. 303
    ssp. *mollis* (Trin.) Hult. 303
    *chatangensis* Roshev. 293
    *interior* Hult. 305
    *mollis* Trin. 303
        ssp. *mollis* 303
        ssp. *interior* (Hult.) Bowden 305
        ssp. *villosissimus* (Schribn.) A. Löve 304
    *sibiricus* L. 287, 300
    *villosissimus* Scribn. 304
*Elytrigia* Desv. 90, 299
    *dasystachya* (Hook.) A. et D. Löve 299
    *Gmelinii* (Schrad.) Nevski 300
    *jacutorum* (Nevski) Nevski 299–300
    *repens* (L.) Nevski 299, 302
    *villosa* (Drob.) Tzvel. 299
*Equisetum* L. 25
    *alpestre* Wahlb. 33
    *arcticum* Rupr. 33, 35
    *arvense* L. s. l. 27–28, 33, **35**
    *arvense* L. s. str. 27–28, 33
        ε *E. arcticum* Rupr. 35
        var. *arctica* Trautv. 35

ssp. *boreale* (Bong.) Rupr. 27–28, 35
var. *boreale* Milde 35
var. *boreale* Lange 35
β *E. boreale* Bong. 35
*boreale* Bong. 33, 35
*campestre* Schulz 33
*fluviatile* L. 30
   *simplex* Rupr. 30
*heleocharis* Ehrh. 30
   var. *limosum* 30
*hiemale* L. 26, 28
*limosum* L. 26, 30, **31**
   f. *Linnaeanum* Döll 30
*palustre* L. 27, 31, **32**
*pratense* Ehrh. 27–28, 32, **34**
*scirpoides* Michx. 26, 29, **30**
*silvaticum* L. 27–28, 31, **33**
*variegatum* Schleich. 26, 28, **29**

*Festuca* L. 97, 264
   *altaica* Trin. 265–66, **268**
   *arenaria* Osbeck 269
   *auriculata* Drob. 267–68, 274, **275**
   *baffinensis* Polunin 276
   *Beckeri* (Hack.) Smirn. 271
   *brachyphylla* Schult. 266–67, 275, **276**
      f. *vivipara* **277**
   *brevifolia* R. Br. 275
   *capillaris* Liljebl. 257
   *cryophila* Krecz. et Bobr. 269
   *egena* Krecz. et Bobr. 269
   *elatior* L. 268
   *eriantha* auct. 269
   *hyperborea* Holmen 276
   *kolymensis* Drob. 267, 272, **274**
   *Kryloviana* Reverd. 275
   *lenensis* Drob. 273
   *ovina* L. 266, 271, **272**
      ssp. *Beckeri* Hack. 271
      f. *vivipara* 272, **273**
      ssp. *brevifolia* Hack. 276
      var. *brevifolia* Lynge 276
   *polesica* Zapal. 266, 271
   *pratensis* Huds. 265, 268
   *pseudosulcata* Drob. 272
   *Richardsonii* Hook. 269
   *rubra* L. 265–66, 269, **270**
   *sabulosa* Lindb. 271
   *sulcata* (Hack.) Nym. 269, 273
   *supina* auct. 271
   *vivipara* (L.) Smith. 264, 271

*Glyceria* R. Br. 95, 237
   *angustata* (R. Br.) Fries 253
   *arctica* Hook. 251
   *borealis* (Nash) Botch. 237
   *Borreri* auct. 259
   *distans* (L.) Wahlb. 261
      auct. 260
      var. *pulvinata* Fries 258
   *glumaris* (Trin.) Griseb. 191
   *gracilis* Palib. 253
   *fulva* (Trin.) Fr. 230
   *intermedia* Klinggr. 257
   *kamtschatica* Kom. 237
   *Kjellmanii* Lge. 249
   *Langeana* Berlin 247
   *lithuanica* (Gorski) Lindm. 237
   *maritima* (Huds.) Wahlb. 243
   *maritima* auct. 244
   *maxima* subsp. *grandis* (S. Wats.) Hult. 237
   *orientalis* Kom. 237
   *paupercula* Holm 247
   *pendulina* Laest. 230
   *pulchella* (Nash) K. Schum. 237
   *reptans* Krok 244
   *spiculosa* (Fr. Schmidt) Roshev. 237
   *striata* ssp. *stricta* (Scribn.) Hult. 237
   *tenella* Lge. 247
      f. *pumila* Lge. 258
   *vaginata* Lge. 257
      var. *contracta* Lge. 253
   *vilfoidea* (Anders.) Fries 244
   *Vahliana* (Liebm.) Fries 249
*Graphephorum fulvum* (Trin.) A. Gray 230
   *Fisheri* A. Gray 227
   *psilosantha* A. Gray 226
   *pendulinum* (Laest.) A. Gray 230
*Gymnocarpium* Newm. 3, 17
   *continentale* (Petr.) Pojark. 17–18
   *dryopteris* (L.) Newn. 17
   *Robertianum* (Hoffm.) 18

*Helictotrichon* Bess. ex Roem. et Schult 96, 171
   *dahuricum* (Kom.) Kitagawa 171
      var. *kamtschatica* Kom. 172
   *Krylovii* (N. Pavl.) Henrard 171
   *planiculme* (Schrad.) Pilger 172
*Hierochloë* R. Br. 92, 100
   *alpina* (Liljebl.) Roem. et Schult. 100, **101**
   *borealis* Roem. et Schult. 103
   *odorata* (L.) Wahlb. 100, 103

*pauciflora* R. Br. 100, **102**
*racemosa* Trin. 102
*Holcus odoratus* L. 103
× *Horderoegneria chatangensis* (Roshev.) Tzvel. 293
*Hordeum* L. 89, 306
    *jubatum* L. 306

*Juniperus* L. 65
    *alpina* Clus. 65
    *communis* L. 65
        var. *nana* Schrenk 65
        var. *depressa* 65
        f. *montana* Hulten 65
    *nana* Willd. 65
    *Niemannii* Wolf 65
    *sibirica* Burgsd. 65

*Koeleria* Pers. 94, 175
    *asiatica* Domin 175–76, **177**
    *asiatica* auct. 178
        var. *lanuginosa* Domin 176
        var. *leiantha* Domin 176
        var. *sublanuginosa* Kryl. 176
    *atroviolacea* Domin 176
    *Cairnesiana* Hult. 176
    *caucasica* (Trin. ex Domin) Gontsch. 176–77
    *cristata* auct. 175
        var. *seminuda* Trautv. 175
    *Delavignei* Czern. ex Domin 177
    *glauca* DC. 178
        var. *Pohleana* Domin 178
    *Gorodkovii* Roshev. 177
    *gracilis* Pers. 176
    *gracilis* auct. 178
        ssp. *gracilis* var. *arctica* Domin 175
        ssp. *seminuda* (Trautv.) Domin 175
        ssp. *sibirica* Domin 175
    *janaensis* Petr. ex Karav. 175
    *hirsuta* auct. 176
    *macrantha* (Ledeb.) Spreng. 176
    *Mariae* V. Vassil. 176
    *Pohleana* (Domin) Gontsch. 175, **178**
    *polonica* Domin 177
    *seminuda* (Trautv.) Gontsch. 175
    *sibirica* (Domin) Gontsch. 175

*Larix* Mill. 57, 60
    *archangelica* Laws. 61
    *Cajanderi* Mayr. 63
    *Czekanowskii* Sz. 60, 63
    *dahurica* Turcz. 60–61
    *dahurica* × *sibirica* 60, 63
    *decidua* var. *rossica* Rgl. 60
    *rossica* Rgl. 60
    *sibirica* Ldb. 60
        var. *polaris* Dyl. 61
        ssp. *rossica* (Rgl.) Sucacz. 61
        f. *rossica* Szafer 60
    *Sukaczewii* Dyl. 60
*Lastraea filix-mas* Presl 13
    *fragrans* Presl 14
    *phegopteris* (L.) Borg. 17
    *spinulosa* β *intermedia* Milde 15
× *Leymotrigia Bergrothii* (Lindb. f.) Tzvel. 302
*Leymus* Hochst. 89, 300
    *angustus* (Trin.) Pilger 300
    *arenarius* (L.) Hochst. 301, **302**
    *interior* (Hult.) Tzvel. 301, **305**
    *mollis* (Trin.) Pilger 301, **303**
    *villosissimus* (Scribn.) Tzvel. 301, **304**
*Lycopodium* L. 37
    *alpinum* L. 40, 49, **50**
    *anceps* Wallr. 40, 48
    *annotinum* L. 40, 44
        var. *alpestre* Hartm. 45
        var. *pungens* Desv. 45
    *annotinum* Simmons 45
    *appressum* Petrov 41–42
    *arcticum* Grossh. 42
    *chamaecyparissus* A. Br. 49
    *chinense* Chr. 42
    *clavatum* L. 39, 46
        ssp. *monostachyon* (Grev. et Hook.) Sel. 40, 46, **47**
        var. *monostachyon* Grev. et Hook. 46
    *complanatum* L. 40, 48, **49**
        var. *anceps* 48
        ssp. *chamaecyparissus* Döll 49
        var. *chamaecyparissus* A. Br. 49
    *lagopus* (Laest.) Zinser. 46
    *pungens* La Pyl. 40, 45, **46**
    *sabinaefolium* Willd. 49
    *selaginoides* L. 54
    *selago* L. 38–40, **42**
        var. *appressum* Desv. 39, 41
        var. *appressa* Desv. 43
        ssp. *arcticum* (Grossh.) Tolm. 38, 42, **43**
        var. *dubium* Sanio 41
        var. *laxum* Desv. 39, 41

## INDEX OF PLANT NAMES

var. *patens* Desv. 39, 41
*subarcticum* Vass. 40
*tristachyum* Pursh 40, 49

*Melica* L. 95, 179
   *nutans* L. 179
*Milium* L. 93, 104
   *effusum* L. 104
*Molinia* Schrenk 90, 174
   *coerulea* (L.) Moench 174

*Nardus* 88, 284
   *stricta* L. 284

*Osmunda crispa* L. 21
   *lunaria* L. 24

*Panicum syzigachne* Steud. 173
*Phalaris arundinacea* Trin. 98
   *erucaeformis* L. 173
*Phegopteris dryopteris* (L.) Fée 18
   *polypodioides* Fée 17
*Phippsia* R. Br. 91, 234
   *algida* (Soland.) R. Br. 234, **235**
      var. *concinna* Richt. 236
   *concinna* (Th. Fries) Lindeb. 234, **236**
      ssp. *algidiformis* H. Smith 236
      f. *major* Roshev. 236
   *foliosa* V. Vassil. 234
*Phleum* L. 92, 104
   *alpinum* L. 104–6
      var. *americanum* Hult. 104
      ssp. *commutatum* (Gaud.) Hult. 104
      var. *commutatum* (Gaud.) Mert. et Koch 104
   *commutatum* Gaud. 104, **105**
   *pratense* L. 104, 106
*Phragmites* Adans. 90, 174
   *communis* Trin. 174
      var. *Berlandieri* (Fourn.) Fern. 174
*Picea* Dietr. 57–58
   *abies* (L.) Karst. 58
      ssp. *obovata* Hultén 59
   *excelsa* Link 58
   *excelsa* Keppen 59
   *fennica* Rgl. 58
   *obovata* Ldb. 58
*Pinus* (Tourn.) L. 57, 63
   *cembra pumila* Pall. 64

*dahurica* Fisch. 62
*lapponica* Mayr 63
*Ledebourii* Endl. 60
*orientalis* Ldb. 59
*pumila* (Pall.) Rgl. 63–64
*sibirica* Ldb. 57
*silvestris* L. ssp. *lapponica* Fries 63
*Pleuropogon* R. Br. 89, 179
   *Sabinii* R. Br. 179, **180**
      f. *aquatica* Roshev. 180
*Poa* L. 94, 182
   *abbreviata* R. Br. 183, 207, **208**
   *airoides* Koel. 233
   *Albertii* Rgl. 182
   *alpigena* (Fr.) Lindm. 190, **203**
      var. *colpodea* (Th. Fries) Scholand. 204
      var. *vivipara* (Malmgr.) Scholand. 204
      var. *vivipara* Hult. 204
      f. *vivipara* Roshev. 204
   *alpina* L. 185, 206, **207**
      var. *vivipara* L. 206
   *altaica* Trin. 219
   *anadyrica* Roshev. 187, 219, **220**
   *angustata* R. Br. 253
   *angustifolia* L. 190, 205
   *annua* L. 184, 222
      var. *supina* (Schrad.) Reichb. 222
   *arctica* R. Br. 189, 194, 196, **198**
      ssp. *caespitans* (Simm. ex Nannf.) Nannf. 199
      var. *caespitans* (Simm. ex Nannf.) Hyl. 199
      ssp. *depauperata* (Fr.) Nannf. 197
      ssp. *elongata* (Bl.) Nannf. 197
      ssp. *longiculmis* Hult. 197
      ssp. *microglumis* Nannf. 197
      ssp. *stricta* (Lindeb.) Nannf. 199
      var. *stricta* (Lindeb.) Hyl. 199
      ssp. *tromsensis* Nannf. 197
      var. *vivipara* Hook. 197
      var. *vivipara* auct. 199
      ssp. *Williamsii* (Nash) Hult. 197
   *arenaria* var. γ Trin. 243
   *argunensis* Roshev. 216
   *aspera* Gaud. 219
   *attenuata* Trin. 216
   *attenuata* auct. 215
      var. *stepposa* Kryl. 215
   *botryoides* (Trin. ex Griseb.) Roshev. 187, 217
   *brachyanthera* Hult. 211

*bracteosa* Kom. 193
*bryophila* Trin. 187, 220
*bulbosa* L. 206
*caesia* Sm. 217
*cenisia* All. 197
*cenisia* auct. 196
    f. *caespitans* Simm. ex Nannf. 199
*compressa* L. 183, 212
*conferta* Blytt. 217
*Cusickii* Vasey 222
*deflexa* Rupr. 229
*distans* L. 261
*Eduardii* Golub. 193
*eminens* C. Presl. 187, **191**
*epilis* Scribn. 222
*evenkiensis* Reverd. 217
*fertilis* Host 214
*flavicans* Ledeb. 193
*flavidula* Kom. 209
*filiculmis* Roshev. 187, **216**
*filipes* Lge. 199
*flexuosa* Sm. 208
*flexuosa* Wahlb. 196
    var. *vivipara* Malmgr. 204
*fulva* Trin 229
*Ganeschinii* Roshev. 199, 217
*geniculata* Turcz. 246
*glauca* Vahl 187, 217, **218**
    var. *conferta* (Blytt.) Hyl. 217
*glaucantha* Gaud. 219
*glumaris* Trin. 191
    var. *laevigata* Trautv. 192
*Hauptiana* Trin. ex Krecz. 262
*hispidula* Vasey 193
    var. *aleutica* Hult. 193
*humilis* Ehrh. 205
*infirma* H. B. K. 222
*interior* Rydb. 219
*irrigata* Lindm. 205
*Komarovii* Roshev. 193
    var. *vivipara* Roshev. 195
*kurilensis* Hack. 191
*Laestadii* Rupr. 229
*lanata* Scribn. et Merr. 188, **195**, 196
    var. *vivipara* Hult. 196
*lanatiflora* Roshev. 210, 232
*lapponica* Prokud. 186, 214
*latiflora* Rupr. 229
*laxa* Haenke 208
*laxa* auct. 211
*leptocoma* Trin. 184, 209
*leptocoma* auct. 209
    var. β Trin. 209

*longifolia* Trin. 222
*macrocalyx* Trautv. et Mey. 192–93
*malacantha* Kom. 189, 193, **194**
    var. *vivipara* (Roshev.) Tzvel. 195
*Malmgrenii* Gand. 207
*maritima* Huds. 243
*maritima* Wahlb. 259
*nemoralis* L. 122, 186, 212, **213**
*nivicola* Kom. 194, 209
*occidentalis* auct. 193
*ochotensis* Trin. 215
*palustris* L. 186, **214**
*patens* Trin. 209
*paucispicula* Scribn. et Merr. 184, 209, **210**
*pelligera* Rupr. 227
*remotiflora* Rupr. 229
*pendulina* (Laest. ) Vahl 229
*penicillata* Kom. 193
*petraea* Trin. 196
*petschorica* Roshev. 196
*phryganodes* Trin. 244
*pinegensis* Roshev. 200
*platyantha* Kom. 188, 193
    f. *vivipara* Kom. 193
*poecilantha* Rupr. 229
*pratensis* L. 191, 200, **201**
*pratensis* auct. 203
    ssp. *alpigena* (Fries) Hiit. 203
    var. *alpigena* Fries 203
    ssp. *angustifolia* (L.) Lindb. 206
    var. *angustifolia* (L.) Sm. 205
    ssp. *eupratensis* Hiit. 200
    var. *humilis* (Ehrh.) Griseb. 205
    var. *laxiflora* Lge. 202
    *nudiflora* Ledeb. 221
*pseudoabbreviata* Roshev. 184, **211**
*pseudo-glauca* Roshev. 220
*remotiflora* Rupr. 229
*retroflexa* Curt. 261
*rigens* Hartm. 200
*rigens* auct. 196
*rotundata* Trin. 214
*salebrosa* Roshev. 220
*scabriflora* Hack. 193
*scleroclada* Rupr. 229
*serotina* Ehrh. 214
    var. *botryoides* Trin. ex Griseb. 217
*setacea* Hoffm. 206
*sibirica* Roshev. 183, **221**
*silvicola* Guss. 221
*similis* Rupr. 229
*Smirnowii* Roshev. 196

*Soczawae* Roshev. 210, 217
*sphondyloides* Trin. 216
*stenantha* Trin. 193–94
    var. *leptocoma* (Trin.) Griseb. 209
*stepposa* (Kryl.) Roshev. 187, **215**
*sterilis* auct. 215
*stricta* Lindeb. 199
*stricta* auct. 204
    ssp. *colpodea* Th. Fries 204
*strigosa* Hoffm. 206
*subabbreviata* Roshev. 211
*subcaerulea* Sm. 190, 205
*subfastigiata* Trin. 192
*sublanata* Reverd. 190, **202**
    var. *vivipara* Tzvel. 202
*supina* Schrad. 185, 222
*taimyrensis* Roshev. 209
*Tanfiljewii* Roshev. 186, 213
*tibetica* Munro 192
*Tolmatchewii* Roshev. 189, 199, **200**
    var. *stricta* (Lindeb.) Tzvel. 199
*transbaicalica* Roshev. 216
*Trautvetteri* Tzvel. 189, 192
*trichoclada* Rupr. 229
*trichopoda* Lge. 199
*Trinii* Scribn. et Merr. 191
*trivialis* L. 185, 221
    var. *altaica* Griseb. 221
*turfosa* Litw. 200
*Turneri* Scribn. 193
*udensis* Trautv. et Mey 182
*ursorum* Kom. 118
*Vahliana* Liebm. 249
*vivipara* Willd. 206
*Williamsii* Nash 196
*Wrightii* (Scribn. et Merr.) Hitchc. 250
*Polypodium* L. 3, 22
  *alpestre* Hoppe 20
  *austriacum* Jacq. 15
  *dryopteris* L. 17
  *filix-femina* L. 20
  *filix-mas* L. 13
  *fragrans* L. 14
  *lonchitis* L. 19
  *montanum* Lam. 12
  *phegopteris* L. 17
  *spinulosum* Muell. 16
  *virginianum* L. 22
  *vulgare* L. 22
    ssp. *occidentale* (Hook.) Hult. 22
*Polystichum* Roth 2, 19

  *filix-mas* Roth 13
  *fragrans* Ledb. 14
  *lonchitis* (L.) Roth 19
  *spinulosum* Ldb. 16
*Potamogeton* L. 71
  *alpinus* Balb. 74, 77–78
    ssp. *tenuifolius* (Raf.) Hult. 74, 77
  *borealis* Raf. 75
  *bottnicus* Hagstr. 76
  *digynus* Wall. 74, 80
  *filiformis* Pers. 73, 75
    ssp. *borealis* (Raf.) 75
    var. *borealis* (Raf.) St. John 75
    f. *polaris* Hagstr. 75
  *gramineus* L. 74, 78
  *groenlandicus* Hagstr. 72, 77
  *heterophyllus* Schreb. 78
  *lanceolatus* Sw. 77
  *lucens* L. 74, 79
  *marinus* Ldb. 75
  *natans* L. 74, 78
  *pectinatus* L. 73, 76
  *pectinatus* Scheutz 75–76
  *perfoliatus* L. 74, 79
    ssp. *Richardsonii* (Benn.) Hult. 79
  *Porsildorum* Fern. 72, 77
  *praelongus* Wulf. 74, 78
  *pusillus* L. 73, 77
  *pusillus* Scheutz 76
  *Richardsonii* Fern. 79
  *rufescens* Schrad. 77
  *salicifolia* Wolfg. 77–79
  *sibiricus* A. Benn. 77
  *subretusus* Hagstr. 73, 75
  *subsibiricus* Hagstr. 73, 76
  *tenuifolius* Raf. 77
  *vaginatus* Turcz. 73, 76
*Pteris Stelleri* S. G. Gmel. 22
*Puccinellia* Parl. 97, 237
  *alascana* Scribn. et Merr. 240, 249
  *Andersonii* Swall. 255
  *Andersonii* auct. 255
  *angustata* (R. Br.) Rand et Redf. 241–42, 252, 254, **255**
  *arctica* (Hook.) Fern. et Weath. 251
  *borealis* Swall. 242, 262
    ssp. *neglecta* Tzvel. 262
  *capillaris* (Liljebl.) Jansen 241, 257
    var. *vaginata* auct. 259
  *chinampoensis* Ohwi 256
  *coarctata* Fern. et Weath. 242, 259, **260**